Chromosome Damage
and Repair

NATO ADVANCED STUDY INSTITUTES SERIES

A series of edited volumes comprising multifaceted studies of contemporary scientific issues by some of the best scientific minds in the world, assembled in cooperation with NATO Scientific Affairs Division.

Series A: Life Sciences

Recent Volumes in this Series

This series is published by an international board of publishers in conjunction with NATO Scientific Affairs Division

A Life Sciences B Physics	Plenum Publishing Corporation London and New York
C Mathematical and Physical Sciences	D. Reidel Publishing Company Dordrecht, Boston, and London
D Behavioral and Social Sciences E Applied Sciences	Sijthoff & Noordhoff International Publishers Alphen aan den Rijn, The Netherlands, and Germantown, U.S.A.

Chromosome Damage and Repair

Edited by

Erling Seeberg
Norwegian Defense Research Establishment
Kjeller, Norway

and

Kjell Kleppe
University of Bergen
Bergen, Norway

PLENUM PRESS • **NEW YORK AND LONDON**
Published in cooperation with NATO Scientific Affairs Division

7944716
DLC

11-11-82

Library of Congress Cataloging in Publication Data

Main entry under title:

Chromosome damage and repair.

(NATO advanced study institutes series. Series A, Life sciences ; v. 40)   use

"Proceedings of a NATO advanced study institute and an EMBO lecture course on chromosome damage and repair held at Godøysund Fjord Hotel, near Bergen, on May 27-June 5, 1980" — T.p. verso.

"Published in cooperation with NATO Scientific Affairs Division."

Bibliography: p.

Includes indexes.

1. Deoxyribonucleic acid repair—Congresses. 2. Chromosome replication—Congresses. 3. Mutagenesis-Congresses. I. Seeberg, Erling. II. Kleppe, Kjell. III. European Molecular Biology Organization. IV. North Atlantic Treaty Organization. Scientific, Affairs Division. V. Series.

QH467.C47   574.87,322   81-19871
ISBN 0-306-40886-4   AACR2

Proceedings of a NATO Advanced Study Institute and an EMBO
Lecture Course on Chromosome Damage and Repair held at
Godøysund Fjord Hotel, near Bergen, on May 27-June 5, 1980

© 1981 Plenum Press, New York
A Division of Plenum Publishing Corporation
233 Spring Street, New York, N.Y. 10013

PREFACE

The NATO - EMBO Advanced Study Institute - Lecture Course on "CHROMOSOME DAMAGE AND REPAIR" was held at Godøysund Fjord Hotel outside Bergen, Norway, from May 27th to June 5th, 1980. This book represents the proceedings of this meeting. In addition to the formal lectures, a number of short contributions presented in the discussion sessions following the lectures are also included. The papers have been divided into different groups according to topic, essentially in the same way as they were presented during the meeting. The editors have made few alterations in the manuscript submitted and these were mostly confined to typing style and correction of typographical errors.

We would like to express our appreciation to all the persons who helped in making this meeting possible. Special thanks are due to Tomas Lindahl, Alan R. Lehmann and Erik Boye, who served in the advisory program committee. We would also like to thank our Danish friends and colleagues and others who provided invaluable assistance in an emergency situation. The editors are also grateful for the financial support provided by a number of organizations and institutions. First and foremost of these were the NATO Scientific Affairs Division and the European Molecular Biology Organization. In addition to these the following Norwegian institutions also gave financial assistance to the meeting: The Ministry of Foreign Affairs, The Ministry of Church and Education, The Norwegian Cancer Society, The Norwegian Research Council for Technology and Science, The City of Bergen, The Norwegian Defence Research Establishment and the University of Bergen.

Oslo - Bergen

1980

Erling Seeberg

Kjell Kleppe

v

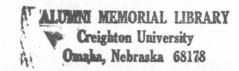

CONTENTS

MECHANISMS OF POSTREPLICATION REPAIR AND REPAIR BY RECOMBINATION

ADAPTATION TO ALKYLATING AGENTS

ACCURACY OF DNA REPLICATION: EFFECTS OF LESIONS IN THE TEMPLATE

ROLE OF ADP-RIBOSE IN DNA REPAIR

LETHAL AND MUTAGENIC EFFECTS OF DNA DAMAGE

INDUCTION OF SISTER CHROMATID EXCHANGES

SPONTANEOUS AND INDUCED DAMAGE TO DNA AND CHROMATIN

DAMAGE TO DNA CAUSED BY HYDROLYSIS

Robert Shapiro

Department of Chemistry
New York University
New York, N.Y. 10003, U.S.A.

INTRODUCTION

The covalent structure of DNA is unstable in aqueous solution. It tends to hydrolyze to its monomeric components, and they
themselves are subject to various hydrolytic reactions. These
processes are slow, when compared to most familiar chemical reactions. However, a reaction that is slow by these standards
may still have great biological significance, if it occurs within
the genetic material of an organism. A single base transformation
within a DNA molecule may be sufficient to cause a mutation, or
inactivate the DNA. Consider a reaction, for example, with a
rate constant of 10^{-10} sec^{-1} at pH 7.4, 37°; it will have a half
life of 220 years. Assume that, within a DNA, it affects two of
the four bases. It will take place once every three hours per
million base pairs of DNA, and thus be a significant source of
damage.

The rates of such reactions cannot conveniently be followed,
however, by conventional chemical methods. They have usually
been examined under conditions where they are greatly speeded up –
by elevated temperatures and acidic or basic catalysis. If reactions have been run at several elevated temperatures, it is usually not difficult to extrapolate down to 37°. Caution must only
be observed when a nucleic acid, rather than a nucleotide, is the
substrate, as changes in secondary structure may take place within
the temperature range considered. An extrapolation from acidic
or basic conditions to neutral solution is more difficult. The
simplest assumption is that the rate will be proportional to the
concentration of H$^{(+)}$ or OH$^{(-)}$ Thus a reaction is assumed to take
place, at pH 7, at a rate 10^{-6} of that measured at pH 1. In some

3

cases, this treatment will give an approximately correct result;
see, for example, the pH-rate profile for the depurination of deoxyadenosine, in Figure 4. Often, however, this approximation
may greatly underestimate the reaction rate at pH 7, as in the
deamination of cytidine in acid. For a better understanding of
the effects of acid and alkali, the dissociation properties of the
reactants must be taken into account.

The function of acid in a hydrolysis reaction is to convert
the substrate to a more reactive protonated form. In the case of
the deamination of cytidine, for example, this takes place as one
decreases the pH from 6 to 3. Once protonation is complete, the
addition of more acid does not speed up the reaction further. The
rate remains constant over the pH range 1-3, affording a plateau
area in the pH-rate profile. Deoxyadenosine and deoxyguanosine,
like deoxycytidine, are converted to cations by aqueous acid, with
pK_a values, respectively, of 3.8, 2.5, and 4.3 at 25°. In concentrated sulfuric acid, or other strongly acidic media, these substances are converted to dications, and thymidine is protonated as
well.

Alkali, like acid, enhances rates of hydrolysis. It does
not do this, however, by increasing the reactivity of the substrate. In fact, it has an opposite effect upon two of the four
deoxynucleosides in Figure 3, thymidine and deoxyguanosine. It
converts them to a less reactive anionic form. The pK_a values are
9.8 and 9.3 respectively. The enhancement of rates observed in
alkaline solution is due to the substitution of an effective nu-
cleophilic reagent, hydroxide ion, for a relatively ineffective
one, water. For the reason given above, deoxyadenosine and deoxycytidine are much more susceptible to degradation by alkali than
thymidine and deoxyguanosine. Again, this is a result which can
not be extrapolated to solutions of pH below 9, as the latter nucleosides will no longer be in anionic form.

DEAMINATION OF CYTOSINE, ADENINE AND GUANINE

Amino substituents on the bases of nucleic acids hydrolyze,
with loss of ammonia, under acidic and alkaline conditions. The
reaction is called hydrolytic deamination to distinguish it from
the oxidative deamination produced by nitrous acid treatment.
The hydrolytic deamination of cytosine is the most rapid one at
neutral pH, and the only one whose kinetics have been studied in
detail. The rate of deamination of cytosine derivatives is
greatly affected by the presence of buffers (1). However, by extrapolation at each pH to zero buffer concentration, we can obtain
a rate for catalysis by water alone (2). The variation of this
rate with pH for cytidine is given in Figure 1. Four distinct
areas should be noted: a plateau in acid, below pH 3; a sloping
area in weak acid; another plateau in neutral solution (pH 6-8)

and a sharp rise as the solution becomes progressively more alkaline. Although several kinetically equivalent possibilities exist
to explain each of these areas, the simplest scheme is probably
the following: The rate in the first two areas reflects the attack
of water on protonated cytidine, with the observed decrease due to
its dissociation (the pK_a for cytidine at 80° is 3.6). This
pathway declines in importance as the concentration of protonated cytidine falls. In the plateau area, pH 6 to 8, the rate depends on the concentrations of protonated cytidine and of hydroxide ion. One term falls as the other rises, and their product is
constant. In the alkaline range, rates are proportioned to the
concentration of hydroxide ion, and reflect the attack of this
species on neutral cytidine.

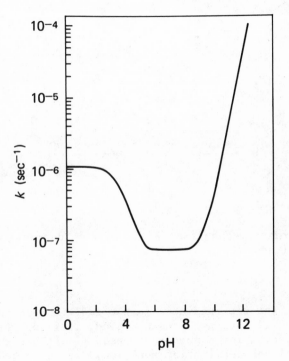

Fig. 1. The pH-rate profile for hydrolytic deamination of cytidine at 80° (2).

Two possible mechanisms have been suggested for the mechanism of deamination (1) (Figure 2), The scheme shown describes
the reaction in acid, but other equivalent ones would exist in

neutral and basic solution. The path, I → III → IV is analogous
to the hydrolysis of an amide, and will be called the direct
route. The alternative is the addition-elimination mechanism,
I → II → V → IV. It postulates addition of water to the 5,6-
double bond of protonated cytidine to afford II. This species is
the cation of a molecule that we will call cytidine hydrate.
Further attack by water at C_4 of II is followed by loss of ammonia, to give V, uridine hydrate. The last step in the reaction
is dehydration of uridine hydrate to uridine (IV).

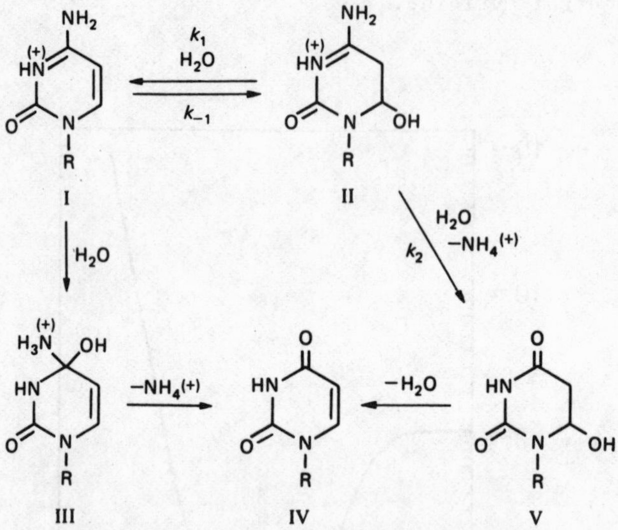

Fig. 2. Alternate mechanisms for hydrolytic deamination of cytosine derivatives.

        Although the intermediates in the addition-elimination me-
chanism (II and V) have not been observed, it is preferred by
workers in the area for a number of reasons. (a) In the related
bisulfite-catalyzed deamination reaction, discussed below, intermediates produced by bisulfite addition to the double bond of cytidine can be observed readily by nuclear magnetic resonance
spectroscopy; the addition-elimination path is firmly established
for this reaction. (b) Other nucleophiles, such as hydroxylamine
and its O-methylderivative, have been shown to add to the 5,6-
double bond of cytidine.(3). (c) Addition of hydroxyl groups to
the 5,6-double bond of pyrimidines has been observed in intramolecular processes; for example in the hydrolytic deamination of
arabinosylcytosine (4,5).

Our understanding of the addition-elimination path has been aided by the fact that the key intermediates, the hydrates of cytidine and uridine, can be prepared by another route: irradiation of the parent nucleosides in aqueous solution (6). For this reason, they have been termed "photohydrates." The availability of these hydrates has permitted the study of their deamination and dehydration processes. For the cytosine derivative most studied, the hydrate of cytidine 3'-phosphate, the rate of reversal under various conditions has always exceeded the rate of deamination by seven to fifty-fold (6). We can therefore deduce that in the scheme for hydrolytic deamination in Figure 2, K_{-1} exceeds K_2. The protonated forms of cytidine (1) and its hydrate (II) are in equilibrium, and the rate determining step is deamination of II to uridine hydrate (V). The last step, dehydration of uridine hydrate to uridine cannot be rate determining.. The minimum rate for this process occurs at pH 6, and increases sharply as the pH becomes more acidic or alkaline (6). The extrapolated rate constant for this dehydration at pH 6, 37°, is approximately 10^{-5} sec^{-1}. This value is still about 10^4 times greater than the overall rate of deamination of cytidine under these conditions.

By an extrapolation from available kinetic data (2,6-8), it can be estimated that the rate of deamination of cytidine is 10^{-4} that of cytidine hydrate at pH 3, 25°. This difference reflects the equilibrium constant for I $\rightleftarrows$ II (Figure 1); in acid, 1 of 10,000 protonated cytidine molecules exists as the hydrate. The equilibrium constant for the conversion of neutral cytidine to its hydrate can also be calculated from this information and the pK_a values at 25° of cytidine (4.1) and its hydrate (5.1). This value is 10^{-5}; 1 in 100,000 molecules of neutral cytidine exists in hydrate form.

Only one kinetic study has appeared on the deamination of cytosine derivatives in alkali (2). Insufficient information is available to determine the details of the reaction, or even to establish whether the direct pathway or the addition-elimination one is operative.

The above discussion has referred to a hypothetical unbuffered solution. Buffer substantces do affect the reaction, however. As they are inevitably present in cellular fluids, it is worth considering the ways in which they can do this. There are two distinct roles that buffers can play in the addition-elimination pathway. The first type is _via_ addition to the double bond, increasing the total equilibrium concentration of partly-saturated cytosine intermediates available for deamination. The catalytic effect in this case should be proportional to the concentration of undissociated acid, HB. The second buffer effect is general base catalysis of the deamination step. In this case, catalysis is proportional to the concentration of buffer anion, $B^{(-)}$.

The most significant catalysis of the first type is that produced by bisulfite. The equilibrium constant is sufficiently
high that in concentrated acidic bisulfite solutions pH 4, virtually complete conversion of cytidine to its bisulfite adduct takes
place (7). This effect, plus the significant ability of sulfite
to catalyze deamination of the cytosine addition product intermediates, have made bisulfite the agent of choice for preparative
deamination of cytosine derivatives, and for modification of biological macromolecules. For example, the half life of cytidine
(or deoxycytidine) in 2.26 M bisulfite, 37°, pH 6.0, is about 19
hours (7). For cytidine in water at that pH, it would be about 50
years. For further information about the use of the bisulfite catalyzed deamination reaction, consult the reviews by Hayatsu (9)
and Shapiro (10). The ability of other buffers to catalyze deamination is significantly less than that of bisulfite.

Hydrolytic deamination of cytosine residues in Eschericia
coli DNA has been studied by Lindahl and Nyberg (11). For single
stranded DNA, the observed rate at pH 7.4 was only slightly slower
than that of the nucleotide, and the pH-rate profile in the range
6-10 was quite similar to that measured for nucleosides (2).
Double-helical DNA, however, was quite resistant to deamination,
and reacted at less than 1% of the rate of the single stranded
form. The reason for this effect is unknown.

The case of 5-methylcytosine derivatives is worthy of special consideration. The product of deamination of this base is
thymine which, unlike uracil, is not excised by DNA-uracil glycosidase. 5-Methylcytosine residues in DNA may therefore be hotspots for deamination (12). Unfortunately, little is known of the
mechanism of hydrolytic deamination of this base. The presence of
the 5-methyl group would be expected to deactivate the 5-6 double
bond of cytosine to addition, for both steric and electronic reasons. In accord with this expectation, we have found that the
rate of deamination of 5-methyldeoxycytidine by 1.0 M bisulfite,
75°, pH 5.25 is about 1/30 of that of cytidine under the same conditions (S. Slae and R. Shapiro, unpublished data). In contrast
to the above, Lindahl and Nyberg (11) have reported that hydrolytic deamination of 5-methyldeoxycytidylic acid in dilute aqueous
buffers, pH 7.4, 95°, proceeds at a rate four times as fast as
dCMP under the same conditions. It is possible that a direct substitution path is operative for the 5-methyl compound, but more
work is clearly needed to determine this.

DEAMINATION OF ADENINE AND GUANINE

Deamination of adenine to hypoxanthine, and of guanine to
xanthine have been reported in acidic and alkaline solution. No
detailed kinetic studies have been reported on the acidic reactions
but from the limited data available (13), it can be estimated that

deamination would take place at 10^{-4} the rate of depurination, in a deoxyribonucleoside. If this ratio of reactivities held at neutral pH, as in acid, then the acid-catalyzed deamination reaction would not be significant, biologically.

The base-catalyzed process may be more important, but the data is limited, and in part, contradictory (14-17). Deamination of adenine nucleosides in alkali takes place, but is slower than the glycosyl cleavage and ring opening reactions, which accompany it. Extrapolation of the data to neutral solution affords a value for adenosine deamination that is about 1% of that of cytidine under the same conditions. Lindahl and Nyberg (18) have estimated that the rate of deamination of adenine in single stranded DNA at pH 7.4 is about 2% of that of cytosine.

It has been reported that deamination in guanine in DNA in 1 N NaOH is more rapid than that of adenine (17). However, this data is very limited, and does not permit an extrapolation to neutral pH.

ACIDIC CLEAVAGE OF N-GLYCOSYL BONDS

The instability of the N-glycosyl bonds of DNA to acid was one of the earliest facts that was recognized in nucleic acid chemistry. The lesser stability of deoxyribonucleosides, compared to ribonucleosides, and of purine nucleosides when compared to pyrimidine nucleosides, has also long been appreciated (19). A pH-rate profile summarizing the behavior of the major deoxyribonucleoside components of DNA in acid is provided in Figure 3. It has been compiled and extrapolated from available data (20-22).

A unified mechanism for deoxyribonucleoside hydrolysis has emerged from the above studies and others (23-26). The methods employed have included the investigation of rates for a variety of sugar and base analogs, secondary deuterium isotope effects, determination of a number of physico-chemical parameters, and product analyses, which established the absence of anomerizations and rearrangement reactions in all but a single case (27). The unified scheme is summarized in Figure 4 and illustrated for the particular case of deoxycytidine cation cleavage in Figure 5.

The pH-rate data of Figure 3 are readily interpreted by the general scheme. The plateau area observed for thymidine is due to cleavage of the neutral molecule (K_0) (20). Only thymidine, of the four major deoxyribonucleosides, has shown this behavior. The enhanced cleavage of thymidine in acid is due to the cation and dication (K_1 and K_2 in Fig. 4). A decline of the cleavage rate in base (above pH 9) would be expected for thymidine, due to conversion of the neutral species to the anion. This behavior was observed for deoxyuridine (20). The pH rate-profile for deoxycy-

6.5
4.5
2.5
0.5
LOG($K \times 10^7$)
DEOXYADENOSINE
DEOXYGUANOSINE
DEOXYCYTIDINE
THYMIDINE
−4.0
−2.0
0
2
4
6
H_0
pH

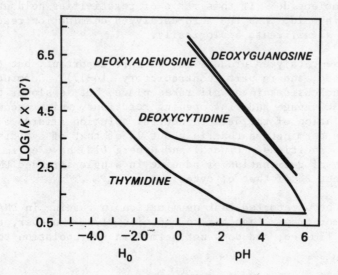

Fig. 3. Dependence of the logarithms of the rate constants (sec^{-1}) on pH and H_o at 95°, for deoxyribonucleoside hydrolysis in acid (21).

tidine at pH values below 0 reflects the hydrolysis of the cation (Figure 3) (21). The upturn in more strongly acidic solution is due to dication hydrolysis (K_2). The plots of the observed log K vs pH for deoxyadenosine and deoxyguanosine are approximately linear, with a slope of -1, over the acidic pH range (22, 24). No plateau area is observed. The situation can be accomodated to the mechanism of Figure 4 if it is assumed that both

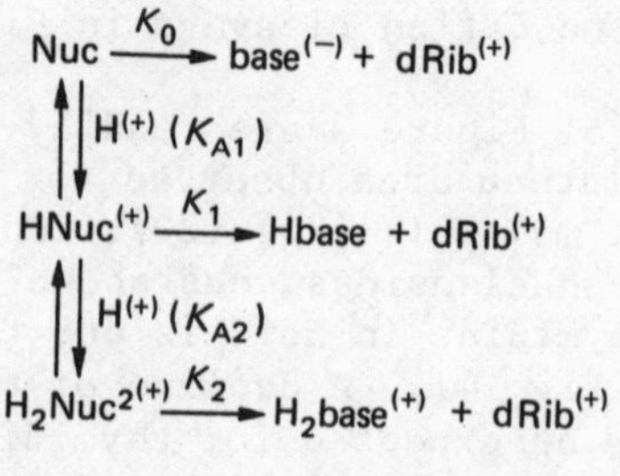

Nuc $\xrightarrow{K_0}$ base$^{(-)}$ + dRib$^{(+)}$
H$^{(+)}$ (K_{A1})
HNuc$^{(+)}$ $\xrightarrow{K_1}$ Hbase + dRib$^{(+)}$
H$^{(+)}$ (K_{A2})
H_2Nuc$^{2(+)}$ $\xrightarrow{K_2}$ H_2base$^{(+)}$ + dRib$^{(+)}$

Fig. 4. Paths for deoxyribonucleoside hydrolysis (21).

monocatonic and dicationic forms of the nucleoside are solvoly-
zing, and the following proportionality exists, $K_1/K_{A1} = K_2/K_{A2}$.
In such a case, the pH-rate profile would be linear through the
region where neutral species, monocation, and dication forms are
present.

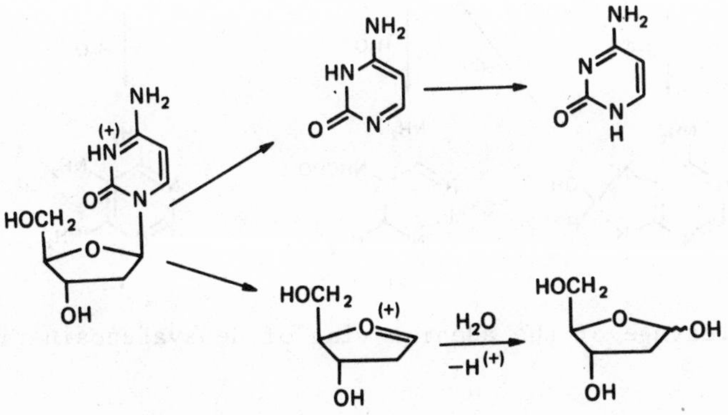

Fig. 5. N-glycosyl cleavage of deoxycytidine cation.

GLYCOSYL CLEAVAGE IN NEUTRAL AND BASIC SOLUTION

While extensive studies have been performed upon the hydrolysis of nucleosides by acid, much less information is available
about the cleavage of their glycosyl bonds in neutral and alkaline solution. Only adenine derivatives have been subjected to
detailed kinetic analysis (14,28). Adenine nucleosides are degraded by alkali to afford adenine directly, and also 4,5,6-tri-
aminopyrimidine. The path to the latter product presumably involves attack by hydroxide at the 8-position of the base, opening
of the imidazole ring, and loss of the sugar to give an observable intermediate, which is most likely the formyl derivative VI
(Fig. 6). Loss of the formyl group then produces the triaminopy-
rimidine, VII, which is further degraded by alkali. A possible
alternative pathway to VII involves loss of the formyl group
prior to cleavage of the sugar.

At any event, both reactions result in cleavage of the N-
glycosyl bond. Their dependence upon hydroxide ion concentration is not simple, however. At lower alkalinity, direct cleavage predominates, while at higher alkalinity, ring opening to

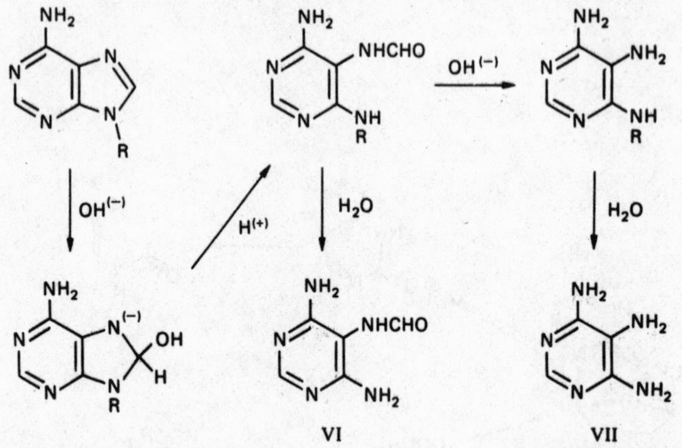

Fig. 6. Cleavage of the adenine ring of deoxyadenosine in alkali.

3,4,5-triaminopyrimidine is more rapid. At the least alkaline
pH reported, ring opening is perhaps 10% as fast as direct cleavage. Taken together, however, the two reactions fortuitously
appears to depend linearly upon alkali concentration (Figure 7).
Guanine nucleosides have been reported to be stable under alkaline
conditions (14), though guanine in DNA may be susceptible to hydrolysis (29) and deamination (17) in alkali. Protection of
guanine derivatives at pH values above 9 would be expected, as
they carry a negative charge. This consideration would not apply
at neutral pH, however.

Depurination rates in Bacillus subtilis DNA at pH 7.4 have
been measured by Lindahl and Nyberg (30). Unlike the case of deamination of cytosine, secondary structure had only a modest effect on the rate, with single stranded DNA depurinating at about
four times as rapidly as the double stranded form. The rate for
single stranded DNA was, however, perhaps 15 times as slow as
that projected for the free nucleosides. The reason for this effect is unknown. The release of guanine from double stranded DNA
was about 1.5 times faster than that of adenine. The result suggests that guanine, like adenosine, is subject to both acid and
base-catalyzed hydrolysis paths. The rate of release purines
from DNA increased as the pH was lowered from 7.4, as expected
from the results with nucleosides (Figure 3).

Depyrimidation of DNA also occurs on heating at pH 7.4, with thymine and cytosine released in roughly comparable amounts. As in the case of the purines, double stranded DNA reacted somewhat more slowly than single stranded DNA. Single stranded DNA was itself perhaps only 1/50 as reactive as the free nucleosides. Because of this effect, the process of depyrimidination of DNA took place at about 5% of depurination, in contrast to the expectation from nucleoside studies (Figure 3). The mechanism of N-glycosyl hydrolysis of deoxycytidine has not been studied at neutral pH, and it is not known whether any process other than the acid-catalyzed path exists.

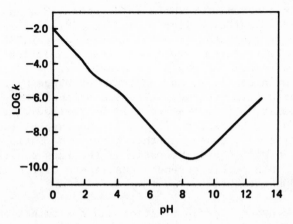

Fig. 7. Dependence of the logarithm of the rate constant (sec^{-1}) on pH, at 37°, for the hydrolysis of deoxyadenosine (23).

DESTRUCTION OF CYTOSINE NUCLEOSIDES BY BASE

In alkaline solution, cytosine nucleosides undergo both deamination and, in a slower process, destruction of the cytosine ring (6,31). The nature of the reaction has not been established, but it has been suggested that opening of the 3-4 bond of the cytosine ring is involved. The stability of the bond connecting the resulting ring-opened fragments to deoxyribose is also undetermined. If the bond were labile, then production of the free sugar might be the ultimate result, and the reaction would be scored as depyrimidination. Cytosine ring cleavage in nucleosides occurs at a rate about ¼ of that of deamination, in strongly alkaline solution (2). At neutral pH, however, its rate is less than 5% of that of deamination. It is not known whether the reaction takes place in single or double stranded DNA.

CLEAVAGE OF PHOSPHODIESTER BONDS

The vacant sites in DNA that result from the hydrolytic release of a base are vulnerable locations for chain breakage. The
deoxyribose residues at such sites exist in part in the aldehyde
form. As such, they can undergo successive β-elimination reactions to afford a degraded sugar residue and adjacent DNA chain
fragments with terminal 3' and 5´ phosphate group (32,33). The
elimination reactions occur considerably more rapidly in alkali,
than in neutral solution. In physiological buffers, pH 7.4, the
rate can be cut in half by omission of Mg$(++)$ or increased up to 25
times by addition of diamines such as lysine or 1,4-diaminobu-
tane.

The rate constant for chain breakage at a vacant site in
DNA (32) in physiological buffers, pH 7.4, 37°, is about 30,000
times as fast as the rate for depurination (depyrimidination can
be negelected, within the accuracy of the calculation). If one
starts with an intact DNA chain, however, there will initially be
no vacant (apurinic) sites, and no chain breakage by this mechanism. As depurination proceeds, the rate of chain breakage will
rise. Eventually, a steady state will be reached where the rate
of formation of vacant sites will equal the rate of their loss
by β-elimination reactions. At that point, the ratio of purines
to vacant sites will equal the ratio of the rate constants:
30,000 to 1. In a double stranded DNA, there will be 1 vacant
site per 30,000 base pairs.

Chain breakage via depurination and β-elimination is a well
recognized path leading to chain breaks in DNA, but not necessarily the only one. In a recent study (34) it was noted that the
rate at which T7 phage DNA accumulated single-strand breaks at
neutral pH was several times faster than the rate of depurination.
The difference was ascribed to "spontaneous destruction of deoxyribose moieties or direct hydrolysis of phosphodiester bonds."
The latter explanation seems more likely in terms of known chemical reactions. The existence of a direct path for phosphodiester
hydrolysis in DNA would also provide an explanation for an observation of Lindahl (35) that the rate of chain breakage in poly
(dA) is only three times as fast as that in poly (dC). A much
greater difference would be expected if the depurination path
were the only one existing. The depurination path was found to
be predominant, however, for the cleavage of phage PM2 DNA which
contained 14 apurinic sites per molecule. The rate of breakage
of this DNA was 30 times larger than that of control PM2 DNA (32).
This data can be understood if we calculate the steady state concentration of apurinic sites in the control, which is less than
one per molecule. The rapid rate of cleavage of the first-men-
tioned DNA, with its high level of apurinic sites, does not pre-
clude the existence of a second path for cleavage in the control.

SUMMARY

    Estimated rate constants for the various processes in nu-
cleosides and in DNA are summarized in Table 1. This information
has been extrapolated from limited data in many cases, and is only
meant to be an approximation. The same figures have been converted to a different form in Table 2. The time needed for a single event, on the average, to take place in a single stranded DNA
of 2 million bases, or a double stranded DNA of 1 million base
pairs has been computed (Table 2).

Table 2. Time Needed for One Event (pH 7.4, 37°)

Reaction	ssDNA(2×10^6 bases)	dsDNA(10^6 base pairs)
Depurination	2.5 hours	10 hours
Depyrimidination	50 hours	200 hours
Deamination of C	2.8 hours	700 hours
Deamination of A	140 hours	?
C ring opening	70 hours	?
Chain break, after depurination	?	10 hours
Chain break, direct	?	2.5 hours

The most significant event for double stranded DNA appears to be
chain breakage, either by a direct mechanism, or via prior depurination. Single stranded DNA is more vulnerable to degradation than the double stranded form (36). Insufficient data are
on hand to estimate chain breakage rates at physiological pH. Depurination and deamination are the most significant of the reactions whose rates are listed.

    Half-lives for degradation of the major nucleoside compo-
nents of DNA and RNA can also be computed from the data of Table
1. They vary from 22 years for deoxyguanosine to 3,5000 years
for uridine. These hydrolyses are too slow to be significant as a
factor in metabolism, but they may play a role in another context.
Many theories concerning the origin of life on Earth have suggested that organic compounds accumulated in the oceans during the
early history of this planet: "It appears quite likely that the
primitive ocean was indeed rich in dissolved organic compounds

Table 1. Estimated Rate of Hydrolysis of DNA (sec^{-1}, pH 7.4, 37°)

Reaction	Nucleoside	ssDNA	dsDNA
Depurination	G: 1×10^{-9}, A: 0.8×10^{-9} } 1.8×10^{-9}	1.2×10^{-10}	3×10^{-11} (G:A, 1.5:1)
Depyrimidination	T: 1.5×10^{-10}, C: 1.0×10^{-10} } 2.5×10^{-10}	6×10^{-12} (T=C)	1.5×10^{-12} (T=C)
Deamination of C	4.4×10^{-10}	2×10^{-10}	8×10^{-13}
Deamination of A (G?)	3×10^{-12}	4×10^{-12}	?
A ring opening	10% contributor to A depurination		?
C ring opening	1.7×10^{-11}	8×10^{-12}	?
Chain break, after depurination	-	?	3×10^{-11}
Chain break, direct	-	?	6×10^{-11}

which may have included many or all of the basic building-block
molecules we recognize in living cells today." (37). The first
replicating nucleic acid was formed by random processes in this
prebiotic "soup," and life began. This process presumably took a
few million years, or longer. The data listed in Table 1 makes
it unlikely that any build-up of nucleic acid components could
take place over a long period of time in an aqueous environment.
The half-lives would be lengthened if the temperatures were lower,
but the presence of catalytic buffers, or a change of pH, would
shorten them.

It appears unlikely, then, that the life arose on Earth by
the process outlined above. Some possible alternatives are that
nucleic acids were produced in a non-aqueous environment, that
the synthesis of the first replicating nucleic acid took place
within a few years, and that nucleic acids were not the first hereditary material, but were made later in evolution, with the aid
of enzymes.

ACKNOWLEDGEMENT

The work cited in this review that was performed in the laboratory of the author was supported by grants ES-01033, GM-11437,
and GM-18589 from the National Institutes of Health, U.S.A.

REFERENCES

1. Shapiro, R. and Klein, R.S. (1966) Biochemistry 5:2358-2362
2. Garrett, E.R. and Tsau, J. (1972) J. Pharm Sci. 61:1052-1061
3. Budowsky, E.I., Sverdlov, D.D., Shibaeva, R.P., Monastyrkaya,
    G.S. and Kochetkov, N.K. (1971) Biochim. Biophys. Acta 246:
    300-319
4. Notari, R.E., Chin, M.L. and Wittebort, R. (1972) J. Pharm Sci.
    61:1189-1196
5. Dudcz, E., Darzynkiewicz, E. and Shugar, D. (1977) Acta Bio-
    chim. Polon. 24:207-214
6. Fisher, G.J. and Johns, H.E. (1976) In: S.Y. Wang (ed) "Photo-
    chemistry and Photobiology of Nucleic Acids", Vol. I, Aca-
    demic Press, New York, pp 169-224
7. Shapiro, R., DiFate, V. and Welcher, M. (1974) J. Amer. Chem.
    Soc. 96:906-912
8. Slae, S. and Shapiro, R. (1978) J. Org. Chem. 43:1721-1726
9. Hayatsu, H. (1976) Progr. Nucleic Acid Res. Mol. Biol. 16:75-
    124
10. Shapiro, R. (1977) Mutation Res. 39:149-176
11. Lindahl, T. and Nyberg, B. (1972) Biochemistry 11:3610-3618
12. Coulondre, C., Miller, J.H., Farabaugh, P.J. and Gilbert, W.
    (1978) Nature 274:775-780

13. Jordan, D.O. (1960)"The Chemistry of Nucleic Acids," Butterworth's, Washington, D.C., p 65
14. Jones, A.S., Mian, M. and Walker, R.T. (1966) J. Chem. Soc. (C):692-695
15. Kammen, H.O. and Spengler, S.J. (1970) Biochim. Biophys. Acta 213:352-364
16. Ganguli, P.K., Reiner, A. and Gyenes, L. (1971) Biochim. Biophys. Acta 254:167-171
17. Ullman, J.S. and McCarthy, B.J. (1973) Biochim. Biophys. Acta 294:396-404
18. Lindahl, T. and Nyberg, B. (1979) Progr. Nucleic Acid Res. Mol. Biol. 22:135-192
19. Capon, B. (1969) Chem. Rev. 69: 407-498
20. Shapiro, R. and Kang, S. (1969) Biochemistry 11:1806-1810
21. Shapiro, R. and Danzig, M. (1972) Biochemistry 11:23-29
22. Zoltewicz, J.A., Clark, D.F., Sharpless, T.W. and Grahe, G. (1970) J. Amer. Chem. Soc. 92:1741-1750
23. Garrett, E.R. and Mehta, P.J. (1972) J. Amer. Chem. Soc. 94: 8532-8541
24. Zoltewicz, J.A. and Clark, D.F. (1972) J. Org. Chem. 37:1193-1197
25. Hevesi, L., Wolfson-Davidson, E., Nagy, J.B., Nagy, O.B. and Bruylants, A. (1972) J. Amer. Chem. Soc. 94:4715-4719
26. Romero, R., Stein, R., Bull, H.G. and Cordes, E.H. (1978) J. Amer. Chem. Soc. 100:7620-7624
27. Cadet, J. and Teoule, R. (1974) J. Amer. Chem. Soc. 96:6517-6519
28. Garrett, E.R. and Mehta, P.J. (1972) J. Amer. Chem. Soc. 94: 8542-8547
29. Greer, S. and Zamenhof, S. (1962) J. Mol. Biol. 4:123-141
30. Lindahl, T. and Nyberg, B. (1972) Biochemistry 11:3610-3618
31. Jones, A.S., Mian, A.M. and Walker, R.T. (1966) J. Chem. Soc. (C):1784-1786
32. Lindahl, T. and Anderson, A. (1972) Biochemistry 11:3618-3623
33. Brown, D.M. (1974) In: P.O.P. Ts'o (ed) "Basic Principles in Nucleic Acid Chemistry," Vol. II, Academic Press, New York, pp 1-90
34. Crine, P. and Verly, W.G. (1976) Biochim. Biophys. Acta 442: 50-57
35. Lindahl, T. and Ljungquist, S. (1977) In: P.C Hanamawalt and R.B. Setlow (eds) "Molecular Mechanisms for Repair of DNA," Part A, Plenum Press, New York, pp 31-38
36. Eigner, J. Boedtker, H. and Michaels, G. (1961) Biochim. Biophys. Acta 51:165-168
37. Lehninger, A.L. (1975)"Biochemistry,"Second Edition, Worth, New York, p 24

METHYLATION OF CELLULAR DNA BY THE CARCINOGEN N-METHYL-N-
NITROSOUREA AND REMOVAL OF METHYLPURINES - COMPARISONS OF
Escherichia coli, MOUSE AND HUMAN CELLS

Philip D. Lawley

Pollards Wood Research Station
Institute of Cancer Research
Chalfont St.Giles, Bucks, U.K. HP8 4SP

INTRODUCTION

The spectrum of methylation products in DNA depends on the
reactivity of the methylating agent and has a marked influence on
the biological effects.

An important secondary factor is the ability of cells to remove methylation products from methylated DNA before this is used
as a template.

It is now clear that some products are potentially "cytotoxic
lesions" (such as 3-methyladenine) and some are "promutagenic"
(such as O^6-methylguanine and O^4-methylthymidine). Carcinogenic
potency of methylating agents is associated with ability of agents
to induce mutation through direct miscoding of the promutagenic
methylated bases.

Thus, methyl methanesulphonate (MMS) and dimethyl sulphate(DMS),
for which the ratio O^6-methylguanine : 7-methylguanine in DNA
methylation products is relatively low (0.004), are correspondingly
weak mutagens (in the sense specified); whereas N-methyl-N-nitrosourea (MNUA) giving a higher ratio O^6-methylguanine : 7-methyl-
guanine (0.11) is a much more efficient mutagen and carcinogen (see
e.g. Lawley, 1980; Newbold, et al., 1980).

Studies of methylation of DNA by both types of carcinogen in
cultured cells (Lawley and Thatcher, 1970) and in mice in vivo (Frei
et al., 1978) showed that the initial spectrum of methylation pro-

ducts was identical with that found with isolated DNA in vitro.
However, using E.coli cells, it was immediately apparent that some
methylation products could be removed from the methylated cellular
DNA, much more rapidly than would result from hydrolysis at neutral
pH in vitro (Lawley and Orr, 1970). Thus, for cells methylated for
a mere 15 min. at 37^0C, the isolated DNA was found to be markedly
deficient in 3-methyladenine (for which the ratio to 7-methylguanine
initially would be 0.21 for DMS and 0.14 for MNUA). Also, subsequent
incubation of methylated cells caused removal of $\underline{O}^6$-methylguanine
over a somewhat longer period of about 1 hr.

The interpretation was that "DNA repair" enzymes could remove
the methylated bases, and the subsequent indications were that
different enzymes were involved, for removal of 3-methylpurines on
the one hand, and for $\underline{O}^6$-methylguanine on the other, with the latter
dependent on ability of cells to synthesise protein (see Warren and
Lawley (1980) for some recent work and a summary of previous
relevant reports).

A particularly interesting outcome of these studies was that,
whereas all cells examined were efficient in removal of 3-methyl-
adenine, in accord with the concept that this product confers
cytotoxicity, there were marked species- and tissue-dependences of
removal of the promutagenic base $\underline{O}^6$-methylguanine. This leads to
the deduction that resistance or sensitivity to the mutagenic or
carcinogenic action of methylating carcinogens would be observed
according to the relative ability or lack of ability of target cells
to remove this base.

An important area of research into the causation of human cancer
concerns the inherited susceptibility of certain individuals to
radiation. It is also a current belief that environmental chemicals
(such as dimethylnitrosamine, a carcinogen that methylates DNA
through metabolic conversion to the same methylating intermediate,
$CH_3N_2^+$, as MNUA) may be significant contributors to cancer causation.

Studies of the ability of individuals to remove potential promutagenic, and hence procarcinogenic, molecular lesions (such as
$\underline{O}^6$-methylguanine) from DNA can thus contribute to detection of
carcinogen-susceptible individuals.

The earlier studies, with cultured mouse and hamster cells
(Lawley and Thatcher, 1970), did not enable detailed studies of this
kind because of lack of sensitivity of methods. Improved methods
for analysis of methylated DNA, involving high performance liquid
chromatography, have now been developed (Frei, et al., 1978) and have
been applied to study the ability of cells treated with MNUA in vitro
to remove $\underline{O}^6$-methylguanine from DNA, and some typical results will
be reported to illustrate the potential of the method.

RESULTS AND DISCUSSION

The methods for preparation of DNA and its analysis have been described in detail elsewhere (Frei, et al., 1978; Lawley and Warren, 1980; Warren et al., 1979). Briefly, [^{14}C]methyl-labelled MNUA is used to methylate cells, DNA is isolated after various times of incubation of methylated cells, and hydrolysed (0.1M-HCl, 70^{0}C) to liberate purines and methylpurines (together with a trace of the minor pyrimidine product O^2-methylcytosine). The extents of methylation at N-1 of adenine, N-3 of adenine (or guanine, in absence of ^{14}C-incorporation into adenine), N-7 of adenine and guanine, and O-6 of guanine can then be determined.

Table 1 presents typical data to illustrate how this procedure can be used for "rapid screening" of the ability of cells to remove O^6-methylguanine. Since the half-life of the reagent MNUA is about 0.25 hr. at 37^{0}C in most media used for cellular suspensions, the methylation reaction is effectively over in about 1 hr. During this time, the cells most proficient in this ability, such as E.coli (in general), and "normal" (i.e. not virally transformed) human lymphocytes or cultured fibroblasts, will have removed a significant part of the O^6-methylguanine. This effect can be recognized by the observation that the ratio O^6-methylguanine : 7-methylguanine is less than 0.11, the "in vitro-chemical" value. Another useful guide, for human cells, is that (as with mammalian cells in general) the rate of removal of 3-methyladenine is less than with E.coli, so that the ratio O^6-methylguanine : 3-methyladenine will be less than unity in proficient cells (Medcalf and Lawley, 1980).

Table 2 summarizes comparative proficiencies of various cells. A relevant comment is that, in general, the proficient cells are more easily distinguished at lower extents of methylation; although the rate of removal of O^6-methylguanine increases as methylation increases, it does not generally appear to do so in proportion to methylation, so that a decreased efficiency as dose increases is apparent. This does not occur for removal of 3-methyladenine.

Comparisons between species have so far indicated that in vivo proficiency parallels that in vitro. However, it must be remembered that in vivo proficiency is known in some cases to be also tissue-dependent, e.g. mouse liver is more proficient than mouse lymphoid tissue. With regard to E.coli, whereas uvr strains are proficient (Warren and Lawley, 1980), methylation-sensitive mutants have been isolated that are deficient in ability to remove O^6-methylguanine (ada, Jeggo, 1979).

An important corollary of the findings is that human cells would be expected by these criteria to be relatively resistant to the mutagenic and carcinogenic action of methylating carcinogens if

Table 1. Removal of methylpurines from DNA - interspecies comparison

Ability of cells to remove from methylated DNA a
promutagenic methylpurine ($\underline{O}^6$-methylguanine) or a
cytotoxic methylpurine (3-methyladenine), as indicated
by the ratios (methylpurine : 7-methylguanine) in
cellular DNA, 1 h (or 0.75 h for E.coli) after treatment
of cells with $\underline{N}$-methyl-$\underline{N}$-nitrosourea at 37^0C.
Significant lack of removal of $\underline{O}^6$-methylguanine in
this time is indicated by *.

	Extent of methylation: (μmol 7-MeGua) (mol DNA - P)	$\dfrac{\text{3-methyladenine}}{\text{7-methylguanine}}$	$\dfrac{\underline{O}^6\text{-methylguanine}}{\text{7-methylguanine}}$
DNA methylated In vitro	21000	0.13	0.11*
E.coli AB1157	1502	~0.00	0.023
E.coli AB1886(uvr⁻)	100	~0.00	0.01
E.coli AB1157 + chloramphenicol	1838	~0.00	0.11*
Mouse (Balb/c,f) Lymphocyte (Thymus)	25	0.10	0.12*
Human (m) Lymphocyte	20	0.090	0.055
Human (m) Lymphocyte	25	0.090	0.054
Human (m) Lymphocyte	78	0.089	0.109*

Table 2. Ability of cells to remove $\underline{O}^6$-methylguanine from DNA
          methylated by carcinogens

Relatively proficient	Relatively deficient
E.coli ($\begin{array}{l}\text{except } \underline{ada}^- \\ \text{including } \underline{uvr}^-\end{array}$)	E.coli with inhibition of protein synthesis
	E.coli $\underline{ada}^-$
Human fibroblasts or lymphocytes	Mouse lymphocytes (and cells of other organs except liver)
Mouse liver	Chinese hamster V79 lung

compared with mouse cells (other than mouse liver). Since most
carcinogens are tested in mice, it is perhaps reassuring that man
is unlikely to be more susceptible to this type of carcinogen than
the mouse, but this assessment cannot necessarily be extrapolated
to other carcinogens.

This leads to consideration of the possibilities of inherited
sensitivity to chemical carcinogens (comparable with the well-known
examples of radiation-sensitivity). Comparative studies of inbred
mouse strains have already shown differences in sensitivity to MNUA
as a carcinogen (Frei and Lawley, 1980), but so far no clear
indication of a correlation between susceptibility and defects in
DNA repair has emerged. Using the methods reported here, further
comparative studies of mouse and man are under way.

ACKNOWLEDGEMENTS

I wish to acknowledge the collaboration in this work of Mr.A.S.
C. Medcalf and Mr. W. Warren of this Institute, and of Dr. J.V. Frei
(Dept.of Pathology, University of Western Ontario, London, Canada)
and Dr. G. Harris (Dept.of Pathology, Kennedy Institute of Rheuma-
tology, London, U.K.); also to thank Dr. D.G. Harnden (Dept.of Cancer
Studies, University of Birmingham, U.K.) and Dr. C.F. Arlett (MRC
Cell Mutation Unit, University of Sussex, U.K.) for supplying some
of the human cells used. This work was supported by grants from
the MRC and CRC.

REFERENCES

Frei, J. V., and Lawley, P. D., 1980, Thymomas induced by simple
    alkylating agents in C57BL/Cbi Mice: Kinetics of the dose
    response, J.Natl.Cancer Inst. 64:845.
Frei, J. V., Swenson, D. H., Warren, W., and Lawley, P. D., 1978,
    Alkylation of deoxyribonucleic acid in vivo in various
    organs of C57BL mice by the carcinogens MNUA, ENUA and EMS
    in relation to induction of thymic lymphoma. Some appli-
    cations of high-pressure liquid chromatography. Biochem.J.,
    174:1031.
Jeggo, P., 1979, The isolation and characterisation of Escherichia
    coli K12 mutants unable to induce the adaptive response to
    simple alkylating agents, J.Bacteriol., 139:783-791.
Lawley, P. D., 1980, DNA as a target of alkylating carcinogens,
    Brit.Med.Bull., 36:19.
Lawley P. D., and Orr, D. J., 1970, Specific excision of methyl-
    ation products from DNA of Escherichia coli treated with
    N-methyl-N'-nitro-N-nitrosoguanidine., Chem.Biol. Inter-
    actions, 2:154.
Lawley, P. D., and Thatcher, C. J., 1970, Methylation of deoxy-
    ribonucleic acid in cultured mammalian cells by N-methyl-
    N'-nitro-N-nitrosoguanidine, Biochem. J. 116:693.
Lawley, P. D., and Warren, W., 1980, The detection and measurement
    of alkylation products in DNA, in "Techniques in DNA Repair
    Research", E. C. Friedberg and P. C. Hanawalt, eds., Marcel
    Dekker, New York.
Medcalf, A. S. C., and Lawley, P. D., 1980, Dose-dependent removal
    of O^6-methylguanine from methylated human cells, in
    preparation.
Newbold, R. F., Warren, W., Medcalf, A. S. C., and Amos, J., 1980,
    Mutagenicity of carcinogenic methylating agents is associ-
    ated with a specific DNA modification, Nature (Lond.),
    283:596.
Warren, W., and Lawley, P. D., 1980, The removal of alkylation pro-
    ducts from the DNA of Escherichia coli cells treated with
    the carcinogens N-ethyl-N-nitrosourea and N-methyl-N-nitro-
    sourea: influence of growth conditions and DNA repair
    defects, Carcinogenesis, 1:67.
Warren, W., Crathorn, A. R., and Shooter, K. V., 1979, The stability
    of methylated purines and of methylphosphotriesters in the
    DNA of V79 cells after treatment with N-methyl-N-nitrosourea,
    Biochim.Biophys.Acta, 563:82.

ON THE USE OF A DAMAGE-SPECIFIC DNA BINDING PROTEIN

TO PROBE LESIONS IN DNA

Ross S. Feldberg

Department of Biology
Tufts University
Medford, MA   02155

INTRODUCTION

I never really intended to study a DNA-binding protein at all.
I originally set out to purify a pyrimidine dimer-specific endonuclease from human cells, but when the fractions were all collected
what I found myself with was a DNA binding protein which bound
UV-irradiated, but not nonirradiated, double-stranded DNA. To my
surprise, I found that the protein bound perfectly well to photoreactivated UV-DNA, which indicated that it was recognizing some
lesion other than the pyrimidine dimer[1].   This conclusion was
confirmed by studies on the photochemistry of damage introduction.
Irradiation of DNA with 313nm light in the presence of acetophenone,
a treatment which should result only in triplet state thymine-
derived photoproducts (e.g. thymine dimers), yielded essentially no
protein-recognizable damage. However, irradiation of either DNA or
poly d(A-T) in the presence of acetone, whose triplet state is
sufficiently energetic to give energy transfer to all four bases,
did give a substrate for the binding protein[2]. These data seemed
to suggest that the protein was recognizing a triplet state-derived
adenine (or purine) photoproduct. Life, alas, is never quite that
simple.

RESULTS AND DISCUSSION

The Specificity of Damage Recognition

If indeed this protein is recognizing adenine-derived damage,
it should be possible to demonstrate binding to a copolymer formed
by annealing irradiated poly dA to unirradiated poly dT. I examined
this prediction by using unlabelled homopolymers as competitors for

25

protein binding to irradiated labelled T7 DNA. The nonirradiated
copolymer gave essentially no competition (Table 1), while a copolymer which had been first annealed and then irradiated gave a 17%
inhibition of protein-DNA binding as measured in a filter binding
assay[1]. As I had predicted, poly dA irradiated with 254 nm light
and then annealed to nonirradiated poly dT did indeed act as a competitor. Unfortunately, and to my total dismay, irradiated poly dT
annealed to nonirradiated poly dA was an even better competitor.
That the inhibition was not simply due to filter saturation was suggested by the fact that incubation of protein and T7 DNA followed by
addition of the various copolymers immediately prior to filtration
gave no inhibition of DNA retention. These results suggested that
the protein wasn't simply recognizing some adenine photoproduct. Up
to this point, I had been concerned with identifying the lesion recognized by the protein, but it began to dawn on me that this focus
might be misplaced and that the protein might actually have a rather
broad specificity for DNA damage.

    An examination of substrate specificity seemed in order and a
summary of our results with a variety of damaging agents is presented
in Table 2.   Although the protein does not bind to single-strand
breaks, apurinic sites, single-strand tails, thymine dimers or cross-
links in DNA, the protein does recognize UV- and γ-irradiated DNA
and poly d(A-T) as well as HNO_2- or $NaHSO_3$-treated nucleic acids[3].
In addition, we have found that an enzymatic superoxide ($O_2 \cdot^-$)-gen-
erating system will give protein-recognizable damage in DNA as will
low levels of H_2O_2 and Fe(II). Although it may be that the protein
has an unusually broad specificity for DNA lesions, a more likely
explanation of these data is that the protein is not recognizing the
damage per se, but is rather recognizing some consequence common to a
variety of forms of monofunctional base damage, such as a minor helix
distortion or a non hydrogen-bonded base pair. A good test of this

Table 1. Irradiated Homopolymer Inhibition of Protein-DNA Binding

Homopolymer[a]	Irradiation	% ^{3}H-DNA Bound to Filter
---	---	65
poly (dA) · poly (dT)	---	62
poly (dA) · poly (dT)	254 nm; 100 J/m^2	53
poly (dA)	" "	45
poly (dT)	" "	33

a. Tritium labelled T7 DNA (7000 cpm/assay; 1.73 x 10^{-4} μmoles
    nucleotides/assay) was irradiated to 50 J/m^2 with 254 nm light.
    Homopolymers (4.65 x 10^{-3} μmoles nucleotides per assay) were
    irradiated as indicated above and, if single-stranded, were an-
    nealed to their unirradiated complements before testing them
    as inhibitors.

Table 2:   The Substrate Specificity of the Damage-Specific
DNA Binding Protein

Nucleic Acid	Damaging Agent	Possible Substrate	Binding
T7 DNA	---	---	-
λ DNA	---	single strand tails	-
φX-174RFI	---	superhelical regions	-
T7 DNA	low pH, heat	apurinic sites	-
φX-174RFI	DNase I	5'-P termini	-
"	Micrococ. nuclease	5'-OH termini	-
T7 DNA	psoralen (313nm)	crosslinks, adducts	-
T7 DNA	254 nm	dimers, glycols, ?	+
"	acetophenone(313nm)	thymine dimers	-
"	acetone (313nm)	dimers, ?	+
"	NaHSO$_3$	G-U mismatches	+
PBS2 DNA	---	A-U base pairs	-
"	254 nm	?	+
T7 DNA	HNO$_2$	G-U, X-C, I-T mismatches	+
"	γ-irradiation	glycols, hydroperoxides, ?	+
"	O$_2\cdot^-$, OH·	?	+
"	Fe(II)	?	+
"	H$_2$O$_2$	?	+
Poly d(A-T)	---	---	-
"	254 nm	glycols, ?	+
"	OsO$_4$	glycols	+
Poly dA$_I$:dT	254 nm	A photoproducts	+
Poly dA:dT$_I$	254 nm	T photoproducts	+

hypothesis would be to prepare an undamaged nucleic acid which contains base mismatches. We have attempted to do this using DNA
polymerase under error-inducing conditions, but our results to date
have been ambiguous. Poly d(A-T) with a high level of misincorporated G is a substrate for the binding protein, but poly d(A-T)
with a low level of G incorporated is not. We could perhaps invoke
"creeping" of the mismatched G into the looped ends of the self
annealed poly d(A-T), but there is no direct evidence for this and we
are continuing our work with synthetically-prepared mismatched polymers in order to obtain a clearer answer.

The Radiation Chemistry of Damage Introduction

As UV- and γ-endonucleases have been employed to study lesion
introduction and repair (reviewed in 4), we have used the damage-
specific DNA binding protein to follow monofunctional base damage in
DNA. Particularly with γ-irradiation, in which the large proportion
of strand breaks and alkali labile sites introduced make endonuclease
probes difficult to apply, the filter binding assay provides a

simple and sensitive probe for base damage. For this reason, I chose
to use the binding protein to examine the radiation chemistry of
damage introduction by ionizing radiation.

Unlike UV, ionizing radiation introduces DNA damage primarily
through the mediation of the radiolysis products of water: the
hydroxyl radical (OH·), the aqueous electron (e$\bar{a}$q) and the hydrogen
radical (H·). In aerated solutions, however, both e$\bar{a}$q and H· will
combine with oxygen to generate the superoxide anion radical ($O_2 \cdot^-$).
We can begin to dissect out the contribution of these species to DNA
damage by carrying out the γ-irradiation under varying conditions.
From dose-response curves for filter binding, I have presented in
Table 3 the DNA binding to an excess of binding protein after 30
krads of irradiation. From the Poisson distribution, it is possible
to calculate the mean amount of damage introduced per T7 molecule
per krad. This calculation is based on the assumption that a single
lesion is sufficient to ensure full retention of a DNA molecule. If
this assumption isn't true, then these data will be an underestimate
of the rate of damage introduction. From the results of Table 3, it
is clear that the maximum rate of damage introduction takes place in
air. Under N_2O, in which the OH· level is doubled due to the conversion of e$\bar{a}$q to OH·, very little damage is introduced. This result is
of interest since it has been demonstrated that OH· is responsible
for the introduction of strand breaks in DNA[5]. That OH· is involved
in nucleotide damage is indicated by the inhibition of damage introduction by I^-, a specific OH· scavenger. The crucial difference
between irradiation in air and in N_2O is the presence of $O_2 \cdot^-$ and
indeed its involvement is confirmed by the inhibition caused by NO_3^-,
a specific e$\bar{a}$q scavenger. In the presence of NO_3^-, e$\bar{a}$q will be removed from solution and thus the $O_2 \cdot^-$ levels will drop. Thus, these

Table 3.   Damage Introduction by γ-Irradiation

Irradiation Condition[a]	% binding/30 krad	lesions/krad/T7
air	26.5	.0103
N_2O	6	.0023
air + I^-(100mM)	9	.0033
air + NO_3^-(100mM)	8	.0029
N_2	18.5	b
N_2 + I^-(100mM)	18.5	"
N_2 + NO_3^-(100mM)	6	.0023

a. T7 DNA irradiated at room temperature at 1.3 µg/ml in 0.01 $\underline{M}$
    Tris-HcL, 0.001 $\underline{M}$ EDTA, pH8.

b. Non-linearity at low doses in N_2-irradiations prevents an
    accurate estimate of the rate of damage introduction.

results indicate that in the presence of air it is a <u>combination</u> of
OH· and $O_2^{·-}$ that is critical for nucleotide damage introduction.
Although irradiation under N_2 did not give a linear dose response, so
that it is difficult to estimate a rate of damage introduction, the
observation that the damage introduction was inhibited by NO_3^- but
not by I^- suggests that eāq itself can introduce nucleotide damage.

Our results with γ-irradiated DNA have illustrated the utility
of the damage-specific DNA binding protein as a probe for monofunctional base damage. We have recently begun to extend our studies
to other damaging agents. With a xanthine plus xanthine oxidase
superoxide-generating system we have found that DNA damage is introduced in a reaction dependent on H_2O_2 and trace amounts of iron,
suggesting again a role for hydroxyl radical. We have also found
damage introduction by low levels of H_2O_2 and by low levels of
Fe(II). In the future, I intend to examine a variety of carcinogens
for their ability to introduce protein-recognizable damage and to
continue with a more detailed examination of the radiation chemistry
of γ-irradiation. It will be of interest to see if we can mimic the
effect of hypoxic cell sensitizers in an <u>in vitro</u> system as well
as to extract DNA from <u>in vivo</u> irradiated cells and to examine the
fate of base damage in a repair-proficient environment.

ACKNOWLEDGMENT

This research was supported by grant CA19419 from the National
Cancer Institute, DHEW. The assistance of Ms. Judith Lucas, Ms.
Josephine Carew, and Mr. Andrew Cranin is gratefully acknowledged.

REFERENCES

1. R.S. Feldberg and L. Grossman, A DNA Binding Protein from Human
      Placenta Specific for Ultraviolet Damaged DNA, <u>Biochemistry</u>
      15: 2402 (1976).

2. R.S. Feldberg, Non-Dimer UV Damage in DNA and Poly d(A-T), <u>in</u>:
      "DNA Repair Mechanisms," P.C. Hanawalt, E.C. Friedberg and
      C.F. Fox, eds, Academic Press, New York (1978).

3. R.S. Feldberg, On the Substrate Specificity of a Damage-Specific
      DNA Binding Protein from Human Cells, <u>Nuc</u>. <u>Acids</u> <u>Res</u>., <u>8</u>:
      1133 (1980).

4. M.C. Paterson, Use of Purified Lesion-Recognizing Enzymes to
      Monitor DNA Repair <u>In</u> <u>Vivo</u>, <u>Adv</u>. <u>Rad</u>. <u>Biol</u>., <u>7</u>: 1 (1976).

5. P. Achey and H. Duryea, Production of DNA Strand Breaks by the
      Hydroxyl Radical, <u>Int</u>. <u>J</u>. <u>Radiat</u>. <u>Biol</u>., <u>25</u>: 595 (1974).

DNA CHEMICALLY MODIFIED BY N-ACETOXY-N-2-ACETYL-AMINOFLUORENE :

NATURE OF THE ADDUCTS AND CONFORMATION OF THE MODIFIED DNA

Marc Leng, Evelyne Sage and Pascale Rio

Centre de Biophysique Moléculaire, C.N.R.S., 1A, avenue
de la Recherche Scientifique, 45045 Orléans Cedex,
France

SUMMARY

Two aspects of chemical modifications of DNA by aminofluorene
derivatives are considered. On one hand, it is shown that the covalent binding of AAF residues to the C(8) of guanine residues can
induce a conformational change of poly(dG.dC).poly(dG.dC) from the
B-form to the Z-form or a Z-like form. On the other hand, the presence of a deacetylated adduct in liver DNA of rats treated with
the carcinogen is demonstrated by the use of specific antibodies.

Chemical carcinogens can react with nucleic acids and proteins.
Until now, most of the works have been devoted to the reaction with
DNA. The covalent binding of the carcinogens to DNA can modify the
conformation of DNA and consequently the interactions between the
modified DNA and some proteins. This might be important in the
tumorogenic process. This assumption is the basis of our work on
DNA modified by aminofluorene derivatives.

In the *in vivo* reaction between N-acetoxy-N-2-acetyl-amino-
fluorene (N-AcO-AAF) and DNA, several adducts are formed. Two of
them have been identified as N-(deoxyguanosin-8-yl)-acetylamino-
fluorene and as 3(deoxyguanosin-N-2-yl)-acetylaminofluorene and
they represent about 30 % of the total adducts. The other adducts
(about 70 %) are not yet completely identified. However, it is
known that the acetyl groups (on the nitrogen of the aminofluorene
residues) are removed. The acetylated adducts are also formed *in
vitro* by the non enzymatic reaction between DNA and N-AcO-AAF
(general review 1).

In this paper, we report two aspects of our work. On one hand, it is shown that specific antibodies can be used to identify and to titrate some adducts. On the other hand, the conformation of modified model DNA (poly(dC.dG).poly(dG.dC)) has been studied by circular dichroism.

Antibodies to Modified Guanosine

N-(guanosin-8-yl) acetylaminofluorene (Guo-AAF) was prepared by the reaction between N-AcO-AAF and guanosine (2). By alkali treatment, a part of the product was transformed in Guo-F (the chemical nature of this product is not yet known but an opening of the imidazole ring of guanine residue is assumed to occur and the acetyl group on the nitrogen of the aminofluorene residue is removed, 1,3). Guo-AAF and Guo-F were linked to bovine serum albumin and rabbits were immunized with the conjugate BSA-Guo-AAF and BSA-Guo-F respectively (2). The specific antibodies were purified by affinity chromatography and their specificities were studied in details by   radioimmunoassays. Some cross-reactions were found with some aminofluorene derivatives but in both cases, the antibodies had the largest affinities to Guo-AAF and Guo-F respectively. In the cross-reactions, the affinities of the antibodies were smaller by at least two orders of magnitude.

Rats were injected with acetyl-aminofluorene (26 mCi/mM) and sacrified after 24 hours. Liver DNA was purified and then treated during three hours at 37°C either by S_1 endonuclease from *Aspergillus oryzae* (pH 4.6 in presence of 4 mM $ZnSO_4$) or by deoxyribonuclease I from bovine pancreas (pH 7.0 in presence of 5 mM $MgCl_2$). The amounts of dGuo-AAF and dGuo-F were determined by radioimmunoassays with the anti Guo-AAF and the anti Guo-F antibodies. About the same amounts of dGuo-AAF and dGuo-F were found. The total amount of bound carcinogen is in agreement with the amount deduced from radioactivity measurements (of the order of 5 µmoles of carcinogen per mole of nucleotide).

One can conclude that the two main adducts are dGuo-AAF and dGuo-F.

Circular dichroism of modified poly(dG.dC).poly(dG.dC)

It is known that in the reaction between N-AcO-AAF and poly(dG.dC).poly(dG.dC), the substitution mainly occurs on the C(8) of guanine residues (4) and that there is no strong clustering of the substituted bases for percentages of substitution smaller than 10 % (5).

The circular dichroism of poly d(G.C)-AAF depended upon the amount of modified guanines. As this amount was increased, the intensity of the first positive band decreased and became negative, the intensity of the negative band decreased and became positive. These changes were similar to those observed with poly(dG.dC). poly(dG.dC) alone as the ionic strength was increased (6). From the analysis of the circular dichroism spectra difference, one can conclude that in low salt concentration, the covalent binding of AAF residues induces a conformational change identical to that induced in poly(dG.dC).poly(dG.dC) by the ionic strength or ethanol (6,7).

A new structure termed Z-DNA has been recently proposed from the study of the hexanucleotide $d(C.G)_3$. This fragment crystallizes as a left-handed double helical molecule with Watson-Crick base pairs. The cytidine residues have the anti conformation while the guanine residues have the syn conformation (8). According to Wang et al. (8), the high salt form of poly d(G.C) is the Z-form and the low salt form is the B-form.

Our results show that the bound AAF residues can induce the Z-form. In this form, the AAF residues are outside the double helix which is quite different from the insertion-denaturation model or the base displacement model (general reviews 9,10).

Specific antibodies can be useful to solve some of the problems encountered in chemical carcinogenesis. We have already shown that antibodies to adenosine, to cytidine, to modified guanosine and to modified DNA can discriminate between local defects in DNA induced by 2-aminofluorene derivatives (11). In this work, specific antibodies reveal the presence of a new deacetylated adduct in liver DNA of rats treated by acetylaminofluorene. It has to be noted that small amounts of adduct can be titrated (about 0.5 pmole).

On the other hand, we show that the covalent binding of AAF residues to the C(8) of guanine residues can induce a Z-form or a Z-like form in poly(dG.dC).poly(dG.dC). One can assume that this form can also be induced by the carcinogen in a G-C rich sequence of natural DNA. In the Z-form, the AAF residues are outside the double helix and the modified guanine residues are paired with the cytosine residues. This conformational change might be of importance in many respects for cellular events which follow the chemical modification of DNA and more especially, the DNA repair enzymes.

REFERENCES

1. K.E. Kriek, Aromatic amines and related compounds as carcinogenic hazards to man, in:"Environmental Carcinogenesis", P. Emmelot and E. Kriek, ed., Elsevier/North Holland Biomedical Press, Amsterdam, 1979.

2. M. Guigues and M. Leng, Reactivity of antibodies to guanosine
    modified by the carcinogen N-acetoxy-N-2-acetylaminofluorene,
    Nucleic Acids Res.,6:733 (1979).
3. M. Spodheim-Maurizot, M. Dreux, G. Saint-Ruf and M. Leng,
    Alkaline stability of guanosine and some of its derivatives
    modified by the carcinogen N-acetoxy-N-2-acetylaminofluorene,
    Nucleic Acids Res., 7:2347 (1979).
4. D.H. Harvan, J.R. Hass and M.W. Lieberman, Adduct formation
    between the carcinogen N-acetoxy-2-acetylaminofluorene and
    synthetic polydeoxyribonucleotides, Chem. Biol. Interactions,
    17:203 (1977).
5. E. Sage and M. Leng, Conformation of poly(dG.dC).poly(dG.dC)
    modified by the carcinogens N-acetoxy-N-2-acetylaminofluorene
    and N-hydroxy-N-2-aminofluorene, Proc. Nat. Acad. Sci. USA,
    in press.
6. F.M. Pohl and T.M. Jovin, Salt-induced cooperative conforma-
    tional change of a synthetic DNA : equilibrium and kinetic
    studies with poly(dG.dC), J. Mol. Biol., 67:375 (1972).
7. F.M. Pohl, Polymorphism of a synthetic DNA in solution, Nature,
    26:365 (1976).
8. A.H.J. Wang, G.J. Quigley, F.J. Kolpak, J.L. Crawford, J.H.
    van Boom, G. van der Marel and A. Rich, Molecular structure of
    a left-handed double helical DNA fragment at atomic resolution,
    Nature, 282:680 (1979).
9. M.P. Daune and R.P.P. Fuchs, Structural modification of DNA
    after covalent binding of a carcinogen in: "Réparation du DNA,
    mutagénèse, cancérogénèse chimique", CNRS ed., CNRS Paris (1977).
10. D. Grunberger and I.B. Weinstein, Conformational changes in
    nucleic acids modified by chemical carcinogen, in : "Chemical
    carcinogens and DNA", P.L. Grover, Ed., CRC Press, Inc. Boca
    Raton (1979).
11. E. Sage, M. Spodheim-Maurizot, P. Rio, M. Leng and R.P.P. Fuchs,
    Discrimination by antibodies between local defects in DNA
    induced by 2-aminofluorene derivatives, FEBS Letters, 108:66
    (1979).

HETERODUPLEX DNA IN BACTERIOPHAGE T1 AFTER IONIZING RADIATION

Heidi Martin-Bertram

Institut für Biologie, GSF
Abt. für Strahlenbiologie
D-8o42 Neuherberg

INTRODUCTION

Most of the DNA lesions induced by ionizing radiation are caused indirectly by the action of water radiolysis products. In biological active systems however, hydroxyl- and peroxy radicals are more likely to react with cellular radical scavangers and the formation of DNA lesions may to a greater extent be mediated by the direct action of radiation. The biological and biochemical effects after exposure of a cell to a damaging agent result from the spectrum of lesions and their distribution in the genome. It seems evident, that bases per se are more involved in lesion formation than phosphates and sugars and in addition it is extremely difficult to select individual lesions among the complexity of modified purines and pyrimidines. One common expression of base damages result – depending on their intergenomic location – in helix distortion, interfered with restricted base stacking.

To study the contribution of base lesions on the deformation of the DNA secondary structure at biological active conditions, DNA of λ-irradiated T1 bacteriophages was investigated for homology by thermokinetics and S1 nuclease cleavage. By this approach, the effects of ionizing radiation on the base pairing ability can be investigated.

RESULTS AND DISCUSSION

Base Pairing Ability

If a denatured DNA molecule is subjected to reannealing, its reassociation rate constant k_2 is dependant on the molecular weight and the genomic complexity[1]. Nonhomologous bases, e.g. modified bases and lost bases, affect the rate constant as well, even if the genomic complexity is maintained [2,3]. In Fig. 1, k_2 values of DNA, degraded by DNAase I or from γ-irradiated phages,[2] were compared to the induced breaks per DNA molecule. The slower reannealing of the irradiated DNA by a factor of three to four is due to lesions different from breaks. It should be mentioned that all kinetics were strictly second-order in this experiment. Contrary, if DNA is irradiated in aqueous solutions, no kinetic differences to unirradiated DNA can be observed, indicating a different pattern of lesion formation and distribution in the genome.

Melting Fidelity

In this experiment, DNA was melted two times: before and after reannealing. From the sigmoid melting curves, the average temperature of melting (T_m) was determined graphically. Broken or weakened hydrogen bonds as well as mispaired bases result in a labilization of the helix integrity and interfere with the T_m value, which decreases with increasing lesions. Nonhomologous base pairing,

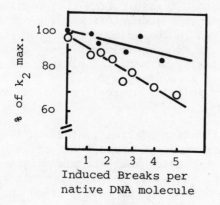

Fig.1. Decrease of the reassociation rate constant k_2 with the induction of breaks and other lesions into DNA. Solid circles for DNAase I digested T1 DNA, open circles for DNA from γ-irradiated T1 bacteriophages. k_2 of 100% corresponds to an experimental value of 6.63 ($mole \times s)^{-1}$. Irradiation conditions as in Table 1; for experimental details see reference 2.

Table 1. T_m Analysis Of DNA From γ-irradiated T1 Phages

Dose[a] kgy	T_m ($^\circ$C) first melting	T_m ($^\circ$C) second melting
0	84.62	84.61
0.5	84.40	84.60
1.0	84.25	84.55
1.5	84.10	84.62
2.0	84.00	84.85
2.5	83.90	84.60
2.5ᵡ	84.55ᵡ	84.55ᵡ

a. Irradiation in N_2 (0°C) with 10^{12} bacteriophages per ml, 0.1 M histidine, 0.165 M NaCl, ^{60}Co-γ at 3.5 kgy/h.
ᵡ after S1 nuclease digestion.

i.e. mismatching during rennealing, also gives a reduced T_m, if the renatured DNA molecule is melted again.

The results in Table 1 indicate no T_m reduction for the re-annealed DNA from γ-irradiated T1 bacteriophages, but a dose de-pendant T_m decrease for the first melting.

The explanation for this phenomenon is as follows: Pre-de-natured regions in the double helix, modified bases and detached base stacking are mediated by the damaging agent and modulate helix distortion to reduced melting temperatures. As no mismatch but a slower reannealing rate is observed, it must be assumed that only homologous bases had matched in the proper manner. During the second melting, those homologous paired bases were melted again. The heterologous sites were obviously not reacting in this approach, i.e. they were overpassed and kept single stranded. This seems to be the reason for the slower kinetics, resulting from enhanced and time consuming nucleation site formation in the irradiated DNA molecule.

Table 2. Molecular weight (MW) Analysis Of S1 Nuclease Digested, Modified DNA On Neutral Sucrose Gradients

Modification	MW (megadalton) native DNA	MW (megadalton) after S1 digestion
None	31.2	25.7
Apurinated	24.4	21.6
DNAase I digested	15.5	15.5
2.5 kgy γ-rays on phage particles	30.3	9.3

Excision with S1 Nuclease

If "heteroduplex" sites are regions of nonpairing bases within double stranded DNA, they should be susceptible to the single strand recognizing enzyme S1 nuclease, which indeed could be shown in the experiment (Fig. 2). The apparent K_M value of 3.2×10^{-9} M (not shown) indicates a high affinity of the S1 nuclease to its substrate, comparable to an endonuclease which recognizes an individual lesion, like apurinic endonuclease with an apparent K_M of $1-10 \times 10^{-9}$ M (reference 4).

The yield of strand breaks cut by S1 nuclease is about the same as the amount induced by γ-irradiation, but the enzyme acts on both strands, e.g. the molecular weight of double- and single stranded DNA is as well reduced. Furthermore, the molecular weight distribution of irradiated and S1 nuclease digested DNA on neutral sucrose gradients (Table 2) is more heterogeneous, compared to DNA degraded by DNAase I (not shown). This indicates, that "heteroduplex" sites are not necessarily associated with radiation induced strand breaks; otherwise a more homogeneous molecular weight distribution on neutral sucrose gradients would be expected.

After S1 nuclease digestion of the "heteroduplex" sites, DNA reanneals like normal DNA, reduced in its size by shearing forces (Table 1); hence all thermokinetical non-reacting sites have been completely excised.

No attempt was made to distinguish individual lesions or the

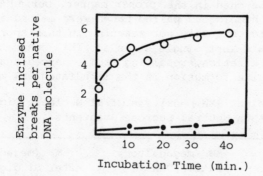

Fig. 2. Time course of S1 nuclease digestion on DNA from γ-irradiated phage particles (open circles) and untreated native T1 DNA (solid circles). Enzyme digestion was performed according to the producers instruction (Boehringer),molecular weights were determined by analytical sedimentation or on neutral sucrose gradients.

size and distribution of "heteroduplex" sites in the genome. It is
likely that these sites are not equally distributed since the excisability and persistance of lesions are probably defined by their
location in the genome[5].

Since the total number of the various base lesions in γ-irradiated DNA is assumed to be three- to fourfold higher than the
number of strand breaks[6], the association of certain types of base
damages within a "heteroduplex", perhaps in a clustered pattern, is
not unlikely.

The formation of bulky lesions in DNA with destroyed base-
pairing ability (e.g. "heteroduplex") by ionizing radiation seems
to be evident[7], but their mechanism of induction remains speculative. "Heteroduplexes" are found in DNA after γ-irradiation in
its specific package form in the T1 phage head, but not in DNA irradiated in aqueous solutions. This might raise the question, whether
these presumed bulky lesions are produced by the direct absorption
of energy, whereas the indirect radiation effect due to water radiolysis in aqueous solutions as well as in the phage head causes a
more uniform distribution of radically induced minor base lesions
in the DNA. Those minor lesions are not likely to interfere with
base pairing, base stacking and helix stability and escape indenti-
fication by thermokinetics.

REFERENCES

1. J.G. Wetmur and N.Davidson, Kinetics of Renaturation of DNA,
   J.Mol. Biol., 31: 349 (1968).
2. H. Martin-Bertram and U. Hagen, Genomic Integrity of T1 DNA
   after γ- and ultraviolet Irradiation, BBA, 561: 312 (1979).
3. J.R. Hutton and J.G. Wetmur, Effect of Chemical Modification on
   the Rate of Renaturation of Deoxyribonucleic Acid. Deaminated
   and Glyoxylated Deoxyribonucleic Acid. Biochemistry, 12: 558
   1973).
4. U. Kuhnlein, E.E. Penhoet and S. Linn, An Altered Apurinic DNA
   Endonuclease Activity in Group A and Group D Xeroderma Pigmentosum Fibrioblasts, Proc. Natl. Acad. Sci, 73: 1169 (1976).
5. P.Cerutti, "DNA Repair Mechanisms", P.C. Hanawalt, E.C. Friedberg, C.F. Fox,eds., Academic Press, New York (1978).
6. J.J. Weiss, " Progress in Nucleic Acid Research and Molecular
   Biology", J.N. Davidson and W.E.Cohn, eds., Academic Press,
   New York (1964).
7. M.C. Paterson, "DNA Repair Mechanisms", P.C. Hanawalt, E.C.Fried-
   berg, C.F. Fox, eds., Academic Press, New York (1978).

INFLUENCE OF ALKYLATING AGENTS ON THE STRUCTURE OF THE BACTERIAL NUCLEOID

Ivar Lossius and Kjell Kleppe

Department of Biochemistry
University of Bergen
Bergen, Norway

SUMMARY

Alkylating agents such as methyl methane sulfonate (MMS) and mitomycin C cause large changes in the nucleoid structure of Escherichia coli. In the presence of high concentrations of these compounds, 25 and 0.1 mM respectively, relaxation of the supercoiled structure takes place. This can be demonstrated both by centrifugation studies as well as by phase contrast microscopy of the cells. Low concentrations of mitomycin C, 1-10 μM, on the other hand cause a rec A dependent increase in the nucleoid mass. During repair, in the absence of the alkylating agents, the number of breaks decreases and supercoiling is introduced again into the nucleoid. The data suggest that consideration of the 3-dimensional structure may be important both during mutagenesis and repair.

INTRODUCTION

In bacteria the chromosome exists in a supercoiled structure called nucleoid[1,2]. The nucleoid body can be directly visualized in bacteria such as E. coli using phase contrast microscopy of living cells embedded in gelatin, a technique originally discovered by Mason and Powelson[3]. In E. coli cells growing in a rich medium one always observes two nuclear bodies per cell while for cells growing in a poor medium 1 or 2 nuclear bodies per cell will be present, Figure 1A,1B. The bacterial cell envelope-free nucleoid can be isolated by carefully lysing the cells with lysozyme, followed by detergents and then centrifugation in a neutral sucrose gradient containing 1 M NaCl. The nucleoid can also be trapped on a membrane filter and examined in a scanning electron microscope[4,5].

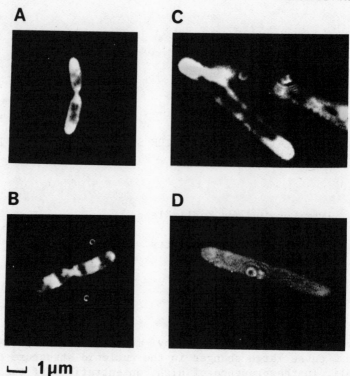

└─┘ **1 μm**

Fig. 1. Phase contrast microscopy of E. coli K-12 DG75 cells.
The phase contrast microscopy was carried out by a
method slightly modified from that described in reference 3.

A. Control, cells grown in minimal medium.
B. Control, cells grown in L-broth medium.
C. Cells treated with 10 μM mitomycin C for
   60 min at 37°C.
   The cells were grown in minimal medium.
D. Cells treated with 100 μM mitomycin C for
   30 min at 37°C.
   The cells were grown in minimal medium.

A fundamental question with regard to the structure of the
nucleoid still remains partly unanswered and this is the relationship between the structure of the in vitro isolated nucleoid and
the in vivo nucleoid. Present evidence suggest that they may not
be the same. In particular the role of RNA in holding the structure together may need reexamination. Possible structures of the
in vitro and in vivo nucleoid are discussed in reference 1. A central theme in both models is, however, the fact that the chromosome
consists of different supercoiled regions or loops.

It is important to examine the fate of the supercoiled structure during mutagenesis and repair. We have studied the effect of
different alkylating agents on the nucleoid structure of E. coli,
using both in vitro and in vivo techniques[6]. The present work
deals with the influence of the monofunctional alkylating agent
methyl methane sulfonate (MMS) and the bifunctional agent mitomycin C.

MATERIALS AND METHODS

Bacterial strains. E. coli K-12 DG 75 F⁻ thy, leu,
E. coli K-12 AB 2463 rec A⁻, thr, leu, his, arg, pro,
E. coli p 3478 F⁻ thy, pol A1, E. coli B and E. coli
AB 1157 thr, leu, his, arg, pro, (isogenic to AB 2463)
were from the laboratory stock.

Growth and lysis conditions. These were essentially as described
in reference 6 and 7. Neutral and alkaline sucrose gradient centrifugation were also carried out essentially as described here.

RESULTS

Mutagen-induced relaxation of the nucleoid. E. coli cells were incubated for various times with different concentrations of alkylating agents and the sedimentation properties of the nucleoid
studied. Incubation of the cells with 25 mM MMS for 30 minutes did
not result in any change in the S-value of the nucleoid. Upon further incubation up to 60 minutes a marked change in the S-value
from 1600 S to approximately 870 S was seen, Figure 2A. In the
case of mitomycin C the presence of small concentrations of this
agent in the medium, 10 μM, caused an unexpected increase in the
S-value to 2000 S, Figure 2B. Upon incubation with higher concentrations a marked decrease was seen, and thus 100 μM caused a decrease to approximately 850 S, Figure 2B.

In the case of high concentrations of MMS and mitomycin C a
number of nicks could be detected in the DNA using alkaline sucrose gradient centrifugations. Incubation with 25 mM for 60 minutes
or 100 μM mitomycin C for 30 minutes resulted in the introduction
of approximately 115 apurinic sites + breaks per chromosome. With
MMS it could further be demonstrated that the breaks most probably
are introduced enzymatically at depurinated sites in the DNA[6].
These results as well as the fact that no change in the DNA:RNA
ratio was found suggested that the decrease in S-value observed
is caused by relaxation of the supercoiled structure. This was
confirmed by ethidium bromide titration of the nucleoid from
the mutagen treated cells, Figure 3. The mitomycin C sample
examined here contained virtually no supercoils. Similar results
were found with 25 mM MMS[6]. The changes in the supercoiled

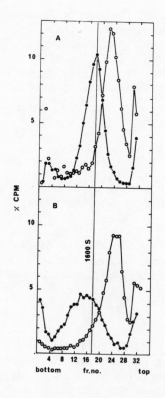

Fig. 2.
A. Sedimentation pattern of the
cell envelope-free nucleoids
from E. coli DG75 cells treated
with MMS. ●——●, control; o——o,
cells treated with 25 mM MMS for
60 min prior to lysis.
B. Sedimentation pattern of cell
envelope-free nucleoids from E.
coli DG75 cells treated with mitomycin C. The cells were grown
in minimal medium in the dark at
37°C. ●——●, cells treated with
10 µM mitomycin C for 60 min;
o——o, cells treated with 100 µM
mitomycin C for 30 min prior to
lysis.

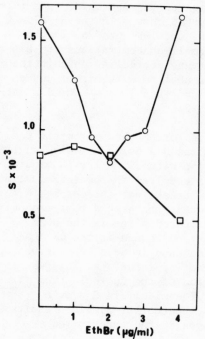

Fig. 3.
Ethidium bromide titration of
envelope-free nucleoids from
cells treated with MMS or mitomycin C. o——o, control; ▫——▫
cells treated with 100 µM mitomycin C for 30 min at 37°C.

structure observed by hydrodynamic methods were also seen by phase contrast microscopy and the result with mitomycin C is given in Figure 1C, 1D. Low concentrations of mitomycin C cause the formation of a large nucleoid, possibly two nucleoids linked together, thus in complete agreement with the increase in S-value observed. With higher concentrations, on the other hand, diffusion of nuclear material throughout the whole cell occurs indicating loss of the supercoiled structure.

Nucleoid studies in repair deficient mutants. We have also employed a number of repair-deficient mutants in these studies. MMS caused complete collapse of the nucleoid structure in pol A⁻ and rec A⁻ strains. Interestingly in the case of mitomycin C the increase in nucleoid mass at low concentrations of this compound was not seen in rec A⁻ strains, Figure 4, suggesting that this increase in S-value is produced by recombination events.

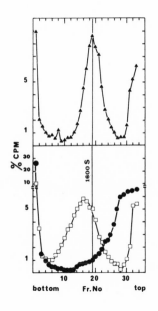

Fig. 4
Effect of low doses of mitomycin C on the sedimentation pattern of envelope-free nucleoids isolated from a rec A⁻ strain and its isogenic parent strain.
▲——▲, control; only rec A⁻ strain shown.
▢——▢ , cell envelope-free nucleoids isolated from E. coli AB 1157 treated with 10 μM mitomycin C for 60 min at 37°C. ●——●, cell envelope-free nucleoids isolated from E. coli AB 2463 rec A⁻ treated with 10 μM mitomycin C for 60 min at 37°C.

Fate of the nucleoid structure during repair. When cells were allowed to repair the mutagen induced damages in the absence of alkylating agents the S-value of the nucleoid increased to the same value as the control, as shown in Figure 5 in the case of MMS. Moreover, there was also a concomitant decrease in breaks as estimated by alkaline sucrose gradients, Figure 5. The supercoiling was also introduced again into the DNA as revealed by titration with ethidium bromide[6].

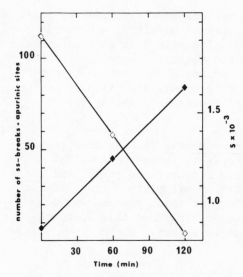

Fig. 5.
Change in S-value and number
of apurinic + single-strand
breaks during repair in MMS-
treated cells incubated in MMS
free medium. E. coli K-12 DG 75
cells treated with 25 mM MMS
for 60 min at 37°C were resuspended in MMS-free minimal medium and incubated at 37°C for
the periods indicated.
◊——◊ number of apurinic sites + single-strand breaks;
◆——◆ , S-value of the cell
envelope-free nucleoid.

DISCUSSION

     The structure of the bacterial nucleoid is thought to be dy-
namic in nature containing a number of supercoiled regions. The
amount of supercoiling must then to some extent depend on the physiological state of the cell. The structure is stabilized by proteins, cell membrane attachment points and possibly also RNA. His-
tone-like proteins have been shown to be present in prokaryotic
organisms[8], but at present there is little evidence to suggest
that nucleosome-type structures are found in the prokaryotic chromosomes. Recent work in this laboratory[9] has indicated that a surprising number of proteins of varying molecular weights normally
are bound to the chromosome and may contribute to the stabilization of certain structures. Relaxation of supercoils takes place during mutagenesis and repair by introduction of breaks in the DNA,
either by a direct attack on the phosphodiester bond or by specific repair endonucleases. Evidence from studies on plasmids[10] and
also from this work suggest that supercoils may be partially restrained in vivo, i.e. one has to introduce a considerable number
of breaks before any measurable changes in structure due to relaxation of supercoils is observed. In the case of the chromosome of
E. coli introduction of up to 30 breaks may not necessarily change
the outward appearance. With approximately 115 breaks per chromosome all supercoiling is lost and drastic changes can be observed
in the structure. For complete repair the supercoiling has to be

introduced again into the chromosomes, presumably by the enzyme DNA gyrase. Results from this laboratory[6] and other groups[11] clearly show that this indeed takes place. Thus consideration of the 3-dimensional structure may be important in DNA repair studies. In the case of repair of double-strand breaks in the E. coli chromosome a study reported in this volume[12] clearly implies that this is the case.

ACKNOWLEDGEMENTS

This study was supported in parts by grants from the Norwegian Research Council for Science and Humanities, and by the Nansen Foundation.

REFERENCES

1. K. Kleppe, S. Øvrebø, and I. Lossius, The bacterial nucleoid, J. Gen. Microbiol. 112:1 (1979).
2. D. E. Pettijohn, Prokaryotic DNA in nucleoid structure, CRC Critical Rev. Biochem. 4:175 (1976).
3. D. J. Mason and D. M. Powelson, Nuclear Division as observed in live bacteria by a new technique, J. Bacteriol. 71:474 (1956).
4. J. Van Ness and D. E. Pettijohn, A simple autoradiographic method for investigating long range chromosome structure: Size and number of DNA molecules in isolated nucleoids of E. coli, J. Mol. Biol. 129:501 (1979).
5. K. Kleppe, unpublished.
6. I. Lossius, P. G. Krüger, and K. Kleppe, Effect of methyl methanesulphonate on the nucleoid structure of Escherichia coli, J. Gen. Microbiol. in press.
7. C. Korch, S. Øvrebø and K. Kleppe, Envelope associated folded chromosomes from Escherichia coli: Variations under different physiological conditions, J. Bacteriol. 127:904 (1976).
8. J. Rouvière-Yaniv, and F. Gros, Characterization of a novel low molecular weight DNA-binding protein from E. coli, Proc. Natl. Acad. Sci. USA, 72:3428 (1975).
9. K. Sjåstad and K. Kleppe, unpublished.
10. D. E. Pettijohn and O. Pfenninger, Supercoils in Prokaryotic DNA restrained in vivo, Proc. Natl. Acad. Sci. USA, 77:1331 (1980).
11. K. M. Ulmer, R. F. Gomez, and A. J. Sinskey, Ionizing radiation damage to the folded chromosome of E. coli K-12: Repair of double-strand breaks in DNA, J. Bacteriol. 138:486 (1979).
12. R. E. Krisch, ^{125}I decay in microorganisms: A prokaryotic model system of examining the role of unrepaired double-strand DNA breaks in cell killing, in: "Chromosome Damage and Repair", E. Seeberg and K. Kleppe, ed., Plenum Press, London, (1980).

NUCLEOSOMAL STRUCTURE OF CHROMATIN : DISTRIBUTION AND EXCISION OF DNA DAMAGE

Peter Cerutti, Motohisa Kaneko and Peter Beard

Swiss Institute for Experimental Cancer Research

CH-1066 Epalinges s/Lausanne, Switzerland

SUMMARY

The implications of the nucleosomal structure of chromatin for the distribution and repair of DNA lesions is of obvious interest. The 'bulky" purine adducts formed in human cells by the highly reactive ultimate carcinogens N-acetoxy-2-acetylaminofluorene (AAAF), benzo(a)pyrene-diol-epoxide I (BPDE I) and benzo(a)pyrene-4,5-epoxide were all introduced in higher initial concentrations in nucleosomal linker- than in core-DNA. In contrast, comparable adduct concentrations in linker- and core-DNA were observed after prolonged exposure of actively metabolizing human lung cells to the procarcinogen benzo(a)pyrene. Since different distribution and repairability of structurally identical lesions may mean different biological potency extrapolation of results of *in vitro* mutagenesis and transformation using ultimate carcinogens to the *in vivo* situation where procarcinogens are metabolized in the target organ has to be done with caution.

No simple rule can be derived for the relative rates of adduct removal from core- and linker-DNA in human cells. While the BPDE I guanine adducts were removed with similar efficiency from both DNA fractions the AAAF adducts disappeared much more rapidly from linker DNA of confluent human fibroblasts. Under certain conditions large fractions of adducts persisted in the DNA over an extended period.

The observed non-random introduction of lesions into chromatin can be exploited to approach basic questions about the dynamic properties of chromatin *in situ*. Using AAAF and repair deficient confluent cultures of *Xeroderma pigmentosum* fibroblasts we have obtained evidence which argues strongly against the existence of constitutive movement of nucleosomes. In a second series of experiments AAAF was used to probe the structure of Simian Virus 40 (SV 40) minichromosomes *in situ* in monkey kidney cells. It was found that the replication origin region of SV 40 DNA was more open to the attack by the agent than the rest of the genome. These results lend direct "*in vivo*" support to the electronmicroscopic observation that the origin region of isolated SV 40 minichromosomes was nucleosome depleted.

INTRODUCTION

The organization of eukaryotic chromatin into nucleosomes is well established. Approximately 140 base pairs (bp) of DNA are wrapped around an octamer core of "inner" histones and these core particles are connected by "linker" DNA of 0 to 70 bp. Nucleosomes are discernible as beads on a string in electron micrographs of unfolded chromatin but *in situ* in the cell they are believed to be packed into higher order structures. Much has been learnt about the nucleosomal structure of chromatin from the selective digestion of nuclei and isolated chromatin with staphylococcal nuclease (SN) and pancreatic DNAse I. SN preferentially digests nucleosomal linker DNA but other regions of chromatin such as certain transcribing genes, replicating DNA and some regions of core DNA may also possess unusual sensitivity to SN [1-4]. The implications of the nucleosomal structure of chromatin for the distribution and repair of DNA lesions induced by chemical and physical damaging agents are of obvious interest [5,6]. In a majority of studies approaching this problem advantage is taken of the selective digestion properties of SN. Two experimental designs are being used which yield different information :   (1) SN is used as a *probe* to distinguish the introduction of patches by repair synthesis into DNA regions of chromatin of higher or lower susceptibility to this enzyme [7-9]. The interpretation of data obtained by this approach in terms of the nucleosomal structure of chromatin is not always straight-forward. In particular this is the case for experiments measuring initial kinetics of the release by SN of nucleotides which had been incorporated by repair synthesis or of DNA lesions [9-11]. As mentioned above a small portion of DNA in core- and linker-regions may possess unusual susceptibility to SN for a variety

of reasons. (2) In the second approach SN is merely used for the
preparation of purified core-DNA. After relative extensive SN-
digestion of isolated nuclei the alcohol or acid precipitable DNA
mostly consists of core DNA [12-18]. In an improved procedure mononucleosomal cores are first purified before the DNA is extracted [19].
The concentrations of lesions or nucleotides incorporated into repair
patches are then compared between purified core-DNA and total nuclear
DNA. The values for linker DNA are calculated from these data. This
experimental design yields results which are directly interpretable
in terms of the nucleosomal structure of chromatin. Since it has not
been possible so far to analyze linker DNA directly the accuracy of
the data is limited, however.

First the principal design of our experiments and a preview of
the major results are given. We have exclusively followed the second
approach outlined above. Benzo(a)pyrene, its ultimate metabolites
7β, 8α-dihydroxy-9α,10α-epoxy-7,8,9,10-tetrahydrobenzo(a)pyrene
(BPDE I) and 4,5-dihydro-4,5-epoxybenzo(a)pyrene (BPE) and N-acetoxy-
2-acetylaminofluorene (AAAF) were used as DNA damaging agents since
they induce a simple well-defined adduct spectrum. BPDE I and AAAF
react preferentially with guanine producing to 70-90% a single adduct
of known structure. BPDE I reacts mostly with the exocyclic amino-
group of guanine under formation of N^2-(7β,8α,9α-trihydroxy-7,8,9,10-
tetrahydrobenzo(a)pyrene-10-yl)deoxyguanosine (BPDE I-dG) [20,21] while
AAAF reacts mostly with carbon-8 to N-(deoxyguanosin-8-yl)-2-acetyl-
aminofluorene or in human fibroblasts to the deacetylated derivative
N-(deoxyguanosin-8-yl)aminofluorene (AF-dG) [19,22,23]. BPE also reacts
preferentially with guanine but also forms significant amounts of
adenine adducts. The structure of the BPE adducts have not yet been
elucidated [17]. For simplicity the DNA adducts formed by the agents
used in this work are referred to as "arylation" products. According
to our basic protocol cultured cells are prelabeled over an extended
period with low levels of ^{14}C-thymidine and then treated with a high
specific activity ^{3}H-labeled DNA damaging agent. Adduct *concentrations*
are determined in core-DNA and total nuclear DNA from the ^{14}C-specific
activity of the DNA and the ^{3}H-specific activity of the damaging
agent. Total amounts of adducts in core- and linker-DNA are calculated
assuming 135 base pairs (bp) for core-DNA and 47 bp for linker DNA in
human cells [24]. No attempt was made to measure the distribution of
minor adducts.

The following general observations were made concerning adduct
distribution and removal. The highly reactive *ultimate* damaging
agents BPE, BPDE I and AAAF all formed adducts preferentially in
nucleosomal linker DNA but their degree of preference varied [13-15,17,19].
In contrast to BPDE I, prolonged treatment of actively metabolizing
human lung cells A549 with the corresponding *procarcinogen* benzo(a)

pyrene (BP) resulted in comparable adduct concentrations in linker-
and core-DNA [16]. No simple rule can be derived for the relative rates
of adduct removal from core- and linker-DNA. While BPDE I-dG was
removed with similar efficiency from both DNA fractions AF-dG disappeared much more rapidly from linker DNA [19]. Under certain conditions large fractions of adducts persisted in the DNA over an extended
period.

In view of the rudimentary knowledge of chromatin structure *in
situ* in the cell it may be premature to speculate in structural terms
about the observed differences in the formation and repair of the
lesions introduced by the agents used in this work. On the other hand,
the observed non-random introduction of lesions into chromatin can
be exploited to approach basic questions about the dynamic properties
of chromatin *in situ*. Using AAAF and repair deficient confluent cultures of *Xeroderma pigmentosum* fibroblasts we have obtained evidence
which argues strongly against the existence of constitutive movement
of nucleosomes. Nucleosomal mobility had been detected in recent *in
vitro* studies but these results cannot be readily extrapolated to
the situation in the cell [25]. Nucleosomal movement had been proposed
as an early step in recent models of excision repair in mammalian
cells [26]. In a second series of experiments AAAF was used to probe
the structure of simian virus 40 (SV 40) minichromosomes *in situ* in
monkey kidney cells. It was found that the replication origin region
of SV 40 DNA was more open to the attack by the agent than the rest
of the genome. These results lend direct "*in vivo*" support to the
electronmicroscopic observation that the origin region of isolated
SV 40 minochromosomes was nucleosome depleted [27-29].

ARYLATION OF DNA IN CULTURED HUMAN CELLS : ADDUCT DISTRIBUTION IN
CHROMATIN

When rapidly growing human lung cells A549 are exposed to BP
(1.3 µM) for 48 h the 7R enantiomer of BPDE I-dG is formed via
cellular metabolism as the major covalent DNA adduct [16]. The same
major adduct is also produced when A549 cells are treated directly
with the ultimate metabolite BPDE I (1.1 µM) [17,20] (Note : the spectrum of minor adducts differs quantitatively and qualitatively under
these two conditions). While the structures of the major adducts are
identical their distribution in DNA is not. Using extensive SN digestion of isolated nuclei for the preparation of core DNA resulting
in 40-50% solubilization of DNA, equal adduct concentrations were
measured in core- and linker-DNA following prolonged treatment with
BP [16] but approximately 10 times higher concentrations in linker-
relative to core-DNA after treatment with BPDE I [17]. Under the first
conditions the adduct distribution reflects a steady state of adduct

formation via metabolic activation of the procarcinogen and adduct removal by repair processes. A trend to equalization of the adduct concentrations between SN-sensitive and -resistant DNA in rat liver nuclei was also observed with increasing time of exposure of rats to the procarcinogen 2-acetylaminofluorene [13]. Under the second conditions the highly reactive ultimate metabolite BPDE I is added at high momentary concentration and reacts rapidly with the exposed sites of chromosomal DNA. During the short time of reaction repair processes are not expected to modify the initial adduct concentration substantially. Extreme preference for reaction with linker DNA was observed for BPE. The initial concentration of BPE-dG was 14 times and BP-dA 33 times higher in linker- than in core-DNA [17].

The observed differences in adduct distribution in chromatin between experiments with procarcinogens and their ultimate metabolites may have important implications. Extrapolation of results of *in vitro* mutagenesis and transformation using ultimate metabolites to the *in vivo* situation where, in general, procarcinogens are metabolized in the target organ has to be done with caution. Different distribution and repairability of structurally identical lesions may mean different biological potency of the respective inducing agents.

EXCISABILITY AND PERSISTENCE OF ARYLATION ADDUCTS IN NUCLEOSOMAL CORE- AND LINKER-DNA

The basic experimental design was as outlined in the "Introduction" and in the preceding section. However, in order to obtain better defined preparations of core DNA we have purified mononucleosomal cores by sucrose gradient centrifugation in recent studies of the repair of AAAF and BPDE I induced adducts in confluent cultures of normal human skin fibroblasts (NF) [19]. Gel electrophoretic analysis using appropriate markers revealed that our core preparations contained three size classes of DNA of 130, 140 and 154 bp, respectively. The relative amount of 154 bp DNA varied somewhat in different preparations. This DNA derives from cores containing short tails of linker DNA which may have been protected from SN digestion by histone H_1. The confluent ^{14}C-thymidine prelabeled NF cultures were treated for 15 min with ^{3}H-AAAF (1.5 μM; specific activity 0.878 Ci/mmole) or ^{3}H-BPDE I (2.0 μM; specific activity 3.76 Ci/mmole) and incubated in conditioned media up to 24 h before total nuclear DNA and nucleosomal core DNA was prepared.

The results obtained with AAAF are summarized first. The concentration of 1.5 μM AAAF is of low toxicity to NF, the residual colony forming ability being 70-80%. The initial adduct concentrations varied between 20 to 38 μMoles adduct per Mole DNA-P in

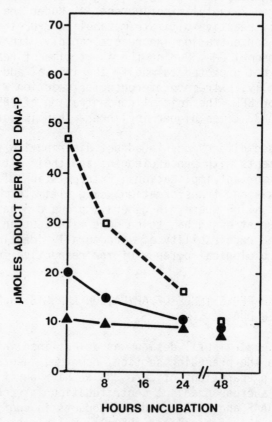

Fig. 1.   Time course of the removal of AAF adducts from total nuclear
          DNA, purified nucleosomal core DNA and nucleosomal linker
          DNA. Normal human fibroblasts CRL 1221 were treated in con-
          fluency with 1.5 μM AAAF. The adduct concentrations were
          determined in total nuclear DNA and in purified nucleosomal
          core DNA as a function of post-treatment incubation. The
          curve for nucleosomal linker DNA was computed from these
          data. ▲——▲, adduct concentration in nucleosomal core DNA;
          ●——●, adduct concentration in total nuclear DNA; □----□ ,
          adduct concentration in nucleosomal linker DNA (from ref.
          19).

different experiments and 85-90% of the adducts consisted of AF-dG, i.e. the deacetylated derivative of the carbon-8 adduct of guanine. Figure 1 shows the kinetics of disappearance of the adducts from total nuclear DNA, nucleosomal core-DNA and linker-DNA for a representative experiment in which the initial adduct concentration was approximately 4 times higher in linker- than in core-DNA. (Note : the curve for linker DNA is calculated from the data for total- and core-DNA). It is evident that adduct removal was rapid during the first 8 h of incubation and occurred 4 times more efficiently from linker- than from core-DNA. After 24 h incubation the rate of adduct removal had decreased to a very low value leaving a substantial fraction of adducts unexcised. In separate experiments it was shown that approximately 50% of the adducts persisted in the DNA after 24 h incubation independent of the initial adduct concentration in the range of 12 to 115 μMoles adducts per Mole DNA-P. Approximately 65% of the persisting adducts were located in nucleosomal cores. It is evident that the excision system accomplishing "early excision" from linker DNA was not saturated at these initial adduct concentrations. This is in agreement with the earlier observation that repair synthesis in human skin fibroblasts was not saturated in the dose range of AAAF used in our work [30,31]. While the rates of adduct removal were much higher in our experiments than in those of Heflich et al. [32] similar levels of persistent adducts were observed. These differences in excision kinetics may be due to the higher initial adduct concentrations of 70 and 120 μMoles adduct per Mole DNA-P in the work by these authors.

In an analogous experiment in which confluent NF were treated with 2 μM BPDE I the initial adduct concentration in total nuclear DNA was 29 μMoles per Mole DNA-P. The adduct concentration in linker DNA was 1.7 time higher than in core DNA. As for the experiments with A549 cells discussed earlier the predominant adduct was 7R BPDE I-dG. Approximately 50% of the adducts were removed within the first 4 h of incubation and the rates of removal from core- and linker-DNA werecomparable. Adduct removal continued at a slower rate and approximately 70% of the initial adducts had disappeared from the DNA within 24 h. Since adduct removal continued at a significant rate over a prolonged time period a persistent fraction of adducts could not readily be discerned. It is interesting to compare these results with earlier work with rapidly growing A549 cells which had been exposed to the procarcinogen BP [16]. As mentioned earlier the initial adduct concentrations were comparable in SN-sensitive and -resistant DNA and, in agreement with the data discussed above for NF, adduct removal occurred with similar efficiency from both fractions of DNA. However, clearly discernible persistent fractions of adducts from 0.40 to 0.55 remained in the DNA of A549 cells over several generation times. Finally, it has been reported that the covalent purine adducts formed by the ultimate carcinogen 7-bromomethylbenzo(a)anthracene were formed at approximately twice the initial concentration in nucleosomal linker-relative to core-DNA but removed with comparable efficiency from the entire DNA in confluent NF cultures [11]. These results are similar to those reported here for BPDE I.

Several interpretations are possible for the formation of large fractions of persistent adducts in DNA which are removed very slowly or not at all as was observed for AAAF treated NF and BP treated A549 cells: (1) The cellular excision system essentially ceases to operate after the initial rapid phase of adduct removal from linker DNA, e.g. a protein required for adduct removal is used up in the reaction. (2) A fraction of adducts is located *ab initio* in regions of DNA which are not freely accessible to the major excision repair system. This fraction of unexcisable adducts is proportional to the initial adduct concentration. (3) Since mass cultures were used in all experiments it is conceivable that different subpopulations of cells removed widely different fractions of adducts, e.g. heavily damaged cells excised none while slightly damaged cells excised all of their adducts. It should be noted that in the case of the persistence of benzo(a)pyrene- and AAF-adducts the majority of the lesions which remain in DNA at prolonged post-treatment incubation were structurally identical to the lesions which were first rapidly removed. A different situation presents itself for the persistency of O^6-alkyl-guanine and the N^2-arylation product of guanine induced by AAF in the brain and liver of rats where under certain conditions a structurally distinct minor fraction of lesions is refractory to excision throughout post-treatment incubation [22,33,34]. These lesions may be irrepairable by the constitutive excision repair system in these organs because of their structural properties, but it cannot be excluded that they could be located in regions of chromatin which are inaccessible to repair enzymes.

N-ACETOXY-2-ACETYLAMINOFLUORENE AS A PROBE FOR THE STRUCTURE OF CHROMATIN IN SITU

The differential reactivity of certain DNA damaging agents with nucleosomal core- and linker DNA can be exploited to study certain aspects of chromatin structure *in situ* in the intact cell [35,36]. We have used AAAF to approach the question of nucleosomal mobility and to probe the structure of the origin region of SV 40.

Is there Nucleosomal Movement ?

The question whether nucleosomes possess inherent mobility allowing slippage along the DNA is of fundamental importance to the understanding of chromatin structure and function. Constitutive or induced nucleosomal movement could conceivably play a role in replication, transcription and repair. Indeed, such a step has been suggested in a recent model of excision repair [26]. *In vitro* experiments using SV 40 minichromosomes indicated that nucleosomes could move into adjacent regions of protein-free DNA [25]. The extrapolation of these results to the *in vivo* situation is difficult, however. Our experiments were based on the following concept. As discussed above AAAF preferentially damages nucleosomal linker regions by a factor of preference from 2 to 4. Constitutive nucleosomal movement in the absence of DNA-replication and -repair would tend to diminish the initial differences in adduct concentration between core- and linker-DNA. Our data obtained with confluent *Xeroderma pigmentosum* fibroblasts of complementation group A which are deficient in the excision of AAF-adducts are summarized in Table 1. The fact that the initial non-random adduct distribution was maintained during 24 h post-treatment incubation indicates that no substantial changes occurred in the nucleosomal register within this time period in the absence of repair and replication. No evidence for repair-induced long-range movement of nucleosomes was obtained from our experiments with confluent NF discussed in the preceding section. The diminution of the initial differences in adduct concentration between linker- and core-DNA during post-treatment incubation can be explained entirely by the observed preferential early adduct excision from linker DNA [19]. At the initial adduct concentration of 20 μMoles per Mole DNA-P in the experiment shown in Figure 1 only every 140 nucleosome on average contains an adduct. Therefore, our data does not exclude the existence of repair-induced *short* range nucleosomal movement.

Hyper-reactivity of the SV 40 origin region to AAAF

The small DNA viruses SV 40 and polyoma represent promising systems for the study of chromatin structure and function, since detailed information is available about their nucleotide sequences and pattern of expression of the genome. For the same reasons these viruses may prove valuable for the study of the interaction of chromatin with carcinogens. SV 40 in the infected cell is present as a minichromosome with nucleosomal organization. *In vitro* studies suggest that the region of the replication origin of SV 40 may have a unique structure and may be free of nucleosomes [27,28]. However, the arrangement of nucleosomes in isolated chromatin is not necessarily stable [29]. We have taken advantage of the differential reactivity of AAAF with linker- and core-DNA to probe the structure of the replicating SV 40 genome *in situ* in the intact cell. Monolayer cultures of BSC-B monkey kidney cells were infected with SV 40 and 40 h later, i.e. when viral replication was very active, ^{3}H-AAAF (9 μM) was added. The viral DNA was isolated, digested with the restriction enzymes Hae III and Kpn I and the restriction fragments analyzed by gel electrophoresis. The sequence of the restriction fragments produced under these conditions is known from previous work[37]. As control relaxed circular SV 40 DNA was treated with ^{3}H-AAAF *in vitro* and analyzed under the same conditions. Since AAAF reacts almost exclusively with guanine the adduct content of each restriction fragment is expected to be proportional to its guanine content if random attack occurs. This was the case for free SV 40 DNA reacted *in vitro*. In contrast, in SV 40 DNA isolated from AAAF treated cells the fragment Hae III F1b/Kpn I which is located immediately to the "late genes" side of the origin of replication contained approximately twice the number of adducts that would be expected from its guanine content. While it was not possible to analyze the small DNA fragment just to the "early genes" side of the origin by our methodology it was evident that the adduct concentrations of the restriction fragments from the rest of the SV 40 genome were proportional to their guanine content. Our data indicate that the DNA of the origin region of the SV 40 chromosome *in the intact cell* is unusually reactive. From these results together with the electron-microscopic observations on isolated SV 40 minichromosomes [27,28] it appears very likely that the SV 40 origin region *in situ* in the infected cell does not possess the usual nucleosomal organization. It cannot be decided whether the DNA in this region is entirely free of proteins or loosely covered with proteins and whether it possesses a unique conformational structure. The preferential binding of AAAF to a region of chromatin believed to have regulatory functions may have important implications for the mechanism of action of AAAF and related carcinogens.

Table 1.   Concentration of Adducts in Nucleosomal Linker- and Core-DNA in AAAF treated *Xeroderma pigmentosum* Fibroblasts [a]

Post-treatment incubation (h)	DNA fraction	Adduct concentration μMoles per Mole DNA-P
0	linker	12.6
	core	29.6
24	linker	12.3
	core	30.9

[a]Confluent cultures; XP25RO belongs to complementation group A.

ACKNOWLEDGEMENTS

M. Kaneko was on leave of absence from the National Cancer Center Research Institute, Tsukiji, Tokyo.

This work was supported by Grant No 3'305.78 of the Swiss National Science Foundation and a Grant from the Swiss Association of Cigarette Manufacturers.

REFERENCES

1. R. Kornberg, Structure of Chromatin, Ann. Rev. Biochem. 46:931 (1977).
2. P. Chambon. Summary : The molecular biology of the eukaryotic genome is coming of age. Cold Spring Harbor Symp. Quant. Biol. 42:1209 (1977).
3. G. Felsenfeld, Chromatin, Nature 271:115 (1978).
4. V. Foe, Modulation of ribosomal RNA synthesis in Oncopeltus fasciatus : An electron microscopic study of the relationship between changes in chromatin structure and transcriptional activity. Cold Spring Harbor Symp. Quant. Biol. 42:723 (1977).
5. See in:"DNA Repair Mechanisms", P.C. Hanawalt, E.C. Friedberg and C.F. Fox, eds., Academic Press, New York (1978).
6. P. Hanawalt, P. Cooper, A. Ganesan, and C. Smith, DNA Repair in Bacteria and Mammalian Cells, Ann. Rev. Biochem. 48:783 (1979).
7. W. Bodell, Non-uniform distribution of DNA repair in chromatin after treatment with MMS, Nucl. Acid. Res. 4:2619 (1977).
8. J. Cleaver, Nucleosome structure controls rates of excision repair in DNA of human cells, Nature 270:451 (1977).
9. M. Smerdon, T. Tlsty, and M. Lieberman, Distribution of ultraviolet-induced DNA repair synthesis in nuclease sensitive and resistant regions of human chromatin, Biochemistry 17:2377 (1978).
10. M. Smerdon, and M. Lieberman, Nucleosome rearrangement in human chromatin during UV-induced repair synthesis, Proc. Natl. Acad. Sci. 75:4238 (1978).
11. F. Oleson, B. Mitchell, A. Dipple, and M. Lieberman, Distribution of damage in chromatin and its relation to repair in human cells treated with 7-bromomethylbenzo(a)anthracene, Nucl. Acid. Res. 7:1343 (1979).
12. R. Ramanathan, S. Rajalkshmi, D. Sarma, and E. Farber, Non-random nature of in vivo methylation by DMN and subsequent removal of methylated products from rat liver chromatin, Cancer Res. 36:2073 (1976).
13. G. Metzger, F. Wilhelm, and M. Wilhelm, Non-random binding of a chemical carcinogen to the DNA in chromatin, Biochem. Biophys. Res. Commun. 75:703 (1977).

14. C. Jahn, and G. Litman, Distribution of covalently bound benzo-(a)pyrene in chromatin, Biochem. Biophys. Res. Commun. 76:534 (1977).

15. P. Cerutti, Repairable Damage in DNA, in:"DNA Repair Mechanisms", P. C. Hanawalt, E. C. Friedberg, and C. F. Fox, eds. pp 1-14, Academic Press, New York (1978).

16. G. Feldman, J. Remsen, K. Shinohara, and P. Cerutti, Excisability and persistence of benzo(a)pyrene DNA adducts in epithelioid human lung cells, Nature 274:796 (1978).

17. G. Feldman, J. Remsen, V. Wang, and P. Cerutti, Formation and excision of covalent DNA adducts of benzo(a)pyrene-4,5-epoxide and benzo(a)pyrene diol-epoxide I in human lung cells A549, Biochemistry 19:1095 (1980).

18. J. Williams,and E. Friedberg, Deoxyribonucleic acid excision repair in chromatin after ultraviolet irradiation of human fibroblasts in culture, Biochemistry 18:3965 (1979).

19. M. Kaneko, and P. Cerutti, Excisability of N-acetoxy-2-acetyl-aminofluorene induced DNA adducts from chromatin fractions of human fibroblasts, Cancer Res., in press.

20. P. Cerutti, F. Sessions, P. Hariharan, and A. Lusby, Repair of DNA damage induced by benzo(a)pyrene-diol-epoxides I and II in human alveolar tumor cells, Cancer Res. 38:2118 (1978).

21. H. Brown, A. Jeffrey, and I. Weinstein, Formation of DNA adducts in 10T 1/2 mouse embryo fibroblasts incubated with benzo(a)pyrene or dihydrodiol oxide derivatives, Cancer Res. 39:1673 (1979).

22. J. Westra, E. Kriek, and H. Hittenhausen, Identification of the persistently bound form of the carcinogen N-acetyl-2-amino-fluorene to rat liver DNA in vivo, Chem.-Biol. Int. 15:149 (1976) and refs. therein.

23. M. Poirier, M. Dubin, and S. Yuspa, Formation and removal of specific acetylaminofluorene-DNA adducts in mouse and human cells measured by radioimmunoassay, Cancer Res. 39:1377 (1979).

24. D. Lohr, J. Corden, K. Tatchell, R. Kovacic, and K. Van Holde, Comparative subunit structure of HeLa, yeast and chicken erythro-cyte chromatin. Proc. Natl. Acad. Sci. 74:79 (1977).

25. P. Beard, Mobility of histones on the chromosome of Simian Virus 40, Cell 15:955 (1978).

26. M. Lieberman, M. Smerdon, T. Tlsty, and F. Oleson, "The role of chromatin structure in DNA repair in human cells damaged with chemical carcinogens and ultraviolet radiation, in:"Environmental Carcinogenesis, Occurrence, Risk Evaluation and Mechanisms", P. Emmelot, and E. Kriek, eds., Elsevier/North-Holland, Amsterdam (1979).

27. S. Saragosti, G. Moyne, and M. Yaniv, Absence of nucleosomes in a fraction of SV 40 chromatin between the origin of replication and the region coding for the late leader RNA, Cell 20:65 (1980).

28. E. Jakobovits, S. Bratosin, and Y. Aloni, A nucleosome-free region in SV 40 minichromosomes, Nature 285:263 (1980).

29. P. Beard, M. Kaneko, and P. Cerutti, unpublished.

30. F. Ahmed, and R. Setlow, Different rate-limiting steps in excision repair of ultraviolet- and N-acetoxy-2-acetylamino-fluorene-damaged DNA in normal human fibroblasts, Proc. Natl. Acad. Sci. 74:1548 (1977).

31. A. Brown, T. Fickel, J. Cleaver, P. Lohman, M. Wade, and R. Waters, Overlapping pathways for repair of damage from ultraviolet light and chemical carcinogens in human fibroblasts, Cancer Res. 39:2522 (1979).

32. R. Heflich, R. Hazard, L. Lommel, J. Schribner, V. Maher, and J. McCormick, A comparison of the DNA binding, cytotoxicity and repair synthesis induced in human fibroblasts by reactive derivatives of aromatic amide carcinogens, Chem.-Biol. Int. 29:43 (1980).

33. R. Goth, and M. Rajewsky, Persistence of O^6-ethylguanine in rat brain DNA : Correlation with nervous system specific carcinogenesis by ethylnitrosourea, Proc. Natl. Acad. Sci. 71:639 (1974).

34. A. Pegg, Formation and metabolism of alkylated nucleosides : possible role in carcinogenesis by nitroso compounds and alkylating agents. Adv. Cancer Res. 25:195 (1977).

35. T. Cech, and M. Pardue, Cross-linking of DNA with trimethyl-psoralen is a probe for chromatin structure, Cell 11:631 (1977).

36. G. Wiesenhahn, J. Hyde, and J. Hearst, The photo-addition of trimethylpsoralen to Drosophila melanogaster nuclei : a probe for chromatin substructure, Biochemistry 16:925 (1977).

37. P. Hale, R. Woodward, and J. Lebowitz, E. coli RNA polymerase promotors on superhelical SV 40 DNA are highly selective targets for chemical modification, Nature 284:640 (1980) and refs. therein.

ELECTRON-MICROSCOPIC VISUALIZATION OF THE AAF BINDING SITES TO DNA

IN CORE PARTICLES AND TRINUCLEOSOMES BY MEANS OF SPECIFIC ANTIBODIES

Robert P.P. Fuchs, M.C. Lang, G. de Murcia, A. Mazen
and M.P. Daune

I.B.M.C., Groupe de Biophysique
15 rue René Descartes, 67084 Strasbourg cedex, France

The covalent binding of the so-called ultimate metabolite of a given carcinogen to DNA is generally believed to be the initial step in chemical carcinogenesis. In the field of aromatic amides or amines carcinogenesis, N-2-acetylaminofluorene (AAF) is the most widely studied compound. In vitro studies are conducted with the model ultimate metabolite N-acetoxy-N-2-acetylaminofluorene (N-Aco-AAF). This compound is known to bind covalently to DNA giving rise to nucleoside – AAF adducts of the same nature as those observed under in vivo conditions[1]. N-2-(deoxyguanosine-8-yl)-acetylaminofluorene (dGuo-C8-AAF), the major adduct, has been shown to induce a structural alteration of the DNA double helix which has been described in this laboratory as the insertion-denaturation model[2,3]. In eukaryotic cells DNA rarely appears free but associated with proteins inside the structure of the chromatin. We asked therefore the question of how this structure would interfere with the covalent binding of N-Aco-AAF to DNA. To study this problem we used a methodology which has been recently developped in our laboratory[4] and which consists in the direct electron-microscopic visualization of the AAF binding sites along the DNA molecule by means of specific antibodies. In fact, native DNA modified with N-Aco-AAF[5] as well as Guo-C8-AAF linked to BSA[6,7] induce in rabbits the synthesis of antibodies that specifically recognize the AAF modified guanine residue in native DNA[7,8]. These specific antibodies have been purified from the crude serum[7,8] and subsequently used to map the AAF binding sites along the DNA fragments isolated from core particles and trinucleosomes.

MATERIAL AND METHODS

N-Aco-(^{14}C)AAF (specific activity 50 mCi/m mol) and erythrocyte chromatin subunits were prepared as previously described[9],[10]. Incubation with N-Aco-(^{14}C)AAF was performed at 37°C during 3 h in a buffer containing 10% ethanol, 0.2 mM EDTA, 0.1 mM PMSF and 4 mM sodium citrate pH 7. Non covalently bound fluorene derivatives were removed by repeated ethyl ether extractions. At this level the integrity of the core particles and of the trinucleosomes was checked by circular dichroism and electron microscopy. This material was subsequently deproteinized and the DNA was precipitated with ethanol. The DNA isolated from both the core particles and trinucleosomes contained covalently bound -(^{14}C)AAF residues. The amount of AAF modification was calculated from the bound radioactivity and from the UV absorbance at 260 nm and expressed as percent of modified bases. In all experiments we used low amounts of modification (0.8% of modified bases in DNA from core particles, i.e. one AAF residue every 125 bases ; 0.36% of modified bases in DNA from trinucleosomes, i.e. one AAF residue every 280 bases). The AAF modified DNA samples were then further incubated with an excess of antibodies against Guo-C8-AAF. The incubation conditions (ionic strength, ratio antibody to antigen) were carefully chosen in order to avoid unspecific binding of antibodies to DNA. The unbound IgG molecules were removed by gel filtration on a Biogel P300 column. The chromatography was monitored by gel electrophoresis or directly by electron microscopy. Spreading and shadowing were performed as described previously[4].

RESULTS

N-Aco-AAF binding to DNA in core particles

Figure 1a is an electron micrograph of AAF modified core particles and figure 1b features the corresponding DNA-AAF (0.8% of modified bases) after deproteinization. It clearly appears that only the extremities of the 145 base pair DNA are labeled with the IgG (fig 1c and 1d). Due to the bivalence of the IgG molecule, some cases of cross-linking between the extremities of two DNA fragments are visible. The histogram (fig 1e) represents the distance between a free DNA extremity and the distal IgG, or between two IgG molecules.

N-Aco-AAF binding to DNA in trinucleosomes

Trinucleosomes being reacted with N-Aco-AAF, and the DNA-AAF fragments (0.36% of modified bases) extracted from these trinucleosomes are shown in figure 2a and 2b respectively. As illustrated in figure 2c the IgG molecules are distributed both between the DNA extremities and at the extremities. The average distance between

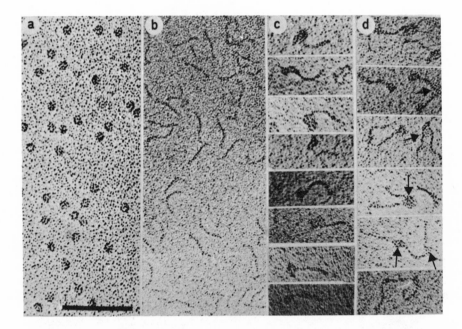

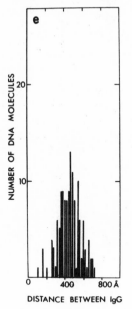

Figure 1 a - Electron micrograph of chicken erythrocyte core particles having 0.8% modified bases. Figure 1b shows the corresponding 145 base pairs DNA-AAF. Figure 1c, 1d : core particle DNA-AAF incubated with anti-Guo-C8-AAF IgG and purified on Biogel P 300. The arrows in figure 1d point to intermolecular cross links resulting from the binding of an IgG to two antigenic determinants of two DNA molecules. The spreading buffer (10 mM Tris, 0.2 mM EDTA, pH 7.4) contained 10 mM NaCl in 1a, 1b and 100 mM NaCl in 1c, 1d (see ref. 2). The bar represents 0.1 μm. The histogram 1e represents the distance between a free DNA extremity and the distal IgG or between two IgG molecules.

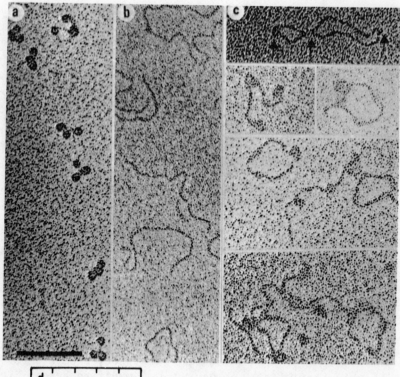

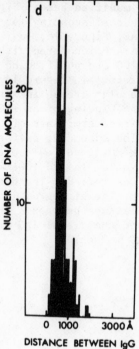

Figure 2a - Electron micrograph of erythrocytes trinucleosomes having 0.36% modified bases. On fig. 2b is the corresponding DNA-AAF. Fig. 2c shows the DNA-AAF fragments after incubation with IgG and gel filtration. Intramolecular as well as intermolecular cross links are visible. The histogram in fig. 2d represents the center-to-center distance between two consecutive IgG. The same spreading conditions as in fig. 1c were used. The bar represents 0.1 μm.

two consecutive antibodies (737 ± 200 Å) is compatible with the average repeat length of erythrocyte chromatin[11] (216 ± 5 base pairs, i.e. 734 ± 17 Å).

CONCLUSION

The preferential binding of AAF to the extremities of the DNA fragments in core particles may reflect the greater affinity of N-Aco-AAF for denatured DNA as compared to native DNA[12] since it is likely to find some denatured DNA at the extremities of the 145 base pair DNA. On the other hand, in DNA extracted from trinucleosomes, one finds that the average distance between two AAF residues (as revealed by the antibodies) is in agreement with the repeat length which is found in chicken erythrocyte chromatin[11]. This result is therefore in favour of a specificity of N-Aco-AAF reaction toward DNA in the linker region, in agreement with results from enzymatic digestion studies on AAF modified chromatin[13] as well as on Benzo(a)pyrene treated chromatin[14].

REFERENCES

1. E. Kriek, Carcinogenesis by aromatic amines, Biochim. Biophys. Acta, 355 : 177 (1974)
2. R. Fuchs and M. Daune, Physical Studies on DNA after Covalent Binding of a Carcinogen, Biochemistry, 11 : 2659 (1972)
3. R.P.P. Fuchs, J.-F. Lefèvre, J. Pouyet, M.P. Daune, Comparative Orientation of the Fluorene Residues in Native DNA modified by N-Acetoxy-N-2-acetylaminofluorene and two 7-halogeno derivatives, Biochemistry, 15 : 3347 (1976)
4. G. de Murcia, M.-C.E. Lang, A.-M. Freund, R.P.P. Fuchs, M.P. Daune, E. Sage and M. Leng, Electron microscopic Visualization of N-acetoxy-N-2-acetylaminofluorene binding sites in Col E$_1$ DNA by means of specific antibodies, Proc. Natl. Acad. Sci. USA, 76 : 6076 (1979)
5. M. Leng, E. Sage, R.P.P. Fuchs and M.P. Daune, Antibodies to DNA modified by the carcinogen N-acetoxy-N-2-acetylaminofluorene, FEBS Letters, 92 : 207 (1978)
6. M.C. Poirier, S.H. Yuspa, I.B. Weinstein, S. Blobstein, Detection of carcinogen-DNA adducts by radioimmunoassay, Nature, 270 : 186 (1977)
7. M. Guigues and M. Leng, Reactivity of Antibodies to guanosine modified by the carcinogen N-acetoxy-N-2-acetylaminofluorene, Nucl. Acid. Res. 6 : 733 (1979)
8. E. Sage, R.P.P. Fuchs, and M. Leng, Reactivity of the Antibodies to DNA modified by the carcinogen N-acetoxy-N-2-acetylaminofluorene, Biochemistry, 18 : 1328 (1979)
9. J.-F. Lefèvre, R.P.P. Fuchs and M.P. Daune, Comparative studies on the 7-iodo and 7-fluoro derivatives of N-acetoxy-N-2-acetylaminofluorene : Binding Sites on DNA and conformational change

of modified deoxytrinucleotides, Biochemistry, 17 : 2561 (1978)

10. G. de Murcia, G.C. Das, M. Erard and M.P. Daune, Superstructure and CD Spectrum as Probes of Chromatin Integrity, Nucl. Acid. Res. 5 : 523 (1978)

11. M.L. Wilhelm, A. Mazen, and F.X. Wilhelm, Comparison of the DNA repeat length in H1 and H5 containing Chromatin, FEBS Letters 79 : 404 (1977)

12. R.P.P. Fuchs, Arylamidation and Arylation by the Carcinogen N-2-fluorenylacetamide : a sensitive and rapid radiochemical Assay, Anal. Biochem., 91 : 663 (1978)

13. G. Metzger, F.X. Wilhelm, M.L. Wilhelm, Distribution along DNA of the bound Carcinogen N-acetoxy-N-2-acetylaminofluorene in chromatin modified in vitro, Chem.-Biol. Interactions, 15 : 257 (1976)

14. C.L. Jahn and G.W. Litman, Accessibility of deoxyribonucleic acid in chromatin to the covalent binding of the chemical carcinogen benzo(a)pyrene, Biochemistry, 18 : 1442 (1979)

STUDIES OF MOLECULAR EVENTS IN ISOLATED NUCLEOLAR CHROMATIN IN RESPONSE TO CHEMICALS AND IONIZING RADIATION

Ole Frederik Nielsen, Bernhard Borkhardt, Mahmut Carin,
Børge F. Jensen, Dieter Jentsch, Johan C. Leer, and
Ole Westergaard

Department of Molecular Biology and Plant Physiology
University of Aarhus
DK-8000 Aarhus C
Denmark

Chromatin is an important target for the effect of chemicals and ionizing radiation in living cells. The structural alterations induced lead to disturbance of the genetic expression, i.e. of the biosynthesis of RNA and of protein. One useful approach to explore the molecular background of various observed end-effects is to study the effect of damage on a specific gene which has been isolated in its chromatin form.

We have isolated the gene (rDNA) coding for precursor rRNA from various species of Tetrahymena as transcriptionally active chromatin. The rDNA exists extra-chromosomally, strongly amplified as palindromic molecules in the nucleoli of the macronucleus. Individual isolated chromatin molecules have been visualized in the electron microscrope as shown in Fig. 1. The palindromic nature of the chromatin molecule is obvious from the existance of two transcriptional units. The transcriptional complexes are seen in Fig. 1A as balls of yarn due to coiling of the RNA side-chains around the RNA polymerase molecules. The direction of transcription from the centre of the molecule towards the ends can be deduced from length-gradient of RNP-fibrils seen in Fig. 1B after the RNA chains have been stretched out by addition of a single-strand binding protein. Also, the non-transcribed distal spacer region of the gene can be seen in Fig. 1A (arrow). The isolated nucleolar chromatin has transcriptional properties which mimic the in vivo process with respect to strand selection, elongation rate and termination (Gocke, et al. 1978).

69

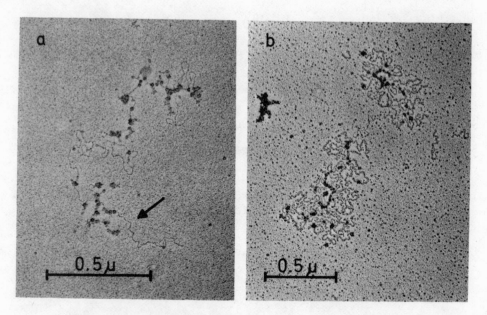

Fig. 1.   Electron micrographs of nucleolar chromatin from T.pyri-
formis. Nucleoli were isolated as described by Gocke et
al. (1978). The nucleolar chromatin was after dissociation
of the nucleolar body spread for electron microscopy by the
method of Dubochet (1971), contrasted with uranyl acetate and
shadowed with platinum-palladium. (a) Untreated nucleolar
chromatin molecule. The arrow points at a distal non-transcribed spacer region. (b) Nucleolar chromatin molecule
with extended RNA side chains after binding of a single-
strand binding protein.

This chromatin is used in the study of the effect of chemicals
such as intercalating dyes on gene expression. One dye, a synthetic
compound Miracil D = lucanthone, has been reported to have a mutagenic effect (see Hirschberg 1975 for references). In addition it
is known to be a potent inhibitor of RNA synthesis in prokaryotes
as well as eukaryotes (Weinstein et al., 1965; Hirschberg et al.,
1968; Hirschberg, 1975). At low concentrations the drug was found
selectively to inhibit the synthesis of rRNA in HeLa cells (Bases
and Mendez, 1969).

Studying the effect of lucanthone in concentrations ranging
from 0-240 µg/ml on isolated nucleolar chromatin we surprisingly
found no reduction in RNA synthesis. At the contrary a stimulation
was observed. Analyses showed that the drug had no effect on the
elongation process (Westergaard et al., 1979). When the size of the
transcript was analysed, it was observed that a high percentage of
of the transcripts (more than 50%) had a molecular weight of 3.0 ×

10^6 Daltons. This indicates that some RNA-polymerase molecules pass
the normal point of transcription and read into the normally non-
transcribed distal spacer. Hybridization of the transcript to restriction fragments of the rDNA confirmed this (Westergaard et al.
1979). It is unknown how lucanthone suppresses normal termination,
but it is plausible that it by intercalation changes specific structures on the DNA template or on the transcript in such a way that
the recognition sites for a termination protein (Leer et al., 1979)
are distorted.

Other chemicals sharing the tricyclic heteroaromatic ring structure with lucanthone have been tested for their effect on termination
of transcription: hycanthone, proflavin, actinomycin D, and acridine
orange. Of these proflavin suppresses normal termination.

The Tetrahymena species T.thermophila has an intervening sequence
of approximately 400 bases in the region of the rDNA that codes for
26S rRNA. The sequence is transcribed and post-transcriptionally
removed from the transcript by a series of events (Cech and Rio,
1979; Din et al., 1979). We have shown that isolated nucleoli have
the enzymatic capacity for cutting out the intervening sequence from
the pre-35S rRNA transcript as accumulation of a small fragment of
0.4 kb,which hybridizes to the intervening sequence of the rDNA, is
observed on incubation. Furthermore, we could show by electron microscopy that the cleaved 35S rRNA is spliced under specific in
vitro-conditions. Fig. 2 shows electron micrographs of R-loop molecules formed by hybridization with pre-35S and 35S rRNA, spective-
ly. In the presence of proflavin, these processes do not take place.
The effect of proflavin might be due to an alteration of the secondary structure of the transcript by intercalation whereby the recognition sites for the endonuclease are altered.

The isolated nucleolar chromatin is also applied in the studies
of the effect of ionizing radiation on chromatin. Damages can be
determined by various methods. Thus, we have used the endogenous
RNA-polymerase activity to monitor the extent of damage as reduction
in transcriptional activity - and to determine the distance between
two damages, using the stripping technique described earlier (Gocke
et al., 1979). The system has also a potential in determination of
various types of template modification, as the RNA polymerase, which
can transcribe in the presence of Mg^{2+}, Mn^{2+}, or Co^{2+}, might have different fidelity in the presence of each of these three divalent cations. Finally, we use the system to localize single-strand breaks
in the nucleolar chromatin. The applied method takes advantage of
the palindromic structure of the rDNA. After denaturation and renaturation (leading to snap-back molecule), a DNA strand, in which a
nick has been introduced, will appear as partially double stranded
- partially single stranded. The single-stranded part can be elucidated in the electron microscope after the addition of single-strand
binding protein (Fig. 3). This method seems very sensitive in the

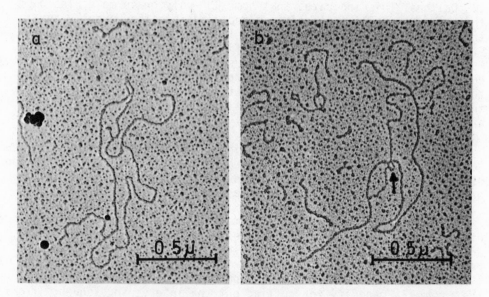

Fig. 2. R-Loop hybrids. The hybrids were formed between snap-back
         rDNA and rRNA from isolated nucleolar chromatin from T.
         thermophila and visualized as described by Din and Engberg
         (1979). The figure shows hybrids between (a) snap-back rDNA
         and pre-35S rRNA and (b) snap-back rDNA and processed 35S
         rRNA (the arrow points at the small intervening sequence
         duplex DNA-loop located at the constriction of the R-loop).

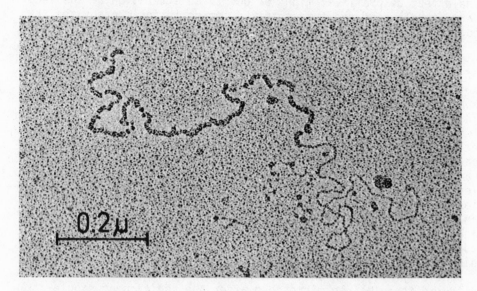

Fig. 3. Visualization of single-strand break in rDNA molecule.

study of accessibility of transcriptionally active versus inactive chromatin, e.g. the distal non-transcribed spacer of the nucleolar chromatin, to radiation.

The isolated transcriptionally active nucleolar chromatin re-represents a very useful system for studies of molecular events on transcriptionally active chromatin of a specific gene.

Bases, R. and Mendez, F., 1969, Reversible inhibition of ribosomal RNA synthesis in HeLa by lucanthone (Miracil D) with continued synthesis of DNA-like RNA, J. Cell Physiol., 74:283.

Cech, T. R. and Rio, D. C., 1979, Localization of transcribed regions of extrachromosomal ribosomal RNA genes of Tetrahymena thermophila by R-loop mapping, Proc. Natl. Acad. Sci. USA, 76:5051.

Din, N. and Engberg, J., 1979, Extrachromosomal RNA genes in Tetrahymena. structure and evolution, J. Mol. Biol., 134:555.

Din, N., Engberg, J., Kaffenberger, W., and Eckert, W., 1979, The intervening sequence in the 26S rRNA gene coding region of T. thermophila is transcribed within the largest stable precursor for rRNA, Cell, 18:525.

Dubochet, J., Ducommun, M., Zollinger, M., and Kellenberger, E., 1971, A new preparation method for dark-field electron microscopy of biomacromolecules, J. Ultrastr. Res., 35:147.

Gocke, E., Leer, J. C., Nielsen, O. F., and Westergaard, O., 1978, Transcriptional properties of nucleoli isolated from Tetrahymena, Nucl. Acids Res., 5:3993.

Hirschberg, E., 1975, Thiaxanthenones: Miracil D and Hycanthone, in "Antibiotics", Vol. III, Corona, ed., Springer Verlag, Berlin-New York.

Hirschberg, E., Ceccarini, C., Osnos, M., and Carchman, R., 1968, Effects of inhibitors of nucleic acid and protein synthesis on growth and aggregation of the cellular slime mold Dictyostelium discoideum, Proc. Natl. Acad. Sci. USA, 61:316.

Leer, J.C., Tiryaki, D., and Westergaard, O., 1979, Termination of transcription in nucleoli isolated from Tetrahymena, Proc. Natl. Acad. Sci. USA, 76:5563.

Weinstein, J. B., Chernoff, R., Finkelstein, J., and Hirschberg, E., 1965, Miracil D: An inhibitor of ribonucleic acid synthesis in Bacillus subtilis, Molec. Pharmacol., 1:297.

Westergaard, O., Gocke, E., Nielsen, O. F., and Leer, J. C., 1979, Effect of lucanthone (Miracil D) on transcription of ribosomal RNA genes from Tetrahymena in vivo and in vitro, Nucl. Acids Res., 6:2391.

Acknowledgement

This study was supported by contract No. 204-76-1 BioDK with EURATOM, CEC, Brussels.

MECHANISMS OF DNA EXCISION REPAIR

BASE EXCISION REPAIR OF DNA

Errol C. Friedberg

Laboratory of Experimental Oncology
Department of Pathology
Stanford University
Stanford, California 94305
USA

INTRODUCTION

The terms base excision repair (BER) and nucleotide excision repair (NER) were first suggested by Duncan et al in 1976 (1), since at that time distinct modes of excision repair appeared distinguishable by a least two criteria.

(i)   In BER, damaged or inappropriate bases are excised as free bases, while in NER such bases are excised as part of nucleotides or oligonucleotide sequences.

(ii)   Mechanistically there is good evidence that BER involves hydrolysis of glycosylic bonds linking bases to the deoxyribose-phosphate backbone of DNA, as a primary enzymatic event during excision repair (2,3).   On the other hand, NER is thought to involve hydrolysis of phosphodiester bonds adjacent to sites of base damage, as a primary enzymatic event. The prototypic example of NER that until recently fulfilled both of these criteria is the excision of pyrimidine dimers from UV-irradiated DNA (4). However, very recent studies from a number of laboratories indicate that at least in M. luteus (5,6) and in phage T4-infected E. coli (7-10), the excision of these lesions requires the action of a specific DNA glycosylase that catalyzes the hydrolysis of the 5' N-glycosylic bond of dimerized pyrimidines.   Phosphodiester bonds associated with the resulting apyrimidinic sites are attacked secondarily by one or more AP (apurinic/apyrimidinic) endonucleases, creating nicks in the DNA 5' with respect to the dimers (see article by Grossman, this volume).   At present, there is no evidence of which I am aware that this 2-step mechanism for the incision of UV-irradiated DNA is more general than the two specific examples just cited. Furthermore, direct analysis of the products of excision catalyzed by the combined action of the T4 enzyme activity and E. coli DNA polymerase I in vitro has demonstrated that pyrimidine dimers are excised as part of an oligonucleotide structure (11). For the purposes of this discussion I define BER only in terms of criterion (i) mentioned above, viz., the structure of the excised product. By this definition therefore the excision of pyrimidine dimers, whether involving the action of a

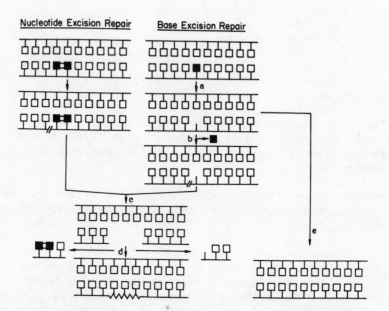

Figure 1 - Diagrammatic representation of base excision repair showing a) DNA glycosylase catalyzed excision of a damaged or inappropriate base (in black), b) AP endonuclease-catalyzed hydrolysis of a phosphodiester bond 5' with respect to a site of base loss, c) excision of the deoxyribose-phosphate residue as part of an oligonucleotide by degradation of the DNA at the site of AP endonuclease-catalyzed incision, d) repair synthesis and DNA ligation. The insertion pathway is designated as e). The nucleotide excision pathway showing endonuclease-catalyzed incision of DNA 5' to a model dinucleotide lesion (e.g., pyrimidine dimer) is shown for comparison.

specific DNA glycosylase or of a direct-acting UV endonuclease, is considered an example of NER and will not be considered further here.

DNA GLYCOSYLASES

At the time of writing, all known examples of BER involve the action of DNA glycosylases. These enzymes have been shown to be ubiquitous in nature and each appears to be specific for a particular form of base damage. Table I lists the 5 DNA glycosylases currently known to be present in cell-free preparations of uninfected E. coli. These enzymes are all monomeric proteins of Mr ~ 30,000, and none have any requirement in vitro for known cofactors (see ref. 3 for a detailed review of the properties of the first 3 DNA glycosylases listed is Table I). The FaPy-DNA glycosylase (12) and the DNA glycosylase recently shown to catalyze the excision of 5,6-dihydroxydihydrothymine and of 5,6-dihydrothymine from DNA (8) are of distinct biological interest, since all of the forms of base damage recognized by these enzymes have been identified in DNA exposed to ionizing radiation (13).

Evidence for a role of some of these enzymes in the excision of damaged or inappropriate bases in living cells has been provided by studies with prokaryotes. Mutants defective in ura-DNA glycosylase have been isolated from strains of E. coli (ung⁻) (14) and B. subtilis (urg⁻) (15). The ung⁻ mutants are spontaneous mutators (16) (presumably due to their inability to excise uracil arising from the spontaneous deamination of cytosine in DNA) and are abnormally sensitive to chemicals known to promote the deamination of cytosine, i.e., nitrous acid (17) and sodium bisulfite (18). Mutants defective in 3-Ma-DNA glycosylase activity (tag⁻) have also been isolated from E. coli (19). At least one of these (tag-1) has been shown to be defective in the structural gene for the enzyme. A second mutation in a strain designated tag-2 has been mapped at a different position in the E. coli genome. Its exact role in the regulation of the expression of 3-Ma-DNA glycosylase remains to be defined. Both mutants are abnormally sensitive to alkylating agents known to produce 3- methyladenine in DNA but show normal sensitivity to UV and ionizing radiation.

As yet no detailed studies have been reported on DNA glycosylase activities from higher organisms. Talpaert-Borlé et al (20) have reported the partial purification of ura-DNA glycosylase activity from extracts of calf thymocyte nuclei. A single activity was found to be associated with a protein of 2.85 S and a Mr ~ 28,700. Sirover (21) has reported the presence of two chromatographically distinct forms of ura-DNA glycosylase in extracts of human lymphocytes. Stimulation of blast transformation by exposure of cells to PHA results in an increase in the level of one of the two forms. Multiple forms of ura-DNA glycosylase have also been identified in human KB cells. One is localized to the cell nucleus; the other is associated with the mitochondrial fraction of cell lysates (22).

TABLE 1

DNA Glycosylases of E. coli

Enzyme	Substrate	Products
Ura-DNA glycosylase	DNA containing deoxyuridine	Uracil + depyrimidinated DNA
3-Ma-DNA glycosylase	DNA containing 3-methyldeoxyadenosine	3-methyladenine + depurinated DNA
Hx-DNA glycosylase	DNA containing deoxyinosine	Hypoxanthine + depurinated DNA
FaPy*-DNA glycosylase	DNA containing a formamidopyrimidine derived from alkylated deoxyguanosine	FaPy residue + depurinated DNA
Endonuclease III	DNA containing 5,6-dihydroxydihydrothymidine or 5,6-dihydrothymidine	5,6-dihydroxydihydrothymine/5,6-dihydrothymine + depyrimidated DNA

*   Formamidopyrimidine (2,6-diamino-4-hydroxy-5-N-methylformamidopyrimidine)

THE REPAIR OF SITES OF BASE LOSS IN DNA (Fig. 1)

Following the enzyme-catalyzed removal of bases, apurinic and/or apyrimidinic sites are left in the DNA. Such sites can apparently be repaired by two distinct enzyme-catalyzed mechanisms, which for convenience we refer to as the degradative and insertion pathways. These pathways presumably also operate at sites of base loss generated by the spontaneous loss of either unmodified or chemically modified (e.g., alkylated) bases.

(i)   The Degradative Pathway. Further degradation of DNA containing sites of base loss can occur by the action of a specific class of endonucleases called AP (apurinic/apyrimidinic) endonucleases. Like the DNA glycosylases, these enzymes are highly ubiquitous in their distribution. In addition, individual species appear to have more than one such enzyme activity. The organism that has been most extensively studied in this respect is E. coli, cell-free extracts of which have been reported to contain a number of apparently distinct AP endonuclease activities. Endonuclease VI is the quantitatively major AP endonuclease of E. coli, accounting for ~ 90% of the total activity on depurinated duplex DNA (23). The remaining activity on this substrate is accounted for by endonuclease IV (24). The literature contains reference to 2 endonucleases of E. coli termed endonucleases III and V which have been reported to attack duplex DNA containing sites of base loss (25,26). Endonuclease III preparations have recently been shown to contain a distinct DNA glycosylase (mentioned above), that catalyzes the excision of 5,6-dihydroxydihydrothymine and of 5,6-dihydrothymine. The AP endonuclease activity reported as endonuclease III may well represent a low level of contamination of this DNA glycosylase with either endonucleases IV or VI. Alternatively it may be an AP endonuclease activity that is physically associated with the DNA glycosylase. The same is possibly true of endonuclease V, an activity reported to attack DNA containing a variety of forms of base damage, including sites of base loss (26).

Recently, a new AP endonuclease activity designated as endonuclease VII of E. coli has been described (2,7; T. Bonura and E.C. Friedberg, MS in preparation). Unlike the other AP endonuclease activities listed in Table I, this enzyme attacks sites of base loss in single stranded DNA or deoxyribopolymers containing sites of base loss. The precise substrate specificity of endonuclease VII in vitro has not yet been clearly established. Thus, for example, depyrimidinated sites in single stranded deoxyribopolymers [poly (dU) or poly (dU.dT)] or single stranded DNA (such as M13 DNA) are attacked by the enzyme, however if such sites are rendered alkaline-stable by reduction with sodium borohydride they are no longer attacked. Depurinated sites created by heat-acid treatment of alkylated poly (dA) are also not attacked by the enzyme and neither are apurinic sites in duplex PM2 DNA. The sensitivity of duplex DNA containing depyrimidinated sites to endonuclease VII is currently under investigation.

More detailed studies on all of the AP endonucleases of E. coli need to be carried out before one can understand the functional significance of their apparent redundancy. It is entirely possible that in the living cell these enzymes discriminate between sites of base loss with respect to the secondary and/or tertiary structure of the DNA, the nature of the base sequences flanking such sites and possibly even whether such sites are apurinic or apyrimidinic in nature. In addition, some may catalyze cleavage of phosphodiester bond 5' with respect to sites of base loss, while others may be 3' specific. Finally, as is possibly true of endonuclease III, some may be physically associated with specific DNA glycosylase activities. Thus, each may in fact play a relatively unique role in the repair of DNA containing missing bases.

The hydrolysis of a phosphodiester bond in close proximity to a site of base loss sets the stage for excision of the deoxyribose-phosphate residue by exonucleolytic degradation of the DNA. The particular enzyme(s) utilized in vivo will in large measure be dictated by the polarity of the phosphodiester bond scission, i.e., if incision is 5' with respect to a site of base loss, excision of the deoxyribose-phosphate residue requires the action of a 5'→3' endonuclease and vice versa. In this regard it is of interest to note that the major AP endonuclease of E. coli, endonuclease VI, has an associated 3'→5' exonuclease activity and in theory might be able to effect an excising function by initiating the degradation of DNA at nicks 3' to sites of base loss. However, in vitro this enzyme reportedly catalyzes incision 5' with respect to apurinic sites in duplex DNA (23). Verly (27) has suggested that the 3' exonuclease activity might serve an anti-ligation function following DNA incision, thereby precluding the rejoining of the sugar-phosphate backbone prior to 5' → 3' excision events. Following incision of apurinic or apyrimidinic DNA on the 5' side of sites of base loss, excision of deoxyribose-phosphate residues can be effected by any one of a number of 5' → 3' exonuclease activities detected in extracts of E. coli, including the 5' → 3' exonuclease function of DNA polymerases I and III, exonuclease VII and exonuclease V. Details of the 5' → 3' enzyme-catalyzed excision of nucleotides and of the repair synthesis and DNA ligation required to complete the degradative repair pathway have been extensively discussed in relation to the later steps of NER (see ref. 4) and will not be further elaborated on here.

(ii)   The Insertion Pathway.   The discovery of a class of enzymes that catalyzes the removal of bases as the free base, immediately suggested the possibility that a different class of enzymes might catalyze the reverse reaction.   Recent studies from two laboratories have resulted in the discovery of purine insertase activity in extracts of E. coli (28) and in extracts of a line of human diploid fibroblasts (29). Neither activity has yet been extensively purified and the details of their mechanism of action remain to be established.   However, studies to date suggest that the eukaryote enzyme has no requirement for a high energy donor and can utilize the appropriate free purine and apurinic duplex DNA as substrates. The activity from E. coli was reported to have a requirement for the appropriate purine as the deoxynucleoside triphosphate (28).   However, more recent studies suggest that this activity too might utilize free bases as substrate (30).

Thus far there have been no reports on the successful isolation of a pyrimidine insertase activity. One would anticipate that such an activity might well exist in certain organisms, particularly thermophiles. In these prokaryotes exposure to low pH and very high ambient temperatures would be expected to promote the spontaneous loss of purines and to a lesser extent pyrimidines from DNA, as well as the spontaneous deamination of cytosine to uracil in DNA. Excision of the uracil by ura-DNA glycosylase would create apyrimidinic sites appropriate for a cytosine-specific DNA insertase activity.

Acknowledgements

Studies in the author's laboratory were supported by research grants from the USPHS (CA 12428), American Cancer Society (NP - 174) and by contract DE-AC03-76SF00326 with the U.S. Department of Energy. The author acknowledges the USPHS for receipt of a research career development award (CA 71005).

References

1. Duncan, J., Hamilton, L., and Friedberg, E.C. (1976). J. Virol.
    19, 338.
2. Friedberg, E.C., Bonura, T., Cone, R., Simmons, R., and Anderson,
    C. (1978). In "DNA Repair Mechanisms", eds. P.C. Hanawalt,
    E.C. Friedberg and C.F. Fox. Academic Press, N.Y., p. 163.
3. Lindahl, T. (1979). Prog. Nuc. Acids, Res. Mol. Biol. 22, 135.
4. Hanawalt, P.C., Cooper, P.K., Ganesan, A.K., and Smith, C.A.
    (1979). Annu. Rev. Biochem. 48, 783.
5. Grossman, L., Riazzudin, S., Haseltine, W.A., and Lindan, C.P.
    (1978). Cold Spring Harb. Symp. Quant. Biol. 43, 947.
6. Haseltine, W.A., Gordon, L.K., Lindan, C.P., Grafstrom, R.H.,
    Shaper, N.L., and Grossman, L. (1980). Nature 285, 634.
7. Friedberg, E.C., Anderson, C.T.M., Bonura, T., Cone, R., Radany,
    E.H., and Reynolds, R.J. (1981). Prog. Nuc. Acids Res. Mol.
    Biol. 26, 197.
8. Linn, S. (1980) (Personal Communication).
9. Radany, E.H., and Friedberg, E.C. (1980). Nature 286, 182.
10. Seawell, P.C., Smith, C.A., and Ganesan, A.K. (1980). J. Virol.
    35, 790.
11. Pawl, G., Slor, H., and Friedberg, E.C. (1975). J. Bact. 122, 341.
12. Chetsanga, C.J., and Lindahl, T. (1979). Nucleic Acids Res. 6,
    3673.
13. Cerutti, P.A. in "Molecular Mechanisms for Repair of DNA",
    (1975) eds, P.C. Hanawalt and R.B. Setlow. Plenum Press
    N.Y., p. 3.
14. Duncan, B.K., Rockstroh, P.A., and Warner, H.R. (1978). J. Bact.
    134, 1039.
15. Makino, F., and Munakata, N. (1977). J. Bact. 131, 438.
16. Duncan, B.K., and Weiss, B. in "DNA Repair Mechanisms", (1978)
    eds. P.C. Hanawalt, E.C. Friedberg and C.F. Fox. Academic
    Press, N.Y. p. 183.
17. Da Rosa, R., Friedberg, E.C., Duncan, B.K., and Warner, H.R.
    (1977). Biochemistry 16, 4934.
18. Simmons, R.R., and Friedberg, E.C. (1979). J. Bact. 137, 1243.
19. Karran, P., Lindahl, T., Ofsteng, I., and Seeberg, E. (1980). J.
    Mol. Biol. 140, 101.
20. Talpaert-Borle, M., Clerici, L., and Campagnari, F. (1979). J.
    Biol. Chem. 259, 6387.
21. Sirover, M.A. (1979). Cancer Res. 39, 2090.
22. Anderson, C.T.M., and Friedberg, E.C. (1980). Nucleic Acids Res.
    8, 875.
23. Verly, W.G., and Rassart, E. (1975). J. Biol. Chem. 50, 217.
24. Lungquist, S. (1977). J. Biol. Chem. 252, 2808.
25. Gates, F.T., and Linn, S. (1977). J. Biol. Chem. 252, 2802.
26. Gates, F.T., and Linn S. (1977). J. Biol. Chem. 252, 1647.
27. Verly, W.G. in "DNA Repair Mechanisms" (1978) eds. P.C.
    Hanawalt, E.C. Friedberg and C.F. Fox. Academic Press,
    N.Y. p. 187.

28.   Livneh, Z., Elad, D. and Sperling, J. (1979). Proc. Natl. Acad. Sci (USA) 76, 1089.

29.   Deutsch, W.A., and Linn, S. (1979). Proc. Natl. Acad. Sci. (USA) 76, 141.

30.   Sperling, J. (1980) (Personal communication)

INCISION OF PYRIMIDINE DIMER CONTAINING DNA BY SMALL MOLECULAR

WEIGHT ENZYMES

R. Grafstrom, N. Shaper, and L. Grossman

Department of Biochemistry
The Johns Hopkins University
School of Hygiene and Public Health
Baltimore, Maryland 21205

And

W. Haseltine
Sidney Farber Cancer Institute
Harvard University

Abstract

In Micrococcus luteus the incision event during repair of DNA
containing pyrimidine dimers has been shown in vitro to be a two-
step enzymatic reaction. The first step results in the cleavage of
the N-glycosylic bond between the 5'-thymine moiety of the pyrimidine
dimer and the deoxyribose generating an apyrimidinic site (Apy) and
a thymine-thymidylate dimer attached to the DNA. The second
catalytic step involves phosphodiester bond hydrolysis 3' to the
Apy site generating a nicked DNA with an Apy site at its 3' terminus and the mixed thymine-thymidylate dimer at the 5' terminus.

Introduction

At the final stage in purification of the pyrimidine dimer
specific endonuclease activity from M. luteus the endonucleolytic
activity was resolvable into two peaks of activity that generated
incised DNA molecules of different structures [1]. It was
considered that these enzymes of different protein and catalytic
properties were functioning in unique regions of the DNA molecule.
These conclusions were based on differences in the conformation of
the DNA sites arising from action by the separated peaks of enzymatic activity.

In order to identify whether the M. luteus dimer specific
endonuclease was functioning in unique nucleotide regions, DNA
fragments of defined sequence were used as both a target for
ultraviolet irradiation and as a substrate for these enzymes
[2,3,4]. Through the use of DNA sequencing techniques [5] it is
possible to determine the distribution of pyrimidine dimers in
irradiated DNA fragments in a diagnostic fashion. The endonuclease
activity catalyzes phosphodiester bond hydrolysis at each pyrimidine
dimer site, thereby generating a population of nicked DNA fragments
which are resolved by polacrylamide gel electrophoresis. From an
analysis of the photosteady states it was demonstrated that the
distribution of dimers is essentially non-random and that the
frequency of dimer formation is affected both by the number of
thymine residues in a potentially dimerizable region and by the two
bases immediately flanking the dimerizable site [4].

One of the unexpected findings of the above experiments was
that the dimer specific endonuclease preparation hydrolyzes the
phosphodiester bond between the two pyrimidine moieties of the
dimer in a two-step reaction: first, cleavage of the 5' N-glycosylic
bond of the pyrimidine dimer by a dimer DNA-glycosylase and then
cleavage of the phosphodiester backbone 3' to the Apy site by an Ap
endonuclease [6]. This two step model for incision of pyrimidine
dimer containing DNA is shown in Figure 1.

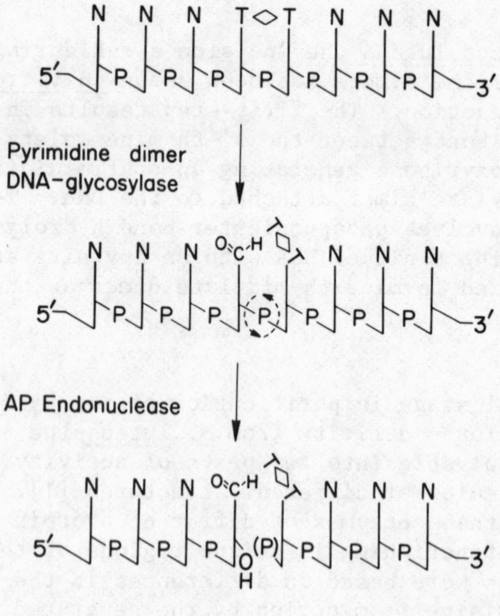

Figure 1

Purification of a Pyrimdine-Dimer-DNA Glycosylase (PyDNG)

As a direct consequence of the above experiments a pyrimidine dimer DNA glycosylase as been purified from the pyrimidine dimer "endonculease" preparation (Table I) by assaying for the appearance of Apy sites in UV irradiated DNA generated by the glycosylase.

Table I. Purification of the Pyrimidine Dimer DNA Glycosylase From (PyDNG) M. luteus

Reaction	Protein mg/ml	Total Units x10^{-7}		PyDNG Specific Activity Units mg^{-1} x10^{-3}	Yield (%)	Ratio PyDNG/ApEndo
		PyDNG	ApEndo			
1) Extract	22.0	3.3	22.3	1.52	100.0	0.16
2) Streptomycin	19.0	1.9	2.7	0.86	54.3	0.71
3) 70% (NH$_4$)$_2$SO$_4$	57.0	1.5	5.3	1.33	43.0	0.28
4) Phosphocellulose	3.6	0.6	0.16	200.0	17.0	3.5
5) Sephadex, G-75	0.09	0.31	0.003	1257.0	9.01	05.0
6) Carboxymethyl-Cellulose	0.004	0.23	0.002	8382.0	7.01	115.0

The enzyme has an isoelectric point of 8.2 and the molecular weight determined by Sephadex G75 chromatography is approximately 15,000 daltons. The enzyme exhibits an optimal activity between pH 6.5 and 8.0 which is stimulated by the presence of 50-100 mm salt. The catalytic activity is 7 times greater in the presence of $MgCl_2$ than in the presence EDTA but was not stimulated by $CaCl_2$, $ZnCl_2$, and $MnCl_2$. The pyrimidine dimer DNA glycosylase activity is unaffected by the presence of ATP in the range of 0.2 to 5.0 mM and no detectible exonuclease or RNase activity could be detected in this enzyme fraction. The DNA glycosylase is specific for pyrimidine dimers since 1) the activity is proportional to the level of pyrimidine dimers in DNA irradiated at 254 nm; 2) it is active against DNA irradiated at 313 nm in the presence of the photosensitizer acetophenone; and 3) is without activity on UV irradiated DNA subsequently treated with the yeast photolyase (generously provided by Dr. Roger Herriott).

Previously it was reported [1] that the Sephadex G-75 fraction could be separated into two distinct "UV endonucleases" by chromotography on DNA cellulose. It has since been found that during elution of the protein from a DNA cellulose column, the DNA glycosylase activity is preceeded by a Mg^{2+}-dependent AP endonuclease and followed by an AP endonuclease that is active in the absence of magnesium and in the presence of EDTA; hence, giving the appearance of two distinct endonucleases.

The enzyme is a DNA glycosylase

Implicit in a dimer DNA glycosylase mechanism is the formation of Apy sites. Evidence for the presence of such sites has been obtained by the predicted sensitivity of UV irradiated ØX174 RFI DNA that had been treated with pyrimdine dimer DNA glycosylase to purified Ap endonuclease from human placenta [6] or E. coli exonuclease III [7]. These sites are also sensitive to alkaline hydrolysis and this alkaline lability can be reversed by $NaBH_4$ reduction of the potential aldehyde site. Another characteristic of the dimer DNA glycosylase reaction is the liberation of free base. However, no free bases are liberated by the pyrimdine dimer DNA glycosylase without subsequent photoreversal of the mixed thymine-thymidylate dimer by either 254 nm light or yeast photo-lyase (see Figure 1). Such treatment yields free thymine in near stoichiometric amounts when chromotographed by PEI-thin layer chromotography.

Mechanism of Action

The mechanism suggested for the dimer DNA glycosylase activity shown in Figure 1 implicates phosphodiester bond hydrolysis 3' to the Apy site generating a nick which is bordered at the 3' end by an Apy nucleotide and a mixed dimer at the 5' terminus. Evidence for these sites of cleavage comes from the DNA sequencing experiments in which the enzymatic digests of the irradiated 5' labelled fragments were heated with alkali or piperidine. Under these β-elimination conditions the digested fragments increased by one nucleotide and their Rf values now corresponded to a break 5' to the dimer site [2]. Conversely, 3' labelled irradiated DNA fragments treated in exactly the same manner after enzymatic activity show no β-elmination dependent mobility shift. Thus, the Ap endonuclease cleaves 3' to the Apy site and the N-glycosylase cleaves the N-glycosylic bond of the 5' pyrimidine of the dimer.

A novel Ap endonuclease activity from M. luteus

In Table I it can be seen that there is a low but detectable level of Ap.endonuclease which co-purifies with the pyrimidine dimer DNA glycosylase. The ratio of glycosylase to Ap endonuclease has never been greater than 100 to 1 even after extensive attempts at further purification. It has a pH optima of 6.5, does not require Mg^{2+} and is not inhibited by 5 mM EDTA. It is heat inactivatable and preliminary results show a Km(app) for Ap sites in UV irradiated DNA generated by the glycosylase of 152 pM. From Sephadex G75 chromatography a molecular weight of 15,000 daltons is estimated. This 3' Ap endonuclease exhibits a much more limited specificity for Ap sites as shown in Table II.

Table II. Substrate Specificities of AP Endonucleases

Initial Rates (Breaks min^{-1} $\mathrm{molecule}^{-1}$) $\mathrm{x10}^{2}$ 1/

Enzyme	DNA Substrates			
	1 Apurinic	2 Reduced Apurinic	3 PyDNG Apyrimidinic	4 Apyrimidinic*
Human AP Endonuclease	2	1	2	1
E. coli Exonuclease III	2	2	2	1
M. luteus AP Endonuclease	3	1	1	1
M. luteus PyDNG AP endonuclease	1	0.2	7	1

*Photolyase product of 3

1/ These values represent initial velocities at 10 μM pRLM
form I DNA. Since the Km's have not yet been determined for
each substrate and enzyme combination, the values are not
necessarily Vmax and comparisons between enzymes are not valid.
However, the values do reflect the relative activity of a
particular enzyme with the different DNA substrates.

Unlike the human Ap endonuclease or exonuclease III from E.
coli, this enzyme exhibits a greater specificity for glycosylase
treated UV irradiated DNA than apurinic DNA or the apyrimidinic DNA
generated by photoreversal of the mixed thymine-thymidylate dimer.
It does not act on NaBH4 reduced Ap DNA. Whether or not the
glycosylase and the Ap endonuclease activities are present on
the same enzyme or are two different enzymes has yet to be determined.

Generality of Dimer Specific DNA Glycosylase Mechanism

The mechanism suggested in Figure I would necessarily invoke
the need for bidirectional excision in which the 3' Ap terminus
requires excision as well as the 5' terminated mixed dimer nucleotide. It is perhaps more than coincidental that those enzymes in
bacteria required for excision of UV damaged regions are bidirectional and are able initiate hydrolysis at both 3' and 5' termini.
For example DNA polymerase I of M. luteus [8], the UV exonuclease
of the same organism [9], a similar enzyme from E. coli, exo-

nuclease VII [10] can initiate hydrolysis at both termini either
on single stranded DNA or those sites which are non-complementary
on incised duplex DNA.

In addition to the already described pyrimidine dimer-DNA-
glycosylase from M. luteus a similar activity has been identified
for the enzyme controlled by the V structural gene of phage T4.
This UV endonuclease has also been shown to act on pyrimidine
dimer containing DNA fragments [11] or DNA to generate a site that
is either sensitive to β-elimination or upon photoreversal liberates
free thymine [12]. Therefore in two specific cases those low
molecular weight enzymes involved in incision of pyrimidine dimers
operate by a two-step enzymatic reaction involving a DNA glycosylase
followed by an AP endonucleolytic activity. At this time there has
been no such evidence for similar dimer DNA glycosylase activity
in E. coli [13]. The large molecular weight proteins isolated from
E. coli by complementation [14] described at these meetings appears
in addition not to contain a pyrimdine dimer DNA glycosylase
activity. Nor have there been reports concerning such an activity
in mammalian sources.

Bibliography

(1) Riazuddin, S. and Grossman, L., J. Biol. Chem 252, 6280 (1977).
(2) Grossman, L, Riazuddin, S., Haseltine, W.A. and Lindan, C.,
       Cold Spring Harbor Symp. on Quant. Biol. 43, 947 (1979).
(3) Haseltine, W.A., Lindan, C.P., D'Andrea, A.D. and Johnsrud, L.,
       Meth. in Enzym. 65, 235 (1980).
(4) Haseltine, W.A., Gordon, L.K., Lindan, C.P., Grafstrom, R.H.,
       Shaper, N.L. and Grossman, L., Nature, 285, 634 (1980).
(5) Maxam, A and Gilbert, W., Proc. Natl. Acad. Sci. 74, 560 (1977).
(6) Shaper, N.L. and Grossman, L., Meth. in Enzym. 65, 216 (1980).
(7) Weiss, B., J. Biol. Chem. 251, 1896 (1976).
(8) Hamilton, L.D.G., Mahler, I. and Grossman, L., Biochem. 13,
       1886 (1974).
(9) Kaplan, J.C., Kushner, S.R. and Grossman, L., Biochem. 10,
       3315 (1971).
(10) Chase, J. and Richardson, C.C., J. Biol. Chem. 249, 4553 (1974).
(11) Haseltine, W.A., personal communication.
(12) Friedberg, E., privileged communication.
(13) Yeung, G.T., unpublished data.
(14) Seeberg, E., Proc. Natl. Acad. Sci. 75, 2569 (1978).

THE USE OF DIRECT PHOTOREVERSAL OF UV-IRRADIATED DNA FOR THE DEMONSTRATION OF PYRIMIDINE DIMER-DNA GLYCOSYLASE ACTIVITY

Eric H. Radany, Jack D. Love and Errol C. Friedberg

Laboratory of Experimental Oncology
Department of Pathology
Stanford University
Stanford, California 94305
USA

A two-step enzymatic mechanism for the incision of UV-irradiated DNA at pyrimidine dimers (PD) by the "UV endonuclease" activity from M. luteus has recently been proposed (1-3). This mechanism is related to the classical base excision repair mode (4), since it consists of the sequential action of a PD-specific DNA glycosylase and an apurinic/apyrimidinic (AP) endonuclease. Such a two-step mechanism was interesting to us for the relevance it might have to our continuing investigations of the repair of UV-irradiated DNA in a variety of biological systems.

The technique originally used in the study of the M. luteus enzymes (1) provides an indirect demonstration of PD-DNA glycosylase activity, and would not be useful for the detection of such activity in crude extracts containing contaminating nucleases. We have therefore developed a more direct assay for the demonstration of PD DNA-glycosylase activity, based on the photoreversibility of PD in DNA. This assay has been used extensively in the identification of a PD DNA-glycosylase activity from T4-infected E. coli (5,6), as well as in the survey of crude extracts from several other biological sources (unpublished observations).

The reversibility of PD formation in DNA has been described in several studies (7,8). If thymine-radiolabeled E. coli DNA is continuously irradiated at 254 nm, the thymine content of PD does not increase beyond 7% of the total thymine content of the DNA, although the theoretical maximum level is much higher. This photosteady state is one of dynamic equilibrium in which the rates of PD formation (which is pseudo zero-order to good approximation) and reversal (which is first-order in PD content) are equal (Fig. 1). The reversibility of PD is convincingly demonstrated by first irradiating the DNA in the presence of Ag^+ (9). The Ag^+ acts as a photosensitizer which increases the rate of PD formation, so that the photosteady concentration of thymine-containing PD exceeds 20%. When the sensitizer is removed and the DNA is re-irradiated, the initial rate of PD reversal exceeds that of PD formation. Thus there is an exponential loss of PD until the original steady-state level (7%) is attained (Fig. 1).

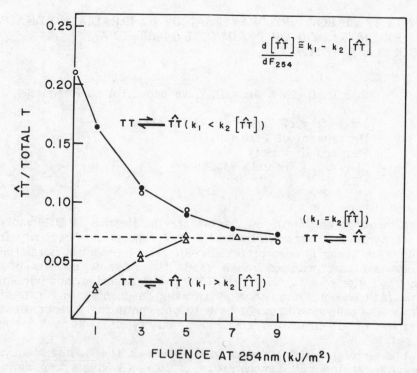

Fig. 1.   Photosteady states for PD formation and reversal in native (unirradiated) DNA (lower curve) and DNA previously irradiated in the presence of photosensitizer (upper curve).

It is evident that if a N-glycosylic bond in a thymine-containing pyrimidine dimer is hydrolyzed leaving the associated base in covalent linkage only with its partner, photoreversal of this species will result in the liberation of free thymine. The qualitative release of thymine from UV-irradiated, ^{3}H-thymine-labeled E. coli DNA into an ethanol-soluble phase has been shown to be mediated by incubation with v-gene product of bacteriophage T4 and subsequent photoreversal (5,6). No soluble thymine is detected prior to photoreversal, or when unirradiated control DNA is incubated with the T4 enzyme preparation and then photoreversed (5,6). These results strongly indicate the existence of a PD-DNA glycosylase activity associated with the v-gene product of T4 (5,6). It is important, however, to support this interpretation by a quantitative demonstration that the free thymine arises by photoreversal of thymine-containing PD specifically, and not from the breakdown of some other photoproduct in the DNA.

If a population of thymine-thymine dimers in DNA uniformly labeled with ^{3}H thymine is photoreversed, exactly half of the radioactivity in the dimers will be released as free thymine base, while half will remain as thymidylate in the DNA. The same prediction holds true for thymine label in cytosine-thymine dimers, provided CT and TC dimers are present in equal numbers in the irradiated DNA, and there is no preference for either species as substrate for PD-DNA glycosylase activity. Thus there should be a 2:1 stoichiometry between the calculated loss of radioactivity in thymine-containing dimers and the release of free thymine, during photoreversal of irradiated, thymine-labeled DNA that was reacted to completion with T4 PD-DNA glycosylase. Such a correlation is shown in the results presented in Figure 2.

Several important conclusions can be drawn from this experiment:

(i)   The free thymine released is derived from thymine-containing PD, since, at the maximum photoreversal fluence used in this experiment, the amount of thymine liberated corresponds to reversal of   60% of thymine-containing PD in the DNA. This is greater than 5% of the total radioactivity in the irradiated DNA. No other known photoproduct that could generate free thymine is present in such large amounts in UV-irradiated DNA.

(ii)   The PD-DNA glycosylase is not a minor contaminant of preparations of T4 "UV endonuclease"; as indicated above, at least 60% of thymine-containing PD in DNA are converted by the enzyme preparation to a form from which free thymine is released following photoreversal.

(iii)   The T4 PD-DNA glycosylase attacks only one of the two N-glycosylic bonds in the dimers. If both were attacked, thymine-containing PD would be lost from alcohol- and acid-insoluble fractions of DNA prior to photoreversal; this is not observed. While our present data do not address the question as to whether uniquely the 3' or 5' glycosylic bond in the dimer is hydrolyzed, previous studies with T4 UV DNA incising activity have shown that nicks in UV-irradiated duplex DNA are catalyzed 5' to PD (10,11). Thus presumably only the 5' glycosylic bond in each dimer is hydrolyzed.

In conclusion, the photoreversal-dependent release of free thymine from UV-irradiated DNA is a reliable and direct assay for PD-DNA glycosylase activity and is currently being utilized in this laboratory for the purification of the T4 v-gene product and for the exploration of similar enzyme activity in other biological sources.

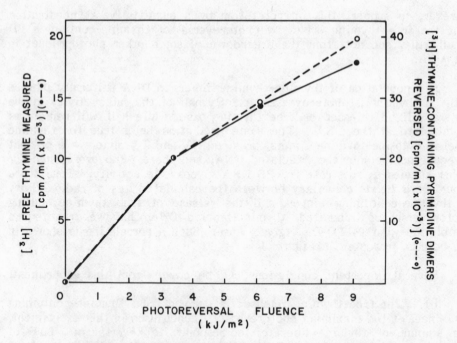

Fig. 2.   Quantitative relationship between the amount of radioactivity
          measured as free thymine and that associated with the reversal
          of thymine-containing PD. The latter parameter is calculated
          from the measured loss of thymine-containing dimers in DNA.

Acknowledgements

These studies were supported by research grants from the USPHS
(CA 12428) and the Foundation-March of Dimes (672) and by contract
DE-AC037-76SF00326 with the U.S. Department of Energy. E.C.F. was
the recipient of a RCDA (CA 71005) from the USPHS. E.H.R. is a trainee
in the Stanford MSTP supported by USPHS training grant GM 07365 and
J.D.L. is a USPHS post-doctoral fellow supported by grant CA-06441.
E.H.R. gratefully acknowledges the American Cancer Society for partial
travel support to the NATO-EMBO course at which this work was presented.

References

1.   Grossman, L., Riazzudin, S., Haseltine, W.A., and Lindan, C. (1979)
      Cold Spring Harb. Symp. Quant. Biol. 43, 947.

2. Haseltine, W.A., Gordon, L.K., Lindan, C.P., Grafstrom, R.H.,
    Shaper, N.L., and Grossman, L. (1980) Nature 285, 634.

3. Grafstrom, R., Shaper, N., Grossman, L., and Haseltine, W.A.
    (1981) (this volume).

4. Friedberg, E.C. (this volume).

5. Radany, E.H., and Friedberg, E.C. (1980) Nature 286, 182.

6. Friedberg, E.C., Anderson, C.T.M., Bonura, T., Cone, R., Radany,
    E.H., and Reynolds, R.J. (1980). Prog. Nuc. Acid Res. Mol.
    Biol. 26, 197.

7. Setlow, R.B. (1966) Science 153, 379.

8. Setlow, R.B. (1968) Prog. Nuc. Acid Res. Mol. Biol. 8, 257.

9. Rahn, R.O. and Landry, L.C. (1973) Photochem. Photobiol. 18,
    29.

10. Yasuda, S., and Sekiguchi, M. (1970) Proc. Natl. Acad. Sci. (US)
    67, 1839.

11. Friedberg, E.C., and Lehman, I.R. (1974) Biochem. Biophys. Res.
    Comm 58, 132.

ENZYMATIC STUDIES OF BASE EXCISION REPAIR IN CULTURED

HUMAN FIBROBLASTS AND IN *Escherichia coli*

Stuart Linn, Bruce Demple, Dale W. Mosbaugh, Huber R.
Warner[+] and Walter A. Deutsch[*]
Department of Biochemistry
University of California
Berkeley, California 94720

INTRODUCTION

With the discovery some six years ago of uracil DNA glycosylase
by Lindahl,[1] it became possible to define a new mode of excision
repair, "base excision repair,"[2] which was hypothesized to occur by
(i) removal of a DNA base by hydrolysis of the glycosylic bond, (ii)
cleavage of a phosphodiester bond adjacent to the resulting apurinic/
apyrimidinic (AP) site by an "AP endonuclease," and (iii) subsequent
excision of the sugar and resynthesis by DNA polymerase. In the
past few years it has been the goal of our laboratory to study the
base-excision processes in depth in order to determine exactly the
sequence and mechanisms of the enzymatic events involved, and
the relation of this repair mode to nucleotide excision repair
processes. This article summarizes some of the recent studies from
our laboratory toward this goal. Space considerations prevent a
review of related studies from other laboratories and such an
omission should not be taken to indicate that our results are particularly unique. Indeed we encourage the reader to consult articles
on similar endeavors that are found elsewhere in this volume as
well as to consult Lindahl's recent comprehensive review.[3]

AP ENDONUCLEASES

Classes of AP Endonucleases

Although many AP endonucleases have been described in the literature, the incision position relative to the baseless sugar has
been determined in relatively few cases. Assuming that the phosphodiester bond cleavage is adjacent to the AP site, there are four
theoretically possible cleavage locations relative to the AP sugar

and the adjacent phosphate residues. Of these, so far as we are aware, only two have been associated with specific enzymes. We should like to define these as "classes I and II" as follows:

B   CHO   B   Class I →   B   CHO   B
                                                                +
---P   P   P   P---   ---P   P   OH   P   P---

Class II →   B   CHO   B
                        +
              ---P   OH   P   P   P---

Class I enzymes produce 3'-deoxyribose- and 5'-phosphomonoester nucleotide termini. As summarized in this article, this class would include E. coli endonuclease III,[4,5] the AP endonuclease associated with preparations of T4 UV endonuclease,[6] possibly M. luteus UV endonuclease[7] and the human fibroblast AP endonuclease described below.

Class II enzymes produce a 3'-hydroxyl-nucleotide terminus and a deoxyribose-5-phosphate at the 5'-terminus. As summarized in this article, this class would include E. coli endonucleases IV[8] and VI,[9] human placental AP endonuclease,[10] and the human fibroblast AP endonuclease II described below.

Class III enzymes might be defined as forming a deoxyribose-3-phosphate terminus, while class IV enzymes would form a nucleoside-3'-phosphate terminus (and deoxyribose at the 5'-terminus). However, no examples of these classes have yet been cited. It might be significant to note that DNA glycosylase activities have been found to be associated with the class I enzyme preparations, but not those of class II enzymes (see below). Whether the pattern will continue to hold will be interesting to follow.

Human Fibroblast AP Endonucleases

Several years ago we reported the existence of two resolvable AP endonuclease activities in extracts from human fibroblasts.[11,12] One of them, which passed through a phosphocellulose column during

chromatography, had a relatively high affinity (Km = 4.6 nM) for
AP sites. We should like to call this enzyme "human fibroblast AP
endonuclease I." The most interesting aspect of this enzyme was
its absence specifically from fibroblasts of the xeroderma pigmentosum complementation group D type.

The other AP endonuclease activity is retained by phosphocellulose and has a relatively low affinity for AP sites (Km =
44 nM). We should like to call this enzyme "human fibroblast
AP endonuclease II." This enzyme generally elutes as a rather
dispersed peak.

Human Fibroblast AP Endonuclease II is a Class II Enzyme. Recently our laboratory has looked into the location of cleavage by
human enzymes by testing whether they can form efficient primer
termini for E. coli DNA polymerase I after having cleaved partially
depurinated, circular duplex DNA molecules.[6] Thus, we have concluded that the human AP endonuclease II is a class II enzyme
cleaving DNA as follows:

B   CHO   B

---P ⌄ P ⌄ P ⌄ P---   Human Fibroblast AP Endonuclease II →

                    B   CHO   B

            ---P ⌄ OH   +   P ⌄ P ⌄ P---

The evidence included:
(i) The enzyme creates nicks that are as efficient as DNA polymerase primer termini as those formed by N. crassa DNase.
Therefore, the 3'-termini are 3'-hydroxynucleotides.
(ii) When partially depurinated duplex DNA circles are nicked
by E. coli endonuclease III, 3'-terminal deoxyribose residues are
formed, which do not prime well for E. coli DNA polymerase I. These
termini are activated by treatment with human fibroblast AP endonuclease II, showing that the latter enzyme breaks on the 5'-side
of the AP site.
(iii) To show that in activating the termini generated by
E. coli endonuclease III, the human enzyme breaks directly adjacent
to the AP site, we have now isolated deoxyribose-5-phosphate from
the combined reaction.[13] In summary, the two enzymes acted as
follows:

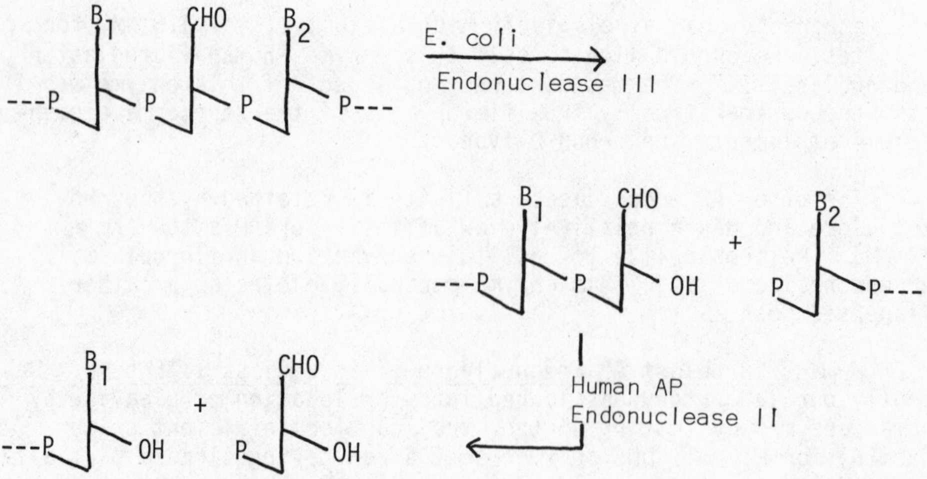

B₁   CHO   B₂

---P   P   P   P---

E. coli
Endonuclease III →

B₁   CHO   B₂
                    +
---P   P   OH   P   P---

Human AP
Endonuclease II

B₁   CHO
       +
---P   OH   P   OH

As a final note it might be observed that AP endonuclease II from xeroderma pigmentosum group D fibroblasts as well as that from HeLa cells are quite similar or identical to the enzyme from normal human cultured fibroblasts.[13]

Human fibroblast AP endonuclease I is a class I enzyme and releases deoxyribose-5-phosphate with AP endonuclease II. When circular DNA is incised by human fibroblast AP endonuclease I, the product is not a good primer for DNA polymerase I, implying that a 3'-deoxyribose might be present.[13] The AP endonuclease I and E. coli endonuclease III together do not create efficient primers; however, nicks generated by human fibroblast AP endonuclease I can be activated by human fibroblast AP endonuclease II. In addition, deoxyribose-5-phosphate is produced by this joint action.[13] This result is observed with both partially depyrimidinated [32P]-labeled DNA and with poly(dA-[32P]dU) which had been partially treated with uracil glycosylase. On the other hand, no labeled sugar-phosphate is released by the AP endonucleases from poly([32P]dA-dU) when similarly treated. Therefore, it appears that the human fibroblast AP endonucleases act efficiently together to release specifically a deoxyribose-5-phosphate in the following manner:

The ability of these two enzymes to carry out together this precise excision of deoxyribose-5-phosphate at an AP site might suggest that base excision repair could proceed in these cells without participation of any exonuclease activity. In addition, the deficiency of AP endonuclease I in the XP group D cells could explain a UV-repair defect in those cells if that enzyme is involved in a base-excision scheme for repairing UV damage (see the following section). The deficiency might also explain the neurological deterioration noted in individuals in this group,[14] if in those cases the non-dividing neuron cells accumulate AP sites for a higher-than-normal length of time, allowing secondary reactions to occur. (It is important to note that the absence of the AP endonuclease I might not totally avoid the specific excision of deoxyribose phosphate, since a site cleaved by AP endonuclease II can undergo the "class I" type of cleavage spontaneously.[3] The rate of this spontaneous cleavage in vivo is, of course, difficult to predict.

Classification of AP Endonuclease Activities Found in E. coli

The AP endonuclease activity of E. coli endonuclease VI (previously termed endonuclease II) has been characterized by Weiss et al.[9] as forming a product consistent with our designation as "class II." As expected, it creates 3'-termini that are efficient primers for DNA polymerase I.[6]

We have characterized E. coli endonuclease III as having class I AP endonuclease activity by virtue of the fact that (i) 5'-termini formed by the enzyme can be phosphorylated by polynucleotide kinase only after alkaline phosphatase treatment, as though it were a nucleoside-5'-phosphate; (ii) the 3'-termini formed by this enzyme are not efficient DNA polymerase primers, but can be activated to be such by either E. coli endonuclease VI[6] or human fibroblast AP endonuclease II[13] (but not I), thus implying that this terminus has a deoxyribose without a base; (iii) when used in conjunction with class II AP endonucleases, E. coli endonuclease II releases deoxyribose-5-phosphate from AP sites.[13]

E. coli endonuclease IV is an AP endonuclease that makes up roughly ten percent of the AP endonuclease activity of E. coli.[8] By virtue of the fact that it makes effective primer termini for DNA polymerase I and that it can activate those termini made by class I AP endonucleases,[6] we have concluded that it is a class II AP endonuclease.

We have noted that preparations of T4 UV endonuclease contain activity against partially depurinated DNA.[6] The 3'-termini produced are not effective DNA polymerase primer termini, but can be activated by E. coli endonucleases IV or VI, but not III. Thus, this activity would appear to be in class I. What the relationship is between this activity on depurinated sites and the UV repair activity is not totally clear (see below).

In summary, E. coli endonucleases IV and VI act as class II AP endonucleases, whereas E. coli endonuclease III and an AP endonuclease activity associated with T4 UV endonuclease are class I activities.

ENZYME PREPARATIONS ACTING BOTH AS AP ENDONUCLEASES AND DNA GLYCOSYLASES TO REPAIR DNA

E. coli Endonuclease III

Extensively purified preparations of endonuclease III had been found to act not only upon AP sites, but also in other sites in DNA exposed to ionizing radiation, ultraviolet irradiation or osmium tetroxide.[4] In the case of UV-irradiated DNA, it was also shown specifically that pyrimidine dimers were not substrate for the enzyme.

A common type of lesion formed by the three damaging agents is of the 5,6-dihydroxydihydrothymine type. To reconcile the enzyme activity on DNA with these lesions as well as upon AP DNA, we reasoned that the enzyme might act first as a DNA glycosylase, then as a class I AP endonuclease on the newly created AP sites:

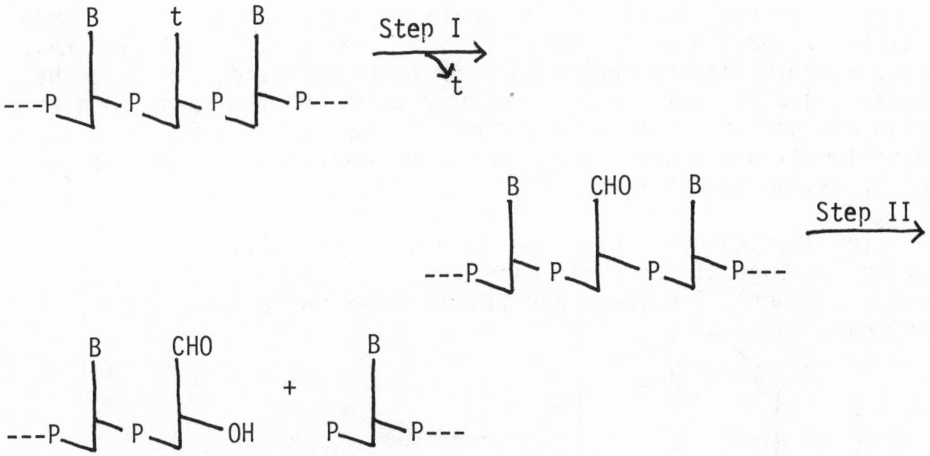

t = 5,6-dihydroxydihydrothymine-type base

Indeed, when DNA was exposed to osmium tetroxide, then to endo-nuclease III, we were able to detect free dihydroxydihydrothymine base by high pressure liquid chromatography.[5] In addition, a small amount of dihydrothymine was detected as being released from untreated DNA, though we are not sure of the origin of this base (tritium decay damage, contamination of labeled thymidine, etc.). Finally, time course studies show that the base release preceeds the endonuclease activity, but that eventually both reach an identical limit. Thus it appears that the above scheme for endonuclease III activity is likely to be correct.

Two questions remain to be answered about E. coli endonuclease III, however. We do not know the exact substrate specificity of the glycosylase activity, particularly in terms of stereoisomeric base derivatives. Further HPLC analysis will be required in this regard. We also have no absolute evidence (beyond apparent inseparability) that the AP endonuclease and glycosylase activities reside in the same protein. We hope to isolate mutants that harbor defects in this enzyme to answer the latter question.

T4 UV Endonuclease

Grossman et al.[7] have suggested that M luteus UV endonuclease activity may, like E. coli endonuclease III, involve the combined action of a glycosylase activity and an AP endonuclease activity. We chose to study whether T4 UV endonuclease might act by a similar mechanism.

By using DNA polymerase I priming efficiency as a test, we indeed were able to show that when cleaving DNA containing pyrimidine dimers with the T4 UV endonuclease, poor primer termini were

formed. However, these termini could be activated by E. coli endonuclease IV or VI[6] or by human AP endonuclease II,[13] implying that
the 3'-termini indeed contained a deoxyribose residue. (That the
5'-terminus, but not the 3'-terminus, contained the phosphomonoester
group was not shown by our experiment. However, the 5'-phosphomono-
ester location has been very clearly demonstrated by Minton et al.[15]
and by Yasuda and Sekiguchi.[16])

The AP site could have been formed by hydrolysis of the 5'-
glycosylic bond of the pyrimidine dimer (as proposed for the M.
luteus enzyme[7]), leaving a pyrimidine dimer nucleotide at the 5'-
terminus:

B   Py   Py   B

---P   P   P   P   P--- $\longrightarrow$

B ·CHO   Py Py   B

---P   P   OH   +   P   P   P---

To test for the existence of this pyrimidine dimer nucleotide,
we have taken advantage of the fact that a subsequent dose of UV
light would release a free pyrimidine base from the hypothetical
enzyme product (though not from its pyrimidine dimer substrate) by
photoreversal as follows:

Py Py   B   Py   B

P   P   P--- $\xrightarrow{h\nu}$ Py +   P   P   P---

Indeed, we were able to detect and to identify thymine base as being
specifically released by a second dose of UV light, but only after
treatment with T4 UV endonuclease.[5] (The enzyme itself releases
no thymine). In addition, at a reaction limit the releasable thymine very nearly equaled the number of thymine-containing dimers
initially present as well as the number of nicks ultimately put in.
Conversely, early in the endonuclease reaction, the number of AP
sites generated is greater than the number of phosphodiester bonds
cleaved,[6] though these can be equalized by the inclusion of exogenous AP endonuclease in the reaction. Apparently, early in the

reaction glycosylase activity preceeds AP endonuclease activity so
transient AP sites exist.

In summary, we have shown the following sequence of reactions
in the test tube with partially purified T4 UV endonuclease:

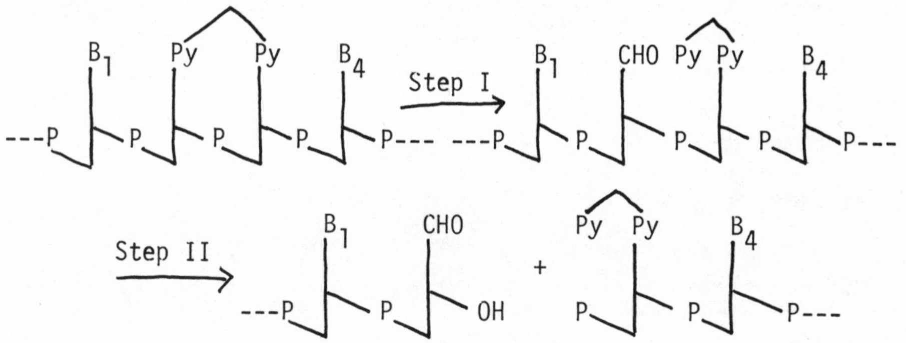

In addition,

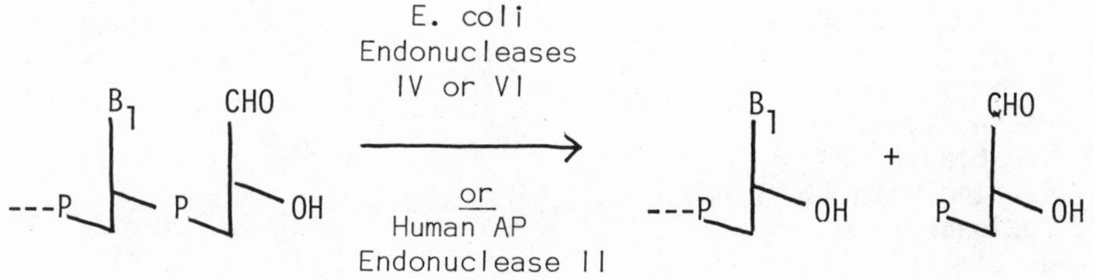

Whether the reactions catalyzed by the T4 UV endonuclease preparation are catalyzed by the same protein species, and whether a
class II AP endonuclease(s) is involved in UV repair in this system
as shown above are questions of immediate interest.

THYMINE DIMERS EXCISED BY E. COLI IN VIVO ARE NOT IN THE FORM OF DIMER NUCLEOTIDES

The glycosylase/AP endonuclease mechanism of the M. luteus and
T4 UV enzymes suggest, of course, that pyrimidine dimer repair may
universally follow such a mechanism. However, this does not appear
to be the case for excision of dimers in vivo in uninfected E. coli
(Table 1). We have extracted acid-soluble oligonucleotides from
cells that had been UV-irradiated, then incubated for 3 hr under
growth conditions, during which time 70% of the pyrimidine dimers
were excised from the cellular DNA. Secondary irradiation of these
oligonucleotides in vitro did not result in the release of
significant amounts of free thymine (less than one-tenth the amount

Table 1. Test for Photoreleasable Thymine in UV
Excision Products from E. coli

Oligonucleotide Source[a]	3600 J/m^2 Photoreleasing UV Irradiation	Pct. Thymine Isolated as Free Base[b]
Irradiated, non-incubated cells (<0.05% thymine as dimers)	− +	0.12 0.07
Irradiated, in-cubated cells (2.9% thymine as dimers)	− +	0.79 0.80
DNase I digest of irradiated T7 DNA (1.6% thy-mine as dimers)	− +	0.41 0.47
DNase digest of irradiated T7 DNA treated with T4 UV endo-nuclease (1.9% thymine as dimers)	− +	0.06 1.04

[a]E. coli JC4583 was irradiated at 2 J/m^2/sec for 30 sec, then incubated for 3 hr where indicated. Acid-soluble oligonucleo-tides were collected from these cells. T7 DNA was irradiated for 100 sec (both DNases) or 300 sec (DNase I only) at 3 J/m^2/sec, digested with T4 UV endonuclease and/or pancreatic DNase, and then acid-soluble oligonucleotides isolated. Thy-mine in dimers was determined as described by Goldman and Friedberg.[17]

[b]Thymine was isolated on polyethyleneimine thin layer plates and quantitated.

expected from the thymine present in dimers). Conversely, secondary irradiation of oligonucleotides formed by a pancreatic DNase diges-tion of UV-irradiated DNA which had been treated with T4 UV endo-nuclease showed an amount of thymine released equal to one-half the thymine initially present in dimers. (In the latter case, omission of the T4 UV endonuclease treatment eliminated the thymine release

as expected). Evidently the dimers excised by the repair mechanism
in uninfected E. coli still contain both glycosylic bonds and are
of the form,

PURINE BASE INSERTION AT APURINIC SITES

In an exploratory study to search for proteins from human
fibroblasts that interact with apurinic sites in DNA, we discovered
a protein that tightly bound specifically to DNA containing apurinic
sites.[18] The protein sediments at 6.4S (molecular weight equivalent
of about 120,000) and separates entirely from AP endonuclease activity which sediments at less than 3S. In fact, the protein has no
apparent effect on AP endonuclease activity, but, conversely, it
does not appear to interact with AP sites once they have been
cleaved by AP endonuclease.

The protein was found to have the ability to make AP sites
alkali-stable in the presence of guanine or adenine bases. Using
labeled bases it was then shown that indeed the bases are incorporated into the partially depurinated DNA and that the reaction was
dependent upon time and protein concentration. It required only
partially depurinated DNA and purine base as substrates and K^+ as
cofactor. It was prevented by heating or freezing of the protein.[18]
The purine incorporation followed Michaelis-Menten kinetics with
respect to purine base,[19] the Km for adenine and guanine each being
5 μM. The latter is equal to the Km of human HGPRT for guanine.[20]
While the free adenine concentration in human fibroblasts is probably
very low, the enzyme binds the base very tightly (<1 μM saturates
base binding) and normal human plasma has 0.6 μM adenine whereas
human erythrocytes have 13 μM adenine.[21] The binding to apurinic
DNA can also be saturated with increasing concentrations of apurinic
sites.

The base substrate is converted to a deoxyribopurine nucleotide
in the DNA, and G is specifically incorporated into partially depurinated poly(dG-dC), whereas A is specifically incorporated into
poly(dA-dT). Both guanine and adenine incorporation are inhibited
by caffeine. The enzyme does not incorporate purine or pyrimidi-
nated bases into either tRNA or partially depyrimidinated PBS2 DNA.

The properties of this protein indicate that it can act as a
partially depurinated DNA:purine base insertase. Obviously a
repair pathway is suggested:

where Pu* is an abnormal purine and Pu is adenine or guanine.

While we have no direct evidence that this pathway exists, Hennings and Michael[22] reported that mouse skin cells exposed to low levels of N-methyl-N'-nitro-N-nitrosoguanidine incorporate label from deoxyguanosine but not from thymidine, leading them to hypothesize a "guanine-specific repair in which a single base apparently replaces a base lost by depurination..." A second indication that this repair pathway might exist is provided by our observation that during host-cell reactivation of SV40 DNA that had been partially depurinated by acid or methyl methane sulfonate, the xeroderma pigmentosum D strains that lack human fibroblast AP endonuclease I are more effective in plaque formation than normal strains.[23] Of course, these cells were at the same time less effective in host-cell reactivation of UV-irradiated SV40 DNA. Perhaps a minor, but abnormally efficient purine insertion repair pathway is exposed when the AP endonuclease is absent as in these cells. Further studies are clearly called for to establish firmly a role of this enzyme in DNA repair rather than in other metabolic functions. (For example, the enzyme might catalyze the exchange of methylated and unmethylated DNA purines.)

Another question about this activity is the energy source of the glycosylic bond that is formed, particularly since a purine insertase activity from E. coli has been reported to use purine dNTPs as cofactors.[24] This question is clearly worthy of future study, particularly with the goal of studying the energetics of DNA purine vs. pyrimidine glycosylic bonds. Of interest in this regard is the inability to detect pyrimidine DNA base insertion, and the observation that uracil DNA glycosylase activity is generally much higher in cells than the purine DNA glycosylases (3-methyladenine and hypoxanthine).[3] Thus further studies aimed at verifying the catalytic character of this protein might also be useful in obtaining some fundamental information on DNA structure.

A NOTE ON EXCISION EXONUCLEASES: HOW IMPORTANT ARE THEY?

Two schemes for DNA repair which did not utilize excision exonucleases were described above. In one case, AP sites were converted to a gap of one nucleotide by the removal of deoxyribose-5-phosphate with a combination of class I and II AP endonucleases. In the other case, depurinated sites were corrected directly by the insertion of the missing base. Whether these processes exist in vivo remains to be seen, but it is clear that a substantial amount of repair might take place without direct "exonuclease" action. Such repair is not likely to be observed as "repair replication" or "unscheduled DNA synthesis."

On the other hand, it is clear that exonucleases often are normally involved in DNA repair processes. In those cases, one must be aware that the exonuclease utilized might be determined directly by the endonuclease that initiates the excision process. Thus one must ascertain whether damaged material remains at or near the 3', 5' or both termini formed by the endonuclease, where the phosphomonoester group resides, and whether such a terminus would be substrate for the exonuclease in question. It would therefore seem reasonable to suppose that "excision exonucleases" will not be fully understood and characterized until the incision endonucleases are identified and can be utilized to prepare appropriate excision exonuclease substrates.

CONCLUSIONS

In an effort to help to understand DNA repair, we have undertaken the detailed characterization of enzymic activities likely to be involved in such processes. Known AP endonucleases were classified into two catalytic types which cleave on either side of an AP deoxyribose such that when acting together deoxyribose-5-phosphate is released. Whether this mechanism operates in vivo remains to be shown. It may be significant that two of the class I AP endonucleases (E. coli endonuclease III and T4 UV endonuclease) are associated with DNA glycosylase activities for specific DNA damage species, whereas the third--human fibroblast AP endonuclease I--has not been examined in this regard. Conversely, no class II AP endonucleases appear at this time to be similarly associated with DNA glycosylase activities, although E. coli endonuclease VI activity is in the same protein as exonuclease III/DNA-3'-phosphatase.[9] Whether a pattern exists of class I enzymes, but not class II enzymes, being associated with DNA glycosylase activities must be answered by further purification of these three, as well as of additional class I enzymes.

The AP endonuclease class I and class II activities acting in concert to correct AP lesions, and the purine base insertion activity each represent repair of DNA damage in a manner that might escape detection if one monitors repair synthesis as indicative quantitatively of DNA repair levels. One should be particularly aware of such effects when assessing or assigning degrees of repair deficiencies in abnormal cell strains by utilizing repair synthesis levels as an indicator. Of course, we have no indication of the extent to which these processes are utilized in the cell, but the neurological defects of the xeroderma pigmentosum group D individuals and the pecularily high ability of fibroblasts cultured from at least some of these individuals to reactivate depurinated SV40 DNA--if related to the absence of AP endonuclease I--are certainly suggestive and call for further study of the subject.

The glycosylase/AP endonuclease mechanism of pyrimidine dimer repair appears not to be ubiquitous, as it is not yet apparent in uninfected E. coli. Nevertheless, it remains to be seen whether it is present in human cells. Given this mechanism, the cell must remove both an AP lesion from the newly formed 3'-terminus and a pyrimidine dimer from the 5'-terminus. Both of these can in theory be carried out without exonuclease action: the former can be done by a class II AP endonuclease, the latter by photoreversal (probably enzymatic), the combination leaving only two nucleotides to be replaced. Alternatively, in bacteria at least, the 5'-terminal dimer nucleotide might be removed by the $5' \rightarrow 3'$ exonuclease activity of a DNA polymerase during a "nick translation" reaction, or, of course, by a $5' \rightarrow 3'$ "excision exonuclease." Most pleasing, perhaps, would be the existence of a dimer glycosylase activity which would cleave the remaining glycosylic bond on the 5'-terminal sugar so as to release a dimerized pyrimidine pair and leave a deoxyribose phosphate which could be removed by a class I AP endonuclease. The search for such an activity is clearly warranted, particularly in M. luteus or phage T4-infected E. coli.

In conclusion, the realization of base-excision repair has allowed the creation of many possible new schemes for DNA repair. Such schemes can be supported by studies of purified enzymes such as those discussed herein. Nevertheless, enzymatic studies alone cannot prove the utilization of a scheme in vivo. Genetic, chemical and cellular studies must accompany biochemical studies in order ultimately to gain a good understanding of the natural phenomena of DNA repair.

ACKNOWLEDGEMENTS

This work was supported by contract DE-AS03-76-SF34 from the Department of Energy, USPHS Grant GM19020 and USPHS Training Grant

ES07075. We are extremely grateful to Dr. Paul Howard-Flanders who, during eight months as a visitor to our laboratory, stimulated much thought, discussion and experimentation on DNA repair processes. We are also grateful to Dr. Rufus Day for suggesting the photo-reversal experiment to us and to Dr. Thomas Lindahl for providing us with E. coli endonuclease IV.

+ Current address: Department of Biochemistry, University of Minn-
  esota, St. Paul, MN   55108
* Current address: Department of Biochemistry, Louisiana State University, Baton Rouge, LA   70803

LITERATURE CITED

1. T. Lindahl, An N-glycosidase from Escherichia coli that re-
    leases free uracil from DNA containing deaminated cytosine
    residues, Proc. Natl. Acad. Sci. USA 71:3649 (1974).
2. S. Linn, Workshop summary: enzymology of base excision repair,
    in: "DNA Repair Mechanisms," P. C. Hanawalt, E. C. Fried-
    berg, C. F. Fox, eds., Academic Press, New York (1978).
3. T. Lindahl, DNA glycosylases, endonucleases for apurinic/
    apyrimidinic sites, and base excision repair, Prog. Nucl.
    Acid Res. and Mol. Biol 22:135 (1979).
4. F. T. Gates III and S. Linn, An endonuclease from Escherichia
    coli that acts specifically upon duplex DNA damaged by
    ultraviolet light, osmium tetroxide, acid, or x-rays, J.
    Biol. Chem. 252:2802 (1977).
5. B. Demple and S. Linn, DNA N-glycosylases and UV repair,
    Nature, in press (1980).
6. H. R. Warner, B. F. Demple, W. A. Deutsch, C. M. Kane, and
    S. Linn, Apurinic/apyrimidinic endonucleases in the repair
    of pyrimidine dimers and other lesions in DNA, Proc. Natl.
    Acad. Sci. USA 77, in press (1980).
7. L. Grossman, S. Riazuddin, W. Haseltine, and K. Lindan, Nucleo-
    tide excision repair of damaged DNA, Cold Spr. Symp. Quant.
    Biol. 43:947 (1978).
8. S. Linquist, A new endonuclease from Escherichia coli acting
    at apurinic sites in DNA, J. Biol. Chem. 252:2808 (1977).
9. B. Weiss, S. G. Rogers, and A. F. Taylor, The endonuclease
    activity of exonuclease III and the repair of uracil-con-
    taining DNA in Escherichia coli, in: "DNA Repair Mechan-
    isms," P. C. Hanawalt, E. C. Friedberg, C. F. Fox, eds.,
    Academic Press, New York (1978).
10. W. S. Linsley, E. E. Penhoet, and S. Linn, Human endonuclease
    specific for apurinic/apyrimidinic sites in DNA. Partial
    purification and characterization of multiple forms in
    placenta, J. Biol. Chem. 252:1235 (1977).
11. U. Kuhnlein, E. E. Penhoet, and S. Linn, An altered apurinic
    DNA endonuclease activity in group A and group D xeroderma
    pigmentosum fibroblasts, Proc. Natl. Acad. Sci. USA 73:1169
    (1976).

12. U. Kuhnlein, B. Lee, E. E. Penhoet, and S. Linn, Xeroderma pigmentosum fibroblasts of the D group lack an apurinic endonuclease species with a low apparent Km, Nucl. Acid Res. 5:951 (1978).

13. D. W. Mosbaugh and S. Linn, Further characterization of human fibroblast apurinic/apyrimidinic DNA endonucleases: the definition of two mechanistic classes of enzyme, Submitted for publication (1980).

14. A. D. Andrews, S. F. Barrett, and J. H. Robbins, Xeroderma pigmentosum neurological abnormalities correlate with colony-forming ability after ultraviolet irradiation, Proc. Natl. Acad. Sci. USA 75:1984 (1978).

15. K. Minton, M. Durphy, R. Taylor, and E. C. Friedberg, The ultraviolet endonuclease of bacteriophage T4, J. Biol. Chem. 250:2823 (1975).

16. S. Yosuda and M. Sekiguchi, Further purification and characterization of T4 endonuclease V, Biochim. Biophys. Acta 442:197 (1976).

17. K. Goldman and E. C. Friedberg, Measurement of thymine dimers in DNA by thin layer chromatography, Anal. Biochem. 53:124 (1973).

18. W. A. Deutsch and S. Linn, DNA binding activity from cultured human fibroblasts that is specific for partially depurinated DNA and that inserts purines into apurinic sites, Proc. Natl. Acad. Sci. USA 76:141 (1979).

19. W. A. Deutsch and S. Linn, Further characterization of a DNA-purine base insertion activity from cultured human fibroblasts, J. Biol. Chem. 254:12099 (1979).

20. J. A. McDonald and W. N. Kelley, Lesch-nyhan syndrome: altered kinetic properties of mutant enzyme, Science 171:689 (1971).

21. G. C. Mills, F. C. Schmalsteig, K. B. Trimmer, A. S. Goldman, and R. M. Goldblum, Purine metabolism in adenosine deaminase deficiency, Proc. Natl. Acad. Sci. USA 73:2867 (1976).

22. H. Hennings and D. Michael, Guanine-specific DNA repair after treatment of mouse skin cells with N-methyl-N'-nitro-N-nitrosoguanidine, Cancer Res. 36:2321 (1976).

23. R. D. Kudrna, J. Smith, S. Linn, and E. E. Penhoet, Survival of apurinic SV40 DNA in the D complementation group of xeroderma pigmentosum, Mut. Res. 62:173 (1979).

24. Z. Livneh, D. Elad, and J. Sperling, Enzymatic insertion of purine bases into depurinated DNA in vitro, Proc. Natl. Acad. Sci. USA 76:1089 (1979).

MECHANISM OF ACTION OF M. *luteus* APURINIC/APYRIMIDINIC ENDONUCLEASES

Jacques Laval and Josiane Pierre

LA 147 CNRS & U 140 INSERM
Institut Gustave-Roussy
94800 Villejuif, France

INTRODUCTION

Apurinic/apyrimidinic sites (AP-sites) occur in DNA. They could have different origin : spontaneous depurination, excision by specific DNA glycosylases of damaged bases. Such lesions are repaired by two independant mechanisms[1]. In the repair by reinsertion, specific insertase inserts the missing purine and regenerates the integrity of DNA. In the repair by base excision, a specific endonuclease for AP-sites (AP-endonuclease) recognises the lesion and incises the DNA backbone near the lesion. The termini generated by AP-endonucleases are of importance as they determine the polarity of the excision of the lesion. Furthermore there is an increasing interest for AP-endonucleases since pyrimidine dimers are excised in M. *luteus* by base excision[2]. In this paper we report the determination of the termini generated by the two different AP-endonucleases extracted from M. *luteus*. We also report experiments showing that these enzymes could be used to detect AP-sites introduced in DNA by γ-rays or by UV irradiation.

RESULTS AND DISCUSSION

The assay measures the release of acid soluble products from depurinated T5 [³H] DNA[3]. Purification procedure involves after lysis of the cells by lysosyme, a phase partition to eliminate the nucleic acids, chromatography on DEAE-cellulose, Sephadex G75 and CM-cellulose, where two peaks of AP-endonuclease activities are separated (AP-endonucleases A and B). The two AP-endonucleases are separatly applied to DNA-Sepharose for further purification[3].

Both AP-endonucleases have the same molecular weight, 35,000 daltons and are monomeric proteins. They are free from DNA glyco-

Table 1 : Selective release of [32P] and [3H] from
d(A[3H]T)$_n$ (A^{32}pU)$_{n'}$ by the combined action of
different enzymes.

	[3H]Acid-soluble radioactivity (%)	[32P]Norit Nonadsorbable radioactivity (%)
uracil-DNA glycosylase + AP-endonuclease A + phosphatase	1.2	17
uracil-DNA glycosylase + AP-endonculease B + phosphatase	1.2	36

sylase, unspecific endonucleases, and phosphatase activities. The
breakage of the phosphodiester backbone is purely endonucleolytic.
There is no exonuclease activity associated as the enzymes do not
release mononucleotides[4].

Study of the action of snake venom phosphodiesterase (VPH) and
spleen phosphodiesterase (SPH) on depurinated DNA treated by AP-
endonuclease A or B shows that [4] i) the treated DNA is substrate
for the VPH, but not for the SPH, suggesting that the AP-endonucleases generate 5'phosphate termini, ii) removal by phosphatase
of the phosphate does not promote the action of SPH. This suggests
that the lesion is 3' to the incision, as AP-site end group is
inhibitor of the SPH[5].

Due to the importance of the incision introduced by the AP-
endonucleases, the above determination is repeated using a different approach. We used a polymer d(A-[3H]T)$_n$, in which every ten
bases 5' [32P]-dUMP is incorporated. The polymer is then treated
with *M.luteus* uracil-DNA glycosylase[6], in order to generate AP-
sites and is used as substrate for AP-endonucleases. Then the
incised polymer is treated by phosphatase. Upon this treatment,
when 1.2% of the [3H] DNA becomes acid soluble (indicating that
there is no exonuclease activity), 17% of the 32P radioactivity
of the polymer nicked by AP-endonuclease A are released as Norit
non-absorbable form and 36% when nicked by AP-endonuclease B
(table I). This result indicates that the cleavage occurs 5' to
the lesion and generates 3'OH and 5'P end groups[4]. If the incised
DNA is treated under alkaline condition similar results are obtained (data not shown) indicating that the AP-sites already incised
by the enzyme is liberated by β-elimination[7].

Table II : Average number of nicks introduced by AP-endonuclease
A or B, by *M.luteus* extracts on native PM_2 DNA and on
UV or γ-rays irradiated PM_2 DNA.

	No enzyme	*M.luteus* extract	AP-endonuclease A	B
Native DNA	0.07	0.094	0.09	0.06
UV irradiated DNA (10 J/m^2)	0.15	1.96	0.12	0.17
UV irradiated DNA (500 J/m^2)	0.094	3.05	0.41	0.31
γ irradiated DNA (2700 rads)	0.19	0.69	0.38	0.31

PM_2 DNA is irradiated with either UV or with γ-rays and then
treated with different amounts of AP-endonuclease A (0.037 unit)
or B (0.08 unit) or by *M.luteus* extract (10 µg of proteins).
Supertwisted and nicked DNA molecules are analysed by agarose
gels. The number of breaks (n) per molecule is calculated by
Poisson's law[4].

   Availability of pure AP-endonucleases prompted us to use them
to detect AP-sites introduced by UV or γ-rays. PM_2 DNA irradiated
with low doses of UV ($10 J/m^2$) is not substrate for AP-endonucleases showing that there is no UV-endonuclease. However with high
doses ($500 J/m^2$) DNA is nicked by our AP-endonucleases. This confirms that in DNA treated with high doses of UV, AP-sites are
generated besides the base lesions (table II). A similar result
is also observed with γ-rays at high doses.
   The results reported above show that in *M.luteus* two AP-endonucleases are present. This is also the case in *E.coli* [8-9]. The
incision of the DNA backbone occurs on the 5' side to the apurinic
site and generates 5'phosphate termini. It has already been shown
that the AP-sites generated by the 3-methyladenine-glycosylase[10]
and by high fluence of UV or γ-rays are substrate for the enzymes.
In *M.luteus* the pyrimidine dimer, introduced by UV light are
repaired by a specific DNA-glycosylase which breaks the N-glycosylic bond between 5'pyrimidine of the dimer and the corresponding sugar. It generates an AP-site having on its 3' side thymine
residue[2]. As both AP-endonucleases described here incise 5' to
the lesion, they could be of importance for the economy of the

cell as the exonuclease could remove at once the AP-sites and the residual dimer. It should be recall that during excision, UV lesions are removed as an oligonucleotides[11,12].

ACKNOWLEDGEMENT
   This work was supported by grants from CNRS, CEA Saclay and INSERM (ATP 77.79.109 and CRL 79-5-172-3-A).

REFERENCES

1. J. Laval and F. Laval , Enzymology of DNA repair, in :
   "Molecular and Cellular Aspects of Carcinogens Screening Tests",
   IARC Scientific publications n° 27 (Montesano, R., Bartsh, H.,
   and Tomatis, L., Éds) 55-73,(1980).

2. R. Grafstrom, N. Shaper, L. Grossman and W. Haseltine.
   Incision of pyrimidine dimer contiaining DNA by small molecular weight enzymes (this volume).

3. J. Pierre and J. Laval, M.luteus endonucleases for apurinic/
   apyrimidinic sites in DNA. Biochemistry, 19, 5018-5023(1980)

4. J. Pierre and J. Laval, M.luteus endonucleases for apurinic/
   apyrimidinic sites in deoxyribonucleic acid. Biochemistry
   19, 5024-5029 (1980).

5. G.P. Margison, P.J. O'Connor and A. Cornish-Bowden, Role of
   Apurinic sites in resistance of methylated oligodeoxyribonucleotides to degradation by spleen exonuclease, Biochem. J.
   151, 249-256 (1975).

6. J. Laval, J.P. Leblanc and B. Martin, Properties of uracil-
   DNA gylosylases from M.luteus, E.Coli and mammalian cells,
   this volume (1980)

7. J.E. Clements, S.G. Rogers and B. Weiss, A DNase for Apurinic/
   Apyrimidinic sites associated with exonuclease III of Hemophi-
   lus influenzae, J. Biol Chem. 253, 2990-2999, (1978).

8. W.G. Verly, and E. Rassart, Purification of E.Coli endonuclease
   specific for apurinic sites in DNA. J. Biol. Chem. 250, 8214-
   8219, (1975).

9. S. Ljungquist, A new endonuclease from E.Coli acting at apurinic sites in DNA. J. Biol. Chem. 252, 2808-2814,
   (1977).

10. J.Laval, Two enzymes are required for strand incision in
    repair of alkylated DNA. Nature, 269, 829-832
    (1977).

11. R.B. Boyce and P. Howard-Flanders , Release of ultraviolet
    light induced thymine dimers from DNA in E.Coli K-12.
    Proc. Natl. Acad. Sci. USA, 51, 293-300 (1964)

12. R.B. Setlow, and W. Carrier, The disappearance of thymine
    dimers from DNA : an error-correcting mechanism. Proc. Natl
    Acad. Sci. USA 51, 226-231 (1964).

PROPERTIES OF URACIL-DNA GLYCOSYLASES

FROM M. *luteus*, E. *coli* AND MAMMALIAN CELLS

Jacques Laval, Jean-Pierre Leblanc and Bernard Martin

LA 147 & U 140 INSERM
Institut Gustave Roussy
94800 Villejuif, France

ABSTRACT

Two Uracil-DNA glycosylases have been purified, one from the nucleus and one from the cytoplasm of rat hepatoma cells (H₄ cells). The nuclear enzyme has been purified more than 1,100 fold. The properties of these two enzymes were compared with the uracil-DNA glycosylases extracted from M.*luteus* and E.*coli*.

INTRODUCTION

The enzymatic degradation of uracil containing DNA was initially established by Wovcha and Warner[1] and by Carrier and Setlow[2]. It was later shown that the uracil is released as a free base by a uracil-DNA glycosylase yielding an apyrimidinic site[3]. This enzyme was purified to homogeneity from E.*coli*[4] and from M.*luteus*[5]. Uracil-DNA glycosylase activity was also reported in human placenta[6] and in calf thymus[7] but not extensively purified. We report the purification and properties of a mammalian uracil-DNA glycosylase and its comparison with the enzymes purified from M.*luteus* and E.*coli*.

RESULTS AND DISCUSSION

Uracil-DNA glycosylase activity was determined by measuring the release of acid soluble products from {³H} uracil PBSI DNA[8]. The mammalian enzymes were obtained from H₄ cells, derived from a rat hepatoma and grown as already described[9]. The purification procedure involved separation of the nucleus, mitochondria and cytoplasm.

Table 1. Purification of Uracil-DNA Glycosylase
from the Nucleus of H_4 Cells.

	Total Protein (mg)	Total[b] units (x 10^{-3})	Specific Activity	Uracil in Acid Soluble products (%)
Cell lysate[a]	103.8	15.5	149	N.D.
Purified nucleus	30.6	11.4	372	94
DEAE cellulose	8.2	11.0	1 360	96
H.A. Ultragel	0.12	6.7	5 600	99
Poly-U Sepharose	< 0.005	1.36	> 170 000	> 99.7

a : determined in an aliquot
b : a unit of enzyme liberates one pmole of uracil per minute
   at 37°C in a medium containing Hepes-KOH 0.007M NaCl 0.04M,
   EDTA 0.002M, PBSI {^{3}H}-DNA 0.52 nmole in 50 µl (final pH 7.2)

The nuclear, cytoplasmic and mitochondrial enzymes account for
70 %, 20 % and 10% respectively of the total activity. Due to
scarcity of the mitochondrial enzyme it was not further purified.
The nuclear and the cytoplasmic fractions were separately chromato-
graphed on DEAE cellulose, HA Ultragel (Pharmindustrie, Paris) and
poly-U Sepharose. This purification procedure yields an enzyme preparation of nuclear origin purified more than 1 100fold with a
recovery of 9 % (Table 1). The preparation of the M. luteus enzyme
has been reported[5]. The E. coli enzyme was purified as a by product
during the purification of DNA ligase according to the procedure of
Modrich et al.[11]. In the fourth step, (DEAE cellulose chromatography), the ligase was adsorbed on the column whereas the uracil-
DNA glycosylase was recovered in the flow through. It was concentrated, purified by chromatography on a second DEAE cellulose at
low ionic strength, and on chromatographied HA Ultragel and poly-U
Sepharose. The enzymes obtained from different sources were neither
contaminated by unspecific endonucleolytic or exonucleolytic activity, by endonucleases specific for apurinic sites[8], nor by
3-methyladenine-DNA glycosylase[12].

Table 2.: Comparison of Uracil-DNA Glycosylases of Different Origin

	M.*luteus*	E.*coli*	H$_4$ nucleus	H$_4$ Cytoplasm
Activity on heat.denatured DNA vs native DNA	2.0	1.9 (2.6[a])	2.0	2.0
Km (M)	7.0 x 10^{-8}	4.0 x 10^{-8a}	1.3 x 10^{-7}	6 x 10^{-7}
Optimum pH	5.6 - 6.0	8.0[a]	7.0 - 7.5	7.0 - 7.5
Optimal ionic strength M NaCl	0.05 - 0.1	0.07[a]	0.04 - 0.05	0.02 -0.05
Effect of Divalent Metals	Inhibitors	Inhibitors and[a]	Inhibitors	Inhibitors
Effect of EDTA (0.05 M)	0	0 (0[a])	0	0
Uracil inhibition (0.001 M)	60 % (Ki = 5.10^{-4}M)	56 % (Ki = 1.2x10^{-v}M) [a]	62 %	N.D.
pI (isoelectric focalisation)	7.0 ± 0.1	N.D.	6.2 ± 0.2	8.9 ± 0.2

a : from Lindahl et al.[4]
N.D. : not determined

The enzyme above described were used to compare their properties.
The results are given in Table 2. All the enzymes tested show a two
fold higher activity on single stranded DNA as compared with double
stranded DNA. They are all activated by relatively low salt concentrations and inhibited by high concentrations. Their Km values lie
in the same order of magnitude. The purification and the characterisation of the cytoplasmic uracil-DNA glycosylase was performed in
order to ascertain if this enzyme was different from the nuclear one
or if it was just leaking from the nucleus. The data presented do
not support any firm conclusions as regarding the nuclear or
cytoplasmic origin the properties so far tested are comparable.
It should also be emphasized that their behaviour is the same on
the various chromatographic support material. Uracil can arise in
DNA by deamination of cytosine and by incorporation of dUMP instead
of dTMP during replication of DNA. If not repaired such a base would
give rise to a transition mutation. The results reported here
suggest that uracil-DNA glycosylase is important for the economy
of the cell and that evolution has preserved the molecule.

ACKNOWLEDGMENTS

This work was supported by grants from CNRS, CEA Saclay
CRL 79-5037-2 INSERM and ATP 77-79.109 INSERM.

REFERENCES

1. Wovcha, M.G., and Warner, H.R.: "Synthesis and Nucleolytic
   Degradation of Uracil-containing Deoxyribonucleic Acid by
   Escherichia coli Deoxyribonucleic Acid Polymerase 1ˣ"
   J. Biol. Chem. 248, 1746.(1973).
2. Carrier, W.L., and Setlow, R.B.: "An Endonuclease from
   M.luteus that acts on uracil, in UV-irradiated BrUra-containing
   DNA." Fed. Proc. 33, 1599,(1974).
3. Lindahl, T. : "An N-Glycosydase from *Escherichia coli* that
   Releases Free Uracil from DNA Containing Deaminated Cytosine
   Residues." Proc. Natl. Acad. Sci., U.S.A., 71, 3649,(1974).
4. Lindahl, T., Ljungquist, S., Siegert, W., Nyberg, B. and
   Sperens, B. : "DNA N-Glycosidase - Properties of Uracil-DNA
   Glycosidase from *Escherichia coli*" J. Biol. Chem. 252, 3286,
   (1977).
5. Laval, J., and Pierre, J. : "Base-Excision repair in Micrococcus
   luteus" in DNA repair Mechanism, Hanawelt, P.C., Friedberg Ec.
   Fox C.F., Eds Academic Press New-York(1978).
6. Sekiguchi, M., Hayakawa, H., Makimo, F.,Tanaka, K., and Okada,Y.
   "A human enzyme that liberates uracil from DNA" Biochem.,
   Biophys. Res. Commun 73, 293, (1976).
7. Talpaert-Borlé,M., Clerici, L., and Campagnari, F. :"Isolation
   and Characterization of a Uracil-DNA glycosylase from Calf

thymus", J. Biol. Chem. 254, 6387, (1979).

8. Pierre, J. and Laval, J. : "Micrococcus luteus endonucleases for apurinic/apyrimidinic sites in deoxyribonucleic acid." Biochemistry 19, 5024-5029, (1980).

9. Laval, F., : "Effect of uncouplers on radiosensitivity and mutagenicity in x-irradiated mammalian cells". Proc. Natl. Acad. Sci. U.S.A., 77,2 702, 1980.

10. Laval, F., Malaise E. and Laval J. : "Heterologous DNA nuclear uptake by mouse DNA fibroblasts in vitro and its early fate". Exptl. Cell. Res. 63, 69, (1970).

11. Modrich, P., Yasuhiro A., and LEHMAN, I.R. : "Isolation and physical characterization of the homogenous enzyme from Escherichia coli".J. Biol. Chem.,248,7495-7501, (1973).

12. Laval, J. : "Two enzymes are required for strand incision in repair of alkylated DNA". Nature, 269, 829-832, (1977).

BASE REPLACEMENT MECHANISMS FOR THE REPAIR OF UNNATURAL AND DAMAGED BASES IN DNA

Zvi Livneh and Joseph Sperling

Department of Organic Chemistry
The Weizmann Institute of Science
Rehovot, Israel

ABSTRACT

A purine insertase activity was found in extracts from *E. coli* H502. This activity brings about the insertion of adenine and of guanine into the appropriate apurinic sites in DNA, using dATP and dGTP as the purine donors, in a Mg^{++}-dependnet reaction. Using extracts prepared from *E. coli* strains KMBL 1719 and 1720 we found a Mg^{++}-dependent insertion of the free purine bases into depurinated DNA. It is possible that induced functions are involved in the process of insertion since active extracts could be prepared from *E. coli* strains (including AB3027 ap endo⁻poll⁻) only if they were grown in the presence of sublethal doses of alkylating agents.

A new method was developed for assaying base replacement mechanisms, based on the use of DNA with labeled AP sites. This method is currently used to look for a pyrimidine insertase activity in *E. coli*.

INTRODUCTION

Depurination of DNA may occur spontaneously by hydrolysis of the deoxyribose-purine glycosylic bond of native[1] or of alkylated purines[2], the latter being released much more rapidly, or enzymaticly, by the action of specific DNA-glycosylases, which release unnatural or modified purines or pyrimidines from the DNA[3]. These enzymes may produce in the DNA both apurinic and apyrimidic sites. It is believed that these AP sites are repaired by excision repair, initiated by an AP endonuclease. Indeed, AP endonucleases have been purified from a variety of sources[3] and the total repair of AP sites in DNA has been demonstrated *in vitro* using bacterial[4] or human enzymes[5].

We were intrigued by the need for two enzymes (a glycosylase and an AP endonuclease) to perform the incision near a lesion in the DNA, while this incision can be performed by a single damage specific endonuclease. One possibility to explain these facts is to assume that the glycosylase action is the initiating event in a new repair mechanism, in which the next step will be the direct insertion of the missing base into the AP site which was created by the glyco- sylase. This suggested mechanism implies the existance of a new "base insertase" enzymatic activity. Indeed, we found in a soluble protein extract from *E. coli* an enzymatic activity which directly and specifically inserts purines into depurinated DNA[6]. A similar activity was found in human fibroblasts by Deutsch and Linn[7,8].

E. coli CONTAINS A PURINE INSERTASE ACTIVITY

Incubation of depurinated DNA from phage PM2 (apDNA, prepared by mild acid treatment) with a soluble enzyme extract from *E. coli* H502 (uvrA endoI) in the presence of Mg^{++} and [8-^{3}H]dATP or [8-^{3}H] dGTP resulted in incorporation of radioactivity into the DNA. Only background incorporation was observed when using [α-^{32}P]-labeled dPuTPs instead of the tritium labeled compounds. The incorporation into the DNA was measured by TCA precipitation, binding of the DNA to DEAE-cellulose paper or by agarose gel electrophoresis. The activity requires Mg^{++}, and it is inactivated in high salt concent- rations (1M NaCl) or after heat treatment (60°C, 10 min). The sub- strate for the activity must be depurinated double strand DNA. Depurinated single stranded DNA from phage φx174 failed to serve as a substrate, while depurinated double strand DNAs from phage PM2 and from calf thymus incorporated the bases with similar efficiencies. The substrate apurinic sites could be induced in the DNA either by mild acid treatment or by alkylation with MMS followed by heat treatment; both DNA preparations promoted incorporation of radio- activity.

The insertion was found to be specific, since no inhibition of the incorporation of labeled dATP or dGTP could be detected in the presence of a high excess of any other dNTP. The specificity of insertion and the fact that double standard DNA is required, suggest that the insertion proceeds through base pairing with pyrimidines which are located on the intact strand opposite the apurinic site.

THE BASE MOIETIES OF THE dNTPs ARE THE INSERTED GROUPS

Only background incorporation was observed using [α-^{32}P]- labeled dPuTPs, a result which eliminates the possibility of poly- merization. In order to prove that the base but not the sugar is inserted into the apDNA, we used [U-^{14}C]dPuTPs, in which both the base and the sugar moieties were labeled with C-14. The proportion of label located at the base or at the sugar varied among various batches. The batch used in the experiments to be described contained 70% of the radioactivity in the sugar and 30% in the base moiety.

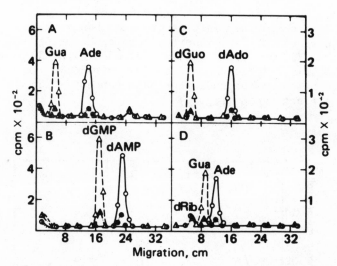

Fig. 1. Chromatographic analysis of acid and enzymatic digests of PM2 apDNA after insertion incubation with [U-^{14}C]dATP or [U-^{14}C]dGTP[7]. o, apDNA incubated with [U-^{14}C]dATP; ● intact DNA incubated with [U-^{14}C]dATP; Δ, apDNA incubated with [U-^{14}C]dGTP; ▲ intact DNA incubated with [U-^{14}C]dGTP.

     DNA isolated after insertion incubation with [U-^{14}C]dGTP or [U-^{14}C]dATP was totally depurinated by acid treatment, and the hydrolysate was analysed by paper chromatography. Only one radioactive peak appeared in each case, comigrating with an authentic sample of adenine or guanine respectively (Fig. 1a). Only negligible radioactivity was detected at the origin, where the depurinated DNA- and hence the deoxyribose moieties - were located. This clearly indicates insertion of the bases into apDNA. Total enzymatic digestion to 5'-mononucleotides of the DNA after insertion incubation followed by paper chromatography, revealed 5'-dAMP and 5'-dGMP as the only radioactive peaks (Fig. 1b). The peaks were extracted from paper, and treated with alkaline phosphatase to remove the phosphate groups. Chromatography of the hydrolysate revealed dAdo and dGuo as the only radioactive materials (Fig. 1c). Since the inserted bases were recovered from the DNA as substances which are chromatographically identical to dAdo (dGuo) or dAMP (dGMP), this indicates that the inserted bases bind to the deoxyribose in the DNA with the formation of the native N9-C1' bond. When the deoxynucleosides obtained after the phosphatase treatment were subjected to mild acid depurination followed by paper chromatography it was found that all the radioactivity comigrated with markers of the free bases (Fig. 1d). These

results imply that the purine bases are the only residues which are
inserted from dNTPs into apDNA.

INSERTION OF FREE BASES; IS INDUCTION INVOLVED IN THE PROCESS?

     Extracts prepared from *E. coli* H502 were incapable of promoting
insertion of free purines into depurinated DNA. However, we found
that free purines could be inserted into apDNA using extracts
prepared from *E. coli* KMBL1720, (F⁻bio⁻endoA101thyA301), and from
E. coli KMBL1719, (F⁻bio⁻endoA101thyA301,uvrE502). The incorporation
of the free bases was dependent on base concentration and on the
amount of extract added (Fig. 2).

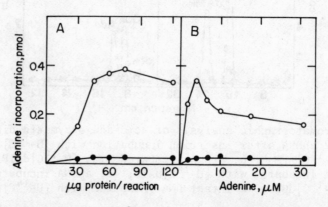

Fig. 2. Incorporation of [2-³H]Ade into depurinated DNA by an
          extract from *E. coli* KMBL1720. o, incorporation into
          PM2 apDNA; •, incorporation into intact PM2 DNA.

     This incorporation does not seem to be polymerization incor-
poration due to the conversion of free adenine to dATP, since the
addition of the other unlabeled dNTPs did not enhance the incorporation, as would have been expected from polymerization incorporation. Moreover, the addition of ATP (0.8mM) to the reaction
mixture, in the presence or in the absence of the other dNTPs
caused nearly total inhibition of the incorporation of adenine.

     Although we could demonstrate insertion of free bases into
depurinated DNA with the extracts mentioned above, we failed to do
so with extracts from some other strains. Trying to find out whether
induced functions are involved in the insertion process, we prepared extracts from *E. coli* cultures which were grown in the
presence of low levels of alkylating agents (MMS or MNNG). The

strains used were *E. coli* 1157 (wt), *E. coli* AB3027 (ap endo⁻poll⁻) and *E. coli* NH5016 (ap endo⁻). The advantage of these strains is a low expected level of excision repair of apurinic sites. Preliminary results show, that while no incorporation of free adenine was found with extracts from bacteria grown in normal cultures, incorporation of free adenine into depurinated DNA occurred with extracts prepared from bacteria which were treated with sublethal doses of alkylating agents. This incorporation was Mg^{++}-dependent, and was nearly totaly inhibited in the presence of 0.4mM ATP (Table 1).

Table 1. Incorporation of [^{3}H]Ade into apDNA
with Extracts from *E. coli* AB3027

Growth Conditions	Additives to the Extract	Incorporation of [^{3}H]Ade pmole (cpm)
Normal	±5mM $MgCl_2$	0.01 (200)
+ MMS	-	0.01 (200)
+ MMS	5mM $MgCl_2$	0.17 (3400)
+ MMS	5mM $MgCl_2$, 0.4mM ATP	0.02 (500)

The involvement of induced functions in the insertion of purines is under further investigation.

AP SITE LABELING METHOD - A NEW ASSAY FOR BASE REPLACEMENT MECHANISMS

We have developed a new method which enables a direct assay of both purine and pyrimidine insertase activities. The assay which is based on the use of DNA with labeled apurinic or apyrimidinic sites, is described in Fig. 3, for the possible assay of pyrimidine insertases. DNA is nick-translated with DNA polymerase I and limited amounts of DNase I, in the presence of unlabeled dATP, dGTP, dCTP and [U-^{14}C]dUTP. The latter is labeled both in the base and the deoxyribose moieties. During the polymerization uracil miscodes for thymine, and is efficiently incorporated into the DNA. Addition of the extract to this substrate DNA results in release of labeled uracil from the DNA, leaving a labeled apyrimidinic sites (exogenous purified DNA uracil glycosylase can be used too). Those labeled apyrimidinic sites which will be repaired by excision repair, initiated by an AP endonuclease, will result in loss of the label from the DNA, thus yielding zero background for excision repair. However, if even a part of the labeled apyrimidinic sites will be repaired by insertion repair, then the label will be fixed in the DNA. In our case, since adenine is located opposite the AP site,

thymine should be inserted and isolation of the DNA after the
reaction, followed by total enzymatic digestion to nucleosides,
should reveal labeled dThd in the DNA.

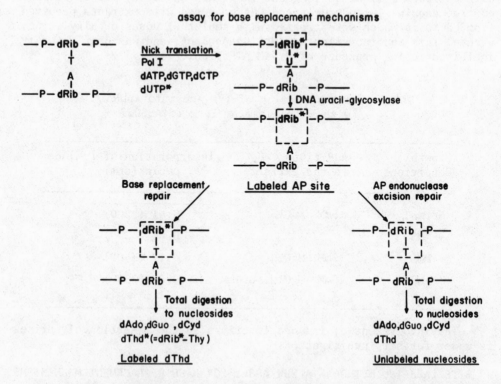

Fig. 3. AP site labeling method for the assay of base insertase
activities.

Thus we start with a DNA labeled in its dUrd residues, and if
base replacement occurs we should end up with a DNA labeled in its
dThd residues. The two big advantages of this assay are: i) We are
measuring directly the true substrate for the insertase, i.e. the
AP site. ii) There is no interference from excision repair which in
other methods may cause a high background. A similar assay was
deviced for labeling apurinic sites, using [U-^{14}C]dITP, which mis-
incorporates for guanine thus creating C-I pairs. The action of a
DNA hypoxanthine glycosylase releases the hypoxanthine from the DNA,
leading to a labeled apurinic site. Base replacement repair will
then lead to the appearance of labeled dGuo in digests of the DNA
after the reaction incubation. These substrates are currently used
to look for base insertase activities in various types of cells.

ACKNOWLEDGEMENT. This research was supported by a grant from the United States-Israel Binational Science Foundation (BSF), Jerusalem, Israel.

REFERENCES

1. T. Lindahl and B. Nyberg, Rate of depurination of native deoxyribonucleic acid, Biochemistry, 11:3610 (1972).
2. P.D. Lawley and P. Brookes, Further studies on the alkylation of nucleic acids and their constituent nucleotides, Biochem. J. 89:127 (1963).
3. T. Lindahl, DNA glycosylases, endonucleases for apurinic/ apyrimidinic sites and base excision repair, Prog. Nucl. Acid Res. Mol. Biol. 22:135 (1979).
4. W.G. Verly, F. Gossard and P. Crine, In vitro repair of apurinic sites in DNA, Proc. Natl. Acad. Sci. USA, 71:2273 (1974).
5. K. Bose, P. Karran and B. Strauss, Repair of depurinated DNA in vitro by enzymes purified from human lymphoblasts, Proc. Natl. Acad. Sci. USA, 75:794 (1978).
6. Z. Livneh, D. Elad and J. Sperling, Enzymatic insertion of purine bases into depurinated DNA in vitro, Proc. Natl. Acad. Sci. USA, 76:1089 (1979).
7. W.A. Deutsch and S. Linn, DNA binding activity from cultured human fibroblasts that is specific for partially depurinated DNA and that inserts purines into apurinic sites, Proc. Natl. Acad. Sci. USA, 76:141 (1979).
8. W.A. Deutsch and S. Linn, Further characterization of a depurinated DNA - purine base insertion activity from cultured human fibroblasts, J. Biol. Chem. 254:12099 (1979).

EXCISION REPAIR OF BULKY LESIONS IN THE DNA OF MAMMALIAN CELLS

R. B. Setlow and Eleanor Grist

Biology Department
Brookhaven National Laboratory
Upton, New York 11973   USA

INTRODUCTION

Although the excision of pyrimidine dimers from
UV-irradiated human cells has been known since 1968 (1) and
excision has been the subject of a number of symposia and recent
reviews (2-4), the details of the process still elude us. We
assume that the excision of pyrimidine dimers is of the
nucleotide excision type, (a) by analogy with bacteria, (b)
because it is of the large patch type, (c) because although few
single strand breaks accumulate during excision in normal human
cells, the numbers that accumulate in excision defective cells
are much less, (d) because the introduction of an exogenous UV
endonuclease enhances repair and survival in excision defective
cells and (e) because inhibitors of the polymerization steps such
as hydroxyurea and cytosine arabinoside result in the
accumulation of single strand breaks (5, 6). (However, the
number of breaks observed is much less than the number of dimers
removed in uninhibited cells, indicating that the various steps
in excision repair act as if they are linked to one another (7).)
The various measures of excision repair of pyrimidine dimers –
the best studied lesion to date because they are easy to identify
and measure in a number of ways – give general agreeement (8)but
different investigators obtain conflicting results in details
such as the dependence on time, dose and the method of
measurement (9-11). Nevertheless, the various techniques
indicate that excision repair of dimers varies widely among cell
lines and strains. For example, rodent cells are low excisers
compared to normal human cells (12), and among humans the cells
of most of the individuals with xeroderma pigmentosum (XP) are
defective in excision repair (13).

131

Table 1:   Ways in Which Some Chemical Damages
Mimic UV Damage in Human Cells

1.   UV-sensitive cells (XP) are more sensitive to the chemical
than normal cells.
2.   Chemically treated viruses show a higher survival on normal
cells than on XP cells.
3.   XP cells deficient in repair of UV damage are also
deficient in excision of chemical damage.
4.   Excision repair of UV and of chemical damage involves long
patches (approx. 100 nucleotides).
5.   XP complementation groups observed for repair of chemical
damage are the same as those for UV damage.

The repair of a number of bulky chemical adducts such as
those derived from N-acetoxy-acetylaminofluorene, benzo(a)pyrene
and dimethylbenzanthracene also seems to be by nucleotide
exicision because their repair mimics in a number of ways the
repair of UV (254nm) damage (14). (See Table I) Thus, we infer
that the many genes, identified as XP complementation groups,
controlling UV excision repair also control the repair of bulky
adducts. The existence of seven complementation groups implies
the existence of rather complicated control mechanisms or enzymic
sequences in the excision repair of bulky adducts. The repair of
UV damage and certain chemical damages must have a number of
steps in common and since for UV it seems as if the rate limiting
step is the endonucleolytic one, it is reasonable to suppose that
it is also rate limiting for the repair of chemical damages.

SATURATION OF EXCISION REPAIR

The kinetics of excision repair is complicated if for no
other reason than that chromatin is not a uniform substance but
is composed of both linker and core regions and repair is
initially more rapid in the linker regions for both UV-irradiated
and AAAF treated human cells (15, 16). Nevertheless, the repair
systems seem to saturate at high doses of UV or chemical
treatment (17) (although some investigators do not observe such
saturation (11, 18)) implying either that UV affects the repair
system directly or indirectly as a result of the accumulation of
photoproducts that inhibit repair or, as seems much more
reasonable, that repair saturates because of the excess of
substrate compared to a rate limiting enzymic step. For example,
at a UV dose that exhibits saturation of repair in human cells,
$20J/m^2$ of 254nm, the effects on known enzymic systems are
negligible. If the same rate limiting step exists for the repair
of UV and bulky chemical adducts one would expect treatment of
repair proficient cells with a combination of agents at high

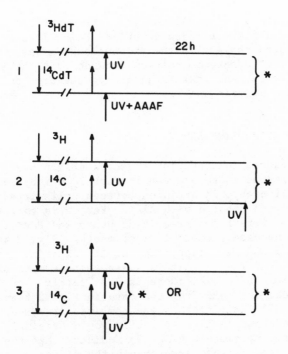

* collect, extract DNA,
treat with UV endonuclease,
measure single strand M_n

Fig. 1. The experimental procedures used to measure (1) the
effect of AAAF on dimer excision (2) excision, and (3)
the precision of the assay. UV: 20 J/m^2 of 254nm; AAAF:
20 μM for 20 min; labeling time before UV: 36h
with ^{3}H-thymidine (0.3 μCi/ml) or ^{14}C-thymidine
(0.04 μCi/ml).

doses would result in no more unscheduled DNA synthesis than from
individual treatments and that the chemical would inhibit the
excision of dimers. Such expectations are not fulfilled. The
results of treating normal human excising strains with
combinations of agents resulted in additivity of repair and no
inhibition of dimer excision, whereas treatment of repair
deficient cells indicated that the combined treatment gave less
repair than either agent separately and the chemical agent
inhibited excision of dimers by greater than 50% (8, 19).We
speculated that although XP heterozygoes fall within the normal
range of excision repair when treated with UV alone, they might
respond in an intermediate fashion and that by use of the
sensitive endonuclease site technique we might detect an
intermediate level of dimer excision between the low inhibition

observed with normal cells and the very large inhibition observed
with XP cells. The experiments described below illustrate
the precision attainable with the endonuclease assay, the fact
that the amount of observed excision depends upon the isotope
used to label cells, and that XP heterozygotes are, on the
average, between normals and XPs.

EXCISION IN XP HETEROZYGOTES

The procedures followed for these experiments are shown in
Fig. 1. The experiment of interest is Part 1, which measures
directly the difference in excision between cells irradiated and
cells irradiatd and treated with AAAF. Part 2 is to estimate the
normal amount of repair observed in 22 hours and Part 3
represents the necessary control experiments. Fig. 2 shows
typical alkaline sedimentation data for the experiments outlined
in Parts 1 and 2 of Fig. 1. It is clear that as a result of the
combined treatment (Fig. 2A) there is less repair than after UV
alone and one can estimate from the data in Fig. 2 that there is
an approximate 35% inhibition by the combined treatment of the
normal excision shown in Fig. 2B. Life would be simple except
that the control experiments (Fig. 3) indicate that the situation
is more complicated because even though the different radioactive
labels do not affect the initial numbers of pyrimidine dimers,
they do affect the amount of observed excision in normal human
cells. Hence, the experiments shown in Part 1 of Fig. 1 were all
done with the labels reversed. Typical data are shown in Fig.
4.The reversal of the labels changes the sedimentation patterns
but the isotope effect is not sufficient to explain the
inhibition in Fig. 2A. We averaged the inhibitions observed in
experiments such as in Fig. 4 and obtained the results shown in

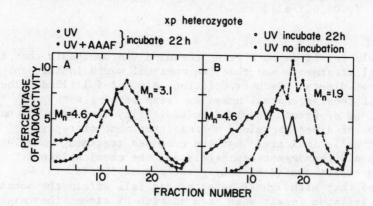

Fig. 2.   Sedimentation profiles, in alkali, of the DNAs from
          Experiments (1) and (2) of Fig. 1.

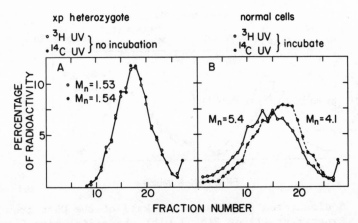

Fig. 3.   Sedimentation profiles in alkali, of the DNA from
          Experiment (3) of Fig. 1.

Table 2. They illustrate that, on the average, XP heterozygotes
show more inhibition of dimer excision by AAAF than do normal
cells although there is a heterogeneity in the heterozygote
population and the heterogeneity follows no clear genetic rules.
Thus, these data indicate that the enzyme systems in XP
heterozogotes act as if they are mixtures of normals and
homozygotes.

TABLE 2.   Inhibition of Dimer Excision ($20J/m^2$, 22h) in
           Presumptive XP Heterozygote Fibroblasts by 20 µM AAAF

Strain	Percentage Inhibition	Remarks
Normals	7	see refs.
CeAr(CRL*1165)	4	mother XPC
ReKo(CRL1202)	3	mother XPD
BeTim(CRL1254)	(7)	mother XPA
BeAr(CRL1167)	18	father XPC
EmAr(CRL1168)	21	brother XPC
DaKam (CRL1278)	21	father XP
LoWen(CRL1159)	30	mother XPD
GM**1631	27	father XP
GM1632	33	mother XP
GM0241	39	mother XP
GM2034	21	mother XP

*   American Type Culture Collection
**   Human Genetic Mutant Cell Repository.

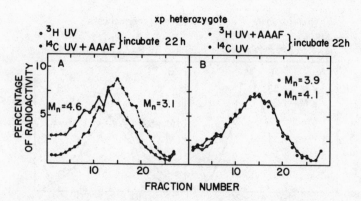

Fig. 4.   Sedimentaton profiles, in alkali of the DNAs from
          Experiment (1) of Fig. 1 with the radioactive labels
          reversed.

Other investigators (20) have not obtained the same results
as we have for normal human fibroblasts and at present we are not
able to explain the discrepancy between the two sets of outwardly
convincing experiments. The difference in excision repair
between cells labeled with ^{3}H-dThd and ^{14}C-dThd depends on the
time at which repair is measured. For example, 6 to 12 h after
UV the difference between the two labels is almost 50%, but as
repair begins to approach completion, the difference drops and
after 20 h only amounts to approximately 10%. We suspect that
the isotope effect results from the difference in radiation
damage from the more numerous and less energetic ^{3}H decays as
compared to ^{14}C and the result of such damage on the progression
of cells through the cell cycle. The faster repair shown by ^{3}H
labeled cells can be mimicked in ^{14}C labeled cells by irradiating
the latter 24 hours before UV with 500 rad of x-rays (500 rads is
approximately the dose received by ^{3}H labeled cells during the 24
hours before UV irradiation) but no effect is observed if the
irradiation takes place immediately before UV.

ACKNOWLEDGMENT

This work was supported by the U. S. Department of Energy.

REFERENCES

1.   J. D. Regan, J. E. Trosko, and W. L. Carrver, Evidence
     for excision of ultraviolet induced pyrimidine dimers
     from the DNA of human cells in vitro, Biophys. J. 8:319
     (1968).

2. P. C. Hanawalt, E. C. Friedberg, and C. F. Fox,
    editors, "DNA Repair Mechanisms," Academic Press, N. Y.
    (1978).
3. L. Grossman, A. Braun, R. Feldberg, and I. Mahler,
    Enzymatic repair of DNA, Ann. Rev. Biochem. 44:19
    (1975).
4. P. C. Hanawalt, P. K. Cooper, A. K. Ganesan, and C.
    A. Smith, DNA repair in bacteria and mammalian cells,
    Ann. Rev. Biochem. 48:783 (1979).
5. A. R. S. Collins, S. L. Schor, and R. T. Johnson, The
    inhibition of repair in UV irradiated human cells,
    Mutat. Res. 42:413 (1977).
6. W. C. Dunn and J. D. Regan, Inhibition of DNA excision
    repair in human cells by arabinofuranosyl cytosine:
    effect on normal and xeroderma pigmentosum cells,
    Molec. Pharmacol. 15:367 (1979).
7. R. B. Setlow and E. Grist, unpublished observation.
8. F. E. Ahmed and R. B. Setlow, DNA repair in xeroderma
    pigmentosum cells treated with combinations of
    ultraviolet radiation and
    N-acetoxy-2-acetylaminofluorene, Cancer Res. 39:471
    (1979).
9. J. I. Williams and J. E. Cleaver, Excision repair of
    ultraviolet damage in mammalian cells. Evidence for
    two steps in the excision of pyrimidine dimers,
    Biophys. J. 22:265 (1978).
10. U. K. Ehmann, K. H. Cook, and E. C. Friedberg, The
    kinetics of thymine dimer excision in ultraviolet
    irradiated human cells, Biophys. J. 22:249 (1978).
11. F. E. Ahmed and R. B. Setlow, Kinetics of DNA repair
    in ultraviolet irradiated and
    N-acetoxy-2-acetylaminofluorene-treated mammalian
    cells, Biophys J. 24:665 (1978).
12. R. W. Hart and R. B. Setlow, Correlation between
    deoxyribonucleic acid excision-repair and life-span
    in a number of mammalian species, Proc. Natl. Acad.
    Sci. USA 71:2169 (1974).
13. E. C. Friedberg, U. K. Ehmann, and J. I. Williams,
    Human diseases associated with defective DNA repair,
    Adv. Radiat. Biol. 8:85 (1979).
14. R. B. Setlow and F. E. Ahmed, DNA repair in human
    cells exposed to combinations of carcinogenic agents,
    in: 13th Jerusalem Symposium on Carcinogenesis:
    fundamental mechanisms and environmental effects,
    D. Reidel, Dordrecht (1980).
15. J. E. Cleaver, Nucleosome structure controls rates of
    excision repair in DNA of human cells, Nature 270:451
    (1977).
16. M. J. Smerdon, M. B. Kastan, and M. W. Lieberman,
    Distribution of repair incorporated nucleotides and

nucleosome rearrangement in the chromatin of normal and
xeroderma pigmentosum fibroblasts, Biochemistry 18:3732
(1979).

17.   F. E. Ahmed and R. B. Setlow, Saturation of DNA repair
      in mammalian cells, Photochem. Photobiol. 29:983
      (1979).

18.   D. A. Scudiero, Decreased DNA repair synthesis and
      defective colony-forming ability of ataxia
      telangiectasia fibroblast cell strains treated with
      N-methyl-N'-nitro-N-guanidine, Cancer Res. 40:984
      (1980).

19.   F. E. Ahmed and R. B. Setlow, Excision repair in ataxia
      telangiectasia, Fanconi's anemia, Cockayne syndrome,
      and Bloom's syndrome after treatment with ultraviolet
      radiation and N-acetoxy-2-acetylaminofluorene,
      Biochim. Biophys. Acta. 521:805 (1978).

20.   A. J. Brown, T. H. Fickel, J. E. Cleaver, P. H. M.
      Lohman, M. H. Wade, and R. Waters, Overlapping pathways
      for repair of damage from ultraviolet light and
      chemical carcinogens in human fibroblasts, Cancer Res.
      39:2522 (1979).

INDUCIBLE EXCISION REPAIR IN Escherichia coli

Priscilla K. Cooper

Department of Biological Sciences
Stanford University
Stanford, California 94305

INTRODUCTION

Excision repair of ultraviolet-irradiation (UV) damage in wild type Escherichia coli results in a bimodal distribution of repair patch sizes.[1] The majority of lesions are repaired with short patches approximately 20 nucleotides in length by a pathway mediated by DNA polymerase I, while a minority are repaired with much longer patches by a $recA^+$-dependent pathway.[2] Excision repair assayed by closure of incision breaks similarly seems to proceed via two pathways, one which is polymerase I-dependent and one which is $recA^+$-dependent. The latter pathway additionally requires the $lexA^+$ genotype and protein synthesis.[3] The long patch pathway and the $recA^+$-dependent closure of incision breaks presumably represent manifestations of a single process, having a number of features in common with the coordinately controlled group of responses to DNA damage known collectively as "SOS functions".[4] Indeed, it has recently been shown that long patch repair is UV-inducible and requires protein synthesis, unlike the constitutive short patch pathway;[5] moreover, the requirement for the $lexA^+$ genotype as well as $recA^+$ for long patch repair has been confirmed.[6] However, the role of the long patch pathway, its relationship to other inducible responses to damage, and its importance for cell survival have yet to be established. We have begun to address these questions through use of the tif mutation[7] to induce the long patch pathway prior to DNA damage.

CHARACTERIZATION OF INDUCED LONG PATCH REPAIR SYNTHESIS

Repair synthesis in the presence or absence of chloramphenicol (CAP) was measured as a function of UV dose in the wild type strain W3110 and a dnaB mutant temperature sensitive for semiconservative

replication (Fig. 1). Identical results were obtained with the two
strains despite the occurrence of extensive replication in W3110 and
none in the mutant. When protein synthesis was inhibited by CAP
during post-UV incubation, there was a striking reduction in the total
amount of repair synthesis. In other experiments a similar reduction
has been obtained with the RNA polymerase inhibitor rifampicin. The
amount of inducible repair synthesis (that which requires new protein
synthesis) is dose-dependent, becoming maximal at 60-80 J/m^2. A similar dose dependence has been found for Weigle reactivation, the enhancement of survival of irradiated λ in hosts irradiated prior to
infection.[9] Repair patch size distributions can be obtained from the
density shift in alkaline CsCl gradients produced by the repaired
regions when the molecular weight of the DNA fragments containing
them is reduced by sonication.[8] Such analysis has shown that the reduction in repair synthesis by CAP and rifampicin is largely due to
almost total inhibition of the long patch component of repair.[6]

We have previously shown that long patch repair is inducible by
UV; that is, the inhibition of repair by chloramphenicol following
40 J/m^2 UV is prevented if the cells had been previously irradiated
with 20 J/m^2 and allowed to carry out protein synthesis for 30 min.[5]
Long patch repair is similarly induced in tif-1 mutants incubated at
43°, a procedure which also results in induction of other SOS func-

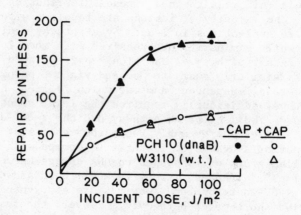

Fig. 1. Repair synthesis in E. coli strains W3110(thy⁻) and PCH10
         (dnaB266 thy⁻). Cultures were prelabelled with ^{14}C-thymine
         in minimal medium plus casamino acids at 32°, irradiated in
         buffer, and incubated with ^{3}H-BrUra at 43° for 45 min, during
         which repair synthesis is completed. Repair synthesis values
         are the ratios of ^{3}H to ^{14}C cpm in parental density DNA puri-
         fied by two sequential bandings in neutral CsCl equilibrium
         density gradients. The detailed procedure has been described
         elsewhere.[8] A portion of each culture received 100 µg/ml
         chloramphenicol (CAP) 15 min prior to irradiation and again
         during post-UV incubation.

tions, e.g., Weigle reactivation, Weigle mutagenesis, and prophage induction.[7] Thus, tif cells which were grown at 32°, then shifted to 43° for one hour prior to irradiation showed no inhibition of total repair synthesis by rifampicin after 60 J/m² UV and no reduction in the long patch component of repair, while long patch repair was inhibited by rifampicin in cells which were continuously incubated at 32° before irradiation (Fig. 2 and patch size analysis, not shown). It is therefore possible by this means to examine the situation in which the capability for long patch repair is already present at the time of irradiation.

In the usual case where induction occurs subsequent to and as a result of irradiation, a minority of lesions (less than 10%) are repaired with long patches, which are at least 200 nucleotides in length in contrast to the short patch size of about 20 nucleotides. (Because of this size differential, however, a large fraction of the repair label is contained in the long patch component). At the earliest times following UV only short patches are made, with both short and long patches being made at intermediate times, while in the final stages of the repair process (between 30 and 45 min after irradiation)

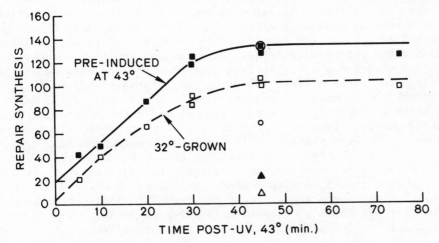

Fig. 2.   Repair synthesis in E. coli strain GC3217 (thy⁻ tif-1 sfiA11). after 60 J/m² UV. The experiment was performed as described in the legend to Fig. 1 except that a portion of the culture was incubated at 43° with 80 µg/ml adenine for the final hour of the prelabelling period. Rifampicin was used in the same way as CAP in Fig. 1. Results from two experiments were normalized by the 30 min and 45 min values. Closed symbols: cells induced prior to UV by 43° incubation. Open symbols: cells incubated at 32° prior to UV. Unirradiated controls▲△; 60 J/m² UV■□; 60 J/m² + 100 µg/ml rifampicin ●○.

only long patches are made.[6] One possible explanation for these observations is that the delay in appearance of long patches is due
solely to the time required for induction to occur and that the small
fraction of sites repaired by long patches is a consequence of the
delay, since short patch repair proceeds rapidly in the initial
phases. However, the results of repair experiments with the pre-in-
duced tif mutant do not support this simple model. While repair synthesis reaches a somewhat higher plateau value in the cells induced
prior to irradiation, it is far less than the approximately 10-fold
increase predicted if long patches can be produced at every incision
site once the pathway is operational (Fig. 2). Furthermore, no
greater proportion of long patches occurs at the earliest times.
These results imply that there is a small defined class of sites
which are repairable with long patches, and they raise the possibility that such sites might arise at certain lesions as a consequence
of events within the cell rather than being created directly by irradiation. Further experiments are required to test this interpretation.

IMPORTANCE OF INDUCED EXCISION REPAIR FOR SURVIVAL

RecA and lexA mutants, which are unable to induce SOS functions
are very UV sensitive, as are wild type strains in which induction
is prevented by protein synthesis inhibition during post UV incubation before determining survival.[4,10] However, it is difficult to
evaluate the possible contribution of induced repair processes to
survival of irradiated cells from such observations because of the
multitude of other effects involved. An alternative approach is to
study the consequences of induction prior to damage. Tif-1 cells
incubated at 43° prior to irradiation showed an appreciable enhancement of survival relative to cells that were not similarly pre-in-
duced (Fig 3, left panel). This inducible enhancement of UV resistance was largely, though not entirely, absent in cells which also
carried the uvrA6 mutation and therefore were incapable of performing
any excision repair,[11] either short or long patch (Fig. 3, right
panel). Thus, there are at least two inducible processes which contribute to the ability of cells to survive UV damage. The major one
of these is uvr[+]-dependent and therefore probably long patch repair.

The increased ability of cells induced prior to irradiation to
survive UV damage resembles Weigle (W) reactivation of irradiated
phage. Accordingly we examined W-reactivation of λ by tif induction
under the same conditions (Fig. 4). There is a striking similarity
between the two phenomena, both in the magnitude of the increase in
resistance and in the extent of its dependence on uvr[+]. If this
uvr[+] dependence of the major component of W-reactivation is indeed
due to a requirement for an excision repair process, then it would
be predicted that W-reactivation of a single strand phage should be
small (similar to that of λ in uvr[−]) and indistinguishable in uvr[+]
and uvr[−] cells. We have measured W-reactivation by tif induction of

the single-strand phage ST-1 in experiments (manuscript in preparation) like those of Fig. 4 and find that this is indeed the case. It would therefore seem, from this result and the data of Figs. 3 and 4, that the inducible repair processes responsible for Weigle reactivation also operate on the bacterial chromosome and that the major one of these is long patch excision repair.

To begin to identify the effects of pre-induction of the long patch pathway which might explain the increased survival, we compared the resumption of DNA replication after irradiation in <u>tif</u> uvr⁺ cells induced by 43° incubation to that in such cells grown only at 32°. The post-irradiation delay in replication is of much shorter duration in the 43°-induced cells, while there is no appreciable effect on the induction on the rate of replication in unirradiated cells(P. Cooper, manuscript in preparation). Thus it may be that the sites which require long patch repair are potent blocks to replication which are for some reason refractory to excision repair by the short patch pathway.

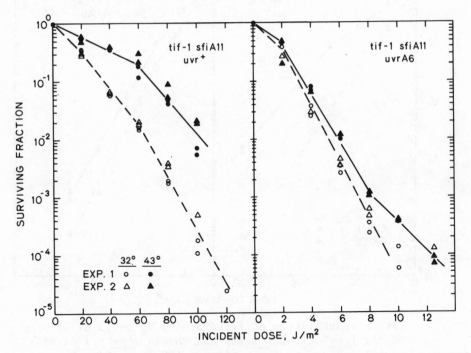

Fig. 3. Colony-forming ability of irradiated cells of strains GC3217 (<u>thy⁻</u> <u>tif</u>-1 <u>sfiA11</u>) and DM1444 (<u>thy⁻</u> <u>tif</u>-1 <u>sfiA11</u> <u>uvrA6</u>). Cultures were induced by 43° incubation prior to irradiation (closed symbols) or not (open symbols) as in legend to Fig. 2. Cells were plated on the same minimal medium plus casamino acids in which they were grown. Plates were incubated at 32°.

DISCUSSION

We have shown that long patch excision repair is one of the inducible "SOS functions" and that there is apparently a small defined
class of sites that are repairable by long patches but not by the
constitutive short patch pathway. The ability to perform this repair
appears to be important for cell survival and also seems to be the
major factor in Weigle reactivation. However, a great many questions
regarding the nature of long patch repair remain. One concerns the
nature of the substrate: what is the special lesion or lesion configuration requiring long patch repair? The possibility that a minor
photoproduct is the substrate has not yet been eliminated but seems
unlikely. A more likely possibility is that long patch repair is
required in particular lesion configurations. These might include
closely-spaced lesions on opposite strands,[12] lesions in the vicinity
of blocked replication forks or in actively transcribing regions, or

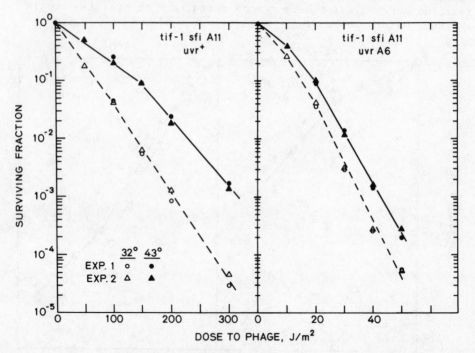

Fig. 4.   Weigle reactivation of irradiated λ by cells of strains
           GC3217 (thy⁻ tif-1 sfiAll) and DM1444 (thy⁻ tif-1 sfiAll
           uvrA6) induced by incubation at 43° as in Fig. 2 legend
           (closed symbols). Open symbols: survival of irradiated λ
           in non-induced cells. Cells grown in minimal medium plus
           casamino acids were induced or not and infected with un-
           irradiated or irradiated λ at a multiplicity of approximate-
           ly 0.01. Infected cells were plated on tryptone medium and
           incubated overnight at 32°.

even lesions in regions important for the maintenance of the ordered structure of the bacterial chromosome. A second question is the nature of the requirement for $recA^+$: are high levels of the recA protein required for a direct interaction with the DNA, or is derepression of one or more other essential genes by the activated recA protease required? The mechanism by which long patch repair proceeds is also as yet largely unknown. Although uvr^+-promoted incision is required,[11] it is not clear whether the subsequent steps involve excision and resynthesis analogous to that in the short patch pathway. An alternative possibility is that some form of incision-promoted recombination with homologous parental DNA is the mechanism, with the long patch resynthesis necessary for completion of the recombinational process.[13] Another question concerns the consequences of repair by the long patch pathway. While we have shown that long patch repair enhances survival, it is not yet known whether it is mutagenic. Although all UV-induced mutations require $recA^+$ and $lexA^+$,[4] it has been reported that mutations presumably occurring during excision repair are not inhibited by CAP under conditions similar to those which block long patch repair.[14] This result is suggestive but not conclusive that long patch repair may be error free.

ACKNOWLEDGEMENTS. I thank Dr. Martine Defais for providing phage stocks, advice, and valuable discussions. This work was supported by contract DE-AC03-76SF00326-7 with the Department of Energy.

REFERENCES

1. Cooper, P. K. and Hanawalt, P. C., Heterogeneity of patch size in repair replicated DNA in Escherichia coli, J. Mol. Biol. 67:1 (1972a).
2. Cooper, P. K. and Hanawalt, P. C., Role of DNA polymerase I and the rec system in excision repair in Escherichia coli, Proc. Nat. Acad. Sci. USA 69:1156 (1972 b).
3. Youngs, D. A., Van der Schueren, E., and Smith, K. C., Separate branches of the uvr gene-dependent excision repair process in ultraviolet-irradiated Escherichia coli K-12 cells, J. Bacteriol. 117:717 (1974).
4. Witkin, E. M., Ultraviolet mutagenesis and inducible DNA repair in Escherichia coli, Bacteriol. Rev. 40:869 (1976).
5. Cooper, P. K. and Hunt, J. G., Alternative pathways for excision and resynthesis in Escherichia coli: DNA polymerase III role?, in: "DNA Repair Mechanisms," P. C. Hanawalt, E. C. Friedberg, and C. F. Fox, eds. Academic Press, New York (1978).
6. Hanawalt, P. C., Cooper, P. K., and Smith, C. A., Repair replication schemes in bacteria and human cells, in: "Progress in Nucleic Acid Research and Molecular Biology", W. E. Cohn, ed., Academic Press, New York (in press).

7. Castellazzi, M., George, J., and Buttin, G., Prophage in-
    duction and cell division in E. coli. I. Further char-
    acterization of the thermosensitive mutation tif-1 whose
    expression mimics the effect of UV irradiation, Mol. gen.
    Genet. 119:139 (1972).
8. Smith, C. A., Cooper, P. K., and Hanawalt, P. C., Measure-
    ment of repair replication by equilibrium sedimentation,
    in "DNA Repair: A laboratory Manual of Research Proce-
    dures", E. C. Friedberg and P. C. Hanawalt, eds., Marcel
    Dekker, New York. (in press).
9. Defais, M., Caillet-Fauquet, P., Fox, M. S., and Radman, M.,
    Induction kinetics of mutagenic DNA repair activity in
    E. coli following ultraviolet irradiation, Mol. gen.
    Genet. 148:125 (1976).
10. Ganesan, A. K. and Smith, K. C., Requirement for protein
    synthesis in rec-dependent repair of deoxyribonucleic
    acid in Escherichia coli after ultraviolet or X-irradi-
    ation, J. Bact. 111:575 (1972).
11. Cooper, P. K. and Hanawalt, P. C., Absence of repair rep-
    lication following ultraviolet irradiation of an ex-
    cision-deficient mutant of Escherichia coli, Photochem.
    Photobiol. 13:83 (1971).
12. Sedgwick, S. G., Misrepair of overlapping daughter strand
    gaps as a possible mechanism for UV induced mutagenesis
    in uvr strains of Escherichia coli, Mutat. Res. 41:185
    (1976).
13. Clark, A. J. and Volkert, M. R., A new classification of
    pathways repairing pyrimidine dimer damage in DNA, in:
    "DNA Repair Mechanisms," P. C. Hanawalt, E. C. Friedberg,
    and C. F. Fox, eds. Academic Press, New York (1978).
14. Bridges, B. A. and Mottershead, R. P., Mutagenic DNA repair
    in Escherichia coli VII. Constitutive and inducible
    manifestations, Mutat. Res. 52:151 (1978).

ON THE NATURE OF THE REPAIR DEFICIENCY IN E.coli uvrE

R. Ben-Ishai and R. Sharon

Department of Biology
Technion-Israel Institute of Technology
Haifa, Israel

ABSTRACT

In uvrE mutants incision is saturated at significantly lower fluences than in the isogenic parent strains. Repair of incisions is delayed and dependent on protein synthesis. As repair takes place at incision sites new dimer sites become susceptible to incision. It is suggested that in uvrE mutants the UV endonuclease complex is altered and remains bound at the site of incision.

INTRODUCTION

Excision repair is a highly coordinated process in which the incision, excision-repair synthesis and ligation steps proceed in rapid sucession[1]. The tight coordination of excision repair results in efficent small patch repair in uvr$^+$ and DNA polymerase I proficient E.coli [2,3] . Mutations in polA, uvrD, uvrE, and recL delay dimer excision and closure of incision breaks[4,5,6].

Recent studies indicate that uvrD, uvrE and recL are alleles of the same gene[7]. Mutations in these alleles confer similar levels of UV sensitivity, ability to carry out host cell reactivation and inhibition of small patch repair [5,6,7]. The role of this uvr product in UV repair is not yet known. It has been proposed that it may control a step subsequent to incision[5,6], regulate polI or polIII[7] or cooperate with polI in the same step of repair[6]. This paper demonstrates that the incision step of repair is impaired in uvrE mutants.

147

RESULTS

Toluene treated cells were used to study the incision step of repair in uvrE 502 mutants. Fig. 1 shows that in the mutant significantly fewer incisions are made than in the isogenic parent strain and that formation of incisions in the mutant decreases gradually with dose. These results indicate that in the mutant the incision enzyme has a low affinity for dimers or that it dissociates slowly from its site of action. Experiments with NMN or various concentrations of ATP indicate that incomplete incision is not due to reversal of incisions by polynucleotide ligase or limitation of ATP, respectively. Results to be reported elsewhere,demonstrated that addition of dimer specific T4 endonuclease V to toluene treated uvrE cells results in dose dependent incisions and repair synthesis, provide further evidence for a defect in incision in uvrE.

In vivo a limited number of incisions accumulate in uvrE mutants within 10 min after irradiation, and these incisions persist until closure commences (Fig.2) The persistance of a constant number of incisions, independent of dose, has been taken to mean that after an initial delay incisions and closure are closely coordinated[6]. To determine whether the incision enzyme of uvrE is capable of functioning after initial formation of incisions, we permeabilized cells after postirradiation incubation and tested their ability to incise.

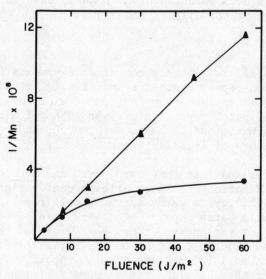

Fig.1   Formation of ATP dependent incision breaks in toluene treated E.coli uvrE502 (●-●) and its isogenic uvrE[+] parent strain (▲-▲).

Fig. 3 shows that in cells permeabilized 10 min after irradiation almost no new incisions, additional to those present before permeabilization are made even if such cells are reirradiated after permeabilization. However, capacity to incise is regained when dimer excision and break closure start occuring. The closure of breaks in vivo and the reappearance of the capacity to incise in permeabilized cells are inhibited if chloramphenicol is present during postirradiation incubation.

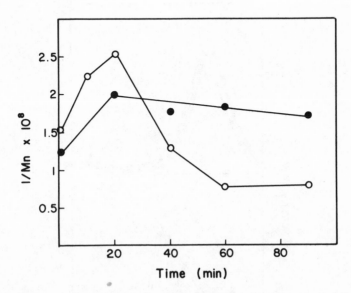

Fig.2   Number of single strand breaks in DNA at different times during incubation after exposure to a dose of $15J/m^2$; (•-•) and (o-o) denote incubation in the presence and absence of chloramphenicol, respectively.

DISCUSSION

The present study provides evidence for a defect in incision in uvrE mutants. This conclusion is based on the following observations: i) incision in toluene - treated uvrE mutants is saturated at significantly lower fluences than in the isogenic parent strains; ii) the number of incision breaks persisting after irradiation in vivo is similar to the saturation level of incisions in toluene treated cells; iii) after initial formation of incisions in vivo almost no additional incisions are formed after permeabilization, additional incisions are also not made if cells are reirradiated after permeabilization; iv) capacity to incise is regained when break closure is occuring; v) the presence of chloramphenicol during postirradiation incubation prevents both break closure and regained capacity to incise.

In E.coli the gene functions coded by uvrA[+], uvrB[+] and uvrC[+] are jointly required for the incision step [9]. A plausible interpretation of our results is that in uvrE mutants the UV endonuclease complex is altered and remains tightly bound to its site of action. Accordingly, the uvrE function may control the formation (or dissociation) of the incision complex or a function that is part of such a complex.

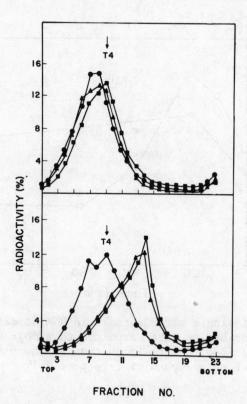

FRACTION   NO.

Fig.3   E. coli uvrE were exposed to UV ($12.5J/m^2$) and permeabilized after 10 min (Fig. 3a) or 60 min (Fig. 3b) incubation. Following a second exposure to UV light permeabilized cells were incubated with ATP and assayed for formation of incision breaks by alkaline sucrose gradient centrifugation. Sedimentation profile of DNA immediately after permeabilization (■-■) after 20 min incubation in the presence of ATP with (●-●) and without (▲-▲) a second exposure to UV light.

Rothman[10] has presented evidence that repair in recL mutants occurs in long patches. We have observed that repair of incisions in uvrE mutants is dependent upon protein synthesis and occurs after induction of the recA protein (unpublished results). These results are in agreement with recent findings that long patch repair is UV-inducible and recA dependent [11],[1 2]. Our observation that regained capacity to incise is coupled to repair of incisions may imply that recA or a recA controlled function facilitates displacement of the UV endonuclease complex from incised DNA in uvrE mutants.

REFERENCES

1. R.B. Setlow, Repair of DNA in "Regulation of nucleic acid and protein biosynthesis," p. 51-62. V.V. Koningsberger and L. Bosch, ed. Elsevier, Amsterdam. (1967).
2. P.K. Cooper, and P.C. Hanawalt, Role of DNA polymerase I and the rec system in excision repair in E.coli, J. Mol.Biol.67:1(1972)
3. R. Ben-Ishai and R. Sharon, Patch size and base composition of ultraviolet light-induced repair synthesis in toluenized Escherichia coli, J. Mol. Biol. 120:423 (1977).
4. J.M. Boyle, M.C. Paterson and R.B. Setlow, Excision-repair properties of an Escherichia coli mutant dificient in DNA polymerase, Nature 226:708 (1970).
5. C.A. Van Sluis, I.E. Mattern and M.C. Paterson, Properties of uvrE mutants of Escherichia coli K12. Effect of UV irradiation on DNA metabolism. Mutn. Res. 25:273 (1974).
6. R.H. Rothman and A.J. Clark, Defective excision and postreplication repair of UV-damaged DNA in a recL mutant strain of E.coli K12. Molec. Gen. Genet, 155:267 (1977).
7. S.R. Kushner, J. Sheperd, G. Edwards and V.F. Maples, UVRD, UVRE and RECL represent a single gene in "DNA repair mechanisms", p. 251-254. P.C. Hanawalt, E.C. Friedberg and C.F. Fox, ed. Academic Press, New York (1978).
8. B.I. Sinzinis, G.B. Smirnov and A.S. Saenki, Repair deficiency in Escherichia coli UV-sensitive mutator strain uvr 502, Biochem. Biophys. Res. Comm. 53:309 (1973).
9. E. Seeberg, Reconstitution of an Escherichia coli repair endonuclease activity from the separated $uvrA^+$ and $uvrB^+/uvrC^+$ gene products, Prod. Natl. Acad. Sci. USA 75:2569 (1978).
10. R.H. Rothman, Dimer excision and repair replication patch size in a recL 152 mutant of Escherichia coli K-12. J.Bact.133:444(1978).
11. P.K. Cooper and J.G. Hunt, Alternative pathways for excision and resynthesis in Escherichia coli: DNA polymerase III role? in "DNA repair mechanisms", p. 255-260. P.C. Hanawalt, E.C. Friedberg and C.F. Fox, ed. Academic Press, New York (1978).
12. R. Ben-Ishai, E. Pugravitsky and R. Sharon, Conditions for constitutive and inducible gap filling of excision and postreplication repair in toluene treated E.coli, in "DNA repair mechanisms". 267-270. P.C. Hanawalt, E.C. Friedberg and C.F. Fox, ed. Academic Press, New York (1978).

REPAIR OF UV-DAMAGED DNA TRANSFORMED INTO E. coli

Peter Strike[1] and R. John Roberts[2]

Department of Genetics, University of Liverpool
Brownlow Street, P.O. Box 147, Liverpool L69 3BX
Merseyside, England

ABSTRACT

The ability of repair proficient and deficient strains of E. coli to repair transforming plasmid DNA has been investigated, using calcium treatment to render the cells competent. Excision repair pathways are particularly important for transforming DNA, but recA mediated pathways play only a minor role. The presence of a homologous resident plasmid in the recipient cell is shown to enhance the survival of incoming damaged plasmid. This enhancement depends on the host uvrA+ and uvrB+ gene products, but not on host recA functions. Explanations of the enhanced survival are considered, based on the known interactions of plasmid molecules.

INTRODUCTION

Calcium treatment is widely used to render cells of Escherichia coli capable of uptake of exogenous DNA, particularly in association with the in vitro techniques for DNA manipulation. Despite this widespread use, the mechanisms underlying the uptake process are not well understood, and it was partly in an attempt to probe these mechanisms that the present study was carried out. Concurrently we hoped to develop a picture of the way in which repair mechanisms acted on transforming plasmid DNA, and so gain an insight into the interactions of repair processes. The plasmid we used routinely in this work was NTP16, a small non-transmissible plasmid (MW 5.6 x 10^6) carrying resistance to ampicillin and kanomycin. Details of the plasmid, transformation procedure and selection for transformants, have been described previously.[1,2]

1. This work was supported by an MRC grant to P.S.
2. R.J.R. acknowledges the receipt of an SRC postgraduate studentship.

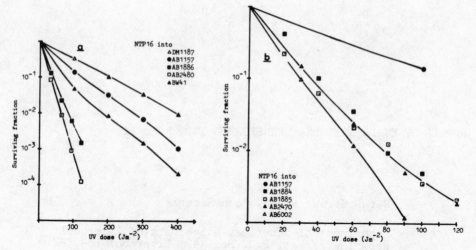

Fig. 1. Survival of UV irradiated plasmid NTP16 transformed into:
(a)   DM1187 tif-1 sfi-A11 lexA3 spr-51;   AB1157 uvr⁺ rec⁺;
      AB1886 uvrA6 rec⁺;   AB2480 uvrA6 recA13;   BW41 uvr⁺
recA13. (b)   AB1157 uvr⁺ rec⁺;   AB1884 uvrC34 rec⁺;
      AB1885 uvrB5 rec⁺;   AB2470 uvr⁺ recB21;   AB6002
uvrA6 recB21.

RESULTS AND DISCUSSION

        Transformation of UV irradiated plasmid NTP16 DNA into excision
proficient (Uvr⁺) calcium-treated E. coli results in a much better
survival than when the same DNA is transformed into an excision
deficient strain (Figure 1). In contrast, mutations within the recA
gene have only a small effect on the survival of NTP16 in either Uvr⁺
or Uvr⁻ backgrounds, indicating that recA-mediated repair pathways
are not particularly active on transforming DNA. However, the small
decrease in plasmid survival observed in recA mutants compared to
wild-type is mirrored by a small increase in survival over wild-type
levels in a lexA spr strain, showing quite clearly that the recA gene
product is involved in some minor pathway(s) for the repair of such
DNA molecules. Also noteworthy is the effect of a mutation in the
recB gene on the survival of NTP16 - such a mutation is almost as
severe as a uvrA mutation in reducing plasmid survival. In the
double mutant uvrA recB, survival of NTP16 is only slightly less than
that observed in uvrA or recB single mutants, implying that the uvrA
and recB gene products must act principally in a single pathway for
the repair of transforming DNA. It may be that the excision repair
which occurs in plasmid DNA is of the long-patch type, since this is
the only in vivo repair pathway known which involves these two gene
product.[5] Alternatively, repair may be of the short-patch type and
still involve both of these gene products, as had been observed in
toluene treated cells.[4]

The efficiency with which excision repair acts on transforming DNA implies very strongly that such DNA is taken up in the double stranded form since, as we have pointed out previously,[1] excision repair is unable to act on single stranded DNA. This is a very different situation from transformation in such naturally competent species as B. subtilis or H. influenzae, where uptake is known to involve the conversion of DNA to the single stranded form. The low efficiency with which recA mediated pathways act on transforming plasmid DNA might be expected under the conditions used, since the DNA concentration is slightly less than saturating, and in consequence the majority of uptake events will involve only a single DNA molecule.[2] Any recombination repair of plasmid damage must therefore be limited to intramolecular events, either during replication in the case of a plasmid monomer, or possibly prior to replication in the case of multimeric plasmid molecules.

In order to investigate whether the availability of homologous undamaged plasmid molecules could improve the survival of damaged transforming DNA, NTP16 was transformed into recipient strains containing a point-mutant of NTP16 which had lost the ability to code for ampicillin resistance. Selection for ampicillin resistance thus allowed survival of NTP16 to be followed. The results of such an experiment, using AB1157 uvr+ rec+ (NTP16 Ap^S) as the recipient strain are shown in Figure 2a. Clearly the presence of the resident

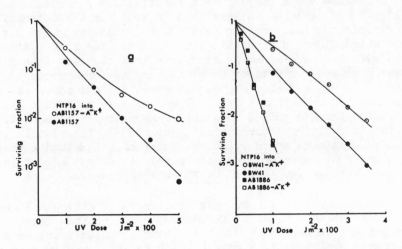

Fig. 2. Survival of UV irradiated plasmid NTP16 transformed into
(a)   AB57 - A⁻K⁺(A⁻K⁺ indicates that this strain contains the Ap^S derivative of NTP16, which normally specifies both ampicillin and kanamycin resistance).   AB1157 uvr+ rec+.
(b)   BW41 uvr+ recA13;   BW41 - A⁻K⁺;   AB1886 uvrA6 rec+;   AB1886 - A⁻K⁺.

homologous plasmid enhances the survival of the incoming damaged transforming molecule. However, an investigation of the host gene products involved in the enhancement reveals that the increased survival is independent of the recA gene product, and is completely dependent on the uvrA and uvrB gene products (Figure 2b). We have tested many mutants showing defects in repair processes for their ability to show this enhancement, and have so far identified only the uvrA and uvrB gene products as being essential.

Investigation of the recombination ability and UV sensitivity of strains carrying NTP16 has ruled out the possibility of a plasmid coded recA analogue, and we have also ruled out the possibility of transposition since restriction analysis of the DNA in the survivors shows no increase in plasmid size. We have also considered two other possible explanations. Firstly we have considered the possibility that we are observing a situation analogous to liquid holding recovery i.e. replication of the damaged transforming molecule is delayed in the cell which contains the resident plasmid, and this delay permits more complete excision repair than is normally possible. Such a delay in replication might be caused by the copy number control system to which plasmids with relaxed replication are susceptible.[5] The usual copy number for NTP16 is 12 copies/cell,[6] and if this level were controlled by a negative control protein as had been suggested for other plasmids,[5] this would provide a mechanism for delaying the replication of an incoming plasmid DNA molecule. If this were the case, it might be expected that plasmids sharing a common copy number control mechanism would display this reactivation phenomenon between themselves. Figure 3a shows that this is true for NTP16 transformed into a recipient strain containing NTP1, a partially homologous plasmid with which NTP16 is incompatible. In fact, the reactivation observed with NTP1 is rather greater than that observed with NTP16 Ap[s], and this may be a reflection of the higher copy number of 18 for NTP1 compared to 12 for NTP16. However, not all pairs of incompatible plasmids show this effect, as can be seen in Figure 3b. The colE1 related plasmid pBR325 shows no enhanced survival when transformed into a recipient strain containing pBR313, the plasmid from which pBR325 was made by in vitro manipulation. Indeed, testing other pairs of plasmids we have been unable to detect significant enhancement between any pair of colE1 related plasmids.

The second possibility we have considered is shown in Figure 4, and involves the interwinding of two homologous plasmid molecules, one damaged the other not. Such a reaction is known to be catalyzed by the enzyme synaptase,[7] an enzyme which can be found in both Rec[+] and Rec[-] bacteria. Such interwinding would have the effect of separating adjacent dimers on opposite strands, and pairing them with undamaged DNA. In this way, lesions which were previously refractory to excision repair would become susceptible to it. As a further step, resolution of the interwound plasmids into two monomers could follow, or equally, conversion of the complex into a

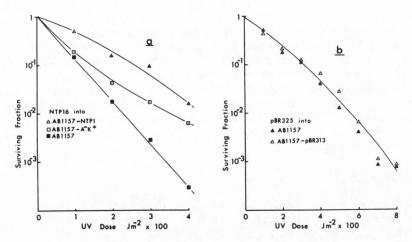

Fig. 3.   (a) Survival of UV irradiated NTP16 transformed into AB1157
          uvr⁺ rec⁺ ■ and derivatives containing either NTP16 Apˢ ▣
          or NTP1 △ .   (b) Survival of UV irradiated pBR325 trans-
          formed into AB1157 uvr⁺ rec⁺ ▲ and a derivative containing
          pBR313 △ .

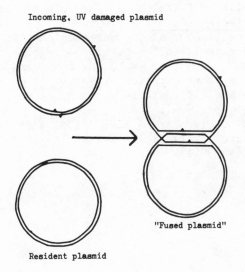

Fig. 4.   Possible first step in resident plasmid enhanced repair.
          The interwinding of an incoming damaged transforming DNA
          molecule with an undamaged homologous resident plasmid
          creates a fused plasmid in which adjacent dimers on opposite
          strands have been separated, and thus become reparable by
          excision repair.

dimeric molecule. Restriction analysis of DNA of survivors from cell
containing NTP1 resident plasmid shows bands characteristic of both
NTP1 and NTP16, implying that in this case either incompatibility is
not complete, or that many of the NTP16 survivors exist as part of a
multimeric plasmid which also contains NTP1. The presence of such a
chimeric plasmid would not of course be proof of synaptase mediated
repair, since dimer formation could occur equally well following the
replication-delay enhanced repair discussed previously. We are
currently working to determine the exact nature of resident-plasmid
enhancement of repair.

REFERENCES

1. P. Strike, G. O. Humphreys and R. J. Roberts, Nature of trans-
      forming deoxyribonucleic acid in calcium-treated Escherichia
      coli, J. Bacteriol. 138:1033 (1979).
2. G. O. Humphreys, A. Weston, M. G. M.,Brown and J. R. Saunders,
      Plasmid transformation of Escherichia coli, in: "Transformatio
      1978", S. W. Glover and L. O. Butler, eds., Cotswold Press,
      Oxford (1979).
3. L. Grossman, A. Braun, R. Feldberg and I. Mahler, Enzymatic
      repair of DNA, Ann.Rev.Biochem. 44:19 (1975).
4. P. Strike, DNA synthesis and degradation in UV-irradiated toluene
      treated cells of E. coli K12: The role of polynucleotide
      ligase, Molec.gen.Genet. 157:99 (1977).
5. A. J. Twigg and D. Sherratt, Trans-complementable copy-number
      mutants of plasmid ColE1, Nature 283:216 (1980).
6. A. Weston, G. O. Humphreys, M. G. M. Brown and J. R. Saunders,
      Simultaneous transformation of Escherichia coli by pairs of
      compatible and incompatible plasmid DNA molecules, Molec.gen.
      Genet. 172:113 (1979).
7. H. Potter and D. Dressler, A biochemical assay designed to detect
      recombination intermediates formed in vitro, Proc.natl.Acad.
      Sci. 76:1084 (1979).

AN ENDONUCLEOLYTIC ACTIVITY FROM M. luteus DIRECTED TOWARDS

8-(2-HYDROXY 2-PROPYL)-PURINES IN DNA

Zvi Livneh and Joseph Sperling

Department of Organic Chemistry, The Weizmann Institute
of Science
Rehovot, Israel

ABSTRACT

An endonuclease directed towards 8-(2-hydroxy 2-propyl)
purines in double stranded DNA was discovered in extracts of
Micrococcus luteus. The activity does not require divalent cations
and it is fully active in 1 mM EDTA, while ATP and caffein cause
inhibition. Extracts of mutant M. luteus lacking pyrimidine-dimer
"endonucleolytic" activity were found to contain normal levels of
the endonuclease. The activity is not inhibited by single stranded
DNA, nor by irradiated or depurinated single stranded DNA. Native
or depurinated double stranded DNAs do not cause inhibition whereas
UV-(254nm)-or γ-irradiated double stranded DNAs do inhibit the
activity hinting at the possibility of a common type of damage in
these damaged DNAs. The conformation of 8-substituted purines at
the nucleoside and nucleotide level, namely their non-regular syn
conformation, suggests that it is the distortion in the DNA which
is recognized by the endonuclease. After incision we could
demonstrate the specific excision of the 8-alkylated purines from
the damaged DNA.

INTRODUCTION

Excision repair is believed to be the main mechanism for the
repair of damaged DNA.[1,2] Two types of excision repair have been
described according to their initiating incision events; base
excision repair in which incision of the DNA requires the action
of a specific DNA-glycosylase followed by an AP endonuclease[3], and
nucleotide excision repair in which incision is performed by a
damage specific endonuclease.[1]

So far only two types of damaged directed endonucleases have been described: A pyrimidine-dimer endonuclease was purified from E. coli,and was shown to be composed of the products of genes uvrA,uvrB and uvrC,[4] and another endonuclease acting at pyrimidine dimers was isolated from calf thymus.[5] The only other class of endonucleases which has been found in a variety of sources is the apurinic/apyrimidinic (AP) endonuclease.[3]

Many other repair endonucleases have been isolated from various sources on the basis of their incision activity on UV-irradiated, γ-ray irradiated or chemically damaged DNA substrates.[1,2] The damage specificities of these enzymes are unknown, mainly due to the fact that the damaged DNA substrates usually contain a multiplicity of products, part of them uncharacterized. This difficulty interfers with the characterization of repair endonucleases and to the study of their role in repair mechanisms in vivo.

We approached these problems by the development of selective UV-induced free radical alkylation reactions of the various nucleic acid bases.[6] The photoalkylations of adenine and of thymine with 2-propanol are representatives of these reactions, and are shown in Fig. 1. It has been found that in mixtures of purines and pyrimidines,[7] in heterolygonucleotides[8,9] or in DNA[10-12] the reaction was selective for the purines, while the reactivity of the pyrimidines was supressed to an extent dictated by their involvement in base-stacking interactions.[9]

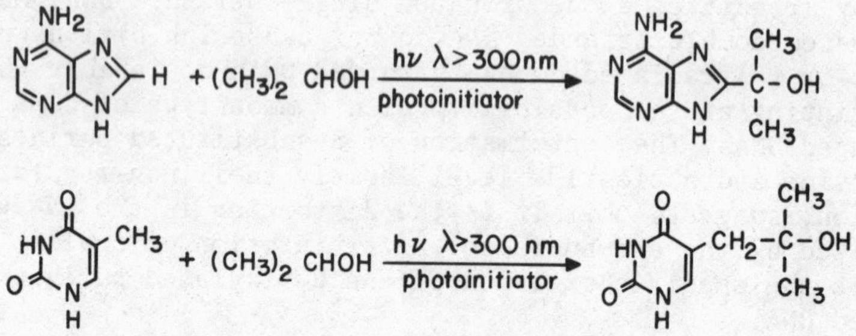

Fig. 1.   Photoalkylation of adenine and of thymine with 2-propanol.

Utilizing this type of reactions we have prepared DNA from phage PM2 which contains 8-(2-hydroxy 2-propyl)-purines as the only type of damage (8-hpPu-DNA). This preparation was shown to be free of single or double strand breaks, interstrand crosslinks,AP sites, pyrimidine dimers, any other pyrimidine damage,and purine damage

other than the 8-substituted purines.[13] Using this DNA as a substrate,
we discovered in M. luteus an endonuclease which is directed
towards 8-(2-hydroxy 2-propyl)-purines in DNA.[14]

M.LUTEUS EXTRACTS CONTAIN AN ENDONUCLEASE WHICH IS DIRECTED TOWARDS 8-(2-HYDROXY 2-PROPYL)PURINES IN DOUBLE STRANDED DNA.

Incubation of 8-hpPu-DNA with a protein fraction from M.luteus
resulted in the formation of single strand breaks in the damaged
DNA, whereas negligible nicking of the untreated DNA occurred (Fig.
2). The activity was measured by its ability to convert circular

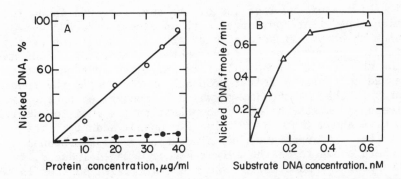

Fig. 2.   Incision of PM2 8-hpPu-DNA by protein fraction from
M. luteus[14]. ○, 8-hpPu-DNA; ●, intact DNA.

superhelical 8-hpPu-DNA from phage PM2 to the circular relaxed
form. The relaxed and the superhelical forms of DNA were separated
usually by denaturation-renaturation followed by filtration on
nitrocellulose filters or by agarose gel electrophoresis. In a few
cases alkaline gradient centrifugation was used too.

The plot of the rate of incision as a function of the substrate
DNA concentration reaches a plateau, and at saturation concentrations
of the substrate the rate of incision increases linearly with
increasing protein concentration (Fig. 2). The activity is
inactivated by heat (60°, 10 min) and by high salt concentration
(1 M NaCl), and is partially inhibited by ATP, caffeine and EDTA.
The activity does not require divalent cations, and it is fully
active in 1 mM EDTA (higher EDTA concentrations cause inhibition).

We could not find any release of free modified bases from the
DNA under various incubation conditions. This indicates that we are
observing a true endonucleolytic activity, rather than a combination
of a specific glycosylase followed by an AP-endonuclease. We also
found that extracts prepared from M.luteus DB7, a strain lacking

"endonucleolytic" activity towards pyrimidine dimer, contain normal
levels of the incision activity. This is not surprising in view of
the recent finding that those pyrimidine-dimer "endonucleases" are
actually a combination of pyrimidine dimer glycosylases and AP
endonucleases.

Three lines of evidence suggest that the activity is indeed
directed towards 8-(2-hydroxy 2-propyl)-purines in the damaged
DNA substrate:
1. The substrate contains 8-(2-hydroxy 2-propyl)purines as the
only type of damage in the DNA, and the incision activity was found
to increase linearly with the number of alkylated purines per DNA
molecule.
2. No inhibition of the incision of the substrate was found in
competition experiments performed with an excess of a DNA
preparation which was irradiated under the same condition as the
substrate(di-tert-butyl peroxide as photoinitiator, UV, λ>305 nm)
but in the absence of 2-propanol. This indicates that the attack of
the free radical derived from 2-propanol on the DNA is the event
which forms in the DNA the damage which is recognized by the
endonuclease and makes the DNA susceptible to its incision activity.
Since we have shown that the thymine-2-propanol adduct is not
present in the DNA substrate, and since the purine-2-propanol
adducts where shown to be the only types of damaged induced, these
further support the conclusion that the 8-alkylated purines in the
DNA are the substrates for the endonuclease.

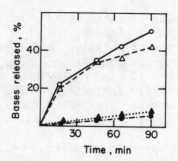

<u>Fig. 3.</u>   Specific excision of 8-alkylated purines from 8-hpPu-DNA
        by an extract from <u>M</u>. <u>luteus</u>[14]. ○, 8-hpAde; ●, Ade; △,
        8-hpGua; ▲, Gua.

3. Following the incision step we could demonstrate the specific
excision of the modified purines nucleotides from the DNA (Fig. 3)
indicating that incision occurred near the 8-alkylated purines in
the DNA.

WHAT IS THE SPECIFICITY OF THE ENDONUCLEASE?

The endonuclease does not seem to be a single-strand specific
endonuclease since it was not inhibited in competition experiments
with a large excess of single stranded DNA from phage φX174. The
same results were obtained by using various damaged single stranded
DNA preparations [UV (260 nm)-irradiated-, γ-ray irradiated-, and
depurinated ssDNAs, and ssDNA containing 8-(2-hydroxy-2-propyl)-
purines]. We therefore conclude that the substrate must be a double
stranded DNA. Competing double stranded DNA or depurinated dsDNA
did not inhibit the incision activity, however, UV (254 nm)-
irradiated or γ-irradiated dsDNA preparations inhibited the activity
to a maximum extent of 60%. If we assume that the inhibition was
competitive, this raises the possibility that lesions that appear
in UV- or γ-irradiated DNA and in 8-hpPu-DNA share structural or
conformational features which are recognized by the endonuclease.
The question of the possible nature of this lesion brings us to the
question of the specificity of the endonuclease. The fact that it is
directed towards 8-(2-hydroxy 2-propyl)purines in dsDNA does not
necessarily imply that it is specific only for this damage. Moreover,
it seems unlikely that nature has designed an endonuclease specific
only for this damage. It is reasonable to assume that what is
recognized by the endonuclease is some conformational change induced
in the DNA by the alkylation of the purines at the 8-position. In
purine nucleosides or nucleotides substitution at the 8 positions
with bulky groups causes steric hindrance with the sugar ring and
with the phosphate group. Consequently these nucleosides or
nucleotides rotate around their glycosylic bond from the native anti
conformation to the syn conformation (Fig. 4).[15] For the case of

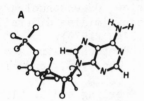
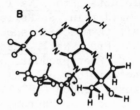

Fig. 4. (A) AMP, anti conformation; (B) 8-hpAMP   syn conformation.

8-(2-hydroxy 2-propyl)-AMP or 8-(2-hydroxy 2-propyl)-Ado this change in conformation was shown by x-ray diffraction studies[16] and from CD and NMR spectra.[17] The question whether 8-substituted purines acquire the syn conformation in DNA is still open, but this might well be the case in light of various studies which hinted towards the existance of syn conformation in N-acetoxy-N-2-acetylaminofluorene (AAAF)modified DNA[18] (this carcingen substitutes mainly guanine at the 8 position). The discovery of the Z-DNA conformation, in which unmodified purines were found to be in the syn conformation[19]further strengthens this possibility, especially in light of a recent report that substitution of the 8 position of guanine in poly dG-dC with AAAF induces the appearance of Z-DNA conformation in the polynucleotide.[20]

It is possible that the syn conformation, if it is induced in the DNA by the 8-substitution of its purines, is the distortion which is recognized by the endonuclease. Since syn conformation can be induced in the DNA by a variety of agents (including, possibly, γ-radiation which produces 8-hydroxyadenine in DNA),[21] it may define a class of leasions which serve as substrates for the described M.luteus endonuclease.

ACKNOWLEDGEMENT

This research was supported by a grant from the United States-Israel Binational Science Foundation (BSF), Jerusalem, Israel.

REFERENCES

1. L. Grossman, A. Braun, R. Feldberg and I. Mahler, Enzymatic repair of DNA, Ann. Rev. Biochem. 44:19 (1975).
2. E.C. Friedberg, K.H. Cook, J. Duncan and K. Mortelmans, DNA repair enzymes in mammalian cells, in: "Photochemical and Photobiological Reviews", K.C. Smith, ed. Plenum Press, New York (1977).
3. T. Lindahl, DNA glycosylases, endonucleases for apurinic/apyrimidinic sites and base excision repair, Prog. Nucl. Acid Res. Mol. Biol. 22:135 (1979).
4. E. Seeberg, Reconstitution of an E. coli repair endonuclease activity from the separated uvrA+ and uvrB+/uvrC+ gene products, Proc. Natl. Acad. Sci. USA 75:2569 (1978).
5. E.A. Waldstein, S. Peller and R.B. Setlow, UV-endonuclease from calf thymus with specificity toward pyrimidine dimers in DNA, Proc. Natl. Acad. Sci. USA 76:3746 (1979).
6. D. Elad, Photoproducts of purines, in "Photochemistry and Photobiology of Nucleic Acids", S.Y. Wang, ed, Academic Press, London (1976).
7. A.A. Frimer, A. Havron, D. Leonov, J. Sperling and D. Elad, Ultraviolet and γ-ray-induced free-radical reactions of nucleic acid constituents. Selectivity of some reactions for

purines. Suppression of the reactivity of pyrimidines, J. Am. Chem. Soc. 98:6026 (1976).

8. A. Havron, J. Sperling and D. Elad, Reactivity and Selectivity in light induced free radical reactions of 2-propanol with purine and pyrimidine nucleotides and dinucleoside triphosphates, Nucleic Acids Res. 3:1715 (1976).
9. E. Livneh (Noy), D. Elad and J. Sperling, Light-induced free radical reactions of nucleic acid constituents. The effect of sequence and base-base interactions on the reactivity of purines and pyrimidines in ribonucleotides, Biochemistry 17:3128 (1978).
10. R. Ben-Ishai, M. Green, E. Graff, D. Elad, H. Steinmaus and J. Salomon, Photoalkylation of purines in DNA, Photochem. Photobiol. 17:155 (1973).
11. J. Salomon and D. Elad, Selective photochemical alkylation of purines in DNA, Biochem. Biophys. Res. Commun. 58:890 (1974).
12. Z. Lorberbaum, J. Sperling and D. Elad, The selective photo-initiated alkylation of purines in DNA with 2-propylamine, Photochem. Photobiol. 24:389 (1976).
13. Z. Livneh, D. Elad and J. Sperling, in preparation.
14. Z. Livneh, D. Elad and J. Sperling, Endonucleolytic activity directed towards 8-(2-hydroxy 2-propyl) purines in double-stranded DNA, Proc. Natl. Acad. Sci. USA 76:5500 (1979).
15. P.O.P. Ts'o, Bases, nucleosides and nucleotides, in:"Basic Principles in Nucleic Acid Chemistry", P.O.P. Ts'o, ed., Academic Press, New York (1974).
16. G.I. Birnbaum and D. Shugar, A. purine nucleoside unequivocally constrained in the syn form, Biochim. Biophys. Acta 517:500 (1978).
17. M. Ikehara, S. Uesugi and K. Yoshida, Studies on the conformation of purine nucleosides and their 5'-monophosphates, Biochemistry 11:830 (1972).
18. R.P.P. Fuchs, J.F. Lefevre, J. Pouyet and M.P. Daune, Comparative orientation of the fluorene residue in native DNA modified by N-acetoxy-N-2-acetylaminofluorene and two 7-halogeno derivatives, Biochemistry 15:3347 (1976).
19. A.H.J. Wang, G.J. Quigley, F.J. Kolpak, J.L. Crawford, G. van der Marel and A. Rich, Molecular structure of a left-handed double helical DNA fragment at atomic resolution, Nature 282:680 (1979).
20. M. Leng, E. Sage and P. Rio. This volume.
21. P.A. Cerutti, M. Kaneko, and P. Beard. This volume.

MECHANISMS OF POSTREPLICATION REPAIR AND REPAIR BY RECOMINATION

REGULATORY AND ENZYMATIC FUNCTIONS OF recA PROTEIN

IN RECOMBINATION AND POSTREPLICATION REPAIR

P. Howard-Flanders, E. Cassuto and S. C. West

Dept. Molecular Biophysics & Dept. Therapeutic Radiology

Yale University, P.O. Box 6666, New Haven, CT 06511

ABSTRACT

Investigations into the genetics of recombination, repair and prophage induction in E. coli, have uncovered a remarkable regulatory system involving the lexA gene product as repressor and the recA gene product. This system normally produces lexA and recA proteins at low constitutive levels. The recA protein is a highly specific, single stranded DNA dependent protease capable of cleaving lexA protein and also phage lambda repressor. Postreplication gaps or other damage that produced single stranded regions in the bacterial DNA, may cause recA protein to bind to the DNA. The resulting increase in protease activity leads to the lexA protein being cleaved and thus turns on the transcription of the genes including recA that are under negative control by lexA.

RecA protein is an ATPase and requires ATP for most reactions. It does not appear to have any nuclease activity. It catalyses the annealing of single strands and homlogous duplex DNA to form D loops. It binds to duplex DNA in complex ways. If the duplex contains single stranded regions, recA protein binds cooperatively and loads the duplex molecules, covering the duplex as well as the single stranded regions. It prepares duplex DNA containing one or more single stranded gaps for hydrogen bonding to intact homologous duplex DNA and promotes the formation of hydrogen bonded joint molecules, but but does not cut the intact homolog in preparation for joining in recombination and postreplication repair.

1. Genetic analysis and characterization of recA and lexA mutants.

Following the isolation by Clark (1965) of recA recombination
deficient mutants of E. coli, the properties of these mutants were
investigated in many laboratories. Witkin (1976), Radman (1975) and
others have noted the pleiotropic nature of the recA gene of E. coli
and listed a series of functions under recA control:
   1. Genetic recombination (not site-specific recombination).
   2. Postreplication repair, resistance to UV light, X-rays, etc.
   3. Induction of prophage lambda by UV irradiation.
   4. UV mutagenesis.
   5. Colicin induction.
   6. Weigle reactivation (survival UV'd phages on UV'd hosts).
   7. Delay in cell division after UV irradiation.
   8. Synthesis of RecA protein (formerly called protein X).
The genetic analysis of recombination and the location of recA in
the genetic map of the bacterial chromosome proved to be unusually
difficult because recA mutations interfere with the recombination
needed for the process of genetic mapping. The problem is further
complicated by the wide range of phenotypes of mutants that map in
the recA gene. These include mutations such as tif, zab from
Castellazzi et al. (1972) and lexB from Morand et al. (1977), in
which recombination is almost normal, as well as recA1 and recA13
which exhibit a recombination deficienct phenotype. The position of
recA between serA and srl at 57 mins in the E. coli map (Fig. 1) was
established by cotransduction and deletion analysis by Willetts et
al, (1969); McEntee, et al., (1976) and McEntee (1977). A detailed
physical map of the recA region was constructed by restriction
mapping by Ogawa et al. (1978) and Sancar and Rupp (1979), and by
sequencing (Horii et al., 1980; Sancar et al., 1980). This work may
help us to understand the properties of mutant recA proteins.
Figure 1 also shows the region near lexA. As we will see later,
lexA has important regulatory functions in repair, mutagenesis and
prophage induction.

2. Identification of RecA protein.

The genetic analysis just described opened the way for the
investigation of the biochemistry of genetic recombination and
postreplication repair. McEntee et al., (1976), McEntee, (1977)
isolated a lambda transducing phage carrying srl and recA. A
similar transducing phage was isolated by Ogawa et al, (1978), while a
multicopy plasmid carrying recA was constructed from the plasmid
bank of Carbon and Clark (1976) by Sancar and Rupp (1979). McEntee
et al. (1976) identified the recA gene product as a protein of about
40,000 MW.   From independent work in four laboratories, McEntee
(1977), Emmerson and West (1977), Gudas and Mount, (1977) and Little
and Kleid (1977), we know the RecA protein to be the same as the
radiation-induced protein X, that had been described some years
earlier by Inouye and Pardee (1970) and by Gudas and Pardee (1976)

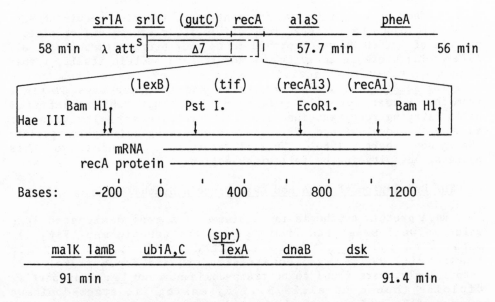

Fig. 1. Genetic maps of the <u>recA</u> and <u>lexA</u> regions of the <u>E. coli</u>
chromosome (McEntee, 1977a; 1977b) and restriction map of
the <u>recA</u> gene. The <u>recA</u> gene has been sequenced through the
1400 base pair region shown (Horii et al. 1980; Sancar et
al, 1980). The mutation <u>recA441</u> (<u>tif-1</u>) increases the
protease activity of RecA protein, <u>lexB30</u> permits
recombination but not UV prophage induction, <u>recA13</u> is
dominant, while <u>recAl</u> is recessive. The mutation <u>lex spr</u>
appears to be an operator mutant which produces no
detectable lexA repressor and permits a high constitutive
synthesis of RecA protein. Uncertainty in position is
indicated by parentheses.

and characterized as a protein which binds to single stranded DNA
and bacterial membrane. A good yield of recA protein can be
obtained by growing <u>E. coli</u> carrying <u>recA</u>⁺ on a multicopy plasmid and
inducing it with nalidixic acid, following which the level of recA
protein may rise to several percent of the total protein of the
cell. The purified protein has a molecular weight of 40,000 by SDS
gel electrophoresis and an ATPase activity dependent on single
stranded DNA (Ogawa et al., 1978; Roberts et al., 1978; Weinstock et
al, 1979).

3. <u>The protease activity of RecA protein</u>.

The induction of <u>E. coli</u> cells lysogenic for prophage λ occurs
in exponentially growing cells following exposure to UV light. This
<u>recA</u> dependent process has been studied by Roberts and Roberts
(1975) who showed that induction occurs following cleavage of the

λ repressor protein into two fragments in a reaction requiring ATP. They purified the protease responsible for cleavage and obtained a protein of 40,000 MW which proved to be recA protein (Roberts et al. 1978). The protease activity was due to recA protein itself, rather than a contaminating activity, since the protease from cells carrying recA441 (tif) was more active and in other ways distinct from the protease of wild type (recA$^+$) cells, while that from cells carrying recA1 was inactive. The protease activity requires single stranded DNA and ATP. It is slow acting and highly specific (Craig and Roberts, 1980). We will return to a discussion of this protease activity in the following section.

4. The Regulation of RecA and LexA Protein Synthesis.

RecA protein synthesis is regulated by a gene designated lexA which is well separated from recA on the genetic map (Fig. 1). Mutants carrying lexA were isolated on the basis of their sensitivity to X-rays (Howard-Flanders and Boyce, 1966) and, surprisingly, were found to be trans-dominant over lex$^+$ in partial diploids (Mount et al.1973). By analogy to trans-dominant non-inducible lacI mutants which produce an altered repressor, it was suggested that lexA might be a regulatory gene producing a repressor protein that controlled recA (Mount et al., 1975; Gudas and Pardee, 1975; Mount, 1977; Gudas and Mount, 1977). In support of this idea, radiation resistant revertants of lexA3 designated tsl were selected. They mapped in the lexA region and partially suppressed the radiation sensitive phenotype of certain recA mutants (Mount, 1975).

Subsequent work was based on the hypothesis that the lexA gene product was the repressor for the recA gene and was also inactivated when lambda repressor was cleaved during the process of induction. In a strain of E. coli carrying recA441 (formerly tif-1), the lambda prophage inducing function and the RecA protease activity were expressed constitutively at 42 C (Roberts et al.,1975; 1978). This strain also carried sfi which suppressed filamentation and loss of viability at high temperature. Into this strongly inducing strain was introduced the transdominant mutation lex3 to provide lexA repressor protein resistant to inactivation. The resulting triple mutant was UV sensitive at 42 C, indicating the resistance of lexA3 repressor to inactivation under these conditions. Using this triple mutant carrying recA441 sfi lexA3, Mount (1977) selected mitomycin C-resistant revertants, designted spr, which have the important characteristic that recA functions appear to be switched on constitutively at all temperatures, rather than being inducible. The spr mutations map in or near lexA and behave as operator mutations that block the synthesis of lex protein. These findings support the concept that lexA acts as repressor for recA and the recA gene product inactivates the lexA gene product. In fact, lexA regulates its own rate of synthesis (Brent and Ptashne, 1980) and

that of several other genes. The synthesis of colicin in cells
carrying Col E is induced by UV-irradiation, the production of uvrA
(but not uvrB) gene product is increased if cells carry spr (Rupp et
al., 1981). Several other loci including a mutator gene may be
controlled in this way (Kenyon & Walker, 1980; Ginsburg et al.,
1980).

A.   Intact duplex DNA

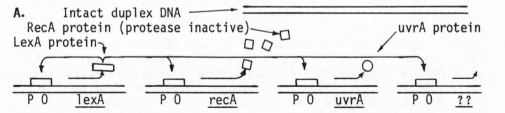

B.   DNA with postreplication gap at UV induced pyrimidine dimer

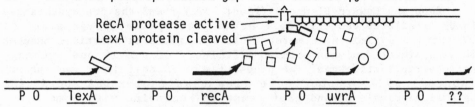

Fig. 2. The lexA regulatory system of E. coli K12.
           A. Quiescent, lexA repressed, constitutive transcription.
           B. UV induced, repression lifted, high recA transcription.
           LexA protein acts as repressor and binds to the operator
           regions of lexA, recA and co-regulated genes. In the
           presence of ATP, recA protein binds cooperatively to single
           stranded regions formed in duplex DNA following irradiation
           and becomes an active protease, cleaving lexA protein, thus
           lifting repression. In E. coli lambda lysogens, the phage
           repressor protein is cleaved by recA protein in a similar
           reaction and prophage induction occurs. (Mount, 1977; Gudas
           and Mount, 1977; 1977; McEntee, 1977; Emmerson and West,
           1977; Roberts et al, 1978; Craig and Roberts, 1980; Little
           et al., 1980; Brent and Ptashne, 1980; Radman, 1975; Rupp et
           al., 1981; Kenyon & Walker, 1980; Witkin, 1976).

     This regulatory system is important in connection with
ultraviolet mutagenesis,   which is blocked in lexA and in recA
mutants (Witkin, 1976). Strains carrying mutations thought to be
operator mutants of recA have been isolated in E. coli B (Volkert et
al., 1979) and K12 (Ginsburg et al. 1980), and found to produce high
constitutive levels of recA protein even when they also carried lexA.

Evidently lexA protein failed to repress recA protein synthesis. UV mutagenesis occured with lexA⁺ but not lexA⁻,   showing that constitutive synthesis of recA protein was not sufficient to ensure UV mutagenesis. At least one other gene product under lexA control was needed.

Biochemical evidence for lexA recA regulation has been obtained through the identification of the lexA gene product as a protein of 24,000 MW (Little and Harper; 1979). Little et al., (1980) demonstrated in extracts of recA⁻ cells   that lexA protein was cleaved when RecA protein was added, and that cleavage occured in the presence of ATP, the conditions employed being similar to those used for cleaving purified phage lambda repressor (Roberts et al., 1977; Little et al., 1980). A summary diagram of this scheme for lexA recA regulation is shown in Figure 2.

5. Postreplication repair and cutting in trans.

Rupp and Howard-Flanders (1968; 1971) found the DNA synthesized in UV irradiated bacteria to contain postreplication gaps at the UV-induced pyrimidine dimers. This DNA initiated genetic exchanges between sister duplexes with an efficiency of nearly one exchange per postreplication gap. The properties of E. coli hyper rec mutants (Konrad, 1977) support the view that duplex DNA containing unrepaired single strand gaps tends to recombine with homologous DNA.

In another attempt to define the DNA structures involved in genetic exchanges, studies were made of reactions occuring in phage or prophage DNA treated with a DNA crosslinking agent. This treatment substantially increased the frequency of genetic exchanges between phage lambda DNA molecules infecting E. coli(λ) lysogens. Ross and Howard-Flanders demonstrated that infection with suitably treated phages resulted in the cutting of intact circular homologous phage DNA present in the cell. This phenomenon, referred to as cutting in trans, was found to occur in response to the presence of crosslinked DNA that had been incised by bacterial enzymes. Since it was dependent on recA⁺ and took place only between homologs,   it appeared that cutting in trans might be equivalent to the cutting step of genetic recombination (Ross and Howard-Flanders, 1977; Cassuto et al, 1978). These experiments, which were performed both in vivo and in vitro, showed that duplex DNA containing unrepaired single strand breaks or gaps, tends to initiate genetic exchanges, and provided a rationale for the experiments with RecA protein next to be discussed.

6. RecA protein promotes homologous pairing.

RecA protein might play any one of a number of roles in genetic recombination and postreplication repair. It might unwind duplex

DNA making single strands available for pairing, it might promote pairing between DNA strands or duplexes, it might be an endonuclease that cuts one strand following homologous pairing, it might have other functions or it might have an indirect or regulatory function.

Suggestive evidence that RecA protein promoted the pairing of homologous DNA molecules was obtained by Holloman and Radding (1976). D loops (paired single strands and covalent circular supercoiled ϕX phage DNA) were converted to recombinants following the transfection of bacterial spheroplasts without need for the recA gene product. However, this experiment is open to other interpretations. More direct evidence was soon obtained in two laboratories. Weinstock et al. (1979) purified recA protein and found it to catalyse the pairing of complementary DNA stands in an ATP dependent-reaction. Significantly, the purified protein from a cold sensitve mutant carrying recA629 failed to promote pairing at restrictive temperature, thus demonstrating pairing to be a recA controlled function. Shibata et al., (1979) found that wild type RecA protein greatly increased the speed of pairing between single strands and superhelical circular duplex molecules to form D loops. In this case the protein was needed in stoichiometric rather that catalytic amounts, sufficient protein being required to bind all the single stranded DNA with one protein monomer per five single strand DNA bases. However, superhelicity was not essential as RecA protein catalysed the pairing of single strands with relaxed duplex DNA, the needed energy presumably coming from the ATPase activity of the protein (McEntee et al., 1979; Shibata et al., 1979).

7. __Binding to single stranded, duplex and gapped duplex DNA.__

At low concentrations, RecA protein was found to bind strongly to single stranded DNA without need for ATP, but to have little affinity for duplex molecules (Inouye and Pardee, 1970; Gudas and Pardee, 1975; McEntee et al, 1979). At higher concentrations (above 40 μg/ml), it bound cooperatively to linear duplex DNA in the presence of ATP, GTP or gamma-S-ATP (McEntee et al, 1979). It did not exhibit nicking closing (topoisomerase) activity, but reduced the number of spiral turns in nicked circular DNA in the presence of gamma-S-ATP and single stranded DNA fragments (Cunningham et al. 1979). It has not been reported to unwind DNA to the point where the strands can separate. In the presence of single strand binding (SSB) protein (Weiner et al., 1975) and ATP, strand separation occurred with Rep protein (Scott et al., 1977) and DNA helicase (Abdel-Monen et al. 1977), but no separation was induced by recA protein (West et al., 1981).

As remarked, not only does recA protein bind to single stranded DNA, but its ATPase activity is dependent upon such binding, duplex DNA being ineffective. In the presence of duplex DNA containing single stranded gaps, therefore, recA protein might be expected to bind to the single stranded regions and to exhibit only the corresponding ATPase activity, the rest of the protein remaining unbound. However, this is not the case. West et al. (1980) found that duplex molecules containing single stranded gaps stimulated the ATPase activity almost as effectivly as an equal weight of single stranded DNA. An example of this effect is seen in Figure 3, which shows the rates of hydrolysis of ATP by recA protein in the presence of equi-molar amounts of single stranded, duplex or gapped duplex circular DNA. Surprisingly, the rate of ATP hydrolysis with the

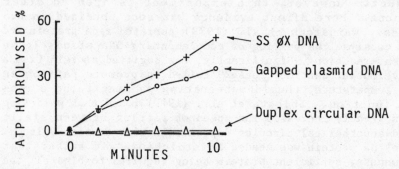

Fig. 3. The hydrolysis of ATP by recA protein in the presence of
equal   equal weights of (a) single strand øX DNA, (b) plasmid
         (pBR322) duplex circular DNA containing 4 to 6 gaps of about
         30 nucleotides each, or (c) intact plasmid (pBR322) duplex
         circular DNA. Reaction conditions were as in West et al.
         1980.

gapped DNA was nearly as great per nucleotide as with single stranded DNA, even though not more than 2% of the gapped DNA was single stranded. As intact Form I and II plasmid DNA without gaps did not stimulate signifiant ATPase activity, we conclude that recA protein may have held the gapped duplex in configurations differing from normal Watson-Crick structure and permitting a high ATPase activity through single strand-like contacts.

        Since recA protein promotes homologous pairing between gapped and intact duplex molecules (see below), recA protein might hold gapped duplex DNA in a special configuration for this purpose. Consistant with this idea, recA protein in the presence of gamma-S-ATP, bound cooperatively to duplex DNA containing single standed regions, forming thick rigid rods readily seen by electron microscopy. Interestingly enough, these were seen only on duplex molecules with single stranded regions, and may reflect cooperative binding starting in the single strands and spreading over the duplex regions (West et al., 1980).

8. <u>Formation of joint molecules by homologous pairing between intact and gapped duplex DNA.</u>

Since gapped duplex DNA caused sister exchanges in postreplication repair and also promoted high ATPase activity in recA protein, Cassuto et al. (1980) investigated whether recA protein could promote pairing between two duplex molecules if one of them was gapped. Circular duplex plasmid DNA molecules with several single stranded gaps was incubated with recA protein, Mg^{++} and ATP, and found to aggregate into a compact tangled mass. Each

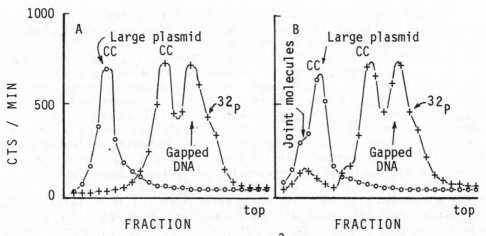

Fig. 4   Detection of joint molecules. ^{3}H-labeled DNA from the large plasmid pKC16 and ^{32}P-labeled gapped DNA from the small plasmid pBR322 were incubated 30 mins with: A, Mg^{++} and ATP; B, recA protein, Mg^{++} and ATP, and sedimented in neutral sucrose. The gapped DNA consisted half of intact covalent circular DNA (CC) and half of form II DNA in which the nick was widened with exonuclease III. (Cassuto et al., 1980).

gapped duplex tended to bind to several others in a multivalent manner and was specific, since it affected homologous molecules only. On the other hand, Cassuto et al. (1980) and Cunningham et al. (1980) found that duplex molecules containing only one gap were monovalent in their ability to bind, and they produced mainly dimeric structures. Cassuto et al., (1980) showed that RecA protein also promoted pairing with chimeric molecules containing heterologous as well as homologous regions. Dimeric strucures were detected when gapped plasmid DNA was incubated with intact DNA from a chimeric plasmid carrying a non-homologous insert of phage genes. Figure 4 shows that joint molecules could be detected in neutral sucrose sedimenting at the leading edge of the main peak (Cassuto et al., 1980).

Joint molecules formed with recA protein appeared to be hydrogen bonded rather than covalently bonded, as they separated readily in alkali. Since recA protein showed no detectable nuclease activity and failed to cut intact molecules so that covalently joined recombinants could be formed, at least one additional enzyme is needed. Potter and Dressler (1980) described an enzyme from E. coli that cut and joined plasmid DNA in homologous regions and did not require recA protein. It remains to be determined how this activity is related to recA promoted recombination.

9. Prospect.

E. coli responds to ultraviolet irradiation and to other DNA damaging treatments through the recA-lexA regulatory system, which controls the levels of recA, uvrA and certain other needed proteins. In addition to being a regulatory protease, RecA protein is the enzyme of homologous pairing, an aspect of recombination now studied in several laboratories. Duplex DNA molecules containing one or more single stranded gaps (gapped duplexes) are recognised by recA protein and are key intermediate structures in genetic recombination. Preliminary evidence indicates that homologous pairing in bacteria may take place by a mechanism differing from that of D-loop formation (Radding 1978). One alternative mechanism involves hydrogen bonding between four complementary polynucleotide chains (McGavin, 1971). Efforts are being made to identify the enzymes of cutting and joining in recA-dependent recombination and to determine the detailed structure of recA protein by X-ray crystalographic methods (McKay, 1980).

This work was supported by grants from the National Institutes of Health AMK69397, CA06519 and GM11014.

Abdel-Monem, M., Lauppe, H.F., Kartenbeck, J., Durwald, H. & Hoffmann-Berling, H. (1977). Mol. Biol. 110, 667-685.
Brent, R. & Ptashne, M. (1980). Proc. Natl. Acad. Sci. U.S. 77, 1932-1936.
Castellazzi, M., George, J. & Butin, G. (1972). Mol. Gen. Genetics. 119, 139-152 and 153-174.
Cassuto, E., Mursalim, J. & Howard-Flanders, P. (1978). Proc. Natl. Acad. Sci. U.S. 75, 620-624.
Cassuto, E., West, S.C., Mursalim, J., Conlon, S. & Howard-Flanders, P. (1980). Proc. Natl. Acad. Sci. U.S. 77,
Clark, A.J. & Margulies, A.D. 1965. Proc. Natl. Acad. Sci. U.S. 53, 451-459.
Craig, N.L. & Roberts, J.W. (1980). Nature 238, 26-30.
Cunningham, R.P., DasGupta, C. Shibata, T., & Radding, C.M. (1980). Cell. 20, 223-235.
Cunningham, R.P., Shibata, T., DasGupta, C. & Radding, C.M. (1979). Nature 281, 191-195.

Emmerson, P.T. & West, S.C. (1977).   Molec. Gen. Genetics 155,
    77–85.
Ginsburg, H., Edmiston, S. & Mount, D. (1980). Proc. Natl. Acad.
    Sci. U.S. In press.
Gudas, L.J. (1976).   J. Mol. Biol. 104, 567–583.
Gudas, L.J. & Mount, D.W. (1977). Proc. Natl. Acad. Sci. U.S. 74,
    5280–5284.
Gudas, L.J. & Pardee, A.B. (1975). Proc. Natl. Acad. Sci. U.S. 72,
    2330–2334.
Gudas, L.J. & Pardee, A.B. (1976).   J. Mol. Biol. 101, 459–477.
Holloman, W.K. & Radding C.M. (1976). Proc. Natl. Acad. Sci. U.S. 73
    3910–3914.
Horii, T., Ogawa, T. & Ogawa, H. (1980). Proc. Natl. Acad. Sci. U.S.
    77, 313–317.
Howard-Flanders, P. & Boyce, R.P.   (1966). Radiat. Res. Suppl. 6,
    156–184.
Inouye, M. & Pardee, A. (1970). J. Biol. Chem. 248, 5813–5819.
Kenyon, C.J. & Walker, G.C. (1980). Proc. Natl. Acad. Sci. U.S. 77,
    2819–2823.
Konrad, E.B. (1977).   J. Bacteriol. 130, 167–172.
Little, J.W., Edmiston, S.H., Pacelli, L.Z. & Mount, D.W.   (1980).
    Proc. Natl. Acad. Sci. U.S. 77
Little, J.W. & Harper, J.E. (1979). Proc. Natl. Acad. Sci. U.S. 76,
    6147–6151.
Little, J.W. & Klied, D.G. 1977. J. Biol. Chem. 252, 6251–6252.
McEntee, K. (1976).   Virology 70, 221–222.
McEntee, K. (1977).   J. Bacteriol. 132, 904–911.
McEntee, K. (1977).   Proc. Natl. Acad. Sci. U.S. 74, 5275–5270.
McEntee, K. (1978).   In: DNA repair mechanisms. Eds: Hanawalt, P.
    C., Friedberg, E.C. & Fox, F.C. pp349–360. Academic Press, New
    York.
McEntee, K., Hesse, J.E. & Epstein, W. (1976).   Proc. Natl. Acad.
    Sci. U.S. 73, 3979–3983.
McEntee, K., Weinstock, G.M. & Lehman, I.R. (1979). Proc. Natl.
    Acad. Sci. U.S. 76, 2615–2619.
McEntee, K., Weinstock, G.M. & Lehman, I.R. (1980).   Proc. Natl.
    Acad. Sci. U.S. 77, 857–861.
McGavin, S. (1971).   J. Mol. Biol. 55, 293–298.
McKay, D.B., Steitz, T.A., West, S.C. & Howard-Flanders, P. (1980).
    J. Biol. Chem. 255, 6662.
Morand, P., Goze, A. & Devoret, R. (1977). Mol. Gen. Genetics 157,
    69–82.
Mount, D.W. (1977). Proc. Natl. Acad. Sci. U.S. 74, 300–304.
Mount, D.W. & Kosel, C. (1975). J. Bacteriol. 121, 1203.
Mount, D.W., Low, K.B. & Edmiston, S.H. (1973).   J. Bacteriol. 112,
    886–893.
Ogawa, T., Wabiko, H., Tsujimoto, T., Horii, T., Masuka, H. & Ogawa,
    H. (1978).   Cold Spring Harbor Symp. 43, 909–915.
Potter, H. & Dressler, D. (1980). Proc. Natl. Acad. Sci. U.S. 77,
    2390–2394.

Radding, C.M. (1978). Ann. Rev. Biochem. 47, 847-880.

Radman, M. (1975). In: Molecular mechanisms for repair of DNA. Part A. Eds. Hanawalt, P.C. & Setlow, R.B. pp 355-367.

Roberts, J. W. & Roberts, C.W. (1975). Proc. Natl. Acad. Sci. U.S. 72, 147-151.

Roberts, J. W., Roberts, C.W. & Craig, N.L. (1978). Proc. Natl. Acad. Sci. U.S. 75, 4714-4718.

Roberts, J. W. Roberts, C.W., Craig, N.L. & Phizicky, E.M. (1978). Cold Spring Harb. Symp. 43, 917-920.

Roberts, J. W., Roberts, C.W. & Mount, D.W. (1977). Proc. Natl. Acad. Sci. U.S. 74, 2283-2287.

Ross, P. & Howard-Flanders, P. (1977). J. Mol. Biol. 117, 137-158.

Rupp, W.D. & Howard-Flanders, P. (1968). J. Mol. Biol. 31, 291-304.

Rupp, W.D., Wilde, C.E., Reno, D. & Howard-Flanders, P. (1971). J. Mol. Biol. 61, 25-41.

Rupp, W.D. et al (personal communication).

Sancar, A. & Rupp, D.W. (1979). Proc. Natl. Acad. Sci. U.S. 76,

Sancar, A., Stachelek, C., Konigsberg, W. & Rupp, W.D. (1980). Proc. Natl. Acad. Sci. U.S. 77, 2611-2615.

Scott, J.F., Eisenberg, S. Bertsch, L.L. & Kornberg, A. (1977). Proc. Natl. Acad. Sci. U.S. 74, 193-197.

Shibata, T., Cunningham, R.P. DasGupta, C., & Radding, C.M. (1979). Proc. Natl. Acad. Sci. U.S. 76, 5100-5104.

Volkert, M.R., Spencer, D.F. & Clark, A.J. (1979). Mol. Gen. Genetics 177, 129-137.

Weiner, J.H., Eisenberg, S. Bertsch, L.L. & Kornberg, A. (1977). J. Biol. Chem. 250, 1972-1980.

Weigle, J. J. (1953). Proc. Natl. Acad. Sci. U.S. 39, 628-636.

Weinstock, G.M., McEntee, K. & Lehman, I.R. (1979). Proc. Natl. Acad. Sci. U.S. 76, 126-130.

West, S.C., Cassuto, E., Mursalim, J. and Howard-Flanders, P. (1980). Proc. Natl. Acad. Sci. U.S. 77, 2569-2573.

West, S.C. et al., (1981) In preparation.

Willetts, N. S., Clark, A.J. & Low, K.B. (1969). J. Bacteriol. 97, 244-249.

Witkin, E.W. (1976). Bacteriology Revs. 40, 869-907.

<u>Haemophilus influenzae</u> DOES NOT FIT THEORIES OF POSTREPLICATION

REPAIR

Jane K. Setlow and Nihal K. Notani*

Department of Biology,
Brookhaven National Laboratory,
Upton, New York 11973

Studies with <u>Escherichia coli</u> have provided all the initial
insights into the molecular bases of repair processes. It is
the thesis of this article that especially if we believe that
there are some general mechanisms in nature, it is important to
consider more than one microorganism in arriving at an
understanding of the biology of repair of DNA.

<u>Escherichia coli</u> cells lacking the ability to excise
pyrimidine dimers can nevertheless survive ultraviolet radiation
considerably better than a strain containing two mutations,
which is recombination defective as well as excision defective
(1). The first clue to the mechanism of the repair system
defective in <u>recA</u> strains was provided by the discovery that the
DNA synthesized after UV in an excision-defective <u>E. coli</u> on an
irradiated template was shorter than in unirradiated cells, but
that upon incubation the single-strand size of the DNA increased
towards that of the control cells (2). The ability to
accomplish this size increase was later found to be dependent on
expression of the <u>recA</u> gene (3). All of these phenomena are
also found in <u>Haemophilus influenzae</u> (4), including a mutation
in the <u>rec1</u> gene, somewhat analogous to <u>recA</u>, that causes the
DNA size increase to be eliminated.

A further advance in understanding postreplication repair
came when it was found that much of the increase in size of the
fragments resulted from transfer of stretches of DNA from the
parental to the daughter strand in <u>E. coli</u>, filling in the gaps

*On leave of absence from Biology and Agriculture Division,
Bhabha Atomic Research Centre, Bombay 400085, India.

in the daughter strand (5). Again H. influenzae was also found
to undergo this process, but whereas in E. coli the number of
gaps was approximately equal to the number of UV-induced
pyrimidine dimers (2), in H. influenzae there appeared to be
considerably more sister-strand exchange than could be accounted
for by the number of dimers (6).

When sister-strand exchange takes place in E.coli, not only
are the gaps in the daughter strands filled in, but in the
process some of the dimers become transferred to the daughter
strands (7). In H. influenzae this process is much more rapid
and complete, so that the dimers become equally distributed
between parental and daughter strands in a few minutes (8).
This takes place even in a mutant lacking a functional
recombination system for phage or transforming DNA. Thus it is
clear that the recombination that leads to dimer exchange and
gap-filling in H. influenzae is a different type from the
recombination of transforming DNA or phage DNA.

The phenomena of postreplication repair have aroused great
interest in recent years because of the connection between this
type of repair and induced mutation by UV. It has been shown
that part of the postreplication repair involves an inducible
system in E. coli, and that functional recA and lexA genes are
required for it (9, 10). If this inducible system is eliminated
by treatment of E. coli cells with chloramphenicol before and
twenty minutes after UV irradiation, part of the gap-filling and
all of the UV-induced mutation is lacking (11).

A hypothesis was developed from these and other related
phenomena in the reaction of E. coli to UV, that when cells with
intact recA and lexA genes are irradiated, there is coordinate
induction of a number of functions (lysogen induction, cell
filamentation, Weigle (W) reactivation, gap-filling and
induction of mutations) (9, 10). The discovery of a mutant
(C600 T44, tif-1) that is thermally inducible without UV
treatment, and mimics all these effects, strengthened the
hypothesis (12, 13). Some of the difficulties with the
hypothesis have been that various functions supposed to be
coordinately controlled are clearly separable under certain
circumstances. For example, E. coli carrying a recF mutation is
normally mutable by UV, but W reactivation is reduced (14) and
gap-filling is defective in excision-defective mutants (15,
16). Revertants of strains carrying the tif-1 mutation (zab)
are lacking in W reactivation, but when lysogenized are
UV-inducible (17).

H. influenzae shows many of the same responses to UV which
have been considered to be coordinately controlled in E. coli.
Thus there is UV-induced lysogen induction (18), cell

filamentation (19), W reactivation (18) and gap-filling (4, 6).
Most of these responses have been shown to be recl-dependent
(20). However, in a most important respect, H. influenzae is
different, namely in being immutable by UV (21, and J. K.
Setlow, unpublished experiments).

A recent study of W reactivation in H. influenzae has been
made in an attempt to understand more about this phenomenon
(22). We found that W reactivation could be carried out with
UV-irradiated HP1cl phage when wild type but not recl mutant
cells were treated with either UV or mitomycin C (MMC),
resulting in an increase of viable progeny phage by more than a
factor of ten. Similar treatment of cells infected with
unirradiated phage only caused a decrease in the number of
progeny. If the cells were incubated in growth medium following
UV or MMC treatment, but before infection with UV-irradiated
phage, there was a further increase in the number of progeny (up
to a factor of eight), although the progeny from unirradiated
phage were almost unaffected by the incubation.
Chloramphenicol during this incubation eliminated the
reactivation. Reversions of two temperature-sensitive phage
mutations were measured as a function of conditions giving
maximal W reactivation. There was no dose-dependent increase in
mutations, and all the measured frequencies were well within
previously measured spontaneous reversions at these loci (23).

Conclusions from these data were: (a) H. influenzae has an
inducible repair system dependent on recl function that operates
on UV-irradiated phage, although no reactivation of cells was
observed when they were treated so as to produce maximal induced
reactivation of phage. (In E. coli W reactivation conditions
also do not contribute significantly to chromosome repair
(24).) (b) This repair system, unlike the one in E. coli (25),
does not result in phage mutations.

There are several responses of H. influenzae to UV
radiation that differ from those of E. coli and that could have
relevance to the problem of why H. influenzae is immutable by UV
radiation. Whereas UV-induced stimulation of recombination in
E. coli is readily observed (26), UV does not increase the
genetic recombination of H. influenzae phage or transforming DNA
(27). While part of the gap-filling is changed by
chloramphenicol in E. coli, no chloramphenicol-induced change in
the kinetics of this process could be observed in H. influenzae
(21). We have already noted the much more rapid and widespread
sister-strand exchange that takes place in H. influenzae after
UV (6, 8).

Several hypotheses on the reasons for the UV immutability
of H. influenzae may be considered. Kimball, Boling and Perdue

(21), taking into account the failure of certain agents
including UV to induce mutations in H. influenzae cells, have
proposed that H. influenzae lacks the inducible repair system.
However, it is now clear that this bacterium does have such a
system, dependent on the recl gene. Since the recl gene is
known to affect postreplication repair (4, 6), it is reasonable
to assume that postreplication repair is somehow involved with
the inducible repair system in H. influenzae. The question
remains how such an inducible system should be error free,
unlike that of E. coli.

A second hypothesis is that UV causes induction of an
inhibitor of proofreading function of a polymerase (28) in E.
coli but not in H. influenzae. The evidence for such a function
in E. coli is rather indirect. The hypothesis also explains the
mutations resulting from UV irradiation of E. coli host cells
subsequently infected with unirradiated phage (29, 30), and the
lack of such mutation in H. influenzae phage. However, there
are indications that the mechanism of such indirect mutagenesis
in E. coli may be entirely different from that observed in the
direct mutagenesis of W reactivation (30), providing some
evidence against the hypothesis.

A third hypothesis is that in E. coli the inducible part of
gap-filling forces synthesis past a pyrimidine dimer (31),
particularly where there are overlapping daughter strand gaps,
whereas the sister-strand exchange in H. influenzae results in
an inserted piece of parental DNA that not only fills in the gap
but also extends past the dimer responsible for the gap. Hence
there is no synthesis possible on the dimer-containing template,
and thus no mutation.

REFERENCES

1. Howard-Flanders, P. (1968). Ann. Rev. Biochem. 37, 175.
2. Rupp, W. D. and Howard-Flanders, P. (1968). J. Mol. Biol.
    31, 291.
3. Smith, K. C. and Meun, D. H. C. (1970). J. Mol. Biol. 51,
    459.
4. LeClerc, J. E. and Setlow, J. K. (1972). J. Bacteriol. 110,
    930.
5. Rupp, W. D., Wilde, C. E., III, Reno, D. L. and
    Howard-Flanders, P. (1971). J. Mol. Biol. 61, 25.
6. Small, G. D. (1975). J. Bacteriol. 124, 176.
7. Ganesan, A. K. (1974). J. Mol. Biol. 87, 103.
8. Waldstein, E., Setlow, J. K. and Santasier, L. (1979).
    Cold Spring Harbor Symp. Quant. Biol. 43, 1059.
9. Radman, M. (1975). In "Molecular mechanisms for repair of
    DNA" (P. C. Hanawalt and R. B. Setlow, eds.), p. 355,
    Plenum Press, N. Y.

10. Witkin, E. M. (1975). In "Molecular mechanisms for repair of DNA" (P. C. Hanawalt and R. B. Setlow, eds.), p. 347, Plenum Press, N. Y.

11. Sedgwick, S. G. (1975). Proc. Natl. Acad. Sci. (USA) 72, 2753.

12. Castellazzi, M., George, J. and Buttin, G. (1972). Mol. Gen. Genet. 119, 139.

13. Witkin, E. M. (1974). Proc. Natl. Acad. Sci. (USA) 71, 1930.

14. Rothman, R. H., Margossian, L. J. and Clark, A. J. (1979). Mol. Gen. Genet. 169, 279.

15. Ganesan, A. and Seawell, P. C. (1975). Mol. Gen. Genet. 141, 189.

16. Rothman, R. H. and Clark, A. J. (1977). Mol. Gen. Genet. 155, 279.

17. Castellazzi, M., George, J. and Buttin, G. (1972). Mol. Gen. Genet. 119, 153.

18. Harm, W. and Rupert, C. S. (1963). Zeit. f. Vererbungsl. 94, 336.

19. Setlow, J. K., Brown, D. C., Boling, M. E., Mattingly, A. and Gordon, M. P. (1968). J. Bacteriol. 95, 546.

20. Setlow, J. K., Boling, M. E., Beattie, K. L. and Kimball, R. F. (1972). J. Mol. Biol. 68, 361.

21. Kimball, R. F., Boling, M. E. and Perdue, S. W. (1977). Mutat. Res. 44, 183.

22. Notani, N. K. and Setlow, J. K. (1980). J. Bacteriol. 143, 516.

23. Boling, M. E. and Kimball, R. F. (1976). Mutat. Res. 37, 1.

24. Salaj-Smic, E., Petranovic, D., Petranovic, M. and Trgovcevic, Z. (1979). Mol. Gen. Genet. 117, 91.

25. Weigle, J. J. (1953). Genetics 39, 628.

26. Jacob, F. and Wollman, E. (1955). Ann. Inst. Pasteur 88, 724.

27. Boling, M. E. and Setlow, J. K. (1969). J. Virology 4, 240.

28. Villani, G., Boiteux, S., Radman, M. (1978). Proc. Natl. Acad. Sci. (USA) 75, 3037.

29. Bleichrodt, J. K. and Verheij, W. S. D. (1974). Mol. Gen. Genet. 135, 19.

30. Ichikawa-Ryo, H. and Kondo, S. (1975). J. Mol. Biol. 97, 77.

31. Sedgwick, S. G. (1976). Mutat. Res. 41, 185.

RecA-MEDIATED ASYMMETRIC REPAIR OF LETHAL DNA LESIONS IN

E. coli[*]

H.E. Kubitschek and C.N. Newman[§]

Division of Biological and Medical Research, Argonne
National Laboratory, Argonne, Illinois 60439, U.S.A.

ABSTRACT

Mechanisms of action of DNA lesions produced by decay of
genetically incorporated ^{125}I or by 313-nm photolysis of incorporated 5-bromouracil (5BU) in exponential phase cultures of Escherichia coli were studied by inducing lesions after subsequent
periods of DNA replication. The analogues were incorporated during
brief labeling periods to provide prelesional substrates for later
induction of the lesions, and label concentrations were chosen to
give cell survivals of about 10% when lesions were induced immediately after labeling. When the same number of lesions were induced
at intervals during the first generation after labeling survival
levels decreased. After replication of the labeled regions, however,
survival levels increased sharply in the wildtype strains. Three
very different patterns of recovery were observed.

1. 5BU photolysis, wildtype strains. Recovery was essentially
complete.

2. [125]I decay, wildtype strain. Approximately half of the
cells recovered at the end of the first round of DNA replication.

---

[*]This work was supported by the U.S. Department of Energy under
contract No. W-31-109-ENG-38.
[§]Present address: Battelle Pacific Northwest Laboratories, Radio-
logical Physics, Richland, WA 99352, U.S.A.

3. [125]I decay, <u>recA</u> strain. No significant increase in recovery occurred at the first generation, but recovery increased
gradually during the second generation, as expected for production
of irreparable damage in labeled regions.

The results are consistent with <u>recA</u>-mediated asymmetric
repair of ^{125}I decay-induced lesions.

INTRODUCTION

Lethal effects of DNA lesions produced by incorporation of
base analogues usually have been studied either in cells labeled
continuously for many generations or in cells tested immediately
after pulse-labeling. We have employed a new approach in which
pulse labeled cells were grown for various periods before inducing
the DNA lesions[1,2]. In this way, modifying effects of cell growth,
DNA replication, nuclear separation, and cell division can, in
principle, be examined. Our results do indeed show that cell survival depends upon the intervening growth period between labeling
and lesion induction.

METHODS

Cultures of <u>Esherichia coli</u>, strains B/r K (wildtype), B/r TT
(<u>thy trp</u>), or B/r WP10 (<u>thy trp recA</u>), were grown overnight to
exponential phase in minimal medium (M9-salts plus 2 mg/ml glucose).
For both auxotrophs the growth medium was supplemented with thymine
(20 µg/ml) and tryptophan (50 µg/ml, and arginine, methionine, and
proline were also added to 100 µg/ml for cultures of strain WP10.
Doubling times of these cultures ranged from 40 to 87 min. Cultures
were labeled for 2-4 min with 5-bromouracil (5BU) or with
5-iodo[^{125}I]-2'-deoxyuridine, washed, and resuspended in growth
medium. There was no detectable change in growth rate of the
labeled cells.

Periodically during culture growth, samples were removed and
DNA lesions were induced in 5BU-labeled cultures by exposure to
313-nm radiation at a fluence of 5 x 10 J/m^2. Survival was determined from colony counts after plating labeled cells and those
from an unlabeled control culture on nutrient agar. With ^{125}I,
samples of the labeled culture and an unlabeled control were frozen
in liquid nitrogen to permit accumulation of decay-induced DNA
lesions. Samples were thawed and plated after a period sufficient
to give approximately 10% survival for cells frozen immediately
after labeling. Further details were described earlier[1,2].

RESULTS AND DISCUSSION

The surviving fraction after lesion induction was small immediately after labeling and during the first generation of growth.
In the repair—proficient strains this was followed by an abrupt
increase in cell survival at the first generation after labeling.
For the repair deficient strain no such increase was observed. More
specifically, Fig. 1 shows the following responses.

1. E. coli B/r TT, 5BU photolysis. Full recovery (100%) was
observed at the end of the first round of DNA replication.

2. E. coli B/r, strains K and TT, ^{125}I decay. Survival increased
abruptly to 50% and this level was maintained throughout the second
generation.

3. E. coli WP10 recA, ^{125}I decay. There was no abrupt increase
in survival at the end of the first generation, but a gradual increase after that time.

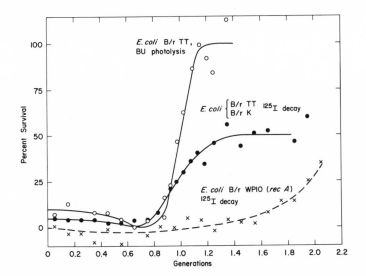

Fig. 1.   Survival patterns after BU photolysis or ^{125}I Decay. Exponentially growing cultures were labeled with 5BU or ^{125}IUdr
for 2–4 min. Samples of the growing cultures then were
removed at intervals for exposure to 313–nm radiation at
5×10^5 J/M^2, or for storage at liquid nitrogen temperature
to accumulate ^{125}I decay-induced lesions. The data points
are averages over periods of 0.1 generation, or over periods
of 0.05 generations in the region of abrupt recovery near
1 generation. Values for the repair proficient strains
exposed to ^{125}I decay are averages of data for both strains.

The sharp increase in survival and complete recovery of resistance to BU photolysis at the end of the first generation (Fig. 1) is consistent with the survival requirement of replication of the labeled duplexes to produce unlabeled homologous duplexes. On the basis of this requirement, we predicted and observed that survival to BU photolysis would decrease and recover again during the second generation after labeling[1].

The sharp increase in cell survival to BU photolysis also provided a measurement of the variability of the replication fork interval[1]. The coefficient of variation of this interval was observed to be 9.3%. This value is approximately half as large as the coefficient of variation for the cell inter-division period, showing that DNA replication is much more tightly controlled than cell division in these bacteria.

Cell survival also increased abrutply at the first generation after exposure to ^{125}I decay in both repair proficient strains B/r K and B/r TT (Fig. 1) but in contrast to the results with 5BU, only half of the cells recovered. Moreover, there was no abrupt increase at the end of the first generation in the recA strain, and recovery occurred only gradually thereafter, consistent with survival of totally unlabeled cells. These different responses show that cell survival following ^{125}I decay at the first post-labeling generation requires the recA gene function as well as the presence of unlabeled homologous duplexes.

The failure of 50% of the cells to recover from ^{125}I decay at the end of the first round of DNA replication shows that repair is controlled by yet another factor. It is unlikely that this value of 50% simply represents repair of half the induced lesions because the same value was observed at more than one fluence. Furthermore, while the requirement for the recA function suggests that some form of recombination is required for repair, it is also very unlikely that the value of 50% recovery can be explained on the basis of random recombination between labeled and unlabeled homologous duplexes. Grivell et al.[3] showed that random recombination in E. coli occurs at frequencies much too low (0.01-0.1%) over stretches far too short (<30 nucleotides) to account for the required 50% repair of the very long labeled stretches (~10[5] nucleotides) produced in our experiments.

We have proposed that the observed recovery of 50% is due to asymmetric strand repair[2]. Because of the symmetry of labeling of daughter duplexes after the first round of DNA replication, the mechanism of asymmetric strand repair is not evident. It may be, however, that there is asymmetry of DNA strand attachment to the cell membrane and that this is the source of asymmetric strand repair.

Very recently we have obtained preliminary evidence support-
ting the hypothesis of asymmetric strand repair (H.E. Kubitschek
and D.M. Williams, unpublished results). If asymmetric strand
repair has high efficiency, we would expect to observe that the
dose response of cells after the first generation of growth following labeling would not decrease below 50%. Initial evidence has
been obtained to support that prediction.

REFERENCES

1. C.N. Newman, and H.E. Kubitschek, J. Mol. Biol. 121:461 (1978).
2. C.N. Newman, and H.E. Kubitschek, Submitted to J. Mol. Biol.
3. A.R. Grivell, M.B. Grivell, and P.L. Hanawalt, J. Mol. Biol.
       98:219 (1975)

^{125}I DECAY IN MICROORGANISMS: A PROKARYOTIC MODEL SYSTEM FOR EXAMINING THE ROLE OF UNREPAIRED DOUBLE-STRAND DNA BREAKS IN CELL KILLING

Robert E. Krisch

Department of Radiation Therapy, School of
Medicine, University of Pennsylvania,
3400 Spruce St., Philadelphia, Pa. 19104

ABSTRACT
    Iodine-125, when incorporated into the DNA of micro-
organisms as iododeoxyuridine, has been demonstrated to
induce double-strand DNA breaks (DSBs) with approximately
100% efficiency when it undergoes radioactive decay. In
the biological dose range approximately 2/3 of these
breaks are repaired in a wild type ($\underline{rec}^+$) strain of $\underline{E.}$
$\underline{coli}$ K12, but none are repaired in the corresponding recombination deficient ($\underline{rec}$A) strain. Each unrepaired DSB
appears to produce a lethal event in either strain. Survival experiments imply that maximum repair of DSBs in
$\underline{rec}^+$ cells can occur in the presence of only a single
genome per cell and therefore does not require a recombinational event. However, experiments with ^{125}I-labelled
small intracellular closed circular DNA mulecules (superinfecting phage λ, MW=33x10^{6}d) demonstrate no repair of
decay-induced DSBs, suggesting that the superstructure
of the large cellular chromosome plays an important role
in the repair of DSBs, even in the absence of recombination.

INTRODUCTION
    Unrepaired double-strand DNA breaks (DSBs) are
widely believed to constitute a major and perhaps the
dominant lethal lesion from ionizing radiation (e.g. see
Hutchinson, 1978, and following papers in the same volume
for a review of recent data). However, these lesions have
been difficult to study and particularly difficult to

correlate with lethal or other biological effects, for a
number of reasons. Among the most important of these are:
1) the much larger numbers of single-strand DNA breaks
(SSBs) and non-breakage DNA lesions induced simultaneously
with DSBs by most forms of ionizing radiation; 2) the difficulty of measuring DSBs and their repair in intact cellular chromosomes using prevailing neutral gradient sedimentation techniques and doses in the range of biological interest, which result in DNA fragments of about 10^9
daltons for prokaryotes and even larger for eukaryotes.

In the present studies the decay of ^{125}I incorporated
into the DNA of viable microorganisms as 5-iododeoxyuridine
(IdUrd) has been used as a tool to study DSBs and their
biological repair in ways that minimize these problems.

IdUrd is an excellent analogue for thymidine (dThd),
and is incorporated as such by thymine-requiring strains
of E. coli and by various coliphages. ^{125}I decays by electron capture, with most decays accompanied by two successive vacancy cascades with large numbers of electrons ex-
pelled, characteristic x-rays, a large residual charge on
the daughter tellurium atom, and a possibly significant
nuclear recoil energy, all combining to make up the so-
called Auger effect. Carlson and White (1963) showed that
this decay mode causes extensive fragmentation of at least
99% of the affected molecules when ^{125}I decay occurs in
a small molecule, such as ethyl iodide, in the gaseous
phase. They ascribed this effect, which they described
as a "molecular explosion", to the Auger effect. Their
findings suggested to myself and to others that ^{125}I decay
in the genetic material might have drastic biological
effects and, in particular, that this decay mode might be
a highly efficient source of DSBs relatively uncontaminate
by other kinds of DNA lesions. The possibility of infecting unlabelled bacteria with ^{125}I-labelled temperate bacteriophage also raised the possibility of studying the
repair of DSBs in small intracellular DNA molecules and
avoiding the difficulties of sedimenting huge DNA molecule
in neutral gradients.

RESULTS AND DISCUSSION

In the experiments to be described, organisms with
uniform ^{125}I labelling of their DNA were generally suspended in a glycerol-containing medium that maintained
their viability and were stored at -196^0C during radioactive decay ($T\frac{1}{2}$=60 days). Dose-response curves were
obtained by periodically thawing aliquots and assaying
for the desired biological and molecular parameters.

In experiments with ^{125}I-labelled coliphage T4 (Krisch and Ley, 1974; Krisch and Sauri, 1975) lethality, SSBs and DSBs were all followed as a function of ^{125}I decay. We found that, within the limits of experimental uncertainty, each decay of a DNA-bound ^{125}I atom causes exactly one DSB and is a uniformly lethal event. The frequency of occurrence of DSBs is nearly independent of whether the DNA is stored in a compact configuration in an intact phage head or in an extended form after DNA extraction, suggesting that DSBs are caused directly by local effects of decay rather than by the electrons and x-rays emitted, since the radiation dose is very dependent on DNA configuration, as demonstrated by measurements of SSBs in the same experiments.

Experiments were carried out next with uninfected bacterial cells to investigate the biological repair of ^{125}I-induced DSBs (Krisch et al., 1976). In these experiments frozen ^{125}I-labelled cells were thawed, spheroplasted, then gently lysed on the surface of neutral sucrose gradient centrifuge tubes; and the extracted DNA was sedimented at low speeds to determine molecular weight profiles, from which numbers of DSBs were calculated. Repair of DSBs was examined by incubating thawed, labelled cells for an hour in growth medium at 37°C prior to the above procedure and comparing with corresponding unincubated cells. In these experiments, wild type (rec$^+$) and otherwise isogenic recombination-deficient (recA) strains of E. coli K12 were compared. These experiments, like the phage experiments already reviewed, revealed an approximate one-to-one correlation between the numbers of ^{125}I decays and DSBs in unincubated cells of either strain. However, incubation in growth medium of decay-exposed rec$^+$ cells with small numbers of DSBs per genome resulted in the disappearance of more than 70% of all DSBs, presumably representing biological repair. Incubation of corresponding recA cells resulted in no detectable repair of DSBs. Parallel assays of colony-forming ability indicated that approximately 1/3 of ^{125}I decays are lethal for rec$^+$ cells while nearly every decay is lethal for recA cells. Thus, for either repair-proficient or -deficient strains of E. coli there was an approximate one-to-one correlation between unrepaired DSBs and lethal events from ^{125}I decay.

In order to determine whether this recA dependent repair of DSBs by E. coli requires actual recombination between a damaged and undamaged copy of the bacterial genome, we compared the relative sensitivity to lethality from ^{125}I decay: 1) between rec$^+$ and recA stationary phase cells averaging just over one unreplicated genome per cell; and 2) between exponential phase cells containing more than three genomes per cell (Krisch and Darby 1977). These experiments demonstrated that rec[+] cells were three times as resistant to killing as were corresponding recA cells under either set of growth conditions. This strongly implies that actual recombination is not necessary for repair of DSBs, since stationary rec[+] cells remained radioresistant even though they did not contain the duplicated genomes necessary for recombination. These findings are consistent with those of Bridges (1971), who studied the x-ray survival of comparable mutant cell lines with varying DNA content governed by growth rate. Our findings appear to conflict with those of Krasin and Hutchinson (1977), who studied DSB repair directly, using the same neutral gradient techniques that we did, in slowl growing cells with fewer than two genomes per cell, and reported a big decrease in such repair compared with cells with more than two genomes per cell. In our own laborator it was very difficult to obtain reproducible results in such repair studies on stationary or very slowly growing cells (unpublished observation).

The simplest interpretation of our findings is that repair of DSBs in E. coli occurs by end-to-end rejoining of the broken DNA molecule, rather than by side-to-side recombination with another DNA molecule, and that the recA gene product is essential for this non-recombinational rejoining process.

Because of the technical problems already alluded to with neutral sedimentation studies of very large DNA molecules, our recent experiments have focussed on the induction and possible repair of DSBs in the small DNA molecule of the temperate bacteriophage lambda superinfecting E. coli cells (Boye and Krisch, 1980). These DNA molecules assume three distinct configurations under neutral isotonic conditions: Species I is a covalently closed circular supercoiled form, representing an unbroken DNA molecule; Species II is a circular molecule with one or more SSBs; Species III is a linear molecule containing one or more DSBs. The three species form three distinct peaks when sedimented at high speed, permitting high resolution studies of the induction and repair of SSBs and DSBs simultaneously. Such studies have failed to demonstrate repair of ^{125}I or x-ray induced DSBs in such small intracellular DNA molecules, even in wild type E. coli cells, under conditions where x-ray induced SSBs are readily repaired.

These results, together with those already reviewed for chromosomal DSBs, suggest that repair of chromosomal DSBs involves ligation of broken ends held in apposition by the tertiary structure of the bacterial chromosome and that DSBs in small intracellular plasmid DNA molecules are not repaired because of the lack of any such stabilizing tertiary structure.

ACKNOWLEDGEMENTS

This research was carried out at the Argonne National Laboratory, Chicago, and was supported by the U.S. Dept. of Energy.

REFERENCES

Boye, E. and Krisch, R.E. (1980), Int. J. Radiat.
        Biol. 37,119
Bridges, B.A. (1971), J. Bacteriol. 108, 1944
Carlson, T.A. and White, M.J. (1963), Chem. Phys.
        38,2930
Hutchinson, F. (1978), In: DNA Repair Mechanisms
        (P.C. Hanawalt et al., eds.), p.457
Krasin, F. and Hutchinson, F. (1977), J. Molec.
        Biol. 116,81.
Krisch, R.E. and Ley, R.D. (1974), Int. J. Radiat.
        Biol. 25,21.
Krisch, R.E. and Sauri, C.J. (1975), Int. J. Radiat.
        Biol. 27,553
Krisch, R.E., Krasin F., and Sauri, C.J. (1976) Int.
        J. Radiat. Biol. 29,37
Krisch R.E. and Darby, D.M. (1977), Int. J. Radiat.
        Biol 32,603

PROTEIN "X" SYNTHESIS IN MAMMALIAN CELLS

Udo Mallick, Hans J. Rahmsdorf, Helmut Ponta and
Peter Herrlich
Kernforschungszentrum Karlsruhe, Institut für Genetik
und Toxikologie, and Institut für Genetik der Univer-
sität Karlsruhe, P.O. Box 3640, D-7500 Karlsruhe 1

INTRODUCTION

The unimpeded progress of proliferation and replication
should be monitored precisely by each cell. In E. coli, such a
monitoring system has indeed been detected. Upon inhibition of
proliferation, it generates signals which cause the induction of
"protein X" (= recA product) (Inouye and Pardee, 1970; Gudas and
Pardee, 1976; McEntee, 1977). This protein then elicits the
appearance of "SOS functions" (for reviews see Witkin, 1976; and
Devoret, 1978). It seems that some of these SOS functions may
also exist in eukaryotic cells, e.g. Weigle reactivation
(Sarasin and Hanawalt, 1978; Lytle and Goddard, 1979) and pro-
virus induction (Links et al., 1977). We examine here whether
mammalian cells respond to an inhibition of replication by induction of new proteins[*].

ARRESTING REPLICATION CAUSES CHANGES IN GENE EXPRESSION IN MURINE AND HUMAN CELLS

We chose to inhibit the proliferation of eukaryotic cells by
inhibiting replication, and to screen for induced gene products
by pulse-labeling with ^{35}S-methionine and resolving the proteins
by two-dimensional (2-D) polyacrylamide gel electrophoresis
(O'Farrell, 1975; O'Farrell et al., 1977). We examined various
cultured cells of murine and human origin: lymphoid cell lines,
lymphoid hybridoma cell lines, peripheral lymphocytes, primary
skin fibroblasts.

Arresting replication indeed caused the appearance and disappearance of protein spots in the autoradiograms indicating distinct changes in gene expression (Figs. 1,2 and 3). In murine cells, the induction of a 35 k dalton protein (XM2) was detectable in one-dimensional gels (Fig.1). The 2-D technique revealed an additional induced protein (XM1) of 40 k dalton in the basic pH range (Fig. 2). An even more basic protein of 18 k dalton was totally repressed at the same time (Fig. 3). XM1 and the 18 k protein labeled almost as strongly as actin did. In human cells, a similar response was observed: two proteins were induced, one was repressed. Their approximate sizes were 70 k, 68 k and 18 k.

To show that the appearance and disappearance of radioactive spots really indicated changes in synthesis, pulse-chase experiments were performed. The prelabeled proteins did not change their electrophoretic behavior after inhibiting replication. The

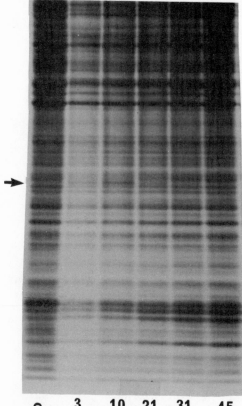

3   10 21 31   45

hrs. after UV

control

Fig.1. Induction of protein XM2. The murine lymphoid hybridoma cell line HPC-52 was irradiated with UV (17 J/m^2) and pulse labeled with 50 μCi/ml ^{35}S-methionine in MEM medium (3 % FCS) for 2 hours. Pulse midpoints are indicated. Sample preparation, electrophoresis and autoradiograms were as described (Laemmli, 1970).

observed results thus relate to changes in the rate of protein
synthesis, not to posttranslational modification.

Both, induction and repression, occurred no matter whether
replication was inhibited by irradiating with low doses of
ultraviolet light (UV), by ionizing radiation or by treatment
with hydroxyurea. Although the physical agents also cause DNA
damage, mutagenesis and inhibition of transcription, the effect
on replication is the only mechanism in common with hydroxyurea.
It is, therefore, likely that signals produced by an arrested
replication fork, influence gene expression.

FUNCTIONAL ASPECTS

To order the observed gene products into a possible functional context, we determined their intracellular fate after synthesis. After a brief chase period to permit migration from the
site of synthesis, XM1 and the 18 k protein were located in the
nucleus. The location of XM2 has not yet been defined. The cytoplasmic fraction did not contain these or other induced or repressed gene products. One induced and secreted protein of about
30 k dalton was, however, found in the culture medium. Its
possible identity with the plasminogen activator (Miskin and
Reich, 1980) needs confirmation.

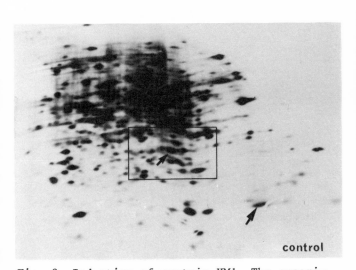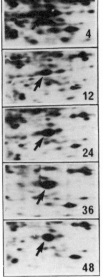

Fig. 2. Induction of protein XM1. The experimental protocol was as in Fig. 1, except that the basic as well as the
acidic proteins were resolved by 2-D gel electrophoresis
(O'Farrell et al., 1977). Electrophoresis in the first D
was 2000 V hr. The autoradiograms are oriented with the
basic proteins to the right. Arrows: XM1 and 18 k protein.

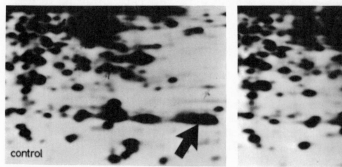

Fig. 3. Repression of synthesis of the 18 k protein. The cells
were pulse-labeled from 2 to 4 hours after UV. The area
shown corresponds to the lower right hand corner of 2-D
gels (see also Fig.2).

From comigration experiments and from the fact that all proteins were strongly labeled with ^{35}S-methionine, we consider it
very unlikely that one of the obwerved proteins was a histone.
Histone synthesis is also correlated with DNA synthesis (Nadeau
et al., 1978), but the repression of the 18 k protein was more
dramatic than the reduction of histone synthesis in our experiments.

The location in the nucleus suggests that the proteins may
act on DNA. Speculations as to a functional similarity of XM1 or
XM2 with the E. coli recA product seem premature but certainly
point at possible experimental approaches.

KINETICS OF SYNTHESIS

The synthesis of the 18 k protein decreased to virtually
zero with a half time of less than one hour. This speaks for
either a control on the level of translation, or for a rapid
mRNA turnover after stop of further transcription. Also the inductions of synthesis were immediate responses to the arrest of
replication. XM1 and XM2 were enhanced after brief lag periods
(Fig. 4). Maximum rates were reached at 10 (XM2), respectively
25 hours (XM1) after irradiation. Thereafter, the rate of synthesis returned to lower levels, approximately in parallel to the
reappearance of the 18 k protein.

The dose response curves using ultraviolet light (Fig. 5)
resemble those obtained for the synthesis of recA product in
E. coli (Fig. 4 of ref. Little and Hanawalt, 1977) and for the
secretion of the plasminogen activator (Fig. 1 of ref. Miskin

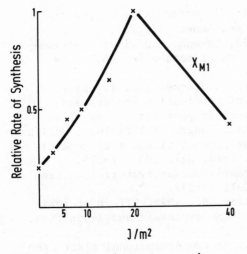

Fig. 4. Kinetics of protein
        induction.

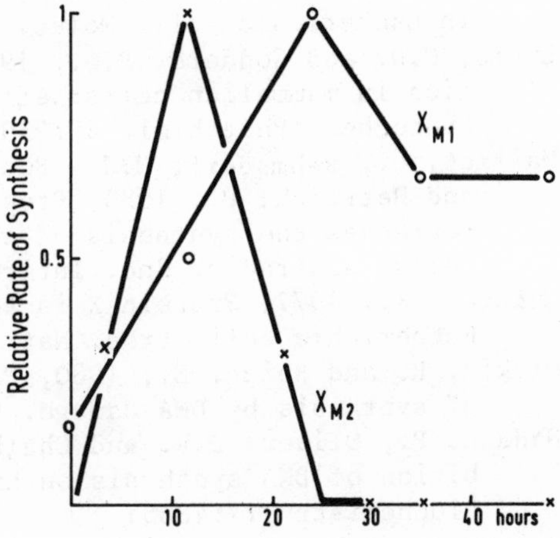

Fig. 5. UV-dose dependence of
        protein induction.

and Reich, 1980). There was a gradual increase of rate up to 20 J/m^2 yielding a "quasi-sigmoidal" curve. We interpret this to indicate the existance of soluble intermediates between the site of pyrimidine dimerization and the activation of XM1 synthesis.

In conclusion, the statistically distributed and arrested replication forks apparently cause, in the majority of cells , a homogeneous biochemical response: both human and murine cells induce at least two proteins and repress the synthesis of one protein. The nuclear location of two of these proteins suggests an enzymatic role on nucleic acid.

REFERENCES

Devoret, R., 1978, Inducible error-prone repair: one of the cellular responses to DNA damage. Biochimie 60:1135.
Gudas, L.J. and Pardee, A.B., 1976, DNA synthesis inhibition and
     the induction of protein X in Escherichia coli. J. Mol. Biol.
     101:459.
Inouye, M. and Pardee, A.B., 1970, Changes of membrane proteins
     and their relation to DNA synthesis and cell division of
     E. coli. J. Biol. Chem. 245:5813.
Laemmli, U.K., 1970, Cleavage of structural proteins during assembly of the head of bacteriophage T4. Nature 227:680.

Little, J.W. and Hanawalt, P.C., 1977, Introduction of protein X
    in Escherichia coli. Molec. gen. Genet. 150:237.

Lytle, C.D. and Goddard, J.G., 1979, UV-enhanced virus reactiva-
    tion in mammalian cells: effects of metabolic inhibitors.
    Photochem. Photobiol. 29:959.

Mallick, U., Rahmsdorf, H.J., Ponta, H. Coquerelle, T., Eife, R.
    and Herrlich, P., 1980, Stop of replication induces and
    represses the synthesis of specific proteins in mammalian
    cells. Abstracts, 2nd. Internat. Congr. Cell Biology, Berlin.

McEntee, K., 1977, Protein X is the product of the recA gene of
    Escherichia coli. Proc. Natl. Acad. Sci. USA 74:5275.

Miskin, R. and Reich, E., 1980, Plasminogen activator: induction
    of synthesis by DNA damage. Cell 19:217.

Nadeau, P., Oliver, D.R. and Chalkley, R., 1978, Effect of inhi-
    bition of DNA synthesis on histone synthesis and deposition.
    Biochemistry 17:4885.

O'Farrell, P.H., 1975, High resolution two-dimensional electropho-
    resis of proteins. J. Biol. Chem. 250:4007.

O'Farrell, P.Z., Goodman, H.M. and O'Farrell, P.H., 1977, High
    resolution two-dimensional electrophoresis of basic as well
    as acidic proteins. Cell 12:1133.

Sarasin, A.R. and Hanawalt, P.C., 1978, Carcinogens enhance sur-
    vival of UV-irradiated simian virus 40 in treated monkey
    kidney cells: induction of a recovery pathway? Proc. Natl.
    Acad. Sci. USA 75:346.

Witkin, E.M., 1976, Ultraviolet mutagenesis and inducible DNA
    repair in Escherichia coli. Bact. Rev. 40:869.

ADAPTION TO ALKYLATING AGENTS

DNA METHYL TRANSFERASE ACTING ON

O^6-METHYLGUANINE RESIDUES IN ADAPTED E. coli

Tomas Lindahl

Department of Medical Biochemistry
Gothenburg University
400 33 Gothenburg, Sweden

INTRODUCTION

Highly mutagenic alkylating agents such as N-methyl-N-nitroso-urea (MNUA) or N-methyl-N'-nitro-N-nitrosoguanidine (MNNG) produce three major purine adducts in DNA: 7-methylguanine, 3-methyl-adenine, and O^6-methylguanine (Lawley and Thatcher, 1970). These lesions are handled in different ways by living cells. The most abundant alkylation product, 7-methylguanine, is relatively harm-less in that it does not cause an appreciable amount of miscoding, and in E. coli it does not seem to be actively removed by any repair pathway (Lawley and Orr, 1970). In contrast, 3-methyl-adenine is rapidly removed from DNA by an excision-repair process which is initiated by a specific DNA glycosylase that does not act on either 7-methylguanine or O^6-methylguanine residues (Riazuddin and Lindahl, 1978; Karran et al., 1980). The fate of O^6-methyl-guanine in DNA has been less clear. While indications of active removal by a DNA repair process were obtained ten years ago by Lawley and Orr (1970), a DNA glycosylase catalyzing the release of such residues has not been found and may not exist. Further, complete removal of the relatively large quantities of O^6-methyl-guanine present in E. coli cells that have been treated with a high dose of MNNG is conspicuously less effective than the simultaneous liberation of 3-methyladenine from DNA.

Replication of a DNA template strand containing O^6-methyl-guanine residues is accompanied by a high frequency of miscoding, since O^6-methylguanine may direct the incorporation of either cytosine or thymine residues into the daughter strand (Mehta and Ludlum, 1978; Abbott and Saffhill, 1979). The difference between strong mutagens, such as MNNG and MNUA, and weak mutagens, such as dimethyl sulfate or methyl methanesulfonate, is largely due to the greater ability of the former agents to produce O^6-methylguanine residues in DNA. Moreover, the relative persistence of O^6-alkylated guanine residues in DNA has been implicated as a major contributing factor in tumor formation in animals exposed to methylating or ethylating agents (Goth and Rajewsky, 1974). For these reasons, we have been interested in investigating the cellular repair pathway(s) for removal of O^6-methylguanine from DNA.

THE ADAPTIVE RESPONSE

Our present understanding of the fate in E. coli cells of O^6-methylguanine residues in DNA is based on the discovery by Cairns and his associates of a specific, inducible repair process, tentatively termed "the adaptive response" (Samson and Cairns, 1977; Jeggo et al., 1977 and 1978; Schendel et al., 1978; Schendel and Robins, 1978; Robins and Cairns, 1979; Cairns, 1980). They showed that cells treated or "adapted" with a low dose of an alkylating agent such as MNNG acquire an increased ability to resist the mutagenic and lethal effects of a higher dose of the same or similar agents. This induction process required protein synthesis and was inhibited by chloramphenicol treatment. The adaptive response could be induced in E. coli lexA and recA mutants as well as in wild type cells, so this pathway was clearly different from the inducible, $recA^+$-dependent, error-prone form of repair termed SOS repair (Radman, 1974). In investigations of the fates of the major DNA purine adducts in adapted cells, it was observed that while 7-methylguanine remained in DNA and 3-methyladenine disappeared efficiently, both in adapted and unadapted cells, the former cells had acquired an increased ability to remove O^6-methylguanine (Schendel and Robins, 1978). It is not known if adapted cells also have an increased capacity to deal with certain minor alkylation lesions, such as O-alkylated pyrimidines.

The repair mechanism dealing with O^6-methylguanine shows some unusual features. Thus, adapted cells differ from unadapted cells by having the capacity to remove up to about 3000 O^6-methylguanine residues per cell very rapidly after challenge with a high dose of MNNG. Any remaining O^6-methylguanine is removed slowly and inefficiently, as in unadapted cells. From these and similar observations, it could be concluded that each molecule induced to remove O^6-methylguanine only acts once (Robins and Cairns, 1979).

DISAPPEARANCE OF O^6-METHYLGUANINE FROM ALKYLATED DNA INCUBATED WITH CELL-FREE EXTRACTS OF ADAPTED E. coli

In order to clarify the biochemical details of the adaptive

response, we have searched for a suitable in vitro system. It was observed (Karran et al., 1979) that the O^6-methylguanine in alkylated DNA was lost when the latter was incubated with cell-free extracts from adapted E. coli. On the other hand, DNA exposed to extracts from unadapted cells retained its O^6-methylguanine content (Fig. 1). The activity present in extracts from adapted cells did not show any requirement for Mg^{2+} or other cofactors, an observation which has subsequently been confirmed with chromato-graphically purified enzyme fractions. Thus, the reaction proceeds efficiently in EDTA-containing reaction mixtures, in which inter-ference with the assay by non-specific, Mg^{2+}-dependent, nucleases present in cell extracts is prevented. A striking feature of the reaction is that although no radioactive material is released in an ethanol-soluble or acid-soluble form from DNA that contains O^6-methylguanine residues that have been labeled in the methyl group with either ^3H or ^{14}C, O^6-methylguanine can no longer be found in acid hydrolysates of the DNA. This shows that the adaptive

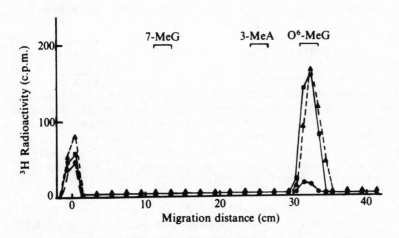

Fig. 1.   Disappearance of O^6-methylguanine from alkylated DNA after incubation with a cell extract from adapted E. coli (Karran et al., 1979). The DNA substrate was treated with ^3H-MNUA, incubated at 80° and pH 7.4 to release 7-methylguanine and 3-methyladenine by hydrolysis, and dialysed before use. This DNA was incubated without cell extract (▲), with cell extract of unadapted E. coli (■), or with extract of adapted E. coli (●) under the conditions of Karran et al. (1979). The O^6-methylguanine content of the DNA was subsequently analyzed by acid hydrolysis followed by paper chromatography. The radioactive material remaining at the origin contains phosphotriesters and O-alkylated pyrimidine nucleotides. (Reproduced with permission of MacMillan Journals, Ltd.).

response is neither due to the induction of a DNA glycosylase, which would liberate free O^6-methylguanine, or to a DNA demethylase, which would release free methanol, or to an endonuclease, which would catalyze the cleavage of a phosphodiester bond adjacent to an O^6-methyl-dGMP residue. Instead, an alteration process has apparently occurred whereby the methyl group remains covalently attached to a macromolecule, but it is no longer bound at the O^6-position of a guanine residue.

The finding that only extracts from adapted cells had a detectable ability to remove O^6-methylguanine residues from alkylated DNA indicated that the in vitro system could be used as an assay for the repair activity induced during adaptation. Confirmatory evidence has come from studies with cell extracts of various E. coli mutant strains that are defective in the adaptive response pathway. Two classes of mutants have been isolated. One type of mutant, ada, which maps at about 50 min on the E. coli K-12 map, has lost the ability to respond by adaptation to a low dose of alkylating agent. Several such mutants have been isolated from the E. coli K-12 strain AB1157 (Jeggo, 1979). The ada mutations have properties consistent with their being located in the structural gene for the enzyme that serves to remove O^6-methylguanine. Another type of mutant, isolated from E. coli B cells by B. Sedgwick, expresses the adaptive response in a constitutive fashion. When growing in normal media, such cells exhibit the enhanced ability to remove O^6-methylguanine normally found only in E. coli that have been induced by exposure to an alkylating agent. The results obtained with cell-free extracts from such mutants (Table 1) are in agreement with the in vivo studies. Thus, extracts from two ada mutants, exposed to a standard adaptation procedure prior to harvesting of the cells, failed to remove a detectable quantity of O^6-methylguanine from DNA, while the parent ada^+ strain was able to do so after adaptation. In contrast, extracts from the constitutive strains E. coli BS11, BS21, BS31, and BS41, grown in a peptone-yeast extract medium containing the required growth factors thymine and histidine but no alkylating agent, efficiently catalyzed the disappearance of O^6-methylguanine from DNA. In fact, the level of activity in extracts of these strains was 3- to 4-fold higher than that seen in other E. coli B strains after adaptation by the standard procedure. These data demonstrate that our in vitro assay measures the adaptive response, as defined by in vivo experiments. It should also be emphasized that the availability of strains that express the enzyme activity in a constitutive fashion greatly simplifies biochemical work. While it has been cumbersome to collect large quantities of adapted cells, the constitutive strains may readily be grown at pilot plant scale. We have obtained 5 kg E. coli BS21 from an 800 liter culture in this fashion. These useful strains are described in the companion article by Sedgwick and Robins (1981).

Table 1. Cell extracts from E. coli strains with mutations affecting the adaptive response pathway have different abilities for removing O^6-methylguanine from DNA.

Experiment a. E. coli K-12 strains. Thirty µg protein were added in each standard reaction mixture.

Source of extract	$[methyl-^3H]O^6$-methylguanine remaining in DNA (cpm)
None	810
E. coli AB1157, ada$^+$, unadapted	748
_"_ , adapted	102
E. coli ada-5, adapted	734
E. coli ada-6, adapted	728

Experiment b. E. coli B strains.

Source of extract	Protein added (µg)	O^6-methylguanine remaining in DNA (cpm)
None	–	928
E. coli B/r, unadapted	20	915
_"_ , adapted	20	170
_"_ , _"_	5	502
_"_ , _"_	1	894
E. coli BS11, unadapted	20	182
_"_	5	169
_"_	1	732
E. coli BS21, unadapted	20	180
_"_	5	198
_"_	1	410
E. coli BS31, unadapted	20	179
_"_	5	235
_"_	1	652
E. coli BS41, unadapted	20	171
_"_	5	197
_"_	1	564

Table 1.   (continued)
Bacterial cultures were treated by the two step procedure
of Schendel et al. (1978) for adaptation, employing
0.075 µg/ml and 0.75 µg/ml MNNG, respectively. No growth
inhibition occurred. Cell extracts were prepared and
assayed for their ability to transfer the methyl group
from O^6-methylguanine in DNA as in Fig. 1.

TRANSFER OF METHYL GROUP TO AN ACCEPTOR PROTEIN

A methyl group, enzymatically transferred from the O^6-position
of guanine, remains bound to a macromolecule in a form that is
precipitable with cold dilute acid or with ethanol. In preliminary
experiments with crude extracts of adapted cells and a radioactively
labeled substrate, it seemed that the methyl groups remained in
DNA (Karran et al., 1979). However, experiments with partly
purified enzyme fractions of greater activity have now shown that
the radioactive methyl groups are transferred to an acceptor
molecule that is susceptible to digestion with proteolytic enzymes
(Table 2). After incubation of the alkylated DNA substrate with
a 30-fold purified enzyme fraction from E. coli BS21, the radioactive methyl groups could be released in an acid-soluble form by
treatment with either pronase or proteinase K.

The results in Table 2 suggest that the enzyme activity induced in adapted cells serves to transfer a methyl group from the
O^6-position of a DNA guanine residue to an amino acid side chain
in an acceptor protein. Four different amino acids have previously
been found to occur in methylated form in various enzymatically
methylated proteins: lysine, arginine, histidine, and glutamic
acid (Paik and Kim, 1975; Springer et al., 1979). The three former
are N-methylated, while glutamic acid is O-methylated. In order to
release and identify the residue that contains the methyl group
which is transferred from O^6-methylguanine in DNA, the partly
purified E. coli protein carrying the methyl group was digested
with proteinase K (see Table 2) followed by digestion with amino-
peptidase M. Analysis of this hydrolysate by paper chromatography
in two systems showed that the radioactive moiety was different
from either ε-N-monomethyllysine, N^G-monomethylarginine, or 3-N-
methylhistidine. Its chromatographic properties were, on the
other hand, similar to those of the glutamic acid methyl ester.
For example, in a n-butanol/pyridine/water (1:1:1) system, the
unknown radioactive compound migrated as a distinct spot with the
same R_f value as the γ-glutamyl methyl ester. Further analysis of
the methylated residue on an automatic amino acid analyzer showed,
however, that the radioactive compound was eluted as a single

Table 2. Transfer of methyl group from O^6-methylguanine in DNA to an acceptor molecule that is sensitive to proteolytic enzymes.

Reagent Enzyme	No E. coli Enzyme			+ E. coli Methyl transferase	
	TCA Prec.	TCA Sup.	(cpm)	TCA Prec.	TCA Sup.
None	147	19		116	23
Proteinase K	135	23		36	144
Pronase	101	20		23	149

The methyl transferase activity was purified 30-fold by phosphocellulose chromatography from extracts of E. coli BS21. Reaction mixtures (Karran et al., 1979) containing DNA with O^6-methylguanine residues radioactively labeled in the methyl group and 1 µg protein each were incubated at 37° for 10 min and then supplemented with either proteinase K (0.1 mg/ml) or pronase (0.5 mg/ml). After continuing the incubation for an additional 30 min, the reaction mixtures were chilled, precipitated with an equal volume of cold 1% trichloroacetic acid, and centrifuged. The radioactivities of the precipitates and the supernatant solutions were determined.

symmetrical peak slightly but significantly earlier than the glutamic acid methyl ester. Again, it was clearly separated from the methylated basic amino acids. Further characterization of this methylated moiety is in progress.

THE PROTEIN IS CONSUMED IN THE REACTION

It was shown by Robins and Cairns (1979) that the enzyme activity present in adapted cells acts only once on its substrate, and consequently appears to be expended in the reaction. We have recently obtained similar results in vitro (Fig. 2). When a limiting amount of active E. coli cell extract was added to the standard reaction mixture (sufficient to transfer about 65% of the methyl groups from O^6-methylguanine after 10 min at 37°), for a typical enzyme reaction it would have been expected that the process would proceed at an even rate during this time interval. Instead, the reaction kinetics clearly demonstrate that the methyl group transfer occurs within the first minute, and no further

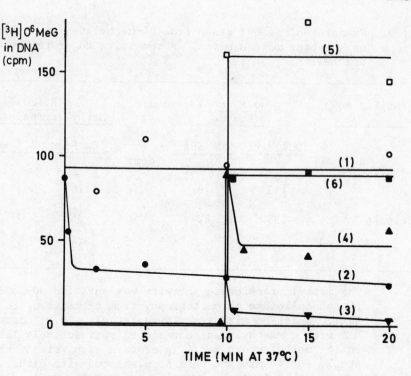

Fig. 2. Loss of methyl transferase activity during the reaction
with alkylated DNA. Standard reaction mixtures (Karran
et al., 1979) were incubated with 1 µg each of a cell
extract from E. coli BS21. The reaction was stopped at
different times and the DNA analyzed for its O^6-methyl-
guanine content as in Fig. 1.

(1) No E. coli protein added (o).

(2) Complete reaction mixture (●).

(3) The complete reaction mixture supplemented with an
     additional 1 µg aliquot of protein from E. coli BS21
     after 10 min (▼).

(4) A 10 min incubation of a reaction mixture containing
     E. coli BS21 protein but no DNA, and DNA subsequently
     added (▲).

(5) Reaction mixture with no E. coli protein and an
     additional amount of DNA (60% of initial input)
     added after 10 min (◻).

(6) Complete reaction mixture, additional DNA
     (60% of initial input) added after 10 min (◼).

significant transfer takes place during the following 9 minutes.
Since addition of more <u>E. coli</u> protein after 10 min results in
rapid removal of the methyl groups from the remaining O^6-methyl-
guanine residues, these residues were not present in a state protected from the action of the enzyme. Moreover, the rapid
cessation of methyl group transfer does not seem to be due to heat
inactivation of the enzyme in the reaction mixture. Thus, when
the enzyme was first incubated for 10 min in a reaction mixture
without DNA, and the latter subsequently added, methyl group
transfer occurred as efficiently as when DNA and protein were
present together at the start of the incubation period. These
data strongly indicate that the methylated protein acceptor
molecule cannot be regenerated, at least under our assay conditions,
and that it is expended in the reaction.

The results obtained do not distinguish between the
possibilities of a methyl transferase activity that moves the
methyl group to a separate acceptor protein   versus the
methylation and accompanying inactivation of the transferase
itself. However, our results from enzyme purification support
the latter alternative. No separation of the transfer activity
from the acceptor activity has been observed by chromatography
on either phosphocellulose, hydroxyapatite, heparin-agarose, or
DNA-cellulose. It would thus appear that the enzyme inactivates
itself during the process of transferring the methyl group from
O^6-methylguanine. The activity associated with this apparently
wasteful reaction does not quite correspond to that associated
with an enzyme, the latter being formally defined as an agent
catalyzing a particular reaction without being itself consumed
in the process. This suicide inactivation may be represented,
on the other hand, as a special case of irreversible substrate
inhibition of an enzyme.

DISCUSSION

The development of an <u>in vitro</u> assay for the adaptive
response (Karran et al., 1979) and the isolation of <u>E. coli</u>
mutant strains which express this function in a constitutive
fashion (Sedgwick and Robins, 1981) makes it possible to predict
that the adaptive response to the mutagenic effect of alkylating
agents will soon be well understood in molecular terms, at least
with respect to the reaction mechanism of the disappearance of
O^6-methylguanine from DNA. On the other hand, the details of
the induction process are presently unclear, and the nature of
the intracellular inducer is unknown. It seems plausible that
a methylating agent which can cause adaptation reacts first with
a molecule in the cell, whose alkylated form then serves to induce
the response. This hypothetical methylated inducer could be an
excisable residue in DNA, or an apparently unrelated compound,

such as an RNA molecule. Alternatively, the presence of methylated inactivated transferase molecules which result from repair of O^6-methylguanine residues could serve, in an auto-regulatory fashion, as a signal to trigger the synthesis of an unmethylated active form of the protein.

A somewhat puzzling aspect of the adaptive response is the difference observed between adaptation to killing vs. mutagenesis caused by alkylating agents. In contrast to adaptation against mutagenesis, the response against killing requires a functional $polA^+$ gene product, and the induction kinetics are different (Jeggo et al., 1977 and 1978). One possible explanation would be that two different enzymes are induced during the adaptive response, one being the methyl transferase for O^6-methylguanine residues described here, and another an enzyme that serves to initiate excision-repair of a different lethal (but not highly mutagenic) alkylation lesion.

An inducible repair mechanism to remove O^6-methylguanine from alkylated DNA by methyl group transfer may also occur in rat liver, according to the in vitro experiments with liver extracts performed by Pegg (Pegg, 1978; Montesano et al., 1980). It remains to be seen if any alkylation product other than O^6-alkylguanine is handled in an analogous way in either E. coli or mammalian cells. In this context, in vivo experiments (Warren and Lawley, 1980) suggest that the corresponding ethylated derivative, O^6-ethylguanine, apparently can be repaired in E. coli both by the pathway described here and by an excision-repair process initiated by the uvr^+ endonuclease function. In contrast, O^6-methylguanine is not removed to any detectable extent by uvr^+-dependent excision-repair (Altamirano-Dimas et al., 1979; Warren and Lawley, 1980). The ability of E. coli cells to cope with this highly mutagenic lesion thus seems to depend entirely on a repair mechanism that is very different from any previously known form of DNA repair.

ACKNOWLEDGMENTS

I thank Barbara Sedgwick and Penelope Jeggo for bacterial strains, John Cairns for helpful discussions, and Anita Jacobsson for excellent assistance. This work was supported by the Swedish Council for Planning and Coordination of Research and by the Swedish Medical Research Council.

REFERENCES

Abbott, P.J., and Saffhill, R. (1979). Biochim. Biophys. Acta 562, 51-61.
Altamirano-Dimas, M., Sklar, R., and Strauss, B. (1979). Mut. Res. 60, 197-206.

Cairns, J. (1980). Nature 286, 176-178.
Goth, R., and Rajewsky, M.F. (1974). Proc. Natl. Acad. Sci. U.S.
    71, 639-643.
Jeggo, P. (1979). J. Bact. 139, 783-791.
Jeggo, P., Defais, M., Samson, L., and Schendel, P. (1977).
    Mol. Gen. Genetics 157, 1-9.
Jeggo, P., Defais, M., Samson, L., and Schendel, P. (1978).
    Mol. Gen. Genetics 162, 299-305.
Karran, P., Lindahl, T., and Griffin, B.E. (1979).
    Nature 280, 76-77.
Karran, P., Lindahl, T., Ofsteng, I., Evensen, G.B., and
    Seeberg, E. (1980). J. Mol. Biol. 140, 101-127.
Lawley, P.D., and Orr, D.J. (1970). Chem. Biol. Interact.
    2, 154-157.
Lawley, P.D., and Thatcher, C.J. (1970). Biochem. J. 116, 693-707.
Mehta, J.R., and Ludlum, D.B. (1978). Biochim. Biophys. Acta
    521, 770-778.
Montesano, R., Brésil, H., Planche-Martel, G., Margison, G.P.,
    and Pegg, A.E. (1980). Cancer Res. 40, 452-458.
Paik, W.K., and Kim, S. (1975). Adv. Enzymol. 42, 227-286.
Pegg, A.E. (1978). Biochem. Biophys. Res. Commun. 84, 166-173.
Radman, M. (1974). In "Molecular Mechanisms for Repair of DNA
    Part A", ed. by P.C. Hanawalt and R.B. Setlow, pp. 355-367,
    Plenum Press, N.Y.
Riazuddin, S., and Lindahl, T. (1978). Biochemistry 17, 2110-2118.
Robins, P., and Cairns, J. (1979). Nature 279, 74-76.
Samson, L., and Cairns, J. (1977). Nature 267, 281-283.
Schendel, P.F., Defais, M., Jeggo, P., Samson, L., and Cairns, J.
    (1978). J. Bact. 135, 466-475.
Schendel, P.F., and Robins, P.E. (1978). Proc. Natl. Acad. Sci.
    U.S. 75, 6017-6020.
Sedgwick, B., and Robins, P. (1981). This volume.
Springer, M.S., Goy, M.F., and Adler, J. (1979).
    Nature 280, 279-284.
Warren, W., and Lawley, P.D. (1980). Carcinogenesis 1, 67-78.

ISOLATION OF MUTANTS OF <u>Escherichia coli</u> WITH INCREASED

RESISTANCE TO ALKYLATING AGENTS

Barbara Sedgwick and Peter Robins

Imperial Cancer Research Fund
Mill Hill Laboratories
Burtonhole Lane, London, NW7 1AD, England

ABSTRACT

A search was made for strains constitutive for the adaptive response by isolating mutants with increased resistance to killing by methylating agents. N-methyl-N'-nitro-N-nitrosoguanidine (MNNG) resistant strains were obtained, but these proved to be less readily methylated by MNNG because they had a low intracellular thiol content (thiols convert MNNG to the proximal mutagen, methylnitrosamine (5)). Four strains with increased resistance to N-methylnitrosourea (MNU) proved to be constitutive for the adaptive response. After exposure to (^{3}H) MNNG, they showed the same low ratio of O^6-methylguanine (O^6MeG) to 7-methylguanine (N^7MeG) as adapted wild type bacteria. When exposed to very high MNNG doses, they were even more resistant to killing and mutation than the adapted wild type. The constitutives may therefore have an even greater ability than the adapted wild type to repair O^6MeG.

INTRODUCTION

When <u>E.coli</u> is exposed to methylating agents the adaptive response is induced (6). This removes the mutagenic lesion, O^6MeG, from the cellular DNA (7). The induction of this DNA repair pathway results in an increased resistance to killing and mutation by several alkylating agents (2). In order to study the genetic control of this pathway it is necessary to isolate mutants which are deficient or constitutive for the response. Jeggo (3) has isolated <u>ada</u> mutants which cannot be adapted. We describe here

the isolation and characterisation of two types of mutants with
increased resistance to alkylating agents : mutants low in
cellular thiols and mutants constitutive for the adaptive response.

RESULTS AND DISCUSSION

In the search for strains constitutive for the adaptive
response, mutants with the following characteristics were sought
(a) increased resistance to killing by methylating agents, (b)
a low ratio of O^6MeG to N^7MeG in the cellular DNA after exposure
to (^3H) MNNG.

MNNG Resistant Mutants

Cultures were mutagenised with 20µg MNNG per ml for 10 mins,
and then exposed once per day for 5 days to 50µg MNNG per ml for
10 mins to enrich for the presence of resistant mutants. Single
colony isolates with increased resistance were obtained from
several separate cultures (Table 1). On exposure to (^3H) MNNG,
however, the ratio of O^6MeG to N^7MeG in the cellular DNA of the
isolates was the same as in the wild type. These mutants are,
therefore, not constitutive for the adaptive response. Overall

Table 1.   Characterization of MNNG Resistant Mutants

Strain	% survival after 50µg MNNG per ml.	a. methylation of cells by (^3H) MNNG (TCA insoluble cpm)	b. acid soluble thiols (µ moles per 10^{12} cells)
AB1157 (wild type)	5	760	11.4
" BS53	94	300	0.5
W3110 (wild type)	3	620	9.9
" BS55	87	230	0.2
" BS56	57	300	2.6
" BS57	60	320	0.4

a. 0.1 ml cultures (3 x 10^8 cells per ml) were exposed to 100µg
    MNNG per ml (19 µCi/mg) for 10 min. The TCA acid insoluble
    cpm were measured.
b. Acid soluble thiols were measured according to Lawley and
    Thatcher (5).

methylation of the mutants by MNNG was reduced, and this presumably accounts for their increased resistance. Thiols, such as glutathione and cysteine, are known to stimulate methylation by MNNG (5) A low cellular thiol content could therefore result in a reduced level of methylation. Glutathione deficient mutants have been isolated previously but their sensitivity to MNNG was not tested (1). All the MNNG resistant strains were, indeed, found to be low in cellular thiols (Table 1).

Methylation by MNU is not stimulated by thiols (8). Thus, in order to avoid the selection of thiol deficient strains, MNU resistant mutants were selected in the search for a strain constitutive for the adaptive response.

Mutants Constitutive for the Adaptive Response

Cultures of F26 mutagenised by MNNG were repeatedly exposed to 2mg MNU per ml to enrich for the presence of MNU resistant mutants. Resistant isolates were obtained from 4 separate cultures. On exposure to (^{3}H) MNNG the ratio of O^6MeG to N^7MeG in the cellular DNA of these isolates was very low, and was similar to the ratio in the adapted wild type (Table 2). These mutants may be constitutive for the adaptive response, and therefore able to remove O^6MeG from their DNA without previous induction of the response. An in vitro assay for the removal of O^6MeG from

Table 2.   Characterisation of MNU Resistant Mutants

Strain	% surivial after MNU (2mg/ml)	methylation of cellular DNA by 50µg (^{3}H) MNNG per ml for 10 min		Ratio of $\dfrac{O^6\text{MeG}}{N^7\text{MeG}}$ x100
		N^7MeG (cpm)	O^6MeG (cpm)	
F26 (wild type)	10	15800	2280	14.4
F26 (wild type) adapted		14680	110	0.8
F26 BS11	92	11170	10	0.09
" BS21	62	21230	110	0.5
" BS31	74	7460	30	0.4
" BS41	88	11650	120	1.0

*100ml cultures of $3x10^8$ cells per ml were concentrated 100 fold and exposed to 50µg MNNG per ml (0.76mCi/mg) for 10 min. Estimation of N^7MeG and O^6MeG was as described previously (7).

methylated DNA by extracts of adapted cells has been developed (4).
This activity is present constitutively in extracts of the MNUA
resistant mutants (Lindahl, pers. commun.).

Killing and Mutation Induction by MNNG

When challenged with very high doses of MNNG, the mutants
constitutive for the adaptive response were even more resistant
to killing and mutation than the adapted wild type (Fig. 1). At
MNNG doses greater than 50µg per ml for 5 min the adapted wild
type accumulated his$^+$ revertants, whereas the constitutives
developed few his$^+$ revertants even when challenged with 400µg per
ml (Fig. 1A). It seemed likely, therefore, that the constitutives
contain an even higher level of the activity which repairs O^6MeG
than adapted F26. This has now been demonstrated in vitro.
Extracts of strain BS21 contain a 4 fold greater activity for the
removal of O^6MeG from methylated DNA than extracts of the adapted
wild type (Lindahl, pers. commun.).

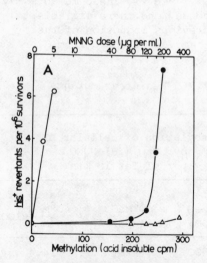 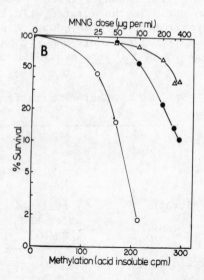

Fig. 1.   Mutation frequency (panel A) and survival (panel B) of
          F26 and the strains constitutive for the adaptive response
          after exposure to various MNNG doses for 10 min. Strains
          BS11, BS31 and BS41 gave the same results as BS21 in that
          few his$^+$ revertants were induced by any of the doses used.
          The effective dose of MNNG is presented as the amount of
          methylation of the cells (estimated by treating with
          (^{3}H) MNNG), because at high MNNG doses the amount of
          methylation was not linearly related to the MNNG
          concentration. (F26 -o-; F26 adapted -●-; BS21 -Δ-).

Spontaneous Mutation Frequency

Strain BS21 and F26, had the same spontaneous mutation frequency of 5×10^{-10} his$^+$ revertants per generation when grown in rich media. In these conditions, therefore, alkylation to produce O^6 alkylguanine does not appear to be contributing to the spontaneous mutation frequency.

Reversion to MNNG Sensitivity

Stock slopes of the constitutives BS21 and BS31 stored for several months at room temperature were found to contain MNNG sensitive revertants. These revertants were even more sensitive to the killing and mutagenic effects of MNNG than the wild type (data not shown). When a revertant of BS21 was exposed to 0.1µg MNNG per ml it accumulated mutations for at least 2 hours (data not shown). The wild type develops very few mutations under these conditions because the adaptive response is induced. The constitutives therefore appear to be reverting to strains which are unable to adapt. The isolation and characterisation of such mutants has been described previously (3). Presumably, under certain conditions the strains constitutive for the adaptive response have a lower growth rate or a lower viability than these revertants.

REFERENCES

1. Apontoweil, P. and Berends, W. (1975) Biochim. Biophys. Acta 399, 1.
2. Jeggo, P., Defais, M., Samson, L. and Schendel P. (1977) Molec. gen. Genet., 157, 1.
3. Jeggo, P. (1979) J. Bact. 139, 783.
4. Karran, P., Lindahl, T., and Griffin, B. (1979) Nature (lond.) 280, 76.
5. Lawley, P.D. and Thatcher, C.J. (1970) Biochem. J. 116, 693.
6. Samson, L. and Cairns, J. (1977) Nature (Lond.) 267, 281.
7. Schendel, P.F. and Robins, P. (1978) Proc. Natl. Acad. Sci., (USA) 75, 6017.
8. Wheeler, G.P. and Bowdon, B.J. (1972) Biochem. Pharmacol., 21, 265.

EFFECT OF RIFAMPICIN ON N-METHYL-N-NITROSOUREA-INDUCED

MUTATION FREQUENCY IN Escherichia coli

B.T. Bacon and S. Neale

Courtauld Institute of Biochemistry
The Middlesex Hospital Medical School
London, W1P 7PN, U.K.

SUMMARY

Rifampicin enhances by 4-6 fold the induced mutation frequency
of N-methyl-N-nitrosourea (MNUA) in E.coli WP2. The magnitude of
this increment is dependent on the time of addition of the antibiotic with respect to that of mutagen, and on the prior growth
rate of the cells before treatment. It is not dependent on the
$uvrA^+$ - or the $recA^+lexA^+$-dependent repair capacities of the cell,
under specified growth conditions, but may be modulated by prior
induction of the "adaptive response". Rifampicin inhibits recovery
of RNA and protein synthetic rates in post-incubated cells following MNUA treatment (Neale and Bacon, in press), and also retards,
in vivo, the loss of O^6-methyldeoxyguanosine (O^6-MeG) residues from
DNA of MNUA-treated E.coli. Chloramphenicol enhances the MNUA-
induced mutation frequency by only 3 fold, and its effect on
mutation is compared with that of rifampicin.

MATERIALS AND METHODS

Growth and treatment of cells

E.coli WP2 (trpE) and its repair deficient mutants, WP2uvrA,
WP285 (polA), CM571 (recA) and CM561 (lexA), were grown in 232
salts medium supplemented with glucose 0.5% and tryptophan
30 μgml^{-1} (60 min mean generation time (mgt)) or additionally
supplemented with 1% nutrient broth (30 min mgt). Cultures were
grown to an OD of 0.8, cell density, 2-5 x 10^8 cells ml^{-1}, and
harvested by centrifugation. Cell suspensions ($\approx$0.25-1.0 x 10^{10} cells

ml^{-1}) were treated with MNUA, in the presence and absence of rifampicin, for 20 min at 37°C, pH 7.0, Aliquots of cells were plated on suitably supplemented agar plates.

Unless otherwise stated, MNUA was used throughout at 2 mM, and rifampicin at 50 μgml^{-1}, and cultures were grown at an mgt of 30 min.

Stationary cell cultures were obtained by inoculating into 232 salts, glucose 0.5%, tryptophan 30 μgml^{-1} medium and allowing growth overnight to a cell density of 2 - 5 x 10^{9} cells ml^{-1}.

RESULTS

Effect of rifampicin on the MNUA-induced mutation frequency

When WP2 was grown at an mgt 30 min, and then incubated in the presence of both MNUA, 2 mM and rifampicin 50 μgml^{-1}, the induced mutation frequency was increased by 4-5 fold. This increase in the mutation frequency has been observed throughout the range of the MNUA concentration curve, 0.2 - 3.0 mM (unpublished results).

Prior treatment of WP2 cells with rifampicin, for 30 min at 37°C, before exposure to MNUA, increased the mutation frequency by 2-3 fold (Table 1). Similarly, when cells were treated with MNUA and subsequently resuspended in fresh growth medium for 30 min before exposure to rifampicin, the induced mutation frequency was increased by 2 - 2.5 fold (Table 1).

Table 1   Factors affecting the rifampicin-induced increment in the MNUA-induced mutation frequency

1. Time of addition of rifampicin:	Increment in mutation frequency
30 min before MNUA	x 2-3
during MNUA treatment	x 4-6
30 min after MNUA	x 2-2.5
2. Growth rates of cells before treatment with MNUA and rifampicin	
30 min mgt	x 4-6
60 min mgt	x 2-4
stationary	x 0.2-1.0

(Figures represent increments in the normal level of MNUA mutation frequency, calculated as 5540 $\pm$ 2761 M/10^{9} S.)

Table 1 also shows how the magnitude of the rifampicin-induced increment in the MNUA (2 mM) mutation frequency is dependent on the growth rate of the cells before treatment. The maximum increase, x 6, was obtained when the cells were grown at an mgt of 30 min. An increment of ≃ x 2-3 was observed when cells were grown at an mgt of 60 min, and only a slight increase or often no alteration in the mutation frequency at all, was observed when cells had attained stationary phase growth before treatment.

Influence of the repair capacity of the cell in the rifampicin effect on mutation

The effect of rifampicin on the MNUA-induced mutation frequency was studied in the repair deficient mutants of WP2 (Table 2). The absence of uvrA$^+$-dependent excision repair, or of recA$^+$lexA$^+$-dependent error prone repair, did not affect the magnitude of the rifampicin-induced increment in the mutation frequency. There was no rifampicin effect, however, in the polA mutant and in fact no mutants were found in cells which had been treated with both mutagen and antibiotic. This was principally due to the low survival of these cells (4-7% of control cells).

Table 2 Effect of rifampicin on the MNUA-induced mutation frequency in WP2 and its repair deficient mutants. (Cells grown at 30 min mgt.)

Strain	Mutation frequency M/10^9 S (mean ± SD)	Rifampicin-induced increment (mean ± SD)
WP2	5340 ± 2761	6.19 ± 1.814
WP2 uvrA	9640 ± 3541	5.75 ± 1.571
CM561 lexA	3297 ± 1908	6.36 ± 2.04
CM571 recA	2918 ± 2018	6.93 ± 2.31
WP85 polA	328 ± 88	0

Table 3 O^6-MeG/Pi ratios of DNA from MNUA- and MNUA plus
rifampicin-treated WP2 cells at various times of
postincubation

Strain	Time of post-incubation (min)	MNUA-treated cells	MNUA + rifampicin treated cells
WP2	0	0.95×10^{-4}	1.04×10^{-4}
	30	0.29×10^{-4}	0.74×10^{-4}
	60	0	0
WP2 Rif^R	0	1.08×10^{-4}	1.0×10^{-4}
	30	0.225×10^{-4}	0.21×10^{-4}

Effect of rifampicin on the loss of O^6-MeG residues from MNUA-treated DNA

O^6-MeG is the major mutagenic product produced by MNUA treatment of DNA[1-3] and the formation of this alkylated nucleoside and its loss from DNA was studied by means of a sensitive radioimmunoassay[5].

The level of O^6-MeG in WP2 DNA, quantified as the ratio O^6-MeG/Pi, was the same in DNA from MNUA-treated and MNUA plus rifampicin-treated cells (Table 3). However, the loss of this residue, as calculated at 30 min postincubation following mutagen and antibiotic treatment, was inhibited in the DNA from the MNUA plus rifampicin-treated cells. There was approximately twice as much O^6-MeG remaining in the DNA of these latter cells. In all cases, all measurable O^6-MeG residues had been lost from DNA by 60 min postincubation. In a rifampicin-resistant WP2 mutant, no difference was observed between the rates of loss of O^6-MeG from DNA or MNUA- and MNUA plus rifampicin-treated cell.

Effect of the inducible adaptation pathway on the rifampicin effect

The effect of rifampicin on the MNUA-induced mutation frequency was studied in WP2 cultures in which the adaptive response had been preinduced by growth (at an mgt of 60 min) for 90 min in N-methyl-N-nitro-N-nitrosoguanidine- (MNNG) supplemented medium (0.1 µgml^{-1}).

Table 4   Effect of the adaptive response on the survival and
mutation frequency of MNUA- and MNUA plus rifampicin-
treated WP2.

Unadapted WP2	Mutation frequency $(M/10^9)$		Survival (% of control)	
	Unadapted	Adapted	Unadapted	Adapted
Control (spontaneous)	4	12	100	100
MNUA	4,346	12	86	100
Rifampicin (spontaneous)	0	0	50.4	70.3
MNUA + rifampicin	11,976	0	71.8	68.0
Rifampicin-induced increment in mutation frequency	x 2.76	0		

There was not only no increment in the MNUA mutation frequency, which was itself reduced by a very large extent, but in fact no mutants were obtained at all from the MNUA plus rifampicin treated cells (Table 4).

Effect of chloramphenicol on the MNUA-induced mutation frequency

WP2 was grown and either treated with chloramphenicol 150 μgml^{-1}, 30 min before exposure to MNUA, or treated simultaneously with MNUA and chloramphenicol, 150 μgml^{-1}. The usual mutation conditions were used. Chloramphenicol increased the MNUA mutation frequency by approximately 3 fold, irrespective of the time of its addition to the cells (Table 5). When cells were pretreated with chloramphenicol and then plated with MNUA and rifampicin, the mutation frequency was only increased by 3 fold instead of the usual 4-6 fold.

Table 5 Effect of chloramphenicol on the mutation frequency of
MNUA- and MNUA plus rifampicin treated WP2.

Time of addition of chloramphenicol	Increment in MNUA mutation frequency	Increment in MNUA + rifampicin mutation frequency
30 min before MNUA	x 2.7	x 2.9
during MNUA treatment	x 3.3	nt

DISCUSSION

MNUA-induced mutations, under the conditions of the experiments described are the combined result of misreplication and misrepair (results not shown). However, the enhancement of these
mutations by rifampicin cannot be specifically attributable to
interference in uvrA$^+$-dependent excision repair or in recA$^+$-depen-
dent error prone repair, although it requires a functional DNA
polymerase I. The adaptation pathway is the only repair pathway
which may be affected by rifampicin. It is clear at least that
preadapted WP2 cells are capable of efficiently removing all mutagenic lesions from DNA, and in so doing, remove the factor(s) involved in the rifampicin effect. It is hypothesized therefore that
in unadapted cells, rifampicin inhibits the induction of this
error-free adaption pathway in response to the MNUA dose. Consequently, therefore, rifampicin prevents accurate repair of mutagenic lesions and increases the time that these lesions remain in
the DNA. The overall result is an increase in the induced mutation
frequency.

It is possible that the retardation in the loss of O^6-MeG
residues from DNA of cells which have been treated with MNUA and
rifampicin may be the result of inhibition of the adaptation pathway.

Inhibition of this repair pathway by rifampicin, however, does
not entirely account for the magnitude of the observed increment
in the MNUA mutation frequency. Chloramphenicol, which also inhibits this pathway[4] only increase the MNUA induced mutation frequency to half the extent of the rifampicin increment.

In addition, chloramphenicol pretreatment of cells before the
mutation assay is only capable of partially blocking the rifampicin effect on the mutation frequency. In contrast to chloramapheni-

col, rifampicin must be added to the cells simultaneously with the MNUA in order to maximally increase the mutation frequency. These results imply that factors, other than the inhibition of protein synthesis are important in the expression of the rifampicin effect. The correlation between the maximum rifampicin effect and the fastest growth rate of the cells before treatment may indicate a dependence on the number of replicating forms within the cell. Other factors such as inhibition of the initiation of DNA synthesis, and in particularly, inhibition of RNA synthesis, by rifampicin, are considered to be important with respect to the induction of MNUA mutations.

Acknowledgements

This research was supported by a grant from the Cancer Research Campaign.

REFERENCES

1. P.D. Lawley, Methylation of DNA by carcinogens: Some applications of chemical analytical methods, in: "Screening Tests in Chemical Carcinogenesis", R. Montesano, H. Bartsch, L. Tomatis, eds.,IARC Scientific Publications No. 12 (1976).
2. A. Loveless, Possible relevance of 0-6 alkylation of deoxy-guanosine to the mutagenicity and carcinogenicity of nitro-soamines and nitrosamides, Nature 223:206 (1969).
3. R.F. Newbold, W. Warren, A.S.C. Medcalf, J. Amos, Mutagenicity of carcinogenic methylating agents as associated with a specific DNA modification, Nature 283:596 (1980).
4. P. Robins, J. Cairns, Quantitation of the adaptive response to alkylating agents, Nature 280:74 (1979).
5. S. Kyrtopoulos, P.F. Swann, A radioimmunoassay for O^6-alkyl-guanine derivatives, Br. J. Cancer 38:170 (1978).

THE ENHANCED REPAIR OF $\underline{O}^6$-ALKYLGUANINE IN MAMMALIAN SYSTEMS

P.J. O'Connor and G.P. Margison

[a]Paterson Laboratories,
Christie Hospital and Holt Radium Institute,
Manchester M2O 9BX, U.K.

In Collaboration with: R. Montesano[b], J.D. Buckley[a,e], Y-H, Chu[a,c], D.P. Cooper[a], R.A. Smith[a], N.J. Curtin[d], K. Snell[d], J. Ashby[e], C. Bradbrook[e], and J.A. Styles[e].

INTRODUCTION

The "adaptive" response in bacteria may be defined in terms of a reduction in the toxicity and mutagenicity of certain agents by pre-exposure to low levels of the agent and the effect is intimately associated with an enhanced DNA repair process[1,2,3]. In mammals it is premature to conclude that an equivalent system exists since none of these parameters have been examined simultaneously but there are now a number of systems in which pretreatment with various carcinogenic agents has been shown to enhance the capacity of certain tissues to repair $\underline{O}^6$-alkylguanine lesions in DNA. These systems are outlined here and the possible relationship of this phenomenon to carcinogenicity is considered.

ENHANCED REPAIR OF $\underline{O}^6$-ALKYLGUANINE INDUCED IN DNA BY ALKYLATING AGENTS

[b]Unit of Chemical Carcinogenesis, International Agency for Research on Cancer, 150 Cours Albert Thomas, 69372, Cedex 2, France; [c]Shanghai Cancer Institute, 270 Dong An Road, Shanghai, People's Republic of China; [d]Department of Biochemistry, University of Surrey, Guildford GU2 5XH, U.K.; [e]Imperial Chemical Industries Ltd., Central Toxicology Laboratory, Alderley Park, Macclesfield, Cheshire, U.K.

Methylating Agents:

A phenomenon which apparently resembles the enhanced repair function of the adaptation effect in bacteria has been observed during
investigations of the mechanism of hepatocarcinogenesis in rats by
chronic dimethylnitrosamine administration. It was found that daily
administration of DMN for 8 weeks increased the capacity of the liver
to remove from DNA the O^6-methylguanine produced by a single terminating dose of $[^{14}C]$-DMN in comparison with animals receiving only
the final dose of radiolabelled agent[4]. The levels of 7-methylguanine
and 3-methyladenine were unaffected by pretreatment indicating an
effect specific for O^6-methylguanine (Fig.1). The enhanced repair
capacity develops gradually during the treatment schedule and while
the effect is easily seen after one week it appears to be maximal at
around 3 weeks[5]. In a similar experiment, but with a lower daily pretreatment dose (0.7mg/kg), 4 weeks elapsed before the response was
detected[6]. Experiments using cell-free preparations of liver and
methylated DNA as substrate[5] have shown that in vitro, the capacity
for repair also increased with the period of pretreatment in parallel
with the enhancement found in vivo. In this case therefore, the en-

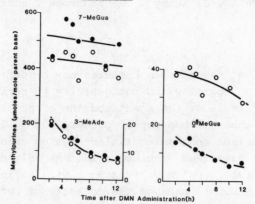

Fig.1.   Methylated purines in the liver DNA of DMN-pretreated rats
         at various times after administration (i.p.) of $[^{14}C]$-lab-
         elled DMN (2mg/kg, sp.act. 5.2mCi/mmole). Pretreatment con-
         ditions: 2mg/kg/day (p.o.) on weekdays for 44 days, O,
         unpretreated control rats ; ●, pretreated rats. (Redrawn
         from Montesano et al., 1979).

Abbreviations: AAF, 2-acetylaminofluorene; AFB_1, aflatoxin B_1; DEN,
N,N-diethylnitrosamine; DMN, N,N-dimethylnitrosamine; DMPT, 3,3-di-
methyl-1-phenyltriazene; MMS, methyl methanesulphonate; MNU, N-methyl-N-nitrosourea.

hanced repair is probably not due to a redistribution of $\underline{O}^6$-alkylation into the cells or regions of DNA from which it is removed at a more rapid rate but to a change in the level of the enzymes concerned.

The enhanced repair is rapid-acting, as it can be detected in the liver of pretreated animals as early as 10 mins after DMN injection[5]. Moreover, the enzyme itself appears to be inactivated during reaction with $\underline{O}^6$-methylguanine and so in these respects it is similar to the bacterial system. Pretreatment with lower doses of DMN also enhanced the removal of $\underline{O}^6$-methylguanine from liver DNA[5,6] and in animals challenged with 2mg/kg [^{14}C]-DMN the extent of the effect was proportional to the $\log_{10}$ of the pretreatment dose. In the lung and kidney of these same pretreated animals, no enhancement was detected, if anything $\underline{O}^6$-methylguanine removal was slower than in the tissues of control animals indicating a possible inhibition of the repair processes[5] but it is already known that the kidney enzyme is easily inhibited[7,8]. However, in this experiment, as in others in which the repair capacity is examined by tracer techniques in animals pretreated with unlabelled alkylating agents, $\underline{O}^6$-methylguanine persisting from the pretreatment doses will compete for the available repair enzymes and could produce an apparent repair inhibition: under the conditions of some experiments this may go undetected. Thus, the effects of chronic administration of DMN on the relative capacity of rat liver, kidney and lung to remove $\underline{O}^6$-methylguanine may explain the earlier observation that chronic treatment with radiolabelled material did not produce an accumulation of $\underline{O}^6$-methylguanine in liver DNA but did in lung and kidney[9].

In another series of experiments, the effects of daily administration of unlabelled MNU and MMS on the removal from rat liver, lung, and kidney DNA of $\underline{O}^6$-methylguanine produced by a terminating dose of [^{14}C]-DMN has been examined. DMN was included as a positive control and the levels of MNU and MMS were selected to produce amounts of 7-methylguanine and (for MNU) $\underline{O}^6$-methylguanine in liver DNA approximately equivalent to the dose of DMN used. Table 1 shows that while DMN pretreatment enhanced $\underline{O}^6$-methylguanine removal from liver DNA, MNU pretreatment caused a slight inhibition. This inhibition was very much more extensive in the kidney and lung, where the removal process was almost completely blocked. MMS had little effect in the liver and kidney but there may have been a slight enhancement in the lung. These results with MNU should be considered with the earlier

Table 1.   Effect of chronic administration of DMN, MNU or MMS on the
amounts of methylated purines in liver, kidney and lung
DNA of rats 6h ater a single dose of [^{14}C]-labelled DMN.

Tissue	Pretreatment	Methylated Purines				
		Amounts (μ mole/mole parent base)			Ratios	
		3-meA	7-meG	O^6-meG	3-meA / 7-meG	O^6-meG / 7 -meG
Liver	Control	8.2	265.8	13.8	0.031	0.052
	DMN	9.3	317.5	8.4	0.029	0.026
	MNU	8.4	279.6	18.1	0.030	0.065
	MMS	6.0	190.3	9.4	0.031	0.049
Kidney	Control	0.52	19.2	0.48	0.027	0.025
	DMN	0.52	18.7	0.51	0.028	0.026
	MNU	0.78	28.1	2.68	0.028	0.096
	MMS	0.54	20.6	0.48	0.026	0.023
Lung	Control	0.87	21.2	0.98	0.041	0.046
	DMN	0.85	20.5	1.17	0.041	0.057
	MNU	0.81	19.6	2.06	0.041	0.105
	MMS	0.85	21.2	0.8	0.040	0.038

Pretreatment conditions:DMN, 1mg/kg/day; MNU, 10mg/kg/day; MMS,
25mg/kg/day and controls given the same volume of saline (2ml/kg/
day) i.p. on weekdays for 14 days. Challenging dose: 1mg/kg/ i.p.
(sp.act. 41mCi/mmole) 24h after the last pretreatment dose.

demonstration[10] that chronic administration of DMN in the drinking
water ($\sim$1.7mg/kg/day) had no effect on the initial level or subsequent loss of $\underline{O}^6$-methylguanine produced in liver DNA by a single dose
of radiolabelled MNU. The inability of each of these agents to enhance the repair of damage produced by the other may be because they
alkylate different, or not completely overlapping cell populations in
the liver, due to a restricted capacity for the metabolism of DMN in
certain cell types. In other tissues, however, the enzyme appears
more sensitive to the inhibitory effects of MNU.

It is interesting to note that the inductive effect has so far
only been detected in mammalian systems using pretreatment with pre-
carcinogens (see also below). The lack of an effect with MMS might
be related to these possibilities but the much lower level of $\underline{O}^6$-
methylguanine produced by this agent may be insufficient to elicit
a response. Although in these experiments, the enhanced repair capacity appears to be specific for $\underline{O}^6$-alkylguanine, the "trigger" remains obscure and it is likely that damage other than $\underline{O}^6$-alkylation
can effect a similar response (see below). Yet another factor in the
ability to detect enhanced repair is the size of the challenging dose:
small increases in repair capacity might be swamped if this is too
high. While enhanced repair, in the case of dialkylnitrosamines, has
so far been detected only in liver, studies with DMPT[11] have suggested
that the effect can occur in liver, ileum, thymus and spleen. How-

ever, in view of the low level of alkylation of DNA in these extra-
hepatic tissues further confirmation of the response at these other
sites is now needed (P. Kleihues, pers.commun).

Diethylnitrosamine:

The effect of chronic administration of DEN in rats is analogous to that produced by DMN treatment. When DEN was given in the
drinking water at a dose of ∿10mg/kg/day and a comparison made with
control animals, there was a relatively lower level of hepatic $\underline{O}^6$-
ethylguanine 12h after a single dose of [^{14}C]-DEN, while 3-ethylade-
nine appears to be unaffected[12]. In this DEN-pretreatment system,
the capacity of the liver to remove $\underline{O}^6$-methylguanine after a single
dose of [^{14}C]-DMN has also been found to be enhanced in comparison
with control animals. Table 2 shows that the $\underline{O}^6$/7-methylguanine ra-
ios for liver DNA 12h after DMN (2mg/kg) are well below the control
values after 1 or 5 weeks of pretreatment. The enhanced repair would
appear to develop more rapidly with DEN, than it does with DMN[5], however, the pretreatment dose of DMN would result in a higher level of
DNA damage which might have, initially, an inhibiting effect[6]. In
a preliminary attempt to explore whether the induced system has the
same capacity to act on both $\underline{O}^6$-methyl- and $\underline{O}^6$-ethylguanine, a terminating dose of DMN of 0.88mg/kg (equivalent to 10mg/kg DEN) was given
to rats pretreated with DEN for 10 weeks. In this experiment, the
amounts of $\underline{O}^6$-methyl and $\underline{O}^6$-ethylguanine removed from the liver DNA of
control animals were very similar and those from DEN pretreated ani-
mals were of similar order (Margison, unpublished). The same enzyme
system therefore could be acting on both types of damage but further
work is needed before a definitive conclusion can be reached.

Table 2.   Effect of pretreatment with DEN on the levels of methylguanines in rat liver DNA 12h after a single dose of [^{14}C]-
labelled DMN.

Group	Period of Treatment (weeks)	Methylguanines (μ mole/mole guanine)		
		N-7	06	Ratio 06/N-7
DEN*	I	433	8.5	0.02
Control	I	423	25.4	0.06
DEN*	5	581	23.2	0.04
Control	5	426	32.7	0.077

*Pretreatment conditions: DEN in the drinking water ∿10mg/kg/day;
challenging dose: 2mg/kg i.p. (Sp.act. 24.5mCi/mmole).

In view of the enhanced O^6-ethylguanine removal which is induced by chronic DEN treatment, the effects of single pretreatment doses of DEN (10mg/kg) on O^6-ethylguanine levels[13] can now be interpreted. For about 6h after pretreatment there is an initial inhibition of the removal of this base produced by a single dose of $[^{14}C]$-DEN (2mg/kg) but afterwards the removal is enhanced. The effects of single doses of DEN are therefore very similar to those produced by single doses of other agents (see below). More recently it has been found that weekly administration of DEN (16.8mg/kg) also enhances O^6-ethylguanine removal and that the extent of this response remained constant for up to 22 weeks of continued treatment (E. Scherer, pers.commun).

Other Species

In order to explore whether similar inductive effects occur in other species a preliminary examination of the effects of weekly administration of DMN on the initial amounts of methylation products formed in the DNA of various Chinese and Syrian hamster tissues following a single dose of $[^{14}C]$-DMN has been made (Table 3). In Syrian hamsters, pretreatment with DMN (3mg/kg/week for 12 weeks) caused

Table 3.   Effects of pretreatment of hamsters with DMN on the levels of methylguanines in tissue DNA at times after a single dose of $[^{14}C]$-labelled DMN.

a)   Syrian hamsters.

Tissue	Group	Methylguanine (μ mole/mole guanine)		
		N-7	O^6	Ratio O^6/N-7
Liver	Control	453	65.2	0.144
	Pretreated	455	75.6	0.166
Kidney	Control	25.5	2.1	0.082
	Pretreated	35.4	5.4	0.153

Pretreatment conditions: 3mg/kg s.c. once weekly for 12 weeks, the challenging dose of DMN was 3mg/kg s.c. (sp.act. 10.3mCi/mmole) and the animals were killed 24h later.

b)   Chinese hamsters.

Group	Survival Time (hrs)	Methylguanines (μ mole/mole guanine)		
		N-7	O^6	Ratio O^6/N-7
Control	12	565	75.3	0.133
	48	285	82.3	0.289
Pretreated	12	656	107.5	0.164
	48	249	87.4	0.351

Pretreatment conditions: 3.5mg/kg i.p. once weekly for 20 weeks, the challenging dose of DMN was 3.5mg/kg i.p. (sp.act. 54mCi/mmole) and animals were killed later as shown.

an increase in the $\underline{O}^6$/7-methylguanine ratios in both liver and kidney (Table 3a). There was an almost identical level of 7-methylguanine in the liver DNA of both groups, whereas the $\underline{O}^6$-methylguanine levels were higher in DMN pretreated animals with a consequently higher $\underline{O}^6$/7-methylguanine ratio. This suggests that the effect may be similar to that caused by DMN in rat lung, i.e. an inhibition or overloading of the repair system. However, in Syrian (and Chinese) hamsters, unlike the rat, the rate of loss of 7-methylguanine after single doses of DMN indicates that active removal of this product takes place[14,15]. An increase in the $\underline{O}^6$/7-methylguanine ratio may therefore reflect an increase in the capacity of these tissues to remove 7-methylguanine. Pretreatment of Chinese hamsters with 3.5mg/kg DMN once weekly for 20 weeks also caused a similar increase in the $\underline{O}^6$-/7-methylguanine ratio in the DNA of liver (Table 3b). Experiments in progress should provide a more precise explanation for these changes in the methylguanine ratios of hamster tissues.

ENHANCED REPAIR OF $\underline{O}^6$-METHYLGUANINE INDUCED IN DNA BY OTHER CARCINOGENS

Chronic Treatments:
     Quite unexpectedly, similar phenomena of enhanced repair of $\underline{O}^6$-methylguanine have been observed in animals pretreated with carcinogens chemically unrelated to the $\underline{N}$-nitroso compounds; these now include AAF and AFB$_1$. Whilst the enhanced repair of $\underline{O}^6$-alkylguanine observed after pretreatment with the methylating and ethylating carcinogens might be envisaged as an 'inductive' or 'adaptive' phenomenon due to the raised level of $\underline{O}^6$-alkylguanine in DNA, this seems less likely in the case of these unrelated chemical carcinogens, unless they also lead to the formation of some adducts at the $\underline{O}^6$-atom of guanine. There are a number of possible explanations at a more general level. These range from effects at the level of a co-factor for the repair enzymes, the induction of a group of repair enzymes as a common response to various types of DNA damage or the induction of enzymes associated with a proliferative response due to the necrogenic action of these agents. Alternatively, an opening up or rearrangement of chromatin structure may make the substrate lesions more accessible and/or cause loss of latency on the part of the enzyme itself, due to the continuous insertion of bulky DNA adducts and subsequent repair occasioned by the pretreatment.

Acetylaminofluorene:
     The initial observations[16] were made for rats that were given single injections of DMN after they had been exposed to a diet cont-

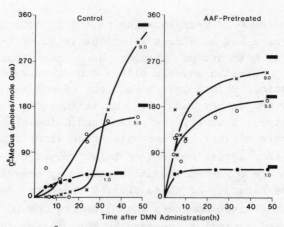

Fig.2.   Removal of O^6-methylguanine (μmoles/moleGua) from the liver
         DNA of control and AAF-pretreated rats at various times
         after administration of DMN. [^{14}C]-Labelled DMN (sp.act.
         5-11mCi/mmole) was given (i.p.) at 1($\bullet$); 5.5 (O) and 9mg/kg
         (X) and the horizontal bars indicate the amount of O^6-meth-
         ylguanine present in DNA 5h after injection. These values
         and the amounts removed from DNA were calculated on the ass-
         umption that the initial ratio of O^6/7-methylguanine in DNA
         is 0.11. Pretreatment conditions: AAF (0.06%) in the diet
         for 3 weeks followed by 1 week of normal diet. This cycle
         was repeated twice and DMN was given at the end of the third
         week on normal diet[16].

aining AAF. The enhanced loss of O^6-methylguanine was seen at three
dose levels of DMN and the data show that the time-course for the repair of O^6-methylguanine was essentially brought forward in the pretreated animals vs the controls. Within the limits of the experiment
the shapes of these removal curves are not materially different and
after 48 hours there is still a similar small proportion of the O^6-
methylguanine residues initially present that remain unrepaired in
the DNA. These experiments (Fig.2) were carried out on animals that
had been exposed to 3 cycles of AAF feeding[16] but the enhanced repair
of O^6-methylguanine was clearly seen after 2 or only 1 cycle of pretreatment (Buckley et al., unpublished). As in the experiments with
the alkylating carcinogens (see above) no effect with these pretreatments were observed for the repair of 3-methyladenine[16]. Also, no
effects were detected for the repair of O^6-methylguanine from kidney
DNA (Buckley et al, unpublished) but it must be noted that the extent
of reaction of metabolites of AAF with the DNA of kidney was almost
certainly very much lower than in the liver.

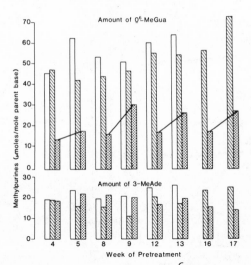

Fig.3.   Preliminary data on amounts of $\underline{O}^6$-methylguanine and of 3-
          methyladenine present in the liver DNA of control and AAF-
          pretreated rats 6h after administration of DMN. Pretreat-
          ment was by gavage on weekdays for the duration of 4, 8, 12,
          & 16 weeks and the effect of pretreatment was assessed by
          giving a challenging dose of [^{14}C]-labelled DMN (2mg/kg;
          i.p.; sp.act. 3-4mCi/mmole) in place of the pretreatment
          dose on the final day. A second group of animals was pre-
          treated in the same way but was then allowed 1 week on nor-
          mal diet before receiving the DMN challenge. Controls were
          ☐ , age-matched untreated animals and ▨ , age-matched but
          given the same volume of corn oil; pretreated animals ▨ ,
          were given AAF, 15mg/kg/day.

In the next series of experiments animals were pretreated with
AAF by gavage at a daily dose of 15mg/kg for 5 days per week over a
period of 16 weeks. One group of animals was challenged on the last
day of treatment and a second group was given DMN after a rest period
of 1 week (Fig.3) in order to assess the effect of the period on normal diet employed in the earlier study[16]. In confirmation of the
previous experiments, an enhanced repair of $\underline{O}^6$-methylguanine from
liver DNA was clearly seen at all stages of pretreatment and at the
earlier times (4 and 5 weeks), the rest-period had very little effect. Later on, however, less enhancement of repair was observed
after a week without AAF-pretreatment. As with the enhancement process itself, it is difficult to give an explanation for this change
but one possibility might be the formation of new populations of
cells that are gradually losing their ability to maintain the level

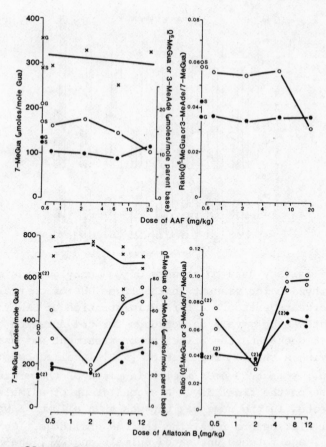

Fig.4. The effect of various doses of AAF and of AFB1 on the amounts
        of methyl purines in rat liver DNA 5 hours after treatment
        with a challenging dose of [^{14}C]-labelled DMN. a) AAF, as a
        suspension in 1.75% gum acacia was injected (i.p.) 24 hours
        before treatment with DMN (lmg/kg; i.p.; sp.act. 12.6mCi/
        mmole); G, gum acacia controls; S, saline controls; (X), 7-me-
        thylguanine; (O), O^6-methylguanine and (●), 3-methyladenine.
        b) AFB1 prepared and administered as above; DMN challenge
        (2mg/kg; i.p.; sp.act. 3.1mCi/mmole).

of enhanced repair and so the inductive effect may be less persistent.
Again, as in the previous study no effects were observed for the repair of 3-methyladenine.

Single Dose Treatments:
        A number of experiments have now been carried out using single
doses of AAF and comparing them with the effects of single doses of
AFB1 (Fig.4). When the challenging dose of [^{14}C]-DMN was given 24

hours after a single treatment with these chemically unrelated carcinogens enhanced repair of $\underline{O}^6$-methylguanine was again observed. This was
seen at 20mg/kg in the case of AAF (Fig.4a) and at 2.0mg/kg with AFB_1
(Fig.4b). At higher doses of AFB_1 the repair of both $\underline{O}^6$-methylguanine
and of 3-methyladenine was inhibited. Although inhibition of these
two repair systems by the higher doses might be due to toxic effects
leading to some tissue damage which can be observed histologically by
this time, it is nevertheless interesting to see that there are circumstances in which the 3-methyladenine repair system can be inhibited
(c.f. the null effect in other systems so far described). Time sequence studies following single doses of AAF are still in progress but
the data for AFB_1 treatment shows that the enhanced repair of $\underline{O}^6$-me-
thylguanine is maintained for at least 5 days.

CONCLUSIONS

     While the repair of $\underline{O}^6$-alkylguanine has been the only parameter
determined in these experiments some general indications are already
evident. 1) Enhancement of the system takes place after treatment
with a wide variety of agents (e.g. DMN, DMPT, DEN, AAF, and AFB_1)
and dose regimes. However, some alkylating agents (e.g. MNU, DMN)
do not mutually enhance the repair of damage produced by each other
and although this might be explained by the alkylation of different
cell populations it also raises the question of whether mammalian
cells may have an inducible system different from that in bacteria
2) In some cases enhanced activity has been confirmed in vitro assays but this has yet to be done for other agents and it is possible
that they act in different ways. 3) Most of the agents which enhance repair also damage hepatic DNA in some way and are both hepa-
totoxic and hepatocarcinogenic. The phenomenon may be related to
any or all of these effects and/or to the proliferative response
caused by cytotoxicity in the liver. 4) In terms of hepatocarcinogenesis and the key role of $\underline{O}^6$-alkylation of guanine (see this volume) it could be proposed that while removal of this lesion is more
rapid, the restorative hyperplasia and attendent DNA replication produced by toxicity may be the decisive factor in producing transformation. In this context the possibility of error-prone repair, should
not be ignored. 5) In animal systems there is no evidence, so far,
to suggest that the enzyme responsible for the removal of 3-methyl-
adenine can be induced but recently, serum-stimulation of WI38 cells
was shown to enhance the activity of the DNA glycosylase for uracil[17].

     Ongoing work in these and other laboratories should help to re-
solve some of these questions.

ACKNOWLEDGEMENTS

    P.J.O'C and G.P.M. gratefully acknowledge support through grant
made to the Paterson Laboratories by the MRC and the Cancer Research
Campaign. We are indebted to Professor P. Kleihues, Dr E. Scherer
and Dr P.F. Swann for personal communications and to E.S. and P.F.S.
for access to data prior to publication.

REFERENCES

1. L. Samson and J. Cairns, A new pathway for DNA repair in Escher-
    ischia coli. Nature (London) 267:281 (1977).
2. P.F. Schendel and P.E. Robins, Repair of O^6-methylguanine in ad-
    apted Escherischia coli. Proc.Natl.Acad.Sci.U.S.A. 75:6017
    (1978) 280:76 (1979).
3. P. Karran, T. Lindahl and B. Griffin, Adaptive response to alky-
    lating agents involves alteration in situ of O^6-methylguanine
    residues in DNA. Nature (London) 280:76 (1979).
4. R. Montesano, H. Brésil and G.P. Margison, Increased excision of
    O^6-methylguanine from rat liver DNA after chronic administra-
    tion of dimethylnitrosamine. Cancer Res. 39:1798 (1979).
5. R. Montesano, H. Brésil, G. Planche-Martel, G.P. Margison and
    A.E. Pegg, Effect of chronic treatment of rats with dimethyl-
    nitrosamine on the removal of O^6-methylguanine from DNA.
    Cancer Res. 40:452 (1980).
6. P.F. Swann and R. Mace, Disappearance of O^6-methylguanine from
    rat liver DNA during chronic dimethylnitrosamine administrat-
    ion. A possible similarity between the system removing O^6-me-
    thylguanine from DNA in rat liver and in E.coli. Chem.-Biol.
    Interact. submitted for publication.
7. A.E. Pegg and J.W. Nicoll, Nitrosamine carcinogenesis: the im-
    portance of the persistence in DNA of alkylated bases in the
    organotropism of tumour induction, in "Screening Tests in Chem-
    ical Carcinogenesis" R. Montesano, B. Bartsch and L. Tomatis
    eds. IARC, Lyon, 1976, p181.
8. A.E. Pegg and G. Hui, Formation and subsequent removal of O^6-
    methylguanine from DNA in rat liver and kidney after small
    doses of dimethylnitrosamine. Biochem.J. 173:739 (1978).
9. G.P. Margison, J.M. Margison and R. Montesano, Accumulation of
    O^6-methylguanine in non-target-tissue DNA during chronic admin-
    istration of dimethylnitrosamine. Biochem.J. 165:463 (1977).
10. G.P. Margison, N. Brésil, J.M. Margison and R. Montesano, Effect
    of chronic administration of dimethylnitrosamine on the excis-
    ion of O^6-methylguanine from rat liver DNA. Cancer Lett.

2:79 (1976).

11. H.K. Cooper, E. Hauenstein, G.F. Kolar and P. Kleihues, DNA alkylation and oncogenesis by 3,3-dimethyl-1-phenyltriazene, Acta Neuropathol (Berl) 43:105 (1978).

12. G.P. Margison, N.K. Curtin, K. Snell and A.W. Craig, Effect of chronic N,N-diethylnitrosamine on the excision of O^6-ethylguanine from rat liver DNA. Br.J.Cancer, 40:809 (1979).

13. E. Scherer, A.P. Steward and P. Emmelot, Kinetics of formation of O^6-ethylguanine in, and its removal from, liver DNA of rats receiving diethylnitrosamine. Chem.-Biol.Interact. 19:1 (1977).

14. G.P. Margison, J.M. Margison and R. Montesano, Methylated purines in the DNA of various Syrian golden hamster tissues after administration of hepatocarcinogenic doses of dimethylnitrosamine. Biochem.J. 157:627 (1976).

15. G.P. Margison, J.A. Swindell, C.H. Ockey and A.W. Craig, The effect of a single dose of dimethylnitrosamine in the Chinese hamster and persistence of DNA alkylation products in selected tissues. Carcinogenesis 1:91 (1980).

16. J.D. Buckley, P.J. O'Connor, and A.W. Craig, Pretreatment with acetylaminofluorene enhances the repair of O^6-methylguanine. Nature (London) 281:403 (1979).

17. P. Gupta and M.A. Sirover, Induction of base excision repair in human cells. J.Cell Biol. 83:5a (1979).

DNA SEQUENCING AND GENE CLONING IN STUDYING MUTAGENESIS AND REPAIR

THE USE OF DNA TUMOUR VIRUSES AS MODEL SYSTEMS FOR STUDYING DAMAGE

AND REPAIR OF EUKARYOTIC CHROMOSOMES: ANALYSIS OF MUTATIONS BY

DNA SEQUENCING

Beverly E. Griffin

Imperial Cancer Research Fund
Lincoln's Inn Fields
London W.C. 2, England

INTRODUCTION

Attempts to define repair processes in eukaryotic systems are
subject to all the difficulties inherent in studying very large and
complex DNA molecules. When the location of a lesion that makes
repair necessary for the efficient functioning of a cell cannot in
itself be precisely defined, it is not surprising that the subsequent repair process can only be described in general terms, usually
related to phenotypic responses. The idea that viruses might be
used as probes for studying repair of DNA in mammalian cells has
been put forward in a number of papers and a limited number of
experiments have been carried out with this aim in mind. The
conclusions from some of these studies, presented and summarised by
Day (1978), do not make very encouraging reading. Nonetheless, it
would seem premature to abandon the approach, particularly when one
considers the remarkable advances made in the past few years in
defining the molecular biology of some of the animal viruses and
the promising efforts to correlate particular regions of viral
genomes with biological activities inside the cell. For the
purposes of this communication, I shall concentrate on the small
DNA tumour viruses, the mouse virus, polyoma, the monkey virus,
SV40, and the human variant of SV40, BKV as being among the better-
characterised and most amenable to study of these viruses.

The DNAs of polyoma virus, SV40 and BKV, are present in their
respective host cells in the form of supercoiled, double-stranded
species carried as minichromosomes. (For review, see Crémisi,
1979). The viruses have limited coding capacities: SV40 is only

249

known to code for five proteins, three of them being involved in
the encapsidation process that leads to the production of viral
particles and the other two, the so-called large and small

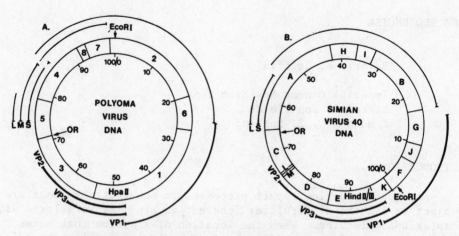

Fig. 1.   (a) A physical map of polyoma virus DNA based on the eight
          fragments produced by cleavage of the DNA with the restric-
          tion enzyme HpaII. The map is divided into 100 units and
          on it are indicated the single EcoRI restriction site, the
          viral origin of replication (O_R) and the location of the
          areas coding for the six known viral proteins, the large
          (L), middle (M) and small (S) T-antigens and the capsid
          proteins VP1, VP2 and VP3. The N-termini of the three T-
          antigens (located near the origin O_R) appear to be iden-
          tical; elsewhere on the genome, a single region codes for
          both middle and large T-antigens making use of two differ-
          ent reading frames. It can also be seen that the C-termini
          of VP2 and VP3, known to share common sequences, overlap
          the N-terminus of VP1, the latter being encoded within a
          different reading frame. (b) A physical map of SV40 DNA
          based on the thirteen fragments produced by cleavage of
          the DNA with restriction enzymes from Haemophilus influenzae
          (Hind II and III). The map is oriented so that homologies
          with polyoma virus are readily apparent. SV40 apparently
          codes for no protein which corresponds to the middle T-
          antigen of polyoma virus. (BK virus shares a similar
          organization with SV40). (For references, see Tooze, 1980).

T-antigens, playing roles in viral DNA replication and/or transformation. Although fewer studies have been made on BKV, the results
seem similar. Polyoma virus is known to encode one additional
protein, designated middle T-antigen (Ito et al., 1977), which
appears to be primarily, if not solely, responsible for the transforming ability of this virus, transformation being defined as the
ability to produce neoplastic growth in a host cell. The primary
DNA sequences of all three viruses have been determined. (For
SV40, see Fiers et al., 1978 and Reddy et al., 1978; for BKV, see
Seif et al., 1979 and Yang and Wu, 1979; for polyoma virus, see
Deininger et al., 1980 and Soeda et al., 1980). In Fig. 1 present
knowledge about the organization of the polyoma virus and SV40
genomes in terms of their coding information is summarised. More
detailed information on these viruses and appropriate references
can be found in The DNA Tumor Viruses (ed. Tooze, 1980). Current
ideas about the role of the polyoma virus middle T-antigen are
discussed in Ito et al. (1980) and Novak et al. (1980) and references found therein. As far as is known, these viruses code for no
replication or repair enzymes of their own. Therefore, for their
survival and their integrity, they must depend to a large extent,
if not wholly, on the relevant functions of their host cells. It
is for these reasons that the viruses should provide excellent
tools for defining host cell repair processes.

APPLICATIONS OF DNA SEQUENCING METHODS TO THE STUDY OF VIRAL
MUTANTS WITH DEFINED BIOLOGICAL PROPERTIES

    The sequencing methods developed by Sanger and collaborators
(1977) and by Maxam and Gilbert (1977) are sufficiently simple and
rapid that intact and altered DNA can now be readily and accurately
compared. For the small DNA tumour viruses, although there are as
yet relatively few data available on the chemistry of the viral
proteins themselves, the DNA sequences allow predictions of the
amino acid sequences and molecular weights of the proteins. Alterations to the DNA in a coding region can thus be extrapolated to
amino acid changes in a particular protein and, conversely, phenotypic changes in the virus correlated with genotypic changes.
Although as yet only limited use has been made of these viruses as
probes into the cellular machinery itself, three examples have been
chosen to show how viral mutants can be used to study replication,
transformation and possibly repair within the cell.

    1. DNA Replication. The origins of replication of polyoma
virus, BK virus and SV40 are remarkably similar, as has been discussed elsewhere (Soeda et al., 1979). Over a region of 42 nucleotides which includes the origins of replication, there is about
80 % sequence homology among the three viral DNAs. Moreover, in all
three cases a portion of these sequences, in addition to the usual
double-stranded species, can form either quite stable hairpin loops

or stereochemically possible four-stranded structures. For SV40, these conformations are shown in Fig. 2. Although such structures may not exist in non-replicating viral chromosomes, they may become important during the actual replication process. The viral large T-antigens appear to bind to DNA at the origins of replication (Tjian, 1978) and may be instrumental in opening up the DNA and allowing such structures to form either prior to or during replication. An important point to be noted here is that not only are the viral origins similar to each other, but they may also be similar to cellular replication origins since it is known that viral infections stimulate replication of host cell DNA. As discussed recently by Jelinek et al. (1980), mammalian genomes contain ubiquitous, interspersed repeated sequences which are very similar to sequences found around the viral origins of replication; they suggest that in mammalian cells this particular class of repeated sequences may serve as replication origins.

In order to examine replication in more detail, Nathans and collaborators have constructed a number of mutants within the SV40 replication origin and assayed the ability of these mutants to replicate. Their procedure for mutagenesis in vitro is relatively simple. It essentially consists of nicking SV40 DNA at a restriction enzyme site which lies within the origin of replication, then generating a small stretch of single-stranded DNA by exonuclease digestion proceeding from the nicked site. Sodium bisulphite is used to deaminate cytosine residues in the single-stranded DNA, and such mutated DNA is then employed to transfect cells, allowing the cell itself to repair the double-stranded viral DNA by gap filling and ligation. Individual plaques were picked and viral DNAs from these plaques analysed by sequencing the DNA and comparing it with the sequence of the wild type virus. (For a detailed discussion of this work, see diMaio and Nathans (1980) and papers cited therein). In Fig. 2 the sites are shown at which six of the mutations (5154-5162) are located and Table 1 gives the properties of the mutants. All mutations except one had some effect upon replication. The single exception (5161) can be seen to lie at the apex of the hairpin loop and at the non-bonded position in the four-stranded structure (see Fig. 2a and b). In another series of experiments, Shortle et al. (1979) have isolated second site revertants of these origin mutants and localised the second mutation on a region of the genome that codes for large T-antigen (see Fig. 1b). Such approaches are extremely promising and should ultimately lead to a more detailed understanding of the initiation of viral DNA replication. Unfortunately, there is as yet no report on the effects these origin mutants have on host cell replication.

2. Transformation. One of the most interesting aspects of DNA tumour viruses is reflected by the in vitro response of cells to viral infection. In certain types of cells, there is a lytic response, that is, viral infection leads to the production of more

virus particles with the concomitant death of the cell. In other
cell types, viral infection results in immortalisation of the cells
and they continue to grow in a manner reminiscent of tumour cells,
i.e. they are "transformed".

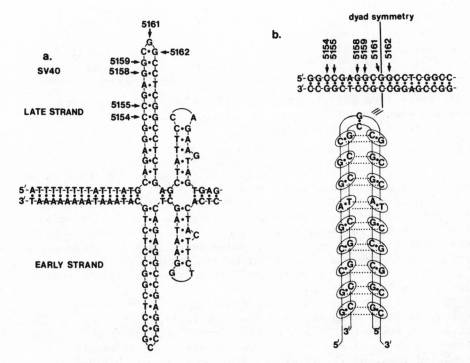

Fig. 2.   Conformations allowed by the DNA sequence located at the
            viral origin of replication. (a) The hairpin loop struc-
            tures and the location of the six mutants (5154–5162)
            discussed in the text. (b) Part of the sequence from the
            major loop in (a) which has a centre of symmetry, as
            indicated. This sequence is presented in the double-
            stranded form with the usual Watson-Crick base pairs
            indicated (·), or, underneath, as a four-stranded struc-
            ture, a composite of the hairpin loop and the double-
            helical structure (see Lim and Mazanov, 1978; Soeda et
            al., 1979). Mutant 5161 which replicates as efficiently
            as the wild type virus (see Table I) maps at the apex of
            the hairpin loop and four-stranded structure and is the
            centre of symmetry in the double-stranded sequence. Its
            biological properties and its location within the allowed
            conformations may not be fortuitous.

Table 1.   Properties of SV40 Point Mutants Derived From the Region
           of the DNA That Encodes the Origin of Replication[+]

Mutation	Location on genome (nucleotide number*) (see Fig. 2)	DNA replication	Plaque morphology
G·C → A·T	5162	< wt	small
G·C → A·T	5161	wt	wt
G·C → T·A	5159	ts	ts
G·C → A·T	5158	ts	ts
C·G → A·T	5155	> wt	small
C·G → T·A	5154	ts	ts

[+]Data taken from Shortle et al. (1979).   *Numbers correspond to
the DNA sequence of Reddy et al. (1978).   Abbreviations: wt = wild
type;   ts = temperature sensitive.

A number of mutants, especially of polyoma virus, have by now
been isolated which allow the area of the viral genome responsible
for cellular transformation to be defined and the transforming
function assigned with some degree of certainty to a particular
viral protein.   For polyoma virus, three classes of mutants have
been identified with lesions that have been assigned to the region
of the genome that is transcribed before DNA replication commences.
This, the so-called "early region", is represented by the restriction enzyme fragments HpaII 5, 4, 8, 7 and 2, see Fig. 1a.   Mutants
of one class, the A mutants, are temperature sensitive and have
lesions in large T-antigen which affect replication and also the
initiation event that leads to transformation, though the latter
effect seems to be an indirect consequence of an accessory large
T-antigen function (U. Novak and B.E. Griffin, unpublished data).
These have been designated tsA mutants.   Another class, the hrt
mutants, represents host range mutants which are non-transforming
and non-tumorigenic.   Sequence analysis, as discussed briefly below,
confirms that they have lesions in middle and small T-antigens.   A
third class of mutants represents those that have lesions in middle
and large T-antigens.   This class should perhaps be designated mlt
mutants, mlt being representative of the viral functions affected,
since the phenotypes of different isolates of such mutants vary
considerably (Griffin and Maddock, 1979;   Bendig et al., 1980).

The properties of the mutant NG-18, the best characterised of
the polyoma virus hrt mutants, became easier to understand once its
DNA sequence had been determined.   It, like other hrt mutants,
proved to be a mutant which affected two viral functions.   Its
deletion of 187 base-pairs, located between 80-83.5 map units

(Fig. 1a) resulted in a frameshift; not only was small T-antigen truncated by the deletion but coding for middle T-antigen was also prematurely stopped by the introduction of a termination codon into the frame containing the information for this protein (Soeda and Griffin, 1978; Hattori et al., 1979). Although the amino acid changes in the mutants can be predicted from their DNA sequences, the phenotypic changes could result from a change in either one (or both) of the two proteins affected.

A study of two of the mlt mutants has been helpful in further delineating transformation. One of the mutants, dl-8, transforms rather better than the wild type virus from which it was derived. The other, dl-23, is essentially non-transforming and non-tumorigenic (Griffin and Maddock, 1979). The former lies between 89.5 and 90 units and the latter between 92 and 94.5 units on the physical map, see Fig. 1a. DNA sequence analysis (N. Smolar and B.E. Griffin, ms. submitted) shows that dl-8 has a deletion of 90 base-pairs which, in addition to its effect upon transformation, also affects replication, the mutant virus being replicated at least 100-fold less efficiently than its wild type counterpart. Replication of the dl-23 mutant, on the other hand, is indistinguishable from that of wild type virus and the effect of its 102 base-pair deletion appears to be mainly on transformation. Protein analysis shows the mutants to have truncated versions of both large and middle T-antigens (Ito et al., 1980). Another mlt mutant, 45, isolated and studied by Folk and his collaborators (Bendig et al., 1980) has a deletion of 66 base-pairs which lies in part within the dl-8 deletion and in part between the dl-8 and -23 mutants. The deletion found in mutant 45 results in no substantial alterations, relative to wild type virus, of either DNA replication or cellular transformation. These results, together with data previously mentioned, emphasise the role of the viral middle T-antigen in transformation. Moreover, the mlt mutants further allow essential and non-essential chromosomal regions for the production of a transformed phenotype to be defined. In addition, the region of large T-antigen deleted in the dl-8 mutant appears to play some role in viral DNA replication, a finding consistent with the second site revertants of SV40 origin mutations discussed above.

3. Mismatch repair? There is evidence to suggest that eukaryotic cells, like prokaryotes, may be able to correct mismatches in DNA by an excision-repair process. Two separate studies have been carried out, one with SV40 (Lai and Nathans, 1975) and the other with polyoma virus (Miller et al., 1976) to determine what happens in vivo to DNA with defined mismatched bases. In the experiments of Lai and Nathans, heteroduplexes were made between non-complementing temperature sensitive (ts) mutants, and cells transfected with the DNA were assayed for plaque formation at the non-permissive temperature. The results from one such study are summarised in Table 2. All four mutants used (TsA 239, 241, 255,

Table 2.   Infectivity of Mutant and ts/ts Heteroduplexes
as Assayed by Production of Plaque Forming
Units at the Non-permissive Temperature[+]

tsA	239	241	255	209
239	0,0	1,0(0)	72,79(86)	74,58(100)
241		0,0	81,83(86)	69,64(100)
255			1,0	34,35(13)
209				0,0

[+]Data taken from Lai and Nathans (1975). Results are
expressed as number of plaques in each of two dishes
of BSC-40 monkey cells infected with 20 ng. of heteroduplexed DNA at 40°C. The number of nucleotides
between the different mutants (Seif et al., 1980) is
given in parentheses (see also Fig. 3).

and 209) had lesions which affected the function of SV40 large T-
antigen and had been mapped in the HindII/III fragment I on the
physical map (see Fig. 1b). Since on its own each of these mutants
was defective in the initiation of DNA replication, the data
suggested that some process other than replication was responsible
for correction of the defects.

In a similar kind of experiment, Miller et al. looked for the
correction of genotypic markers in heteroduplexed polyoma virus DNA.
Their heteroduplexes were made between two mutant DNAs (tsa and CR)
which differed in four distinct sites, one (designated A for tsa
and a for CR) located between 1 and 14 units on the physical map of
polyoma virus, a second (B, b) located within an HhaI restriction
enzyme site at 26 map units, a third (C, c) within an HaeIII site
at 28 map units, and a fourth (D, d) near the viral origin of
replication (see Fig. 1a). In their experiments, in DNA isolated
from sixteen separate plaques, all the markers except two appeared
to segregate independently, with very little apparent bias being
observed in the repair (or correction) process. The two markers
(B, C) and (b, c), on the other hand, always appeared together in
the progeny virus. These markers are the closest of all on the
genome and by DNA sequence analysis (Soeda et al., 1980) can be
shown to be between 85 to 91 nucleotides apart; the next closest
markers, (B, A) and (b, a), would have a minimum distance of about
750 nucleotides, that is, the distance between 14 and 26 map units.
Miller et al. interpreted their data to mean that the markers
closest together on the genome were repaired together, whereas
those further apart were repaired independently.

Recently, Seif et al. (1980) have sequenced the DNA of the A mutants studied by Lai and Nathans. Two of the mutants, tsA239 and 241, proved to be identical and were the result of G:C → C:G changes. (This is somewhat surprising since hydroxylamine was the agent used to generate the mutants and it might have been expected to produce a G:C → A:T change). One other mutant, tsA255, also had a G:C → C:G change, and the fourth, tsA209, the expected C:G → T:A change. DNA sequences are given in Fig. 3. It can be seen from the sequence that conversion of heteroduplex to wild type DNA requires that the mutant markers be repaired independently and on opposite strands. Data for polyoma virus, discussed above, suggest that the further apart the markers lie, the more likely they are to be repaired independently. The SV40 mutants map close enough to each other that segregation would not have been predicted. Nonetheless, all SV40 mutant heteroduplex DNA could recover in vivo to a certain extent the wild type phenotype (see Table 2), the least efficient being the heteroduplex composed of mutations separated by only 13 nucleotides. Two factors must be remembered when considering repair here: one is that in SV40, only one strand appears to be used for coding, the one shown in Fig. 3, and it is therefore the most important one to be correctly repaired. The other is that the position of the mismatch within the codon may be important. In this case, the latter may be particularly significant since one mutation (in tsA209) is in the centre of a codon and cannot be altered without altering the amino acid whereas the other two mutations (in tsA239/241 and 255), although they lie in the wobble position, are part of the only triplet (TGG) that normally codes for tryptophan in mammalian cells.

The data from the studies on polyoma virus and SV40 appear to be somewhat in conflict. It may be, but seems unlikely, that mismatch correction is occurring by two different processes in the two cases. Alternatively, it may be that the studies are actually

```
          C(tsA239/241)
          ↑
5'-TGG·CTA·CAC·TGT·TTG·TTG·CCC·AAA·ATG·GAT·TCA·GTG·GTG·TAT·

GAC·TTT·TTA·AAA·TGC·ATG·GTG·TAC·AAC·ATT·CCT·AAA·AAA·AGA·TAC·

TGG·CTG·TTT·AAA·GGA·CCA-3'
 ↓                      ↓
C(tsA255)            T(tsA209)
```

Fig. 3. SV40 wild type and mutant* DNA sequences (from nucleo-
         tides 3559-3454[+])

[+]Taken from Reddy et al. (1978)
*Taken from Seif et al. (1980)

probing regions which are borderline cases between simultaneous and independent repair, and that the repair patch size is somewhat different in the different (mouse versus monkey) host cells. On the other hand, particularly in the case of the SV40 mutants, it may be that the phenotypic responses reflect alterations in the genome other than those required to restore mutant to wild type DNA in the specific area around the mutations. In the latter case, the tsA209/255 heteroduplex is particularly relevant since if it represents repair, the two lesions are so close together that simultaneous repair (if it occurs in mammalian cases) would appear likely. It cannot of course be ruled out entirely, in either of the two studies, that the effects observed are indicative of high frequency recombination events which may be a property of the viruses themselves or of the transfection techniques used in the studies. Nonetheless, the availability of both mutants and rapid sequencing technology makes this approach a promising one for probing the question of mismatch repair in mammalian cells.

DISCUSSION

The examples discussed above were selected because they represent studies on three different biological processes, all of which probably reflect interactions between the virus and its host cell. In some cases, the virus may act as a switch to turn on cellular processes that were either functioning at a lower level before viral infection or were even essentially dormant. In support of this idea, it is well-documented that an increase in host cell replication accompanies viral replication, and although most mammalian cells in culture undergo some spontaneous transformation, the frequency is considerably lower than observed subsequent to viral infection. The correction of viral DNA mismatched base-pairs may represent a cellular response to a large dose of foreign DNA. Although the available data are consistent with a response concerned with repair, they are as yet too sparse to allow definite conclusions to be drawn about the presence of such a mode of repair in mammalian cells.

In addition to the examples presented here in detail, there are other types of studies concerned with the interaction of small DNA tumour viruses and their host cells which may be relevant to the interaction between the species and possibly to the general field of mammalian DNA repair. For instance, Hale et al. (1980) have shown, using a single-strand specific chemical modifying agent and supercoiled SV40 DNA, that promoters on the viral DNA for E. coli RNA polymerase are selected targets for chemical modification. Their data would suggest that there are regions within the DNA which breathe and may be "hot-spots" for mutagenic agents. To extrapolate from Fig. 2, one such "hot-spot" could be a replication origin. Cornelis et al. (1980) have shown that UV-irradiated SV40 virus survives much better in monkey cells that have been themselves irradiated than in non-irradiated cells. These results are in agreement with earlier work by Das Gupta and Summers (1978) with Herpes simplex, demonstrating an inducible pathway of UV repair in mammalian cells. These studies can and presumably will be extended to examine the cellular functions responsible for viral survival. Finally, Katinka et al. (1980) have sequenced the changes in polyoma virus DNA that allow it to be expressed in undifferentiated embryonal carcinoma cells. The latter cells are not normally "hosts" for polyoma virus, but mutations which involve simple sequence changes near the viral origin of replication are sufficient to allow the virus to grow in the undifferentiated cells. No doubt other examples can be found, but the purpose of this short report is not to provide a full review but rather to try to overcome the hesitation of scientists in one field to using the tools of another field, and to emphasise some routes by which this can be usefully done.

REFERENCES

Bendig, M., Thomas, T. and Folk, W.R., 1980, J. Virol., 33:1215.
Crémisi, C., 1979, in: "Microbiological Reviews", p. 297.
Cornelis, J.J., Lupker, J.H. and van der Eb, A.J., 1980, Mutation Res., 71:139.
Das Gupta, U.B. and Summers, W.C., 1978, Proc. Nat. Acad. Sci. U.S.A., 75:2378.
Day, R.S., 1978, in: "DNA Repair Mechanisms", Hanawalt, Friedberg and Fox, ed., ICN-UCLA Symp. on Molecular and Cellular Biology, IX:531.
Deininger, P.L., Esty, A., LaPorte, P., Hsu, H. and Friedmann, T., 1980, Nuc. Acids Res., 8:855.
DiMaio, D. and Nathans, D.J., 1980, J. Mol. Biol., 140:129.
Fiers, W., Contreras, R., Haegeman, G., Rogiers, R., van de Voorde, A., van Heuverswyn, H., van Herreweghe, J., Volckaert, G. and Ysebaert, H., 1978, Nature, 273:113.
Griffin, B.E. and Maddock, C., 1979, J. Virol., 31:645.
Hale, P., Woodward, R.S. and Lebowitz, J., 1980, Nature, 284:640.
Hattori, J., Carmichael, G.G. and Benjamin, T.L., 1979, Cell, 16:505.
Ito, Y., Brocklehurst, J.R. and Dulbecco, R., 1977, Proc. Nat. Acad. Sci. U.S.A., 74:4666.
Ito, Y., Spurr, N. and Griffin, B.E., J. Virol., in press.
Jelinek, W.R., Toomey, T.P., Leinwand, L., Duncan, C.H., Brio, P.A., Choudary, P.V., Weissman, S.M., Rubin, C.M., Houck, C.M., Deininger, P.L. and Schmid, C.W., 1980, Proc. Nat. Acad. Sci. U.S.A., 77:1398.
Katinka, M., Yaniv, M., Vasseur, M. and Blangy, D., 1980, Cell, 20:393.

Lai, C.-J. and Nathans, D., 1975, Virol., 66:70.

Lim, V.I. and Mazanov, A.L., 1978, FEBS Letters, 88:118.

Maxam, A.M. and Gilbert, W., 1977, Proc. Nat. Acad. Sci. U.S.A., 74:560.

Miller, L.K., Cooke, B.E. and Fried, M., 1976, Proc. Nat. Acad. Sci. U.S.A., 73:3070.

Novak, U., Dilworth, S.M. and Griffin, B.E., 1980, Proc. Nat. Acad. Sci. U.S.A., 77:3278.

Reddy, V.B., Thimmapaya, B., Dhar, R., Subramanian, K.N., Zain, B.S., Pan, J., Ghosh, P.K., Celma, M.L. and Weissman, S.M., 1978, Science, 200:494.

Sanger, F., Nicklen, S. and Coulson, A.R., 1977, Proc. Nat. Acad. Sci. U.S.A., 74:5463.

Seif, I., Khoury, G. and Dhar, R., 1980, Nuc. Acids Res., 8:2225.

Shortle, D.R., Margolskee, R.F. and Nathans, D., 1979, Proc. Nat. Acad. Sci. U.S.A., 76:6128.

Soeda, E. and Griffin, B.E., 1978, Nature, 276:294.

Soeda, E., Arrand, J.R., Smolar, N. and Griffin, B.E., 1979, Cell, 17:357.

Soeda, E., Arrand, J.R., Smolar, N., Walsh, J.E. and Griffin, B.E., 1980, Nature, 283:445.

Tjian, R., 1978, Cell, 13:165.

Tooze, J. (ed.), 1980, "Molecular Biology of Tumor Viruses Part 2: DNA Tumor Viruses", Cold Spring Harbor Lab.

Yang, R.C. and Wu, R., 1979, Science, 206:456.

OLIGODEOXYRIBONUCLEOTIDES AS SITE-SPECIFIC MUTAGENS

Michael Smith and Shirley Gillam

Department of Biochemistry, Faculty of Medicine
University of British Colombia, 2075 Wesbrook
Place, Vancouver, B.C. Canada VGT 1W5

INTRODUCTION

Classical genetic analysis of cells and viruses involves spontaneous or induced mutation together with screening or selection for a changed phenotype followed by assignment of the changed DNA to a specific location in the genome. This approach has been very useful in defining the various functions of DNA both in coding and regulatory regions. The advent of rapid DNA sequence determination methods has created a new situation where extensive regions of DNA sequences can be known without a precise knowledge of the functions of that DNA. Deliberate, defined modification of a DNA at a specific site followed by examination of the biological properties of the modified DNA offers a way out of this dilemma. A number of methods have recently been developed for in vitro modification of the genomic DNA and recombinant plasmids (1,2). These methods in general operate by introduction, using chemical or enzymatic means, of a limited number of changes in the vicinity of a restriction endonuclease cleavage site. The procedure has been called site-directed mutagenesis, reversed genetics or the production of constructed mutants (1,2). Whilst the general locality of the introduced changes are known, their nature or exact position are not defined until the sequence of the mutant DNA is known. Usually more than one type of changed DNA is produced. In this article an alternate strategy for the production of defined point mutations, site-specific mutagenesis, will be described. This method uses a synthetic oligodeoxyribonucleotide,

261

different at only one nucleotide from wild-type, as a
specific mutagen, and involves in vitro integration
of the oligodeoxyribonucleotide into genomic DNA. The
experiments described here were performed with bacteriophage ØX174 DNA; in principle the method is applicable
to any circular genome of a similar size. Thus the
method is appropriate not only for small circular
bacteriophage and viral DNAs but also for parts of
larger genomes cloned as recombinants in bacterial
plasmid vectors.

The attractive and unique features offered by this
approach to the construction of mutants are that (i)
it is not constrained to the neighbourhood of a
restriction endonuclease cleavage (ii) a specific
nucleotide in the DNA is targeted and (iii) the precise
change at that nucleotide (transition, transversion,
insertion or deletion) is programmed.

The Oligodeoxyribonucleotides

Factors which have to be considered in deciding on
an oligodeoxyribonucleotide to be used as a site-
specific mutagen include (i) the ease of synthesis (ii)
the length of the oligodeoxyribonucleotide required
for a stable duplex which includes one mismatched
nucleotide pair (for transition or transversion
changes) or one nucleotide loop out (for insertions or
deletions) and (iii) the length of oligodeoxyribonucleotide required to recognize a unique complementary
sequence for a genome of the size under investigation.

At the moment, there are rapid developments in the
chemical synthesis of oligodeoxyribonucleotides of
defined sequence and it is possible that the chemistry
of oligodeoxyribonucleotide synthesis will shortly be
as accessible to the molecular biologist as is the
chemistry of peptide synthesis or that of DNA sequence
determination by the method of Maxam and Gilbert.
However, at the time of the inception of the studies
discussed in this paper it was decided that a simple
enzymatic method for oligodeoxyribonucleotide synthesis
was required. To this end, conditions were devised so
that E. coli polynucleotide phosphorylase would
catalyse the primer-depended stepwise reaction of
deoxyribonucleoside-5' diphophates (3-10). This reaction
was used to make all of the oligodeoxyribonucleotides
used in the present studies, and is capable of incor-
porating analogs (9). The advantage of a stepwise
enzymatic method is that it uses procedures with which
biochemists are completely familiar. Because the

reaction is kinetically controlled and because of the variability in rates of different phosphodiester forming steps, on occasion yields are low. However, the scale on which the subsequent mutagenic reaction is carried out is of the order of 1 pmole; thus only minute amounts of the synthetic oligodeoxyribonucleotide are required (11,12).

The realization that relatively short oligodeoxyribonucleotides could form stable duplex (Watson-Crick) structures bridging a mismatched nucleotide pair came from a series of studies on oligonucleotide duplex structures using synthetic model oligonucleotides (13-17). The interesting observation was that, even with the mismatch centrally placed in the duplex, stable structures could be obtained with oligodeoxyribonucleotides containing as few as seven or eight nucleotides at high ionic strength and low temperature (17). Clearly, it ought to be possible to utilize such structures, involving relatively accessible synthetic oligodeoxyribonucleotides, to induce specific point mutations.

The length, n, required for an oligodeoxyribonucleotide to recognize a unique sequence in a DNA strand is such that 4^n = number of nucleotides in the DNA strand (14). Thus for small bacteriophages, viruses and plasmid DNAs which contain 5,000-10,000 nucleotides, n is 6 to 7. Since this is a statistical calculation, certain sites may require a slightly longer oligodeoxyribonucleotide for unique recognition. However, the similarity of the length of oligodeoxyribonucleotide demanded by the physical chemistry and by the genetic complexity is both remarkable and gratifying.

Strategy for Oligodeoxyribonucleotide Mutagenesis

The target for these initial studies on oligodeoxyribonucleotide mutagenesis was the genome of bacteriophage ØX174 DNA because its total sequence is known and because of its relatively simple biology (18). Earlier studies on in vitro replication of ØX174 DNA using purified enzymes had shown that, using an oligodeoxyribonucleotide primer, single stranded bacteriophage ØX174 DNA can be converted to infective closed circular double-stranded DNA using E. coli DNA polymerase I and DNA ligase (19-22). Under the conditions used in these experiments, at 20°C, oligodeoxyribonucleotides containing nine or more nucleotides were optimal primers although there was detectable priming with tri- and tetra-deoxyribonucleotides (22). These

results suggested that an oligodeoxyribonucleotide
mismatched at one nucleotide with complementary
bacteriophage ØX174 DNA could be used as a primer for
E. coli DNA polymerase I and then could be integrated
into circular duplex DNA using DNA ligase to yield
a molecule which is a heteroduplex at one position.
Transfection with this DNA should produce progeny
derived from both strands. This basic strategy is
diagrammed in Figure 1. It was evident from studies
on oligodeoxyribonucleotide priming that the 5'-exonu-
clease activity of E. coli DNA polymerase I edits out
most of the primer before ligation (21). It is also
evident that E. coli DNA polymerase I is one of the most
effective DNA synthesizing enzymes when faced with
the need to convert a long single-stranded template to
double-stranded DNA (23). Hence the large, polymerizing
fragment of E. coli DNA polymerase I (Klenow enzyme),
from which the 5'-exonuclease had been removed proteoly-
tically, was used in these experiments (11). Other
experiments now have confirmed that the native E.coli
DNA polymerase I requires a rather long 5'-sequence
of an oligodeoxyribonucleotide mismatched with DNA at
one position if the mismatch is to be effectively
protected against editing by the 5'-exonuclease(24).

If a heteroduplex ØX174 DNA is used to transfect
E. coli spheroplasts, one would expect that 50% of
the progeny phage would be derived from one strand and
50% from the other. Factors which would be expected
to reduce this efficiency, in addition to the 5'-exo-
nuclease discussed above, include the 3'-exonuclease
of the E. coli DNA polymerase I which is present in the
large polymerase fragment of the enzyme. The activity
could edit out the nucleotide mismatch by degradation
of the oligodeoxyribonucleotide from the 3'-end.
Mismatch repair can be influenced by the genetics of
the host bacterium (25) and also by the distance of
the mismatch from the origin of phage DNA replication
(26). These factors, relating mismatch repair to the
efficiency of mutagenesis have yet to be investigated.
One further factor which could influence the yield of
mutant progeny in the case of ØX174 is a consequence
of the high infectivity of single-stranded DNA (19,27,
28). Therefore, a step involving treatment with the
single-stranded specific endonuclease Sl is used to
inactivate residual single-stranded bacteriophage DNA
(Figure 1).

The statistical calculation discussed above predicts
that an oligodeoxyribonucleotide of the appropriate

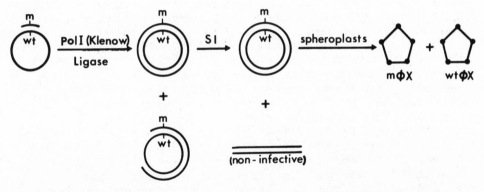

Figure 1.   Strategy for preparation and isolation of a
specific mutation (m) in ØX174. A mutant oligodeoxyribonucleotide is used as primer for E. coli DNA
polymerase I, large fragment [Pol I (Klenow)] with wild-
type bacteriophage DNA as template. The reaction mixture
also contains T4 DNA ligase (Ligase) to convert the
fully extended copy strand to a closed circle. Infective
incomplete molecules are inactivated by the single-
stranded specific endonuclease S1 (S1) before transfection of E. coli spheroplasts to yield a mixed population of mutant and wild-type ØX174.

length should recognize only one site in a complementary strand of genomic DNA. However, it is useful
to be assured that there is not an above average
frequency of occurence of the sequence of interest.
This can be readily established for a known genomic
sequence by using sequence searching techniques (29).
It can also be experimentally established that an
oligodeoxyribonucleotide interacts with only one site
(see later).

Transition Mutations

The first experiments to test the possibility of
using synthetic oligodeoxyribonucleotide as site-
specific mutagens were directed at inducing the two
types of transition change AT→GC and GC→AT, since the
mismatched nucleotide pairs required, viz AC and GT,
should least perturb a DNA duplex (11). The target was
nucleotide 587 of the ØX174 genome. In wild-type

bacteriophage DNA, this nucleotide is G and when this
is replaced by A it generates an amber mutation in the
reading frame of gene E, the lysis function (Figure 2).
The oligodeoxyribonucleotides used in the initial
studies (Figure 2) each contained twelve nucleotides;
there is no sequence in ØX174 bacteriophage DNA complementary to these decadeoxyribonucleotides except at
positions 582 to 593 (excluding the mismatch at 587).
This was established by computer searches of the ØX174
DNA sequence (18,29). That the oligodeoxyribonucleotides
would prime specifically only at the desired site in OX174
DNA was confirmed in a pulse-chase experiment which
accurately measures the number of priming sites and
their distance from adjacent restriction endonuclease
sites (11). Only one site was detected; it was situated
the number of nucleotides from an adjacent Hae III site
which is predicted by the ØX174 DNA sequence (18).

Using the strategy diagrammed in Figure 1, with a
temperature for priming of DNA synthesis of 25°, the
oligodeoxyribonucleotide pGTATCCTACAAA induced a G→A
change at position 587 with about 15% efficiency.
Reversion of this change, A→G, was induced by
pGTATCCCACAAA with a similar efficiency (11). It was
noticeable that in these initial experiments, the
endonuclease S1 had a very beneficial effect, increasing
the fraction of mutated progeny from about 0.1% to
about 15% (11). Presumably this reflects incomplete
elongation of the primer and/or incomplete ligation of
the elongated product. In more recent experiments
using the same primers at the same temperature, but with
a different buffer composition and also with a more
active T4 DNA ligase preparation, the efficiency of
mutant production prior to endonuclease S1 treatment
was 2% to 10% (12). In these experiments, the fraction
of progeny that were mutant after endonuclease S1
treatment was 15% to 22% (12) suggesting that the
effect is due to more efficient production of closed
circular duplexes rather than to a greater efficiency
of production of the specific nucleotide change.

Because the temperature at which the above experiments
were carried out is likely to be close to the melting
temperature of the oligodeoxyribonucleotide duplex, it
was of interest to investigate the effect on yield of
mutant of reducing the temperature at which the
priming reaction was carried out in concert with a
study of the effect of oligodeoxyribonucleotide length
on efficiency of mutant production (12). The results
of a series of studies on the efficiency of produciton

Figure 2. Nucleotides 561 to 600 of wild-type ØX174 bacteriophage DNA and the sequences of the oligodeoxyribonucleotieds complementary to the region around nucleotide 587 (underlined). The mutant complementary oligodeoxyribonucleotide, pGTATCCTACAAA (mismatched nucleotide underlined) induces a G→A change at 587. This produces an amber codon (TAG) in the reading frame of gene E [this is the same mutation as the lysis defective mutation am3 (11, 18)]. The codon change in the reading frame of the overlapping gene D,GTG→GTA does not change the aminoacid (18). The reversion of the mutation at position 587, A→G, is produced by the wild-type complementary oligodeoxyribonucleotide, pGTATCCCACAAA (mismatched nucleotide underlined).

wild-type complementary oligodeoxyribonucleotide

3' A-A-A-C-A-C-C-T-A-T-Gp 5'

mutant complementary oligodeoxyribonucleotide

3' A-A-A-C-A-T-C-C-T-A-T-Gp 5'

wild-type ØX174 DNA   5' -T-G-C-G-T-T-T-A-T-G-G-T-A-C-G-C-T-G-G-A-C-T-T-T-G-T-G-G-A-T-A-C-C-C-T-C-G-C-C-T 3'
                                                570   590   600

gene D   - Cys - Val - Tyr - Gly - Thr - Leu - Asp - Phe - Val - Gly - Tyr - Pro - Arg -

gene E   (Met) - Val - Arg - Trp - Thr - Leu - Trp - Asp - Thr - Leu - Ala -
                                                                                        Ter

Table 1. Effect of length of oligodeoxyribonucleotides and temperature on the change change A → G at position 587 of ØX174 viral DNA (am3 → wild-type) when endonuclease S_1 was used to degrade single stranded phase DNA.

Primer (mismatch is underlined)	33°			25°			10°			0°		
	Total	am^+	%	Total	am^+	%	Total	am^+	%	Total	am^+	%
pGTATCCCACAA	0.5	0.01	2	6	0.9	15	20	7.5	38	27	10	39
pGTATCCCACA	7.7	0.25	3	25	6.6	26	65	24	37	78	27	37
pGTATCCCAC	5.4	2.5×10^{-2}	0.5	20	2.5	13	52	10	19	67	17	25
pGTATCCCA	3.6	2×10^{-3}	5×10^{-2}	9	0.25	3	27	2.7	10	27	5.6	20
pGTATCCC	3.6	3.6×10^{-5}	1×10^{-3}	16	0.03	0.2	18	0.12	0.7	23	0.1	0.5

Total and mutated progeny plaques ($X10^{-6}$) produced at different temperatures

The template viral DNA was am3cs70 and the progeny plaques listed above were produced by 0.1 µg of viral template DNA. In a control experiment with the homologous primer, pGTATCCTAC, the percentage of revertants obtained at 0° was 4×10^{-2}%.

of an A→G change at position 587 of ØX174 DNA is shown
in Table 1; the temperature used for priming DNA synthesis ranged from 0° to 33° and the series of oligodeoxyribonucleotides were heptamer, pGTATCCC, up to the
undecamer pGTATCCCACAA (12). The hoped for results
were obtained: reduction of the temperature of priming
increased the efficiency of mutant production with all
lengths of olideoxyribonucleotide primer. Thus with a
priming temperature of 0°, pGTATCCCACA and pGTATCCCACAA
produced the desired mutation with close to the theoretical efficiency expected of a heteroduplex (50%) and
the octadeoxyribonucleotide was a very efficient mutagen.
Similar results were obtained when the series of oligodeoxyribonucleotides, pGTATCCT to pGTATCCTACAAA were
used to induce the change G→A at position 587 of
ØX174 DNA (12).

Transversion Mutations

The subject of the foregoing section, transition
mutations, is a class of mutational change which can
be induced with reasonable predictability using
nucleoside analogs or chemical reagents (1,2). One of
the potentials of oligodeoxyribonucleotide mutagenesis,
in addition to precise targeting, is the possibility
of programming other types of point mutation. Hence
it was of interest to investigate the possibility of
inducing transversion mutations. The target for these
studies was nucleotides 5276 of ØX174 (30). This
normally is G and can be replaced by T by constructing
a heteroduplex with a purine-purine (GA) mispairing
(Figure 3). Reversion of this change is achieved by a
pyrimidine-pyrimidine (TC) mispairing. The two oligodeoxyribonucleotides constructed to include these
changes were pCCCAGCCTAAA and pCCCAGCCTCAA respectively
and correspond complementarily, to nucleotides 5274-
5284 of ØX174 DNA. These oligodeoxyribonucleotides
contain eleven nucleotides and the mismatched are at
the third nucleotide from the 3'-terminus. These facts
together with the less favorable stereochemistry of
the mismatches suggest that these oligodeoxyribonucleotides provides a more stringent test of the
mutagenic method than the earlier experiments at position 587. In fact, when the G→T transversion at
position 5276 was attempted using a temperature of 25°
for priming DNA synthesis, no mutants were detected
(30). The experiments described in the previous section
which defined optimum priming conditions at 0°, were
next undertaken. When pCCCAGCCTAAA was used as a primer
at 0° on wild-type ØX174 DNA template followed by the

Figure 3. Nucleotides 5256 to 5287 of wild-type ØX174 bacteriophage DNA and the sequences of oligodeoxyribonucleotides complementary to the region around nucleotide 5276 (mismatched nucleotide underlined). The mutant complementary oligodeoxyribonucleotide, pCCCAGCCTAA (mismatched nucleotide underlined) induces a G→T change at position 5276. This produces an amber codon in the reading frame of gene B [the same mutation as am16 (30)]. The codon in the reading frame of the overlapping gene A, TTG→TTT, results in a changed amino-acid. The reversion of the change at position 5276, T→G is induced by the wild-type complementary oligodeoxyribonucleotide (mismatched nucleotide underlined).

wild-type complementary oligodeoxyribonucleotide   3' -A-A-C-T-C-C-G-A-C-C-C-p -5'

mutant complementary oligodeoxyribonucleotide   3' -A-A-A-T-C-C-G-A-C-C-C-p -5'

ØX174 wild-type DNA   5' -G-C-A-A-A-A-G-A-G-A-A-T-G-A-G-A-T-T-G-G-G-C-T-G-G-G-A-A-A- 3'
                                      5260   5270   5280

gene B   - Lys - Lys - Arg - Asp - Glu - Ile - Glu - Ala - Gly - Lys -
                                                                 Ter

gene A   - Lys - Lys - Arg - Glu - Met - Arg - Leu - Arg - Leu - Gly -
                                                                 Phe

general strategy of Figure 1, 13% of the progeny phage were the desired mutant (12,30).

When the oligodeoxyribonucleotide pCCCAGCCTAA was used under the same conditions to revert the nucleotide change at position 5276 (T→G) no changed phage were detected (30). The explanation for this is evident from Figure 4. This shows results of the pulse-chase experiment described earlier which defines the number of priming sites on ØX174 DNA for a particular oligodeoxyribonucleotide. Thus, pCCCAGCCTAAA only has one strong priming site whereas pCCCAGCCTCAA has two. Examination of the ØX174 DNA sequences shows that a complement of the 3'-terminal octanucleotide of this sequences occurs twice starting at position 553 and 4060. The additional priming site defined by fragment B (Figure 4) corresponds to that at position 553 since it is 125 nucleotides from a Hae III site (18). Presumably priming at this second site interfered with the integration of the oligodeoxyribonucleotide at the desired site into a covalently closed circular duplex. Knowledge of the phage DNA sequence adjacent to the 3'-end of the desired priming site allowed the primer to be extended preferentially by DNA polymerase in the presence of only three deoxyribonucleoside triphosphates. Subsequent elongation at 37° prevented reaction at the unwanted site at position 553. Application of the strategy of Figure 1 then resulted in the desired mutant being produced in 19% yield (12,30). Comparison of the results of the experiments at position 5276 and those at position 587 suggests that the yields of mutants in each case relate to the number of nucleotides present in the mutagenic oligodeoxyribonucleotides beyond the 3'-end of the mismatch site. Thus, similar yields are obtained when there are two extra nucleotides. Presumably this reflects the extent of protection from the editing action of the 3'-exonuclease of E.coli DNA polymerase I. The results in Table 1 also suggest that three nucleotides on the 3'-end of a mismatch provide fairly complete protection against exonuclease action.

In vitro Selection of Mutant DNA

The experiments described in the two preceding sections involve the production or reversion to wild-type or nonsense mutations, for which there are convenient screening systems. Many desirable mutational changes do not have convenient screens or selection procedures; some changes may be phenotypically silent.

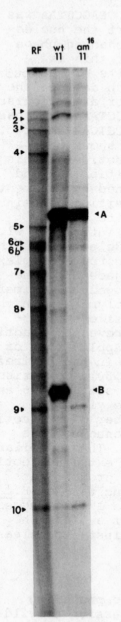

Figure 4. Gel electrophoresis of products which demonstrate specific priming by pCCCAGCCTCAA (wt-11) and pCCCAGCCTAAA (am 16-11) on wild-type ØX174 DNA template (30). The product of pulse (^{32}P)-phase synthesis of DNA was cleaved with Hae III, denatured and separated by gel electrophoresis. The size markers (RF) are fragments of ØX174 DNA produced by Hae III cleavage of ^{32}P-labelled RF DNA followed by denaturation (nucleotide residues in each fragment are 1.1353; 2,1078; 3,872; 4,603; 5,310; 6a,271; 6b,281; 7,234; 8,194; 9,118; 10,72). The fragments A indicate that homologous and heterologous olideoxyribonucleotides prime at the same site. The fragment B indicates that the wild-type oligodeoxyribonucleotide also primes at a second site (see text).

In principle such mutants could be isolated by sequencing individual DNA clones. However, an oligodeoxyribonucleotide mutagen offers the potential for in vitro selection of the desired mutant DNA from a mixture with wild-type DNA. The mutagenic oligodeoxyribonucleotide will form a perfect duplex with the mutant DNA which is more stable than a duplex with one mismatched nucleotide pair (17); the mutagenic oligodeoxyribonucleotide paired with mutant DNA therefore should be a more efficient primer-template combination than is the same oligodeoxyribonucleotide paired with wild-type DNA. The strategy for this experiment is diagrammed in Figure 5 (31).

A temperature of 25° is used for the priming of DNA synthesis because the studies discussed earlier in this

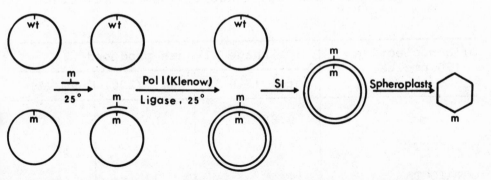

Figure 5. In vitro selection of mutant ØX174 DNA using an oligodeoxyribonucleotide. The oligodeoxyribonucleotide can be the one used to induce the mutation or a homolog (see text). At 25°, the oligodeoxyribonucleotide is a more efficient primer of E. coli DNA polymerase I (large fragment) with homologous (mutant) DNA template than with heterologous (wild-type) DNA. Hence the mutant DNA is selectively converted, after ligation, to closed circular duplex DNA which is resistant to endonuclease S1. Transfection of E. coli spheroplasts with the DNA which survives S1 treatment yields a phage population specifically enriched with mutant.

article have shown that mismatched oligodeoxyribonucleotide priming is reduced in efficiency at this temperature.

The data in Table 2 show that the strategy diagrammed
in Figure 5 causes marked enrichment of a desired
ØX174 DNA (G→A change at 587) as measured by the
enrichment in mutant phage, about seven-fold with the
most effective oligodeoxyribonucleotide. Other experiments directed at enrichment of the oligodeoxyribo-
nucleotide-induced revertant of the change at nucleotide 587 (31) have resulted in enrichment of up to
thirty-fold (Table 3). Clearly, depending on the initial
yield of mutant DNA, one or two cycles of this enrichment procedure is capable of producing a population of
DNA molecules which is close to 100% the desired
mutant, irrespective of phenotype. Sequence determination of the DNA from only one or two clones can
be used to confirm that the desired mutant has been
isolated.

Table 2. In vitro selection of am ØX174 DNA (A at 587)
         using oligodeoxyribonucleotides of different
         length.[a]

Oligonucleotide (20 pmoles)	Phage plaques produced		
	Total (X10^{-5})	am (/200)	am (%)
–	0.85	10	5
pGTATCCT	7	32	16
pGTATCCTA	14	32	16
pGTATCCTAC	5.7	50	25
pGTATCCTACA	4.3	85	42

[a]Experiments were carried out as described in Figure 5.
The mixture of template ØX174 am3sB1 and ØX174 am$^+$sB1
DNAs (0.92 pmoles per experiment) was isolated from
a mixture of phage, assaying as 10% am and 90% am$^+$,
which was produced by mutation of ØX174 am$^+$sB1 DNA
using pGTATCCTAC. Phage plaques resulted from transfection with DNA derived from 0.1 µg of template DNA.

Table 3.   In vitro selection of ØX174 am$^+$ DNA from mixture with ØX174 am^3 DNA using pGTATCCCACA.[a]

am$^+$ DNA (%) in template DNA	am$^+$ Phage plaques produced (%)	
	without oligonucleotide	with oligonucleotide
0.25 pmoles total DNA/experiment		60 pmoles per experiment
0	0	9
1	1.25	18
10	10	60
20	17	80
50	46	100
0.76 pmoles total DNA/experiment		20 pmoles per experiment
0	0	12
1	1.4	33

[a]DNA was used to transfect E. coli spheroplasts and the progeny phage assayed on su$^+$ and su$^=$ hosts. In experiments without oligonucleotide, the template mixtures of ØX174 am3cs70 and ØX174 am$^+$cs70 viral DNAs were used to transfect spheroplasts directly. In experiments with the oligonucleotide, the procedure diagrammed in Figure 5 was followed using the standard conditions for priming of E. coli. DNA polymerase 1 (large fragment), ligation with T4 DNA ligase, enrichment of closed circular duplex DNA with endonuclease S1, spheroplast transfection and progeny phage assay.

Deletion Mutants

In order to develop the full range of oligodeoxyribo-nucleotide-induced point mutations, the deletion of a single nucleotide in ØX174 DNA has been investigated (32). For obvious reasons, an inter-genic target was chosen, nucleotide 2925 which lies between genes G and H (Figure 6). The nucleotide (T) is part of the sequence coding for the ribosome binding site for gene H; its removal would be expected to affect the effici-ency of production of protein H.

Figure 6. Nucleotides 2911 to 2939 wild-type ØX174 bacteriophage DNA and the sequence of the mutant complementary oligodeoxyribonucleotide, pAAATCCCTCA (missing nucleotide indicate by dash) around nucleotide 2925 (the T in the middle of the Hph I recognition sequence). The mutant oligodeoxyribonucleotide induce the deletion of nucleotide 2925 and concommitant inactivation of the Hph I site.

wild type viral DNA

$$5'-C-C-A-C-T-T-A-A-G-\boxed{T-G-A}-G-G-T-G-A-T-T-T-\boxed{A-T-G}-T-T-T-G-G-T-3'$$

2911   Hph I   2939

mutant complementary oligodeoxyribonucleotide

$$3'-A-C-T-C-C- -C-T-A-A-A-p-5'$$

mutant viral DNA

End G   Start H

$$5'-C-C-A-C-T-T-A-A-G-\boxed{T-G-A}-G-G-A-T-T-T-\boxed{A-T-G}-T-T-T-G-G-T-3'$$

Mutagenesis, using the oligodeoxyribonucleotide, pAAATCCCTCA, followed the usual strategy (Figure 1). The efficiency of production of mutant bacteriophage ØX174 DNA could be monitored from the fraction of molecules from which an Hph I restriction endonuclease site had been removed (Figure 6). The initial yield of mutant DNA was low (<5%). However, three cycles of DNA selection and phage isolation produced a population of phage in which four out of four clones had nucleotide 2925 delected (Figure 7). It is of interest that

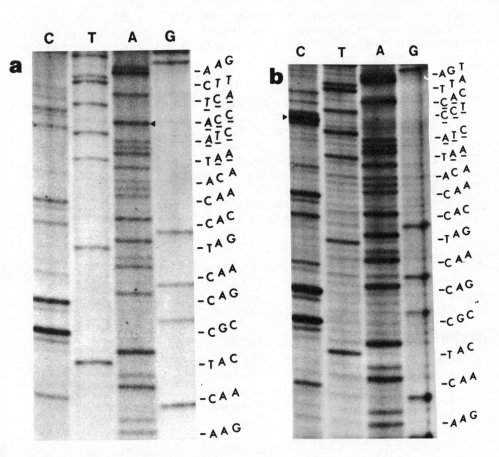

Figure 7. Sequence of the complementary strand of wild-type ØX174 DNA (a) and of the mutant (b) from which nucleotide 2925 (A in the complementary strand) has been deleted. Sequence determination was by the enzymatic terminator method.

growth of phage at several temperatures did not reveal
any differences from the behaviour of wild-type phage
nor did an examination of the in vivo synthesis of
phage protein. Thus, according to these criteria, the
mutation is phenotypically silent. This in turn
emphasizes the unique power of the oligodeoxyribonucleotide selection of mutant DNA.

Conclusions

The experiments described in the foregoing demonstrate the power of the oligodeoxyribonucleotide
method in programming defined point mutations in the
ØX174 genome. The use of this method to produce
defined nonsense codons opens the way to precise
definition of the coding function of regions of DNA
sequence and the definition of the biological functions
of the derived protein (28).

The development of the in vitro selection method
for efficient isolation of mutant DNA opens the way
to the isolation of phenotypically silent mutants.
Apart from use of such mutants in defining the role
of regulatory regions of the DNAs of phage such as
ØX174, the selection procedure will allow specific
mutation of DNA which is not normally expressed in
E. coli, e.g. eukaryote DNA cloned in a plasmid
vector. The mutated eukaryote DNA can be reinserted
in its homologous organism to define the effect of
the mutation.

ACKNOWLEDGEMENT

Research in the authors' laboratory was supported
by a grant from the Medical Research Council of
Canada of which M.S. is a Career Investigator.

REFERENCES

1. Shortle, D., Pipas, J., Lazarowitz, S., DiMajo, D.
   and Nathans, D. (1979). In Genetic Engineering.
   Principles and Methods. Volume I. Setlow, J.K. and
   Hollaender, A. (Eds.). Plenum Press, New York, pp.
   73-92.
2. Weissmann, C,. Nagat, S., Taniguchi, T., Webber, H.
   and Meyer, F. (1979). In Genetic Engineering.
   Principles and Methods. Volume I. Setlow, J.K. and
   Hollaender, A. (Eds.). Plenum Press, New York, pp.
   133-150.
3. Gillam, S. and Smith, M. (1972) Nature New Biol.
   238, 233-234.

4. Gillam, S. and Smith, M. (1974).Nucleic Acids Res.
   1, 1631-1647.
5. Gillam, S., Waterman, K., Doel, M. and Smith, M.
   (1974). Nucleic Acids Res. 1, 1649-1664.
6. Gillam, S., Waterman, K. and Smith, M. (1975)
   Nucleic Acids Res. 2, 613-624.
7. Gillam, S., Jahnke, P. and Smith, M. (1978)
   J. Biol. Chem. 253, 2532-2539.
8. Trip, E.M. and Smith, M. (1978). Nucleic Acids Res.
   5, 1529-1538.
9. Trip, E.M. and Smith, M. (1978). Nucleic Acids Res.
   5, 1539- 1549.
10. Gillam, S. and Smith, M. (1980). Methods in
    Enzymology 65, 687-701.
11. Hutchison, C.A. III, Phillips, S., Edgell, M.H.,
    Gillam, S., Jahnke, P. and Smith, M. (1978).
    J. Biol. Chem. 253, 6551-6560.
12. Gillam, S. and Smith, M. (1979). Gene 8, 81-97.
13. Astell, C. and Smith, M. (1971). J. Biol. Chem.
    246, 1944-1946.
14. Astell, C.R. and Smith, M. (1972). Biochemistry 11,
    4114-4120.
15. Astell, C.R., Doel, M.T., Jahnke, P.A. and Smith, M.
    (1973). Biochemistry 12, 5068-5074.
16. Doel, M.T. and Smith, M. (1973). FEBS Letters 34,
    99-102.
17. Gillam, S., Waterman, K. and Smith, M. (1975).
    Nucleic Acids. Res. 2, 625-634.
18. Sanger, F., Coulson, A.P., Friedmann, T., Air, G.M.,
    Barrell, B.G., Brown, N.L., Fiddes, J.C., Hutchinson,
    C.A. III., Slocombe, P.M. and Smith, M. (1978).
    J. Mol. Biol.l 125, 225-246.
19. Goulian, M, Kornberg, A. and Sinsheimer, R.L. (1967).
    Proc. Nat. Acad. Sci. U.S.A. 58, 2321-2328.
20. Goulian, M., (1968). Proc. Nat. Acad. Sci. U.S.A.
    61, 284-291.
21. Goulian, M. (1968). Cold Spring Harbour Symp. Quart.
    Biol. 33, 11-20.
22. Goulian, M., Goulian, S.H., Codd, E.E. and Blumenfield,
    A.Z. (1973). Biochemistry 21, 2893-2901.
23. Sherman, L.A. and Gefter, M.L. (1976). J. Mol. Biol.
    103, 61-76.
24. Razin, A., Hirose, T., Itaka, K. and Riggs, A.D.
    (1978). Proc. Nat. Acad. Sci. U.S.A. 75, 4268-4270.
25. Glickman, B.W. (1979). Mutation Res. 61, 153-162.
26. Baas, P.D. and Jansz, H.S. (1978). In The Single-
    Stranded DNA Phages. Denhardt, D.T., Dressler, D.
    and Ray, D.S. (Eds.) Cold Spring Harbor Laboratory
    New York, pp. 215-244.

27. Sinsheimer, R.L. (1968). In Progress in Nucleic
    Acid Research and Molecular Biology. Volume 8.
    Davidson, J.N. and Cohn, W.E. (Eds.). Academic
    Press, New York, pp. 115-167.
28. Bhanot, G.S., Kahn, S.A. and Chambers, R.W. (1979).
    J. Biol. Chem. 254, 12684-12693.
29. McCallum, D. and Smith, M. (1977). J. Mol. Biol.
    116, 29-30.
30. Gillam, S., Jahnke, P., Astell, C., Phillips, S.,
    Hutchison, C.A. III and Smith, M. (1979). Nucleic
    Acids Res. 6, 2973-2985.
31. Gillam, S. and Smith, M. (1979). Gene 8, 99-106.
32. Gillam, S., Astell, C.R. and Smith, M. (1980).
    Unpublished results.

CLONING OF $\underline{recA}^+$ AND $\underline{lexA}^+$ AND SOME OF THEIR MUTANT ALLELES;

AN INVESTIGATION OF THEIR MUTUAL INTERACTION

P. T. Emmerson, I. D. Hickson, R. L. Gordon
and A. E. Tomkinson
Department of Biochemistry
The University
Newcastle upon Tyne, UK

ABSTRACT   The $\underline{recA}^+$, recA200 (ts), $\underline{lexA}^+$ and lexA3 alleles
have been cloned into multicopy plasmid vectors. The plasmid
pPE13 ($\underline{recA}^+$) confers only partial UV-resistance on DM49
(lexA3), despite producing an excess of recA protein. This
suggests that lexA has another role in DNA repair in addition
to its role in control of $\underline{recA}$.   A small degree of UV-
sensitivity is conferred upon AB1157 (recA+) by pPE13 ($\underline{recA}^+$).
At 42°, the plasmid pPE23 (recA200, ts) confers a high degree
of UV-sensitivity upon AB1157 (recA+), despite the presence
of the chromosomal recA+ gene.   Purified recA200 (ts)
protein at 42° has a single strand DNA-dependent ATPase
activity which is higher than that of purified recA+ protein.

The $\underline{recA}^+$ gene was subcloned from λprecA (1) by cutting the
phage DNA with endonuclease BstI and ligating a fragment into the
BstI site of pBR322.   Ap^r (Tc^S, UV^r) transformants of AB2463
(recA13) were selected and screened for those that harboured small
plasmids which conferred the ability to synthesise large amounts of
recA protein on induction.   One such plasmid, pPE13 ($\underline{recA}^+$) (4.2 Md;
1.6 Md insert), was transformed into N1462 (spr sfiA ΔrecA).   In
this strain, the $\underline{recA}$ genes on the multicopy plasmids are de-
repressed, because of the spr mutation in lexA, and the recA protein
is synthesised copiously (Figure 1).

The temperature-sensitive allele $\underline{recA200}$ (2) was cloned from
the strain KL399 (recA200).   Chromosomal DNA was cut with BstI, and
inserted into the BstI site of pBR322.   Transformants of KM4196 (ΔrecA)
were selected for Ap^r and screened for Tc^S and MMC^r at 30°.   One
such plasmid, pPE23 (recA200) (4.2 Md; 1.6 Md insert), was used to
transform AB1157 (recA+) and KM4196 (ΔrecA) and transformants examined

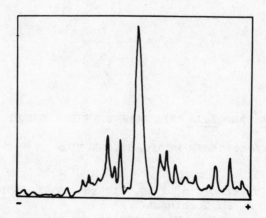

Figure 1. A culture of N1462 (<u>spr sfiA</u> ΔrecA) harbouring pPE13
           (<u>recA</u>⁺) was lysed in SDS buffer and analysed by SDS-PAGE.
           The gel was stained and scanned.   The prominant band
           is the recA protein.

for UV-sensitivity.   The plasmid pPE23 (<u>recA200</u>, ts) confers upon
KM4196 (<u>ΔrecA</u>) almost wild-type UV-resistance at 30° but not at
42° (Figure 2a).

When harboured by AB1157, pPE13 (<u>recA</u>⁺) increases the UV-
sensitivity to a small but significant extent when compared with
pBR322 (Figure 2b).   Thus, an excess of good recA protein in the
cell seems to sensitize it somewhat to UV.   Perhaps the energetic
load on the cell imposed by the high levels of surplus protein
synthesis is generally debilitating to the cell and occurs at the
expense of some of the energy that would be required for repair.
Alternatively, excess recA protein may suppress normal <u>recB/C</u>
nuclease activity in the cell.

Also shown in Figure 2b is that the plasmid pPE13 (<u>recA</u>⁺)
increases the UV-resistance of the strain DM49 (<u>lexA3</u>), but not
up to the level of AB1157 (<u>recA</u>⁺) harbouring pPE13 (<u>recA</u>⁺).   SDS-
PAGE analysis of DM49 harbouring pPE13 reveals an excess of recA
protein.   Thus, the UV-sensitivity of <u>lexA</u> mutants can not be
ascribed entirely to their inability to induce recA protein.   Even
in the presence of an excess of recA protein, defective lexA protein
prevents normal levels of repair.   This suggests that the lexA
protein may control other genes involved in inducible repair, in
addition to <u>recA</u>.

The plasmid pPE23 (<u>recA200</u>, ts) sensitizes AB1157 (<u>recA</u>⁺) to
UV to some extent at 30° and to a much greater extent at 42° (Figure 2
Thus, despite the presence of the chromosomal <u>recA</u>⁺ gene, the

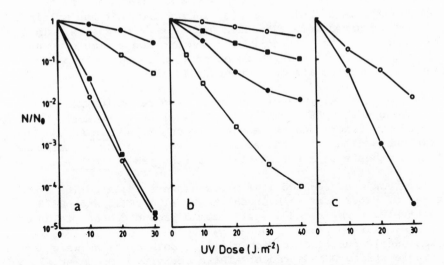

Figure 2. UV survival curves of:
(a) (●) KM4196 (ΔrecA)/pPE13 (recA⁺); (□) KM4196 (ΔrecA)/pPE23 (recA200, ts) 30°; (■) KM4196 (ΔrecA)/pPE23 (recA200, ts) 42°; (O) KM4196 (ΔrecA)
(b) (O) AB1157/pBR322; (■) AB1157/pPE13 (recA⁺); (●) DM49 (lexA3)/pPE (recA⁺); (□) DM49 (lexA3)/pBR322
(c) (O) AB1157/pPE23 (recA200, ts) 30°; (O) AB1157/pPE23 (recA200, ts) 42°.

multiple copies of the defective recA200 (ts) protein at 42° prevent normal DNA repair, possibly by competing with the recA⁺ protein. A similar result was obtained by Ogawa et al. (3) in experiments with a multicopy plasmid carrying the recA1 mutation.

The lexA⁺ gene was cloned from chromosomal DNA. E. coli DNA and pBR322 DNA were digested with EcoRI and PstI endonucleases and a fragment of E. coli DNA inserted between the EcoRI and PstI sites of pBR322 in place of a plasmid fragment. The strain DM961 (tsl lexA3) was transformed and Tcʳ transformants selected at 42° to permit the growth of only tsl⁺ (lexA⁺) transformants. The plasmid pPE24 (lexA⁺) (3.8 Md; 1.8 Md insert) obtained in this way conferred UV-sensitivity upon AB1157 (lexA⁺) (data not shown).

To study the interaction between multiple copies of recA⁺ and lexA⁺ genes and proteins in the same host, the recA⁺ fragment was re-cloned from pPE13 into the compatible plasmid pACYC184 using BstI. The resulting plasmid pPE14 (recA⁺) and pPE24 (lexA⁺) were both transformed into KM4196 (ΔrecA). The presence of both plasmids was demonstrated by their respective drug resistance markers and by agarose gel electrophoresis. However, the normally high level of recA protein expected from pPE24 (recA$^+$) could not be seen on SDS-PAGE analysis of cell extracts, indicating that the multiple recA operator sites were saturated with lexA$^+$ repressors. This titration out of the lexA repressor may provide a simple method of amplifying this protein which is normally only sparcely produced because it regulates its own synthesis (4, 5).

The lexA3 allele was cloned by in vivo recombination between the chromosome of DM49 (lexA3) and pPE24 (lexA$^+$), selecting plasmids that conferred increased UV-sensitivity on AB1157 (lexA$^+$). The plasmid pPE25 (lexA3) obtained in this way also turned off recA protein synthesis in a strain co-harbouring pPE14 (recA$^+$).

The recA$^+$ and recA200 (ts) proteins were isolated (6) from cultures of N1462 (spr sfiA ΔrecA) harbouring pPE13 and pPE23 respectively.   The recA200 (ts) plasmid produced high yields of protein similar to those produced by the recA$^+$ plasmid (Figure 1), and was easily purified in high yield.   Both proteins were assayed for single strand DNA-dependent ATPase activity. Ogawa et al. (3) found that purified recA1 protein was deficient in single strand DNA-dependent ATPase activity.   However, although the plasmid pPE23 (recA200, ts) is similar to a plasmid carrying recA1 (3) in that they both sensitize wild-type hosts to UV, the purified recA200 (ts) protein at 42° was approximately twice as active as an ATPase as the recA$^+$ protein (data not shown).

The cloning of recA200 and lexA described here differs from several other reports of cloning recA and lexA genes in vitro (3, 4, 5, 7), in that here the immediate source of the cloned DNA was the E. coli chromosome rather than a colony bank or a substituted phage.   These studies suggest that chromosomal DNA is a convenient direct source, provided that a positive selection is available for the desired gene.   We have used similar techniques to clone the uvrD-rep and recB/C regions directly from chromosomal DNA.

ACKNOWLEDGEMENTS

We thank Pauline Martin for technical assistance. This work was supported by the Medical Research Council.

REFERENCES

1.   McEntee, K., Hesse, J. E. and Epstein, W. (1976). Proc. Natl. Acad. Sci. USA 73, 3979.
2.   Lloyd, R. G., Low, K. B., Godson, G. N. and Birge, E. A. (1974). J. Bacteriol. 130, 407.
3.   Ogawa, T., Wabiko, H., Tsujimoto, T., Horii, T., Masukata, H.

and Ogawa, H. (1978). Cold Spring Harbor Symp. Quant. Biol. 43, 909.

4.   Little, J. W. and Harper, J. E. (1979). Proc. Natl. Acad. Sci. USA 76, 6147.

5.   Brent, R. and Ptashne, M. (1980). Proc. Natl. Acad. Sci. USA 77, 1932.

6.   Emmerson, P. T., Northrop, F. D., Walker, J. E. and West, S. C. (1979). FEBS Letters 106, 349.

7.   Sancar, A. and Rupp, W. D. (1979). Proc. Natl. Acad. Sci. USA 76, 3144.

PERTURBATION OF THE SOS RESPONSE OF Escherichia coli BY PLASMIDS CARRYING TRUNCATED recA GENES

S.G. Sedgwick and G.T. Yarranton

Genetics Division
National Institute for Medical Research
London NW7 1AA

Several models for the regulation of E. coli recA[+] gene expression and induction of SOS functions envisage that the recA[+] gene is repressed by the lexA[+] gene product. Positive control is thought to occur by "activation" of the basal level of recA protein so that it inactivates both lexA[+] protein and the repressors of the SOS functions[2-5]. Such models predict that an excess of recA promotor-operator sequences would titrate lexA[+] repression of the chromosomal recA[+] gene[2]. The result would be escape synthesis of recA[+] protein, and perhaps easier inducibility of SOS functions. To test these naive predictions recA control sequences were introduced into E. coli by transformation with pBR322 derivatives, carrying the recA control sequence and 100%, 75% and 20% of the recA[+] structural gene (fig.1) pDR1453 and pDR1461 were very generously supplied by Dr. Dean Rupp[6].

Contrary to expectation transformation with pMH1 and pDR1461 carried no increase in the basal level of recA[+] protein (fig.2 lanes 1,3,5) and homologous recombination was normal (table 1). Furthermore scanning densitometry showed that induced amounts of

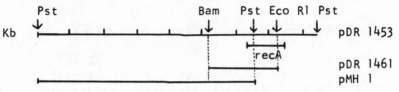

Fig.1.   Restriction fragments of E. coli srl - recA region cloned into pBR322.

recA$^+$ protein following naladixic acid treatment were also similar
to within a factor of two (fig.2, lanes 2,4,6).   DNA degradation
in UV irradiated pDR1461 transformants was normal (table 1) implying
that the recA$^+$ protein induced was functional and able to protect
against nuclease digestion.   In addition to recA$^+$ protein, naladixic
acid treatment of pDR1461 transformants induces synthesis of proteins
with molecular weights of 34,000, 31,000 and 29,000 (fig.2, lane 4).
Little has shown that these proteins are respectively truncated
recA$^+$ protein, β lactamase and processed β lactamase^8.   β lactamase
induction occurs by readthrough from the recA sequence into the
pBR322 bla gene.   Consequently when a shorter recA sequence was
inserted at a different pBR322 site, Pst 1, to generate pMH1, there
was no induction of β lactamase synthesis.   pMH1 should make an
8,000 dalton recA protein fragment but this has not yet been
identified.   Although having little effect on recA$^+$ gene expression,
pMH1 and pDR1461, caused a large increase in radiosensitivity (fig.
3) and reduced or eliminated several of the SOS functions (table 1).
Weigle reactivation and UV induced mutagenesis became undetectable
and λ lysogenic induction required larger doses to induce fewer
cells.   Lysogenic induction, previously thought to be absent in
lexA lysogens^{9,10}, was consistently observed with E.coli lexA3

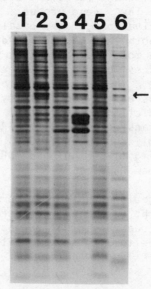

Fig.2.   12.5% SDS - polyacrylamide gel electrophoresis of
           [^{35}S]-methionine labelled proteins from E. coli AB1157,
           lanes 1,2; AB1157 pDR1461, lanes 3,4; AB1157 pMH1,
           lanes 5,6, treated, lanes 2,4,6, or not, lanes 1,3,5,
           with 40μg/ml of naladixic acid for 40 minutes prior
           to labelling.   The arrow marks the position of recA
           protein.

Table 1.   Summary of results

Strain	$recA^+$-dependent process	Parameter	Quantitation
AB1157 (λ)	lysogenic induction[7]	fraction induced	>30%
AB1157pDR1461(λ)			10-15%
AB1157pMH1(λ)			10-15%
AB1157pDR1453(λ)			>30%
AB1157(λ)		optimal UV dose	20-30 J/m^2
AB1157pDR1461(λ)			45 J/m^2
AB1157pMH1(λ)			45 J/m^2
AB1157pDR1453(λ)			20-30 J/m^2
DM49(λ): lexA3		fraction induced	10-20%
DM49pDR1461(λ)			< 0.001%
AB1157	Weigle reactivation[7]	reactivation factor	+50%
AB1157pDR1461			-47%
WP2	UV induced mutagenesis[13]	Trp$^+$ revertants per plate at 10% survival	675,588
WP2pDR1461			4, 1
AB1157	DNA degradation control	% acid insoluble DNA 1 hour after 22.5 J/m^2	98%
AB1157pDR1461			98%
DM938 :recA1			40%
AB1157	homologous DNA recombination	Freq. pro$^+$ recombinants by mating with Hfr C	8.4×10^{-2}
AB1157pDR1461			6.8×10^{-2}
AB1157pMH1			5.1×10^{-2}

and lexA102[11] from several stock collections.   UV induction was optimal at 5-10 J/m^2, a lower dose than wild type, and was completely eliminated by transformation with pDR1461 (table 1). Control experiments with pDR1453 and pBR322 transformants showed no alteration in reduction of the SOS functions.

Disruption of the SOS response by pMH1 and pDR1461 cannot be caused by "readthrough" products of plasmid genes[8] because the inserts cloned in these plasmids are at different sites.   A second possibility is that truncated recA protein prevents the multimeric[12] assembly needed for chromosomal recA$^+$ protein to be active. This explanation seems unlikely as similar effects were seen in

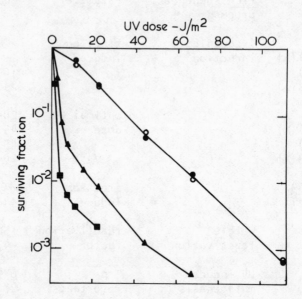

Fig.3.   UV survival of E. coli AB1157 ●   ; and transformants
         harboring pDR1453,   O   ; pDR1461, ▲ ; and pMH1, ■ :

cells with truncated proteins of quite different sizes.   In
addition, not all recA functions were reduced, indicating the
presence of functional recA protein inside the cells.   A third
possibility is that pMH1 and pDR1461 intervene in the SOS response
by both failing to contribute to the basal level of functional recA[+]
protein, and by providing sites at which activated recA[+] protein
acitivity is expended.   Note that an excess of recA - operator
sites per se,such as provided by pDR1453, does not reduce SOS
induction.   Upon induction a competition could develop for activated basal-level recA protein.   If there was a greater affinity
for action at the recA[+] control region, then truncated plasmid
recA genes would be expressed at the expense of reduced expression
of the SOS operons.   E. coli harboring plasmids with full size
recA[+] genes would have normal SOS induction because the increased
number of recA control sequences would be matched by increases in

the basal level of full sized functional recA$^+$ protein^2.   Consistent with this proposal is the observation that lexA3 mutants having less recA$^+$ protein available for activation, were most severely affected by pDR1461.   Conversely tif-1 mutants, having higher levels of recA/tif-1 protein were less subject to UV sensitization than tif$^+$ E. coli (data not shown).   It is concluded that the effects of pDR1461 and pMH1 can be interpreted in terms of titration of activated recA$^+$ protein, and not as predicted, titration of lexA$^+$ protein.

REFERENCES

1.   M. Radman, in: "Molecular and environmental aspects of
       mutagenesis", L. Prakash, F. Sherman, M. Miller, C. Lawrence
       and H.W. Tabor, eds., C.C. Thomas, Springfield, Ill. pp128-
       142 (1974)
2.   K. McEntee, Proc. Nat. Acad. Sci. 74: 5275-5279 (1977)
3.   L. Gudas and D. Mount, Proc. Nat. Acad. Sci. 74: 5280-5284 (1977)
4.   P.T. Emmerson and S.C. West, Molec. Gen. Genet. 155: 77-85 (1977)
5.   S.G. Sedgwick, A. Levine and A. Bailone, Molec. Gen. Genet.
       160: 267-276 (1978)
6.   A. Sancar and W.D. Rupp, Proc. Nat. Acad. Sci. 76: 3144-3148
       (1979)
7.   A. Goze and S.G. Sedgwick, Mutation Res. 52: 323-331 (1978)
8.   J.W. Little, Molec. Gen. Genet. 177: 13-22 (1979)
9.   J. Donch, J. Greenberg, and M.H.L. Green, Genet. Res. 15:
       87-97 (1970)
10. M. Castellazzi, J. George and G. Buttin, Molec. Gen. Genet.
       119: 153-174 (1972)
11. J. Donch, M.H.L. Green and J. Greenberg, J. Bacteriol. 96:
       1704-1710 (1968)
12. T. Ogawa, H. Wabiko, T. Tsurimoto, T. Tsurimoto, T. Horii,
       H. Masukata and H. Ogawa, Cold Spring Harbour Symp. Quant.
       Biol. 43: 909-915 (1978)
13. E.M. Witkin, Proc. Nat. Acad. Sci. 71: 1930-1934 (1974)

PLASMIDS CARRYING THE *uvrA* AND *uvrC* GENES OF *Escherichia coli* K12:

CONSTRUCTION AND PROPERTIES

Cees A. van Sluis and Jourica A. Brandsma

Laboratory of Molecular Genetics
State University of Leiden
64 Wassenaarseweg
NL-2333 AL LEIDEN, The Netherlands

ABSTRACT

From the Carbon-Clarke collection, a plasmid harbouring the *uvrC* gene was identified (pLC13-12) and subsequently subcloned into pBR322. The UvrC$^+$ plasmid pCA32 complemented UvrC deficient bacteria to the wild type phenotype. The *uvrC* gene has been assigned to a 2,000 bp BgIII DNA fragment. Upon introduction into minicells, pCA32 directs the synthesis of a protein of 28,000 dalton, which most likely is encoded by the *uvrC* gene. A novel method is described for the cloning of genes, for which direct selection is impracticable. The technique has been employed in the construction of a plasmid carrying *lexA, uvrA, ssb* on pACYC184.

INTRODUCTION

The incision-step in the major pathway in *E. coli* K12 for the removal of DNA damage induced by ultraviolet (UV) radiation and several mutagens is dependent on the gene products of the *uvrA*, *uvrB* and *uvrC* genes (1-3). Although much effort has been given to the elucidation of the underlying mechanism, insight in the precise reaction mechanism is still incomplete. Recently it was found that the respective *uvrA, B, C* gene products are simultaneously required for the nicking of UV damaged DNA (4,5). To facilitate the purification of the *uvr* gene products several groups isolated plasmids carrying the *uvrA* and *uvrB* genes of *E. coli* K12 (6-12). This report describes the construction and characterization of plasmids carrying the *uvrA* and the *uvrC* gene.

For the cloning of the *uvrA* gene we employed a novel technique which was based on the introduction of a resistance gene

(ampicillin resistance from Tn1) adjacent to the gene to be cloned.
This method is also applicable to other genes for which no good selection is available. Insertion of transposon Tn1 in the *E. coli*
K12 chromosome at 91.3 min enabled us to clone a 21 Mdal fragment
of chromosomal DNA, carrying *ubiA, lexA, uvrA and ssb*.

RESULTS AND DISCUSSION.

The Materials and Methods for the plasmid construction and
the analysis of the inserted DNA fragments has been partially described earlier (6-8) and will be given in detail elsewhere (Van
Sluis *et al.*; Brandsma *et al.*, manuscripts in preparation).

Cloning of the uvrC gene.

The *uvrC* gene is located at 42-43 min on the *E. coli* K12 genetic map, between the *flaD* and *flaI* genetic loci (13). The Carbon-
Clare 'colony bank' plasmid collection contained several plasmids
which carried these genes (14). Plasmid donor strains were therefore conjugated with a UvrC⁻ acceptor strain and assayed for UV resistance. Plasmid strain pLC13-12 harboured both *flaD* and *uvrC*. The
10.5 Mdal plasmid was unstable in RecA⁺ cells and had to be maintained in a RecA UvrC⁻ host to ensure integrity of the inserted
chromosomal fragment. Fig. 1 shows the cellular survival of strain
WDR1984 *recA uvrC* carrying plasmid pLC13-12. The introduction of
the *uvrC* plasmid into the mutant strain resulted in a radiation
phenotype comparable with a UvrC⁺RecA⁻ strain, whereas the colE1
vector plasmid had no influence on the UV phenotype. The expression
of the *uvrC* gene carried on the plasmid, was also observed in host
cell reactivation of irradiated λvir phage (results not shown).

In order to obtain a more stable plasmid, pLC 13-12 was analyzed for suitable restriction endonuclease sites. The inserted DNA
contained several *Pst*I sites and one *Hind*III site approx. 2 Mdal
from the poly dA-dT linker region (see fig. 2). Hybrid plasmids
were constructed by *in vitro* ligation of *Pst*I restricted pLC 13-12
and pBR 322 DNA (15) and tet^r transformants were selected in a *recA*
uvrC strain. One of 51 tet^r amp^s plasmid isolates (pCA32) conferred
UV resistance and contained a 3.5 kb *Pst*I DNA fragment carrying
one *Hind*III site. The latter is in accordance with the properties
of a λ.*uvrC* hybrid phage harbouring a 5 kb *Hind*III fragment of bacterial DNA (Seeberg, personal communication). The pCA32 plasmid
confers UV resistance to all *uvrC* mutant strains investigated whereas the behaviour of UvrA and UvrB mutants is not affected (not
shown).

The response of the standard mutant strain AB1884 *uvrC*34 is
given in fig. 3. The UV phenotype of bacteria carrying the pCA32
plasmid approaches the level of wild type bacteria, whereas the

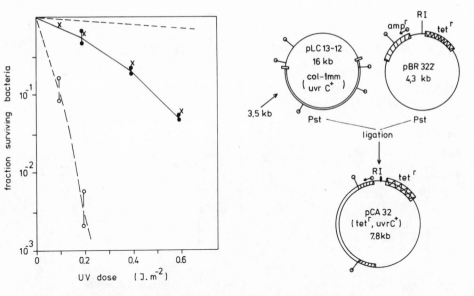

Fig. 1. Survival of WDR1984 harbouring pLC13-12.
(left) (o-o) control; (o-o) with plasmid. For comparison, the
survival of a RecA UvrC⁺ strain also given (crosses).

Fig.2. Schematic construction of pCA32.
(right)The *Pst*I fragment carrying the *uvrC* gene is marked by
'3.5 kb'.
*Pst*I sites are also indicated (o——)

wild type seems to be sensibilized slightly by the plasmid. This
phenomenon is presently under investigation. In all cases UvrC
mutants regained host cell reactivation ability upon introduction
of the plasmid (not shown).

Subsequently a detailed restriction map was prepared which
is depicted in fig. 4. The inserted *Pst*I fragment comprises 3400
+ 100 bp and carries two unique *Bgl*II and one *Bgl*I site. To locate
the DNA segment encoding the UvrC message, two approaches were
used. Firstly the *Bgl*II DNA fragment (from 51-76 map units) was
removed by cutting pCA32 with *Bgl*II restriction endonuclease and
religation. The deletion plasmid pCA33 (5.8 kb) did not carry an
active UvrC gene, indicating that at least part of the gene was
removed on the *Bgl*II fragment. This result could be substantiated
further by the insertion of Tn5 (Km) transposons into the pCA32
plasmid. One insertion plasmid, pCA502, was unable to complement
for UvrC deficiency, whereas two others - pCA520 and pCA524 - behaved similarly compared to the parental plasmid pCA32. The location of the Tn5 insertion in the latter plasmids indicated that
the *uvrC* gene is probably mainly located on the 2 kb *Bgl*II fragment.

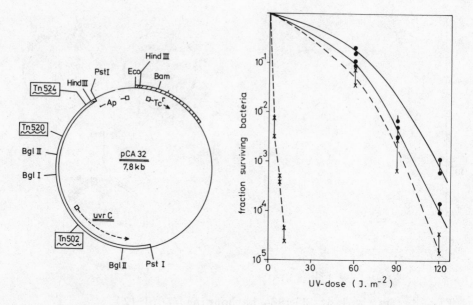

Fig. 3.   Physical map of pCA32.
(left)   Only relevant restriction sites are shown. The direction
          of transcription of the respective genes is indicated by
          an arrow.

Fig. 4.   Influence of pCA32 plasmid on cellular survival of AB1884
(right)   *uvrC*34 (x—x) and AB1157 wildtype (o—o). Vertical bars
          indicate experimental variation.

    Attempts to clone the entire *uvrC* gene on the 2 kb *Bgl*II frag-
ment proved to be unsuccessful, although a very low expression
could be obtained by insertion of the *Bgl*II fragment into the unique *Bam*HI site of pACYC177 ((16), fig. 6; pCA76,78). Therefore the
respective *Bgl*II sites were modified separately in the parental
pCA32 plasmid. Plasmid DNA nicked mainly once by the *Bgl*II restriction endonuclease in the presence of ethidium bromide (17),
was subsequently repaired by DNA polymerase I to blunt termini and
religated. After transformation, two derivative plasmids were recovered, which both lacked a different *Bgl*II site. The result of
the introduction of 4 additional basepairs and hence a frame shift
in the putative reading frame was different in both plasmids (see
fig. 5, left). In plasmid pCA74 the 'removal' of the rightmost
*Bgl*II site had no effect, whereas the pCA72 plasmid lacking the
leftmost site was only marginally active in complementation. The
partially active, incomplete gene product which is presumably encoded by the altered *uvrC* gene present on the pCA72 plasmid is
still sufficiently active to render a UvrC deficient host cell proficient for phage reactivation (fig. 5, right).

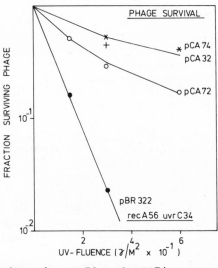

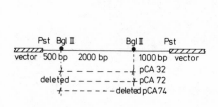

Fig. 5. Mutant plasmids of pCA32.
   left panel: deleted *Bgl*II sites in pCA72 and pCA74.
   right panel: survival of irradiation λvir phage measured
   in *recA uvrA* host cells carrying various plasmids.

   More recently an extended number of derivative plasmids of
pCA32 has been prepared by cloning subfragments into pACYC177 (16).
A schematic survey of derivatives is given in fig. 6, where a preliminary physical map of the *uvrC* region is also shown. Full expression of the UvrC function was only obtained by the plasmids
pCA79 and pCA80, whereas partial expression was observed with
pCA62 and pCA76. Reversal of the orientation of the 2 kb fragment
(pCA78) abolished the *uvrC* expression completely (results not
shown).

   A preliminary identification of the *uvrC* gene product was
carried out by the analysis of plasmid-encoded polypeptides synthesized in *E. coli* minicells (18). Plasmids pCA32, pCA33 and the vector pBR322 were introduced into the minicell producing strain
P 678-54 and ^{35}S-labelled polypeptides were isolated and separated
employing standard methods (20). In fig. 7 the autoradigram is presented showing that the 29 kdal β-lactamase product is absent from
both pCA32 and pCA33, and that a weak 27 kdal tet[r] protein band is
observable which partially overlaps the presumptive *uvrC* gene product, which is absent in pCA32. The estimated molecular weight of
the UvrC protein, based on this and similar experiments, is approx.
26-28 kdal. The protein is also absent in minicells carrying the
insertion plasmid pCA502 (see fig. 4).

   The *in vitro* manipulation of chromosomal DNA present on plasmids allows the deliberate rearrangement of genes which subsequently may be reinserted at the original location on the *E. coli*

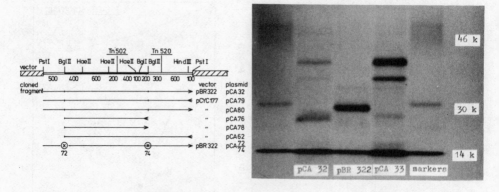

Fig. 6.   Derivative plasmids of pCA32.
(left)   The relative orientation to the ampicillin gene on the
            pBR322 and pACYC177 vectors is indicated by arrows.

Fig. 7.   Autoradiography of ^{35}S-labelled polypeptides from plas-
(right)   mid-containing minicells. Radioactive marker-proteins
            were run on the two outer lanes of the electropherogram.

chromosome. Transposons inserted in the cromosomal fragment concerned, can be reintroduced into the DNA of the host cell by recombination. During this process, plasmid replication is stopped
by thermo-inactivation of DNA polymerase I (PolAts). This experimental approach has been used to introduce transposon Tn5 into the
chromosomal uvrC gene. The molecular configuration present on plasmid pCA502 (uvrC::Tn5, see fig. 5) was transferred to the E. coli
chromosome, resulting in a UV-sensitive phenotype (19). A slightly
modified method was employed for the uvrA gene and is described
below. The technique can also be used to clone mutant genes.

The cloning of the uvrA gene.

     The absence of uvrA in the Carbon-Clarke plasmid collection
(9) and the failure to clone the uvrA gene directly from the chromosome, urged us to employ a method based on indirect selection.
To clone the uvrA gene, the colony bank plasmid pLC44-14 was employed which comprises the genes ubiA, lexA and presumably dnaB
(14,20). Transposon Tn1 (Ap) was introduced into pLC44-14 and a
derivative (pLC44-14::Tn1-3) was used for the transfer of the ampicillin resistance (see fig. 8). PolAts cells carrying the derivative plasmid pLC44-14::Tn1-3 were incubated at the non permissi-
ve temperature (40° C) in selective medium and kept sufficiently
long to loose the plasmid (21). A bacterial culture was obtained
showing reduced colicin EI immunity and stable inheritance of the
resistance to ampicillin. These cells were used to propagate phage
P1. Upon transduction of a UvrA⁻ amp^S recipient strain with this

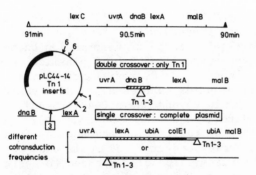

Fig. 8. Method for the insertion of Tnl into the chromosome
top   : part of the *E. coli* K12 genetic map
left   : sites of insertion of Tnl into pLC44-14 ·
center: possible recombination events
bottom: modes of plasmid integration

P1 lysate, we observed that the cotransduction frequency of UvrA[+]
with ampicillin-resistance varied from 54-88%. Presumably in most
cases the entire pLC 44-14::Tnl-3 plasmid had been inserted into
the host chromosome by a single cross-over mechanism (see fig. 8).
This assumption was further sustained by the reappearance of the
original plasmid in the cytoplasm of several transductants during
subculturing.

Chromosomal DNA, isolated from transductant strains which had
been checked again for high cotransduction of UvrA[+] with ampicillin
resistance, was employed for the construction of recombinant

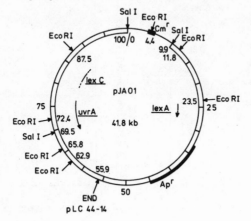

Fig. 9. Physical map of plasmid pJA01.
Arrows at the *UvrA* and *lexA* genes indicate direction
of transcription according to (11) and (20) respectively.
One map unit is approx. 0.42 Kb

plasmids with the pACYC184 vector (16). Mixtures of chromosomal
and plasmid DNA were treated with *Bam*HI, religated and recombinant plasmids were selected by transformation on medium containing
chloramphenicol and ampicillin. Hybrid plasmids were obtained which
unfortunately did not confer UV resistance to the RecA⁻UvrA⁻ recipient strain. All plasmids carried a 3-4 Mdal DNA region extending
approximately 0.15 min beyond the pLC44-14 boundary on the *E. coli*
K12 map, in the direction of the *uvrA* gene (13). The experiment
was repeated with *Sal*I restriction endonuclease and one out of
50 recombinant plasmids conferred UV resistance to a *uvrA* recipient strain. The plasmid (pJA01) appeared to contain 21 Mdal fragment comprising approx. 0.8 min of the bacterial chromosome and
the Tn-1 transposon (fig. 9). In the course of this work cloning
of the *uvrA* gene was reported; together with *uvrA*, *ssb* was cloned
on the pDR2000 plasmid (10,11). The physical map of the pJA01 plasmid given in fig. 9 is in good agreement with the restriction analysis of pDR2000 (11) and pLC44-14 (20). The presence of a *Sal*I
restriction site in pJA01 was recently confirmed. The further cloning of the 13.7 kb *Sal*I fragment comprising *ssb* in pACYC184 and
pBR322 yielded plasmids that did not confer UV resistance to *uvrA*
strains (not shown).

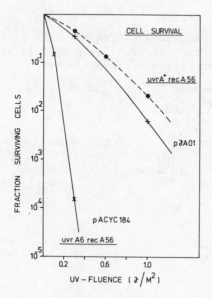 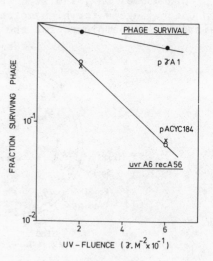

Fig. 10. Complementation of the *uvrA* mutation by pJA01.
      left panel : pACYC184 (x), pJA01 (+). The UvrA⁺ derivative
      is included for comparison.
      right panel: no plasmid (o), pACYC184 (x) pJA01 (●).

The introduction of the pJA01 plasmid into a *uvrA recA* cell restored the UvrA deficiency as illustrated by increased cellular survival and host cell reactivation (figs. 10A and B). All *uvrA* mutants investigated so far, responded to pJA01, whereas UvrB and UvrC mutant strains showed unchanged radiation phenotype upon introduction of the plasmid. These results strongly indicate that pJA01 encodes the *uvrA* gene. The analysis of polypeptides produced by the plasmid is currently under investigation.

The pJA01 plasmid also appeared to carry the *ubiA* and *lexA* genes which are also present on pLC44-14 (20). The pJA01 plasmid is rather unstable in a Rec$^+$ background and strains have to be subcultured on selective media. The impracticability of this large plasmid is presently under investigation and small derivative plasmids carrying the various genes from pJA01 are constructed. Apart from its application in the purification of the *uvrA* gene product and the study of gene structure and regulation, pJA01 and derivatives will undoubtly be valuable in the more precise location of genetic loci on the *E. coli* K12 genetic map (13).

ACKNOWLEDGEMENTS

Thanks are due to Dick Dubbeld for his expert technical assistance. The experimental help of Hans Baak, Johan Twilt and Huib Storm are much appreciated. We are grateful to Hans Pannekoek and Piet van de Putte for continuing advice and encouragement. Drs. E. Seeberg and W.D. Rupp are thanked for providing strains and giving their experimental results prior to publication.

This work was supported by the Commission of European Communities under Contract 194-76 BION.

REFERENCES

1. van de Putte, P. van Sluis, C.A., van Dillewijn, J. and Rörsch, A. (1965) Mut. Res. 2, 97-110.
2. Howard-Flanders, P., Boyce, R.P. and Theriot, L. (1966) Genetics 53, 1119-1136.
3. Hanawalt, P.C., Cooper, P.K., Ganesan, A.K. and Smith, C.A. (1979) Ann. Rev. Biochem. 48, 783-836.
4. Seeberg, E., Nissen-Meyer, J. and Strike, P. (1976) Nature 263, 524-526.
5. Seeberg, E. (1978) Proc. Natl. Acad. Sci. USA 75, 2569-2573.
6. Pannekoek, H., Noordermeer, I.A., van Sluis, C.A. and van de Putte, P. (1977) J. Bacteriol. 133, 884-896.
7. Pannekoek, H. Noordermeer, I. and van de Putte, P. (1979) J. Bacteriol. 139, 48-53.
8. Pannekoek, H., Noordermeer, I. and van de Putte, P. (1979) J. Bacteriol. 139, 54-63.

9. Rupp, W., Sancar, A., Kennedy, W.J., Ayers, J. and Griswold,
    J. (1978). In 'DNA Repair Mechanisms' (P.C. Hanawalt, E.C.
    Friedberg and C.F. Fox, eds) pp. 229-235, Academic Press,
    New York.
10. Sancar, A. and Rupp, W.D. (1979) Biochem. biophys. Res. Commun.
    90, 123-129.
11. Sancar A., Wharton, R.P., Seltzer, S., Kacinsky, B.M., Clarke,
    N.D. and Rupp, W.D. (1981) J. Molec. Biol. (in press).
12. Sancar, A., Clarke, N.D., Griswold, J., Kennedy, W.J. and
    Rupp, W.D. (1981) J. Molec. Biol. (in press).
13. Bachmann, B.J. and Brooks Low, K. (1980) Microbiol. Rev. 44,
    1-56.
14. Clarke, L. and Carbon, J. (1976) Cell 9, 91-99
15. Bolivar, F., Rodriguez, R.L., Green, P.J., Betlach M.C.,
    Heijneker, H.L. and Boyer, H.W. (1977) Gene 2, 95-113.
16. Chang, A.C.Y. and Cohen, S.N. (1978) J. Bacteriol. 134, 1141-
    56.
17. Parker, R.C., Watson, R.M. and Vinograd, J. (1977) Proc. Natl.
    Acad. Sci. USA. 74, 851-55.
18. Reeve, J.N. (1979) In 'Methods in Enzymology' (R. Wu, ed.)
    Vol. 68, pp 493-503, Academic Press, New York.
19. van Sluis, C.A., Dubbeld, D. and Twilt, J.C., unpublished re-
    sults.
20. Little, J.W. (1980) Gene 10,237.
21. Kingsbury, D.T. and Helinski, D.R. (1970) Biochem. biophys.
    Res. Commun. 41, 1538-44.

CLONING OF THE uvrC[+] GENE FROM Escherichia coli ONTO A PLAQUE FORMING PHAGE VECTOR

O.R. Blingsmo, A.L. Steinum, E. Rivedal, and E. Seeberg

Norwegian Defence Research Establishment

P.O. Box 25, N-2007 Kjeller, Norway

INTRODUCTION

Specialized transducing phages are important tools in the characterization of host coded gene functions. Isolation of specialized transducing phages was previously restricted to those carrying host genes located close to the normal phage attachment sites[1]. However, recent advances in techniques of DNA recombination both in vivo and in vitro have made possible in principle the isolation of transducing phages carrying genes from anywhere on the bacterial chromosome[2-4]. This communication describes the isolation of a plaque-forming bacteriophage λ carrying the E.coli uvrC[+] gene.

The uvrC gene in E.coli is functionally closely related to the uvrA and uvrB genes even though these loci are widely spaced on the bacterial chromosome[5,6]. It was recently shown that the uvr[+] genes code for three different gene products which complement each other to produce an ATP-dependent repair endonuclease activity[7,8]. This enzyme activity initiates nucleotide excision repair in E.coli of various kinds of DNA damage, including those induced by UV-light[8], cross-linking agents[9,10] and other alkylating agents producing bulky type of lesions in DNA[11]. The mechanism of action for this complex type of enzyme activity is still unknown, and it seemed likely that cloning of the uvrA, uvrB and uvrC genes would be very useful in further characterization of the uvr gene functions. We have therefore searched for λ-phages carrying uvr[+] genes in populations of phages constructed in vitro from E.coli DNA and a phage vector digested by restriction endonuclease Hind III[4].

This report describes the strategy for isolation of λuvrC[+] and the initial characterization of the gene product(s) encoded by the bacterial segment on the phage genome.

RESULTS

Isolation of $\lambda uvrC^+$. A two-step selection procedure was used for the isolation of $\lambda uvrC^+$. Firstly, the phage population was UV-irradiated and propagated repeatedly, until a marked increase in the survival was observed. It was assumed that a $\lambda uvrC^+$ phage would complement the host cell reactivation deficiency (hcr^-) of the uvrC mutant and thus have a selective advantage towards other UV-irradiated phages which are not reactivated in uvrC host cells[6]. The increase in survival observed after several cycles of irradiation and propagation indicated that the phage population had been enriched for $\lambda uvrC^+$ phages. Phages from single plaques were then isolated, repurified, and tested for UV-survival on the various excision defective hosts. Two out of four isolates showed positive and specific reactivation on uvrC, relative to uvrA and uvrB host cells, and were scored as $\lambda uvrC^+$. The other two phage isolates were not reactivated by any of the uvrA, uvrB or uvrC host bacteria. The UV-survival of the plaque-forming ability of one of the $\lambda uvrC^+$ phages is shown in detail in Table 1. Results from a similar experiment with the parent phage vector are included for comparison. It appears that $\lambda uvrC^+$ is subjected to host cell reactivation in uvrC cells, although not to the same extent as in uvr^+ cells. The parent phage vector is, as expected, not reactivated in any of the excision defective hosts, but fully reactivated by uvr^+ cells.

TABLE 1

Host cell reactivation of $\lambda uvrC^+$ and $\lambda 540$ in excision proficient and deficient E.coli. The table shows surviving frequencies for the phage after a UV-dose of 80 J/m^2.

Host cell	Surviving frequency		Survival ratio
	$\lambda uvrC^+$	$\lambda 540$	$\lambda uvrC^+/\lambda 540$
AB1157(uvr^+)	2.4×10^{-1}	2.3×10^{-1}	1.0
AB1884(uvrC34)	1.9×10^{-2}	3.7×10^{-6}	5130
AB1885(uvrB5)	9.1×10^{-6}	2.7×10^{-5}	0.3
AB1886(uvrA6)	1.6×10^{-5}	9.6×10^{-6}	1.7

We have also searched for λuvrA⁺ and λuvrB⁺ phages in the population of transducing phages using the same approach as for λuvrC⁺,
but with negative results. It is possible that the uvrA and uvrB
genes contain a restriction site for Hind III endonuclease or alternatively that the Hind III fragments from those regions are too
large to be cloned on the λ540 vector.

Insertion of λuvrC⁺ as prophage or as superinfecting homoimmune
phage restores repair endonuclease activity in uvrC mutated cells.
The λuvrC⁺ phage is integration proficient and capable of entering
the host chromosome as a stable prophage. A heterogenote uvrC⁻/uvrC⁺
bacteria could therefore be constructed by lysogenizing uvrC cells
with the λuvrC⁺ phage. Cell extracts from this strain were prepared and subjected to enzyme tests to see if the ATP-dependent
UV-endonuclease activity normally deficient in the uvrC mutant was
restored by the presence of the uvrC⁺ gene on the prophage. As
control was included experiments with extracts from uvrC mutated
cells lysogenized with the phage λ540. **Fig.** 1A shows that the
heterogenote lysogen possesses normal levels of the repair endo-

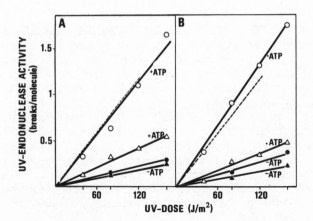

Fig. 1.   Dose dependent cleavage of UV-irradiated ColE1 DNA by cell
          extracts. A. Breakage caused by extracts of strain AB1884
          (uvrC34) lysogenic for λuvrC⁺ (circles) or λ540 (triangles).
          B. Breakage caused by AB1884 lysogenic for λ540 super-
          infected (multiplicity of 5 phages per cell) with λuvrC⁺
          (circles) or λ540 (triangles).
          Essays were performed in the presence (open symbols) or
          absence (closed symbols) of 2.0 mM ATP and reaction condi-
          tions were as previously described[7],[8]. Dashed line shows
          the breakage caused by wild type extracts in the presence
          of ATP.

nuclease activity, while the uvrC control lysogen is deficient. The presence of the uvrC⁻ allele at the uvrC locus on the chromosome of the heterogenote was confirmed by curing the lysogen for the λuvrC⁺ phage. The cured cells were highly UV-sensitive, while the heterogenote lysogen had wild-type resistance (Blingsmo and Seeberg, in preparation).

The enzyme data obtained with the λuvrC⁺ lysogen suggests that the uvrC gene on the phage carries its own promotor, since the phage promotors are repressed in the prophage state[1]. The possibility remained, however, that expression of the uvrC gene carried by the prophage was caused by read-through from a bacterial promotor located close to the prophage attachment site. To test this possibility the uvrC lysogen carrying the phage vector λ_{540} was superinfected with λuvrC⁺ and extracts from the superinfected lysogens tested for the repair endonuclease activity. After superinfection of a homoimmune lysogen λ-DNA will form nonreplicating, covalently-closed circular molecules repressed by the prophage coded repressor[1]. As shown in Fig. 1B, the ATP-dependent UV-endonuclease activity is restored to normal levels by superinfection with λuvrC⁺, but is deficient in uvrC lysogens superinfected with λ_{540}. This confirms that the uvrC gene operon on the λuvrC⁺ phage genome includes its own promotor.

Identification of two proteins encloded by the bacterial segment in λuvrC⁺. Ptashne[12] originally devised a method for specific radioactive labelling of phage coded proteins. The method involves inactivation of the host chromosome by irradiation with heavy doses of UV-light prior to phage infection. Such heavily irradiated cells are incapable of synthesizing proteins encoded by its own chromosome, but still retain the capability of synthesizing phage-coded proteins[12,13]. Upon infection of heavily irradiated cells with λuvrC⁺ and incubation in the presence of ^{14}C-labelled amino acids, newly synthesized proteins were identified after gel electrophoresis and autoradiography of the labelled proteins. On the radiochromatogram two distinct bands specific for the transducing phage appeared. The bands represent proteins of approximately 30,000 and 70,000 daltons in molecular weight.

DISCUSSION

Phage λuvrC⁺ complements uvrC mutated cells in host cell reactivation of UV-irradiated phages. However, as compared to the reactivation observed in uvr⁺ cells, reactivation of λuvrC⁺ in uvrC host bacteria is only partial. One explanation for this could be that the uvrC gene function on the phage is only partially expressed. This seems unlikely, however, since the amount of uvrC gene product produced from λuvrC⁺ as prophage or superinfecting

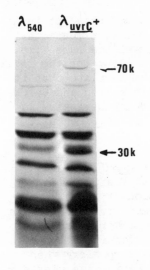

Fig. 2 Autoradiochromatogram of radio-actively labelled phage (λuvrC$^+$ or λ_{540}) coded proteins separated on SDS-polyacrylamide gels.

homoimmune phage appears to be normal, as measured by the ability to restore repair endonuclease activity in the uvrC mutant. Furthermore, uvrC cells lysogenized with λuvrC$^+$ have acquired wild type resistance to UV-light (Blingsmo and Seeberg, in preparation). It thus appears that the uvrC gene on the λuvrC$^+$ phage is fully capable of complementing the uvrC mutation in UV-repair of the host cell and therefore is normally expressed from the transducing phage.

Another explanation for the partial reactivation of λuvrC$^+$ on uvrC host cells is that the UV-exposure of the phage inactivates the uvrC gene itself before entering the cells. A single hit in the transcribed strand of the uvrC operon would be expected to prevent expression of the repair function. The LD_{37} cose for UV inactivation of λuvrC$^+$ on uvrC host cells is 20 J/m^2, which induces in average 12 pyrimidine dimers per phage genome[14]. Assuming that the dimers are evenly distributed, this dose of UV-light produces in average one dimer per single-strand DNA unit, corresponding to 15% of the phage DNA. This size DNA would have a maximum coding capacity of about 2400 amino acids, corresponding to a protein of 300,000 daltons, which seems too large to represent the size of the uvrC product. Gel filtration experiments have indicated that the size of the uvrC$^+$ product is no larger than 70,000 daltons[8]. However, inactivation of the uvrC operon by the UV-exposure of the λuvrC$^+$ phage may still partially account for reduced survival of the transducing phage on uvrC relative to uvr$^+$ host bacteria.

Excision repair enzymes in E.coli are constitutively expressed. When an irradiated phage enters the host cell, repair starts immediately. When irradiated λuvrC$^+$ enters the uvrC host cell, the situation is different. The uvrC$^+$ product must be synthesized before the phage genome can be repaired. At the same time, phage proteins are synthesized which initiate phage replication. This means that repair and replication will start approximately at the same time. There is therefore in this case a greater probability for the replication fork to reach an unrepaired lesion in the template than in the case when repair starts immediately after infection and before onset of replication. It is possible that the delayed repair reponse partially accounts for lower survival of λuvrC$^+$ on uvrC cells than on uvr$^+$ cells (Table 1).

The bacterial DNA segment carried by phage λuvrC$^+$ encodes two polypeptides of about 30,000 and 70,000 daltons. Positive identification of either one of these as the uvrC$^+$ product would be possible by means of a λuvrC$^-$ phage carrying an amber mutation in the uvrC gene. We are presently trying to select an amber mutant from mutagenized λuvrC$^+$ phage stocks. This involves identification of λuvrC mutant phages directly by means of plaque morphology when plated on uvrC bacteria. This is achieved by treating the plates at an early stage of plaque appearance with psoralen plus near UV-light This treatment will kill the centre growth of lysogenic bacteria within λuvrC$^-$ plaques, but not within λuvrC$^+$ plaques, because λuvrC$^-$ lysogens are sensitive and λuvrC$^+$ lysogens are resistant towards psoralen plus light treatment. The mutant phages will then appear as clear plaques while the λuvrC$^+$ plaques will remain turbid. Several hundred mutant phages can then easily be isolated, if necessary, and screened for the amber character.

During the course of this study other groups have also cloned uvr genes either on plasmids[15,16,17] or phage vectors[18]. However, so far plaque forming phages carrying the uvrC gene,have not been reported, nor have biochemical studies of the expression of the uvrC gene product. An extensive genetic characterization of the λuvrC$^+$ phage will be published elsewhere (Blingsmo and Seeberg, in preparation).

REFERENCES

1. M.E. Gottesman and R.A. Weisberg, in: "Bacteriophage Lambda", A.D. Hershey, ed., Cold Spring Harbour Laboratory, New York (1971).
2. K. Shimada, R.A. Weisberg, and M.E. Gottesman, J. Mol. Biol. 63:483 (1972).
3. K. Murray and N.E. Murray, J. Mol. Biol. 98:551 (1975).
4. K. Borck, J.D. Beggs, W.J. Brammar, A.S. Hopkins, and N.E. N.E. Murray, Mol. Gen. Genet. 146:199 (1976).
5. P. van de Putte, C.A. van Sluis, J. van Dillewijn, and A. Rörsch, Mutat. Res. 2:97 (1965).
6. P. Howard-Flanders, R.P. Boyce, and L.Theriot, Genetics 53:1119 (1966).

7. E. Seeberg, J. Nissen-Meyer, and P. Strike, <u>Nature</u> 263:524
    (1976).
8. E. Seeberg, <u>Proc. Natl. Acad. Sci. USA</u> 75:2569 (1978).
9. E. Seeberg, <u>Mutat. Res.</u> (in press).
10. R.S. Cole, D. Levitan, and R.R. Sinden, <u>J. mol. Biol.</u> 103:39
    (1976).
11. H.N. Thielmann, <u>Eur. J. Biochem.</u> 61:501 (1976).
12. M. Ptashne, <u>Proc. Natl. acad. Sci. USA</u> 57:306 (1967).
13. R.W. Hendrix, <u>in</u>: "Bacteriophage Lambda", A.D. Hershey, ed.,
    Cold Spring Harbour Laboratory, New York (1971).
14. J.R. Boyle and R.B. Setlow, <u>J. Mol. Biol.</u> 51:131 (1970).
15. H. Pannekock, I.A. Noordermeer, C.A. van Sluis, and P. van
    de Putte, <u>J. Bacteriol.</u> 133:884 (1978).
16. W.P. Rupp, A. Sancar, W.J. Kennedy, J. Ayers, and J. Griswold,
    <u>in</u>: "DNA Repair Mechanisms", P.C. Hanawalt, E.C. Friedberg
    and C.F. Fox, eds., Academic Press, New York (1978)
17. C.A. van Sluis and R. Brandsma, this volume (1981).
18. J. Auerbach and P. Howard-Flanders, <u>Mol. Gen. Genet.</u> 168:341
    (1979).

DNA REPAIR DEFICIENCY AND HUMAN DISEASE

XERODERMA PIGMENTOSUM - A HUMAN
MODEL OF DEFECTIVE DNA REPAIR

Errol C. Friedberg

Laboratory of Experimental Oncology
Department of Pathology
Stanford University
Stanford, California 94305
USA

A.   Introduction

The human disease xeroderma pigmentosum (XP) has been intensively studied ever since Cleaver first provided evidence of defective repair of DNA damage in 1968 (1). The disease is inherited by an autosomal recessive mode and is clinically characterized chiefly by severe photosensitivity of the skin and eyes, a high frequency of a variety of neurologic abnormalities and an almost 100% incidence of malignant skin tumors.   It is noteworthy that tumors in other sites have not been reported with an abnormal frequency in this disease.   All known cases of XP fall into two major classes, based on the ability of cells in culture (usually fibroblasts) to repair DNA damage caused by ultraviolet (UV) radiation.   So-called typical or classical XP is characterized by defective nucleotide excision repair of pyrimidine dimers in asynchronous fibroblast cultures exposed to UV light at ~254 nm.   XP variant cases are clinically indistinguishable from the typical form of XP, but fail to show defects in any of the currently available biochemical parameters that directly or indirectly measure the excision of pyrimidine dimers in asynchronous fibroblast cultures.   The technique of cell hybridization has been used to classify typical XP cases into a number of apparently distinct genetic groups (complementation groups) defined by their ability to complement the restoration of normal levels of excision repair measured by unscheduled DNA synthesis (UDS) in heterodikaryons. At the time of writing, approximately 100 cases have been designated in the literature by an internationally agreed upon nomenclature (2) and assigned to one of 7 complementation groups.   The clinical symptoms and signs and the variety of genetic, cellular and bio-

chemical defects associated with XP have been extensively reviewed
by a number of authors (3-6) in recent years and the interested
reader is referred to this literature for a more detailed description.
In this article I have specifically focused on some of the literature
published during the past two years and have attempted to draw
attention to selected new aspects of the disease.

B.   Clinical and Genetic Features of XP

Recent additions to the rapidly expanding library of XP cases
stem from reports by Hashem et al (7) of 16 new patients in Egypt
and Arase et al (8) and Keijzer et al (9) demonstrating two new
complementation groups (F&G). Of the 16 new Egyptian cases, 8
fall into complementation group A and 8 into group C. Thus of
106 complemented cases culled from the world literature, groups A
(39.6%) and C (26.4%) collectively account for 66% of the total.
Of the remaining 34%, the XP variant accounts for 15% and XP
group D 13.2%. Only 5% of the total cases are constituted by
groups B (1 case reported), E (2 cases reported), F (1 case reported),
and G (1 case reported). It is of interest to note that cells from
the 8 group C cases recently studied in Egypt showed significantly
higher levels of unscheduled DNA synthesis than did the group A
cases. This is consistent with previous comparisons reported by
others (see ref. 6). Hashem et al (7) also measured DNA strand
breaks induced in UV-irradiated cells in vivo by incubating the cells
in cytosine arabanoside plus hydroxyurea, a treatment that blocks
the completion of excision repair. Relative to normal human fibroblasts, XP group A cells were found to induce strand breaks at a
frequency of about 3%, whereas the XP group C cells induced breaks
at a frequency of 15.6%.

Central nervous system disorders were found in all of the group
A patients and in none of the group C patients from the recent
Egyptian study. This is consistent with a recent report by Andrews
et al (10) indicating a general correlation between sensitivity of XP
cells to killing by UV radiation and the presence of neurologic
abnormalities in the donor. In the latter study neurologic abnormalities were confined to XP group A and D patients and were not
detected in group C or E cases. Unfortunately, the recent Egyptian
study did not report data on UV-induced cell killing and thus it is
still not clear whether in general, the sensitivity of fibroblasts in
culture to UV radiation correlates best with the presence or absence
of neurologic complications, the severity of the excision repair
defect, or the severity of the dermatological symptoms and signs.

C.   Response of XP Cells to DNA Damage by Chemicals

It has been previously reported by numerous authors that XP
cells are abnormally sensitive to a variety of forms of chemical
damage, including damage by 4-nitroquinoline-1-oxide, nitrogen

mustard, aflatoxin B_1 and acetylaminofluorene (see ref. 6 for a recent review). Recently, Hurley et al (11) demonstrated that the antibiotic anthramycin binds to the DNA of normal human fibroblasts. Confluent normal skin fibroblasts were able to remove 86% of the bound anthramycin within 72 hours. However, XP cells of a single patient from complementation group A were only able to remove 49% of the lesions during the same incubation period. Coppey and Nocentini (12) have shown that the survival of typical XP and of XP variant cells was about five times and twice, respectively, more sensitive to formaldehyde treatment than that of skin fibroblasts from healthy and XP heterozygote donors. The capacity of formaldehyde-treated host cells to support herpes virus replication was also more sensitive in XP than in normal cells.

All cell complementation studies have been confined to XP cells treated with UV radiation, and the question as to whether the identical distribution of complementation groups would result using a different form of DNA damage is an important one. Recently, Zelle (13) has shown that in general, the degree to which UDS was reduced after 4-nitroquinoline-1-oxide (4NQO) treatment of 9 different XP cell strains followed the same pattern observed for the repair of UV damage. Furthermore, complementation analysis using 4NQO-treated cells gave identical results for the 9 cells tested as did the use of UV-irradiated cells.

Another interesting observation on the response of XP cells to chemical DNA damage stems from studies reported by Ahmed and Setlow (14,15). These investigators reported (14) that when normal human fibroblasts are exposed to saturating doses of UV radiation plus N-acetoxy-acetylaminofluorene, the total amount of DNA repair measured by three different techniques was additive, suggesting that the biochemical pathways for the repair of these two forms of base damage are not identical. [Such additivity has not been confirmed in similar studies reported by Brown et al (16).] On the other hand, when Ahmed and Setlow (14) tested an XP variant cell line and two typical XP cell lines (complementation groups C and D), an inhibitory effect was observed in the presence of both DNA damaging agents, i.e., the total amount of repair observed was less than that seen with either agent alone (17).

In 1977 Goth-Goldstein reported that a line of XP group A cells (XP 12RO) was defective in the excision of 0^6-alkyl guanine residues, during incubation of intact cells following their exposure to either methyl- or ethyl-nintrosourea. More recent studies by Altamirano-Dimar et al (18) utilized a lymphoblastoid cell line (XP A-3) derived from an XP patient in complementation group C. These cells were found to be defective specifically in the excision of 0^6-methyl guanine relative to normal cell controls following treatment with N-methyl-N'-nitro-N-nitrosoguanidine. A direct demonstration of a defect in the excision of 0^6-ethyl guanine from XP12RO cells has been reported by Bodell et al (19). These workers treated normal and XP group A cells with ethylnitrosurea and monitored the rates and extent of loss of bases ethylated at seven distinct positions. All except 0^6-ethylguanine were excised normally by the XP cell line.

D.   Enzyme Defects in XP Cells.

About four years ago, Linn and his colleagues (20) reported that extracts of XP cells of complementation group D have about 1/6 the normal level of AP (apurinic/apyrimidinic) endonuclease activity. Additionally, extracts of normal human cells have two species of AP endonuclease activity that can be resolved by phosphocellulose chromatography; one with a high Km for apurinic duplex DNA and the other with a low Km for this substrate (21). Extracts of XP group D cells were reported to be lacking the low Km species. In view of these results it would have been anticipated that XP cells from complementation group D would be relatively defective in the repair of SV40 DNA containing apurinic sites. In fact, a significantly higher level of viral infectious centers was observed in the XP group D cells compared to normal human fibroblasts (22). It is argued that these results indicate the existence of a possible alternative (and presumably more efficient) pathway for the repair of apurinic sties in the DNA of normal human cells. An obvious candidate for such a pathway is the DNA-purine insertase activity shown to be present in extracts of a line of normal human fibroblasts (23).

Reports from other laboratories on AP endonuclease activity in normal human and XP cells have provided different results. Moses and Beaudet (24) found no quantitative differences in the specific activity of total AP endonuclease in extracts of normal and XP group D cells. Witte and Thielmann (25) reported that when DNA treated with methylmethanesulfonate was incubated with extracts of normal or XP group A fibroblasts, the number of DNA strand breaks induced was significantly less with extracts of the latter cells. However, when DNA containing only apurinic sites (prepared by heat-acid treatment) was tested, equal levels of AP endonuclease activity were observed. These investigators therefore concluded that XP group A cells may be defective in an enzyme activity required for excision of alkylated bases from DNA, but are not defective in AP endonuclease activity. Ishiwata and Oikawa (26) directly measured the specific activity of uracil-DNA glycosylase and 3-methyl-adenine-DNA glycosylase in extracts of three normal and five XP

cell extracts and showed no differences. All five XP cell lines used
were lymphoblastoid, but unfortunately only two can be positively
identified with respect to established complementation groups; XPL5
is derived from patient XP5SE (complementation group E) and XPL15
is derived from patient XP15SE (complementation group A).

The enzyme(s) in normal human fibroblasts of obvious and
particular interest with respect to XP, is that involved in the excision
of pyrimidine dimers. Previous studies (see ref. 6) have suggested
that cell lines from a number of complementation groups are defective in the incision of UV-irradiated DNA, presumably at sites of
pyrimidine dimers. Zelle and Lohman (27) have recently examined
examples of cells representative of each of the seven known complementation groups as well as the XP variant form, for their ability
to remove from their DNA, sites sensitive to an enzyme activity
from M. luteus known to be specific for pyrimidine dimers [loss of
enzyme-sensitive sites (ESS)]. Their results indicate that XP group
A, B and G cells are totally defective in the loss of ESS, suggesting
a total inability of these cells to effect incision of DNA at dimers.
Cells from groups C and D show a very slight loss of ESS during
32 hours of post-UV incubation. Cells from complementation groups
E and F are distinctly leaky. In general, these results correlate
well with UDS measured in the same cells, with the exception of
the cells from complementation group F, which show only about 10%
residual UDS. The authors suggest two explanations for this discrepancy. One is that the group F cell lines may remove pyrimidine
dimers by a repair process that is associated with shorter patches
of repair synthesis than that observed in normal cells. A second
possible explanation is that in the group F cells the kinetics of
repair synthesis relative to DNA incision are much slower than in
normal cells, so that at early times post-UV (when UDS is measured)
a significant defect is observed in this parameter.

During recent studies in our laboratory (J.I. Williams and E.C.
Friedberg, unpublished observation) we have directly measured the
loss of thymine-containing pyrimidine dimers during post-UV incubation of normal and XP cells exposed to a UV fluence of $40J/m^2$.
Cells from complementation groups A, D and G were found to be
essentially totally defective in this parameter. Those from complementation groups C and E showed some residual capacity for dimer
excision. XP variant cells showed no detectable defect. XP group
B and F cells have not yet been studied.

E.   The Role of Chromatin Conformation in DNA Repair in XP Cells

Some years ago studies from this laboratory demonstrated that
extracts of normal human fibroblasts prepared by sonication of
[^{3}]H-thymine labeled UV-irradiated cells effected the excision of

pyrimidine dimers from endogenous chromatin (28). Extracts of XP cells prepared under identical conditions catalyzed the excision of pyrimidine dimers from UV-irradiated purified DNA added to the incubation, but such extracts of XP group A (26), C (27) and variant cells (27) failed to excise dimers from endogenous chromatin. On the basis of these results we postulated that the excision of dimers and possibly other forms of base damage to DNA in eukaryote cells requires not only the action of specific repair enzymes, but also the action of factors necessary for nucleosome unfolding and/or the dissociation of DNA from histones so as to provide access of sites of base damage to repair enzymes (28,29).

Since then, studies from a number of laboratories have provided evidence suggesting that the rate of excision repair in micrococcal nuclease resistant regions of chromatin (presumably nucleosome cores) is significantly less than in micrococcal nuclease sensitive (presumably nucleosome linker) regions of chromatin (30-32). Comparisons between normal and XP group C, D and E cells have not shown any qualitative differences in the distribution of repair synthesis label between micrococcal sensitive and resistant regions of chromatin (33). Quantitative deviations from normal have been observed in XP group C and D cells, but their biological relevance is currently unclear (33).

A specific aspect of histone modification during DNA repair that has received recent attention is poly (ADP) ribosylation. A number of studies suggest that any perturbation of chromatin associated with DNA strand breakage, results in increased cellular poly (ADP) ribose polymerase activity and decreased NAD content, indicative of increased poly (ADP) ribose synthesis (see ref. 34). XP group A, B, C and D cells fail to show these responses following UV irradiation (35). Based on these results, Berger et al (35) have suggested that in normal human cells enzyme-catalyzed incision of DNA stimulates the synthesis of poly (ADP) ribose, which in turn produces alterations in chromatin conformation required for base damage excision and for repair synthesis of DNA.

F.   Conclusion

Despite intensive efforts over the past dozen years, very little real progress has been made in our understanding of the biochemical basis (or bases) for the multiple cellular defects described in this disease. In my view the key to this understanding lies in unravelling the specific enzymatic and possible non-enzymatic biochemical events required for the repair of base damage in chromatin-associated DNA in eukaryotes. Advances in DNA repair enzymology in recent years (see other articles in this volume) indicate that the majority of repair-specific enzymes from any biological source (I do not consider

DNA polymerases, exonucleases or DNA ligases as repair-specific enzymes) that have readily yielded to the techniques of classical biochemical isolation and purification are relatively simple monomeric proteins that are DNA glycosylases. However the recent biochemical complexity shown with the uvr A, B and C gene products of E. coli (36-38) leads one to suspect that the "repairosome(s)" in eukaryote cells are at least equally as complex. When ultimately isolated and characterized, such putative "repairosomes" may well consist of multiprotein units in which specific interactions between individual proteins effect multiple functions such as base damage recognition, DNA-histone dissociation and DNA incision and excision of base damage, by one or more specific enzymatic mechanisms. When such information is available, direct comparison with "repairosomes" in XP cells will hopefully yield the answers we have sought for so long.

Acknowledgements

Studies in the author's laboratory were supported by research grants from the USPHS (CA 12428), American Cancer Society (NP-174) and by contract DE-AC03-76SF00326 with the U.S. Department of Energy. The author acknowledges the USPHS for receipt of a research career development award (CA 71005).

References

1. Cleaver, J.E. (1968). Nature, 218, 652.
2. Cleaver, J.E., Bootsma, D., and Friedberg, E. (1975) Genetics 79, 215.
3. Cleaver, J.E. and Bootsma, D. (1975). Annu. Rev. Genet. 9, 19.
4. Setlow, R.B. (1978). Nature 271, 713.
5. Arlett, C., and Lehmann, A.R. (1979). Annu. Rev. Genet. 12, 95.
6. Friedberg, E.C., Ehmann, U.K., and Williams, J.I. (1979). Adv. Rad. Biol. 8, 86.
7. Hashem, N., Bootsma, D., Keijzer, W., Greene, A., Coriell, L., Thomas, G., and Cleaver, J.E. (1980). Cancer Res. 40, 13.
8. Arase, S., Kozuka, T., Tanaka, K. Ikenaga, M. and Takebe, H. (1979). Mutation Res. 59, 143.
9. Keijzer, W., Jaspers, N.G.J., Abrahams, P.J., Taylor, A.M.R., Arlett, C.F., Zelle, B., Takebe, H., Kinmont, P.D.S., and Bootsma, D (1979). Mutation Res. 62, 183.
10. Andrews, A.D., Barrett, S.F. and Robbins, J.H. (1978). Proc. Natl. Acad. Sci. (USA) 75, 1980.
11. Hurley, L.H., Chandler, C., Garner, T.F., Petrusek, R., and Zimmer, S.G. (1979). J. Biol. Chem. 254, 605.
12. Coppey, J., and Nocentini, S. (1979). Mutation Res. 62, 355.
13. Zelle, B (1980). Ph.D. Thesis, Erasmus University, Rotterdam.
14. Ahmed, F.E., and Setlow, R.B. (1979). Cancer Res. 39, 471.

15. Ahmed, F.E. and Setlow, R.B. (1980). Chem-Biol. Interactions 29,
         31.

16. Brown, A.J., Fickel, T.H., Cleaver, J.E., Lohman, P.H.M., Wade,
         M.H. and Waters, R. (1979). Cancer Res. 39, 2522.

17. Goth-Goldstein, R. (1977). Nature 267, 81.

18. Altamirano-Dimar, M., Sklar, R., and Strauss, B. (1979). Mutation
         Res. 60, 197.

19. Bodell, W.J., Singer, B., Thomas, G.H. and Cleaver, J.E. (1979)
         Nucleic Acids Res. 6, 2819.

20. Kuhnlein, U. Penhoet, E.E. and Linn, S. (1976). Proc. Natl. Acad.
         Sci. (USA) 73, 1169.

21. Kuhnlein, U., Lee, B., Penhoet, E.E. and Linn, S. (1978). Nucleic
         Acids Res. 5, 951.

22. Kudrna, R.D., Smith, J., Linn, S., and Penhoet, E.E. (1979). Mutation
         Res. 62, 173.

23. Deutsch, W.A., and Linn, S. (1979). Proc. Natl. Acad. Sci. (USA)
         76, 141.

24. Moses, R.E., and Beaudet, A.L. (1978). Nucleic Acids Res. 5, 463.

25. Witte, I., and Thielmann, H.W. (1979). Cancer Lett. 6, 129.

26. Ishiwata, K., and Oikawa, A. (1979). Biochem Biophys. Acta. 563,
         375.

27. Zelle, B., and Lohman, P.H.M. (1979). Mutation Res. 62, 363.

28. Mortelmans, K., Friedberg, E.C., Slor, H., Cleaver, J.E., and Thomas,
         G. (1976). Proc. Natl. Acad. Sci. (USA) 73, 2757.

29. Friedberg, E.C., Rude, J.M., Cook, K.H., Ehmann, U.K., Mortelmans,
         K., Cleaver, J.E., and Slor, H. (1977) in "DNA Repair Processes",
         eds. W.W. Nichols and D.G. Murphy. Symposia Specialists.
         Miami, Fl., p 21.

30. Cleaver, J.E. (1977). Nature 270, 451.

31. Smerdon, M.J., Tlsty, T.D. and Lieberman, M.W. (1978). Biochemistry
         17, 2377.

32. Williams, J.I., and Friedberg, E.C. (1979). Biochemistry 18, 3965.

33. Smerdon, M.J., Kastan, M.B., and Lieberman, M.W. (1979). Bio-
         chemistry 18, 3732.

34. Dukacz, B.W., Omidiji, O., Gray, D.A., and Shall, S. (1980). Nature
         283, 593.

35. Berger, N.A., Sikorski, G.W., Petzold, S.J., and Kurohara, K.K. (1980).
         Biochemistry 19, 289.

36. Seeberg, E., Nissen-Meyer, J. and Strike, P. (1976). Nature 263.

37. Seeberg, E. (1978). Proc. Natl. Acad. Sci. (USA) 75, 2569.

38. Seeberg, E. (1980). Prog. Nuc. Acids Res. Mol. Biol. (In press)

CORRELATION OF POST-UV COLONY-FORMING ABILITIES
OF XERODERMA PIGMENTOSUM FIBROBLASTS WITH DNA
INCISION REACTIONS CATALYZED BY CELL-FREE EXTRACTS[+]

Heinz Walter Thielmann and Adonis Georgiades

German Cancer Research Center
Institute of Biochemistry
Im Neuenheimer Feld 280, 6900 Heidelberg, FRG.

INTRODUCTION

Xeroderma pigmentosum is an autosomal recessive genodermatosis
which is characterized by abnormal dryness, atrophy, pigmentation
abnormalities, and early carcinomas of the skin if exposed to sunlight (1). Cultivated XP fibroblasts show lower levels of DNA excision repair after exposure to UV irradiation or UV-like carcinogens
such as $(Ac)_2ONFln$. There are currently 7 excision repair-deficient
forms, classified as complementation groups A-G (2), which suggests
that several gene products are involved in excision repair of UV-
damaged DNA. It seems generally accepted that XP cells of all complementation groups are defective at the level of DNA incision
(2,3). However, our knowledge regarding the precise enzymic deficiency by which a particular complementation group is characterized
is still incomplete.

In the present study various normal and XP group A fibroblast
lines (the latter from XP patients who also had neurological abnormalities) were exposed either to UV light or to the ultimate
carcinogen $(Ac)_2ONFln$, and the differences in their abilities to
form colonies were studied. In addition, 105 000 x g supernatants
were prepared from the normal and XP cell lines and tested for
their capacities to incise UV-irradiated PM2 DNA. It will be shown
that the XP group A cell lines investigated exhibit both a drasti-

[+]This work was supported by the Deutsche Forschungsgemeinschaft,
SFB 136.
Abbreviations: XP, Xeroderma pigmentosum; $(Ac)_2ONFln$, N-acetoxy-2-
acetyl-aminofluorene; UV light, ultraviolet light; AP sites, apuri-
nic/apyrimidinic sites.

cally diminished colony-forming ability after UV irradiation or treatment with $(Ac)_2ONFln$ and an absence of endonucleolytic activity, i.e. cleaving UV-irradiated double-stranded DNA, from 105 000 x $\underline{g}$ supernatants. In further experiments, the protocol developed for determination of colony-forming ability was applied to a series of fibroblast lines derived from skin biopsies of XP patients currently under clinical observation by Prof. Dr. E. Jung (Dermatologische Klinik der Städt. Krankenanstalten Mannheim).

MATERIALS AND METHODS

Cell cultures

The cell types and their sources were as follows: normal human fibroblasts, N-1 (Dr. E. Jung), GM 500 (The Human Genetic Mutant Cell Repository, Camden, NJ, USA), CRL 1295 (American Type Culture Collection, Rockville, MD, USA). XP fibroblasts, complementation group A with neurological abnormalities, GM 710, GM 544 (The Human Genetic Mutant Cell Repository), CRL 1223 (American Type Culture Collection), XP02MA to XP32MA, referred to as Mannheim collection (Dr. E. Jung).

Measurement of colony-forming ability

Fibroblasts from stocks in passages 3-14 were seeded into each of four 64 cm^2 dishes at a cell density of approximately 30 cells/cm^2 using F-12 medium containing 100 IU/ml of both streptomycin and penicillin, and supplemented with 10% fetal calf serum (medium A). After incubation for 12 h at 37 $^{\circ}C$ in an atmosphere of 5% CO_2/95% air (with more than 95% humidity), the cells were washed with phosphate-buffered saline (PBS) and UV-irradiated with a germicidal lamp. The intensity was measured with a factory-calibrated model J-260 Radiometer (Ultra-Violet Products, Inc., San Gabriel, CA, USA). After irradiation, cells were fed medium A. Treatment with carcinogen lasted for 1 h at 37 $^{\circ}C$, in F-12 medium supplemented with 5% fetal calf serum. Thereafter cells were washed with PBS before being given medium A. The medium was changed once a week. After 3 weeks, cells were fixed with 4% CH_2O, stained with crystal violet, and the colonies (aggregates of 50 and more cells) counted. Colony-forming ability was calculated by dividing the number of colonies of UV- or $(Ac)_2ONFln$-treated cells per dish by the number of colonies from control dishes. Colony-forming efficiency, i.e., the number of colonies grown having had no treatment divided by the number of cells plated, ranged from 8-15%, depending on the cell line.

Preparation of 105 000 x g supernatants

Cells grown to subconfluency in 3 Roux bottles were washed 3 times with PBS, and scraped off the surface into 3.6 ml buffer

(100 mM NaCl, 50 mM Tris-HCl pH 7.4, 0.1 mM EGTA, 0.1 mM 2-mercaptoethanol, 15% glycerol). The suspension was sonicated until cells and nuclei were completely disintegrated. The sonicate was centrifuged at 4 °C for 90 min at 105 000 x g. The supernatants were used immediately. The protein contents of the supernatants varied between 300-1100 µg/ml.

Endonuclease activity

Endonuclease was measured by monitoring the conversion of UV-irradiated (1300 erg/mm$^2 \cong$130 J/m^2) superhelical PM2 DNA (23 pyrimidine dimers (4)) to nicked circles. The incubation mixture (total vol. 40 µl) contained 0.3 nmol DNA in 30 mM Tris-HCl pH 7.4, 60 mM NaCl, 0.06 mM EGTA, 0.06 mM 2-mercaptoethanol, and 0.1-1.7 µg protein. The reaction proceeded for 60 min at 37 °C, and was stopped with 20 µl of a solution having the following composition: 50 mM sodium dodecylsulfate, 20 mM EDTA, 1.4 M sorbitol, 0.25% bromophenol blue marker. Covalently closed circular PM2 DNA was separated from the nicked form by agarose gel electrophoresis (6).

From the fraction of molecules which contained no breaks (α), the average number of breaks per PM2 DNA molecule (N) was calculated using the equation: $N = -\ln\alpha$. In control experiments using ^{32}P-labelled PM2 DNA as test molecule and alkaline sucrose gradient centrifugation for analysis, the number of alkali-labile sites induced by UV irradiation (1300 erg/mm^2) was found to be 0.08 per DNA molecule. This background value was subtracted to give the data shown in Fig. 2.

RESULTS AND DISCUSSION

Colony-forming abilities were investigated in 3 normal and 3 XP group A fibroblast lines (the latter from patients who in addition to skin manifestations had neurological disorders) after UV irradiation of the cells in situ. There was no trypsinization and replating following UV irradiation; instead, medium was simply added and the cells were allowed to grow. The colonies were counted when 3 weeks had elapsed. Fig. 1A shows the colony-forming abilities of the respective fibroblast lines as a function of the UV dose. It appears that the 3 normal lines exhibit positive shoulders which extend to approximately 30 erg/mm^2 (3 J/m^2), followed by exponential portions. These curves do not differ significantly from each other either in their shoulder regions or in their slopes. In contrast, the curves obtained from the 3 XP group A fibroblast lines do not show positive shoulders. Their exponential portions are far steeper than the ones from normal cell lines, indicating that the XP cells have lower capacities to repair UV-induced damage and to restore their DNA to the biologically functional level required to complete repeated cycles of cellular division (5). These curves closely resemble to those published by Andrews et al.

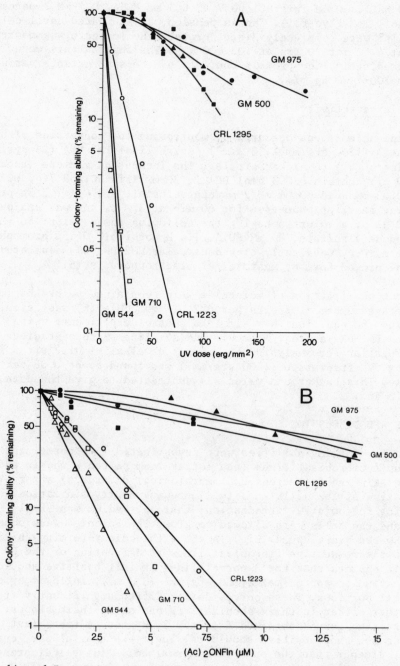

Fig. 1A and B.   Colony-forming ability of 3 normal and 3 XP group A fibroblast lines as a function of UV dose (A) or $(Ac)_2ONFln$ concentration (B) used for cell pretreatment (see Materials and Methods).

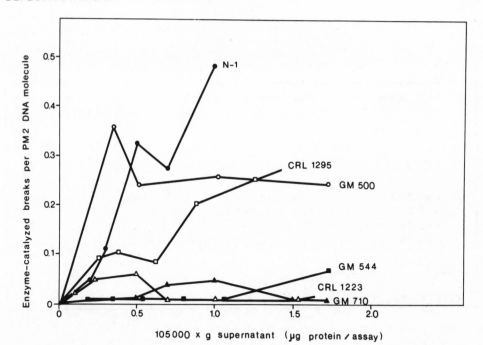

Fig. 2. Comparison of endonucleolytic activity of 105 000 x g
supernatants from normal fibroblast lines with those from XP group
A lines. For details see Materials and Methods.

(5) who followed a protocol involving trypsinization and replating
after UV irradiation. Qualitatively similar results were obtained
when UV light was replaced by the ultimate carcinogen $(Ac)_2ONFln$.
As Fig. 1B shows, normal cell lines again exhibit a positive shoulder; the long exponential portions of the curves are less steep
than the ones from the XP group A cell lines. As is the case after
UV treatment, the XP fibroblasts do not form a shoulder at all.
Table 1 includes the D_0 values calculated from the curves shown in
Figs. 1A and 1B. These D_0 values represent the UV dose or, analog-
ously, the carcinogen concentration required to lower the colony-
forming ability from any point of the straight line to $1/e$ (36.8%)
of that point.

It appears that the D_0 values of the XP fibroblasts are on
average one seventh $((Ac)_2ONFln)$ or one fourteenth (UV) of those
measured with normal cells. Thus the enzymic activity that is lower
or missing in the XP cells investigated was required not only for
the repair of UV-induced DNA damage but also for repair of damage
caused by $(Ac)_2ONFln$.

Attempts were also made to characterize some XP cell lines of
the Mannheim collection with regard to colony-forming ability after
exposure to UV or $(Ac)_2ONFln$. Table 2 includes the extremes of the

Table 1.   Identity of the fibroblast lines studied and
           their sensitivity (reduction of colony-forming
           ability) after UV irradiation or treatment with
           $(Ac)_2ONFln$

Cell line	Complementation group	D_0 values after treatment with	
		UV (erg/mm^2)	$(Ac)_2ONFln$ (μM)
CRL 1295	normal	31	10.8
GM 975	normal	97	12.8
GM 500	normal	66	8.5
CRL 1223	A[a]	6	2.1
GM 544	A[a]	3.5	1.1
GM 710	A[a]	4.0	1.7

[a] with neurological abnormalities

Table 2.   Identity of the XP fibroblast lines of the
           Mannheim collection (Dr. E. Jung) and their
           sensitivity (reduction of colony-forming
           ability) after UV irradiation or treatment
           with $(Ac)_2ONFln$.

Cell lines (Mannheim collection[a])	D_0 values after treatment with	
	UV (erg/mm^2)	$(Ac)_2ONFln$ (μM)
XPO2MA	47	5.4
XPO7MA	49	5.3
XP14MA	50	1.6
XP23MA	49	1.0
XP15MA	10	1.5
XP17MA	18	0.5
XP30MA	6	0.5
XP31MA	6	1.0
XP32MA	9	1.2

[a] complementation groups no determined as yet

D_0 values measured. The first group (XP02MA, XP07MA, XP14MA and
XP23MA) exhibits D_0 values which are close to normal (see Table 1).
These cell lines were derived from biopsies of patients whose only
apparent symptoms were cutaneous. There is a second group (XP15MA,
XP17MA, XP30MA, XP31MA, XP32MA) showing a similar high degree of
UV sensitivity (as well as sensitivity to $(Ac)_2ONF1n$) to that reported characteristic for cell lines belonging to complementation
groups A or D (5). Some of these patients have one or more of the
components of de Sanctis-Cacchione syndrome, in addition to the
characteristic skin manifestations. Given the above selection of
XP cell lines, it seems that the biochemical defect as represented
by colony-forming ability to some extent parallels the severeness
of the clinical status of the patients.

The question was considered as to whether or not an excision
repair defect, as expressed by low colony-forming ability, would be
reflected by a low incision capacity of the respective 105 000 x g
supernatant with UV-irradiated double-stranded DNA as substrate.
Supernatants were therefore prepared from the same normal and XP
cell lines which had been tested for colony-forming ability, and
their capacities for incising UV-damaged DNA were compared. Briefly,
PM2 DNA containing approximately 23 pyrimidine dimers (4) was incubated with increasing amounts of supernatant. At the end of the
enzymic reaction, the product (nicked, double-stranded DNA) was separated from the substrate (dimer-containing superhelical DNA) by
agarose gel electrophoresis. The number of enzyme-catalyzed breaks
(calculated as described in the Materials and Methods section) was
measured as a function of the amount of extract protein in the
assays. As Fig. 2 shows, the supernatants of the XP group A cell
lines tested caused less nicks in the DNA substrate than the supernatants from normal cell lines. These differences are significant
using Fisher's "exact test for four-field tables" as criterion.

The yield of breaks observed is certainly not due to the presence of AP sites, as the small contribution made by these sites
has been carefully determined in control experiments (see Materials
and Methods), and subtracted. Moreover, in a previous study it
was shown that extracts from normal and XP group A fibroblast lines
do not differ in AP endonuclease activity (6). However, the possibility that minor photoproducts other than dimers and AP sites account for the enzyme-catalyzed breaks cannot be excluded. Whatever
their nature, they seem to be recognized by normal supernatants but
not by the XP group A supernatants.

It should be noted that incising activity was measured in the
absence of Mg^{2+} and ATP. These two cofactors were found to be
essential for incision of both UV- and $(Ac)_2ONF1n$-damaged DNA in
nucleotide-permeable E. coli cells (7,8). However, when Mg^{2+} and
ATP were present in the in vitro assays, complete degradation of

the DNA substrate by unspecific DNases was observed.

The data presented in Fig. 2 have to be considered with caution, as, on average, of the 23 dimers present per PM2 molecule, only 0.3 were converted to single-strand breaks (using 1 ug protein per assay). Thus the test reaction appears to be far from optimal. It is conceivable that Mg^{2+} and ATP are necessary for the enhancement of dimer-specific breaks. If so, testing of 105 000 x g supernatants for dimer-specific endonuclease activity will not be possible in the absence of inhibitors which restrain unspecific DNases. Further investigations are needed for the eventual establishment of optimal test conditions.

REFERENCES

1. J. M. Mascaró, Xeroderma pigmentosum, in: Cancer of the skin, Vol. 1, R. Andrade, S. L. Gumport, G. L. Popkin, T. D. Rees (eds.), pp. 573-595, W. B. Saunders Co., Philadelphia, London, Toronto (1976).
2. W. Keijzer, N. G. J. Jaspers, P. J. Abrahams, A. M. R. Taylor, C. F. Arlett, B. Zelle, H. Takebe, P. D. S. Kinmont, and D. Bootsma, A seventh complementation group in excision-deficient xeroderma pigmentosum, Mutat. Res., 62: 183-190 (1979).
3. K. Tanaka, H. Hayakawa, M. Sekiguchi, and Y. Okada, Specific action of T4 endonuclease V on damaged DNA in xeroderma pigmentosum cells in vivo, Proc. Natl. Acad. Sci. USA, 74: 2958-2962 (1977).
4. M. Woodworth-Gutai, J. Lebowitz, A. C. Kato, and D. T. Denhardt, Ultraviolet light irradiation of PM2 superhelical DNA, Nucleic Acids Res., 4: 1243-1256 (1977).
5. A. D. Andrews, S. F. Barrett, and J. H. Robbins, Xeroderma pigmentosum neurological abnormalities correlate with colony-forming ability after ultraviolet radiation, Proc. Natl. Acad. Sci. USA, 75: 1984-1988 (1978).
6. I. Witte and H. W. Thielmann, Extracts of Xeroderma pigmentosum group A fibroblasts introduce less nicks into methyl methanesulfonate-treated DNA than extracts of normal fibroblasts, Cancer Letters, 6: 129-136 (1979).
7. E. A. Waldstein, R. Sharon, and R. Ben-Ishai, Role of ATP in excision repair of ultraviolet radiation damage in Escherichia coli, Proc. Natl. Acad. Sci. USA, 71: 2651-2654 (1974).
8. H. W. Thielmann, Carcinogen-induced DNA repair in nucleotide-permeable Escherichia coli cells, Eur. J. Biochem., 61: 501-513 (1976).

ACKNOWLEDGEMENTS: We thank Dr. L. Edler, Institut für Dokumentation, Information und Statistik, Deutsches Krebsforschungszentrum, for doing the statistical calculation. We also thank Dr. G. Robert N. Jones for critically reading the manuscript and for helpful suggestions. The technical assistance of Frau S. Friemel is gratefully acknowledged.

EFFECTS OF ULTRAVIOLET-IRRADIATION ON DNA LIGASE ACTIVITY OF HUMAN FIBROBLASTS FROM NORMAL AND XERODERMA PIGMENTOSUM DONORS

S. Nocentini and M. Mezzina*

Institut Curie, Section de Biologie, 26 rue d'Ulm
75231 PARIS cedex 05, France
*Institut de Recherches Scientifiques sur le Cancer,
CNRS, B. P. n° 8, 94800 VilleJuif, France

ABSTRACT

Ultraviolet (UV)-irradiation of stationary cultures induces an increase of DNA ligase activity in both normal and complementation group A xeroderma pigmentosum (XP) cells. Extents and doses of induction are cell line-dependent. These inductions can be prevented by the protein synthesis inhibitor cycloheximide, but not by the DNA replication inhibitor fluorodeoxyuridine (FudR).

INTRODUCTION

Direct evidence of the role that DNA ligase is supposed to play in DNA replication and repair in mammals is until now anavailable, essentially because of the lack of lethal conditional mutants defective in DNA ligase or of specific inhibitor of this activity. However, many studies on DNA ligase in different proliferating tissus or cultures have suggested that ligase activity is correlated with the rate of DNA synthesis (for review, see 1). On the other hand, in a precedent report (2) we have shown that DNA ligase activity increases in UV-irradiated monkey kidney CV-1 cells as DNA repair takes place, operationally supporting the current concept that ligation is a necessary step in excision repair process. To further investigate the possible involvement of DNA ligase in DNA repair, we have studied how DNA ligase activity is affected by UV-irradiation in human fibroblasts genetically characterized with regards to their repair capability. Results indicate that UV-irradiation induces new DNA ligase in both normal and A group XP fibroblasts, but the dose-dependences of these inductions are very different.

MATERIELS AND METHODS

Cells. The following fibroblasts lines were used: CRL 1295, a normal diploid strain, and XP 12 BE (CRL 1223), an XP line belonging to complementation group A, were obtained from the American Type Culture Collection, Bethesda M.D. An other normal line, N2, was kindly supplied by Dr Arlett (MRC Cell Mutation Unit, University of Sussex, England). Cultures were grown in RPMI medium 1640 (Eutroph) buffered with 20 mM N-(Tris (hydroxymethyl)- methyl)glycin (Merck) and supplemented with 10 % foetal calf serum (Rehatuin Laboratories). Cells were tested between passage 10 and 30 of subcultivation.

Experimental Procedure. Stationary cultures in 14.5 cm plastic Petri dishes (Greiner) were irradiated, after removal of the culture medium, with a single dose of 254 nm UV-radiation at different time intervals before preparing cellular crude extracts and assaying for DNA ligase according to Olivera's method. Experimental conditions have been published in detail elsewhere (2).

RESULTS

All experiments were carried out with confluent cultures in which there was virtually no mitotic activity. In these conditions, the specific activity of DNA ligase was of the same order in all three cell lines tested (about 2 units/mg of proteins) and its level remained constant in unirradiated cultures through the whole duration of experiments (48 h). The UV-irradiation of cells immediately before their harvesting for DNA ligase assay (t = 0) did not modify the specific activity of the enzyme if compared with unirradiated controls. When post-exposure incubation of cells increased, DNA ligase activity was observed to evolve depending on the dose of irradiation and culture origin. Fig. 1-A shows that in excision repair-proficient fibroblasts (CRL 1295), UV-irradiation over a broad range of doses induced an increase of DNA ligase activity. A maximum two fold enhancement was observed after 48 h in cultures irradiated with about 20 J/m^2. The level of enzyme tended to decrease with time following irradiation with doses $\geqslant 35$ J/m^2. Similar results were obtained with the other normal cell line N2. In XP 12 BE excision repair-deficient cells (Fig 1-B) little fluctuations of DNA ligase level were observed after a dose of 4 J/m^2; higher doses led to a decline of ligase activity whereas doses $\leqslant 2$ J/m^2 induced an increase in its level. Fig. 2 emphasizes the difference of dose-dependence of ligase activity induction between normal and XP cells.

In parallel experiments, cultures were treated with cycloheximide or FudR, inhibitors respectively of protein synthesis and replicative DNA synthesis. Drugs were added at all cultures at time - 48 h and maintained until t = 0. Results show (Fig 1-A and -B)

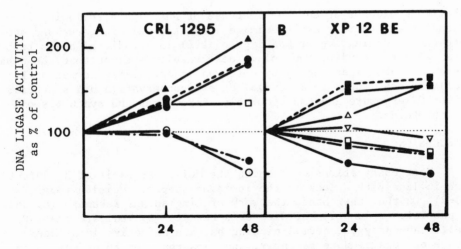

Fig. 1.   DNA ligase activity as a function of the time of post-
exposure incubation in A) normal fibroblasts and B)
complementation group A XP cells irradiated with a dose
of 1 (△), 2 (■), 4 (▽), 9 (□), 17.5 (●), 24 (▲) and 35
(○) J/m^2, in absence ———— or in presence of cycloheximide 5 µg/ml) —·—· or FudR (10^{-5} M) ------- .

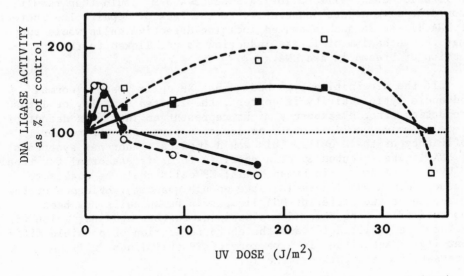

Fig. 2.   Dose-dependence of DNA ligase induction in normal (■,□)
and A group XP (●,○) cells measured 24 h (full symbols)
or 48 h (open symbols) after irradiation.

that FudR did not affect the increase of DNA ligase following UV-irradiation, whereas cycloheximide inhibited this increase and even provoked a diminution of DNA ligase level in irradiated cells below that of corresponding controls. These results suggest that the enhancement of DNA ligase activity observed in irradiated normal or XP cells is not linked to a revival of semiconservative DNA synthesis after UV-exposure and that probably it requires the synthesis of new proteins.

DISCUSSION

Most observations concerning the molecular basis of XP defect are in line with a lack in the incision step of excision repair (3-7). On the other hand, the work of Dingman and Kakunaga (8) and a possible explanation of Cleaver's results (9) indicate that XP cells can carry out normal nicking but are defective in a later stage of repair. More recently, the idea that the XP defect is not simply due to enzyme deficiences is progressing (10,11).

We have shown here that DNA ligase is inducible by UV-radiation in both normal and XP cells but with quite different dose-dependences The increase of activity in XP cells is only observed after low doses of UV-irradiation. In these conditions, DNA ligase activity seems unlikely to represent the limiting factor determining repair deficiency in XP cells and the hypothesis of a defect in an early stage of repair is somewhat strengthened. The decline of ligase activity in irradiated XP cells following doses >4 J/m^2 could then result from degrading events consecutive to the lack of repair. The increase of DNA ligase in such presumed incision-defective cells would suppose that the mechanism of ligase induction is not linked to the introduction of breaks in the DNA.

If the alternative hypothesis that XP cells possess normal endonucleolytic activity is correct, the limited capacity of ligase production after UV-exposure might represent an intrinsic defect of XP cells which could account, at least in part, for the repair deficiency of these cells. This would implicate that new synthesis of DNA ligase, eventually of a specific form, is essential for repair of UV-induced damage in human cells; XP cells could be unable to produce such a DNA ligase in large enough quantity. Evidence of the existence of two different DNA ligases in human cells has been reported (1) making this idea plausible. Further studies, which will include the purification and the characterization of possible different ligase activities in a variety of XP cell lines, will be necessary to test this hypothesis.

ACKNOWLEDGEMENTS

This work was supported by contracts CRL 79.4.179.3 and ATP 77.79.109 from INSERM

REFERENCES

1. S. Söderhäll and T. Lindahl, DNA ligases of eukaryotes, Febs
    Letters, 67:1 (1976)
2. M. Mezzina and S. Nocentini, DNA ligase activity in UV-irradiated
    monkey kidney cells, Nucleic Acids Res., 5:4317 (1978)
3. J. E. Cleaver and J. E. Trosko, Absence of excision of ultra-
    violet-induced cyclobutane dimers in xeroderma pigmentosum,
    Photochem. Photobiol., 11:547 (1970)
4. M. C. Paterson, P. H. M. Lohman and M. L. Sluyter, Use of a UV
    endonuclease from Micrococcus luteus to monitor the progress of
    DNA repair in UV-irradiated human cells, Mutation Res., 19:245
    (1973)
5. K. Tanaka, M. Sekiguchi and Y. Okada, Restoration of ultraviolet-
    induced unscheduled DNA synthesis of xeroderma pigmentosum cells
    by the concomitant treatment with bacteriophage T4 endonuclease
    V and HVJ (Sendai virus), Proc. Nat. Acad. Sci. USA, 72:4071
    (1975)
6. A. J. Fornace, K. W. Kohn and H. E. Kann, DNA single-strands
    breaks during repair of UV damage in human fibroblasts and
    abnormalities of repair in xeroderma pigmentosum, Proc. Nat.
    Acad. Sci. USA, 73:39 (1976)
7. P. R. Cook, I. A. Brazell, S. A. Pawsey and F. Giannelli, Changes
    induced by ultraviolet light in the superhelical DNA of lympho-
    cytes from subjects with xeroderma pigmentosum and normal
    controls, J. Cell Sci., 29:117 (1978)
8. C. W. Dingman and T. Kakunaga, DNA strand breaking and rejoining
    in response to ultraviolet light in normal and xeroderma pigmen-
    tosum cells, Int. J. Radiat. Biol., 30:55 (1976)
9. J. E. Cleaver, Sedimentation of DNA from Human Fibroblasts
    Irradiated with Ultraviolet Light: Possible Detection of Excision
    Breaks in Normal and Repair-Deficient Xeroderma Pigmentosum Cells,
    Radiat. Res., 57:207 (1974)
10. K. Cook, E. C. Friedberg and J. E. Cleaver, Excision of thymine
    dimers from specifically incised DNA by extracts of Xeroderma
    pigmentosum cells, Nature, Lond. , 256:235 (1975)
11. K. Mortelmans, E. C. Friedberg, H. Slor, G. Thomas and J. E.
    Cleaver, Defective thymine dimer excision by cell-free extracts
    of xeroderma pigmentosum cells, Proc. Natl. Acad. Sci USA, 73:
    2757 (1976)

HERITABLE RADIOSENSITIVE AND DNA REPAIR-DEFICIENT DISORDERS IN MAN

M.C. Paterson, N.T. Bech-Hansen, and P.J. Smith

Health Sciences Division
Atomic Energy of Canada Limited
Chalk River Nuclear Laboratories
Chalk River, Ontario K0J 1J0, Canada

INTRODUCTION

The deleterious effects of many environmental agents--ultraviolet (UV) light, ionizing radiation, and polycyclic aromatic hydrocarbons, to cite a few--are primarily attributed to their ability to react with, and thereby structurally modify the deoxyribonucleic acid (DNA) of the living cell[1]. Given the appreciable level of damage believed to occur naturally in DNA[2] and the need to maintain its fidelity as the repository of the genetic script, it is understandable that all living organisms, humans included, should possess multiple cellular processes whose combined actions promote the repair of damage to DNA[3]. An increasing number of genetically transmitted disorders in man is being found to be associated with enhanced sensitivity to extrinsic DNA-damaging agents. Patients afflicted with any one of these particular Mendelian single-gene traits are typically cancer-prone and suffer from progressive neurodegeneration, and, at least for a few disorders, their cells when cultured *in vitro* exhibit anomalies in one or more DNA repair processes[4-8]. The majority of these hereditary diseases can be divided into two broad groups: (i) disorders associated with hypersensitivity to UV light and UV-mimetic chemicals; and (ii) disorders associated with hypersensitivity to ionizing radiation and radiomimetic chemicals. Exemplary diseases of the former group-- namely, xeroderma pigmentosum (XP), Bloom syndrome (BS), and Cockayne syndrome--are reviewed elsewhere in this volume. Here we present an overview of the clinical and laboratory features of the latter group of disorders with emphasis on the prototype disease, ataxia telangiectasia (AT)[7,8], and comment briefly on current insight into the pathogenesis of these disorders and its implications for theories on environmental carcinogenesis and neurodegenerative processes in man.

ATAXIA TELANGIECTASIA: A MULTISYSTEM DISORDER LINKING RADIOSENSI-TIVITY WITH DEFECTIVE DNA REPAIR

Clinical Picture

This complex neurovascular and immunodeficiency syndrome follows an autosomal recessive pattern of inheritance; the disorder is glo-bally distributed, and its occurrence, estimated at 25 per million live births, appears not to be maintained by inbreeding or new mutations (see refs. 7 and 8 for clinical details). The minimal criteria for clinical diagnosis are muscular incoordination (ataxia) due to cerebellar dysfunction, and permanent dilation of blood ves-sels (telangiectasis) of the conjunctivae and skin, especially sun-light-exposed areas. Both symptoms often have their onset in in-fancy and follow a variable but relentless progression, the former typically culminating in severe neurological incapacitation by late adolescence. Thymus, liver, and sex organs are frequently under-developed, possibly because of aberrant tissue differentiation during early embryonic development. The hypoplastic thymus presumably ac-counts for the immunodeficient state which often involves both the cellular system (e.g. delayed skin-graft rejection, reduced response to skin-test antigens) and the humoral system (e.g. reduced levels of serum and salivary IgA and IgE). Impaired immunity is in turn associated with predisposition to recurrent bronchopulmonary in-fection. Afflicted patients are also prone to develop malignancy, notably lymphomas and lymphatic leukemias; the cancer incidence, estimated to be ~10%, is some 1200 fold greater than that in an age-matched control population. Pneumonia, cancer, and ataxia are the major complications leading to premature death. Increased radiosen-sitivity takes the form of a severe (e.g. ulcerative dermatitis, deep tissue necrosis) and sometimes fatal reaction to conventional radio-therapy for tumor treatment. Considerable variation in the appear-ance, age of onset, and severity of such clinical features as immuno-deficiency and hypogonadism is consistent with genetic heterogeneity in the disorder.

Laboratory Picture

The two most consistent laboratory hallmarks of AT are elevated levels of α-fetoprotein in blood serum[9] and numerous signs of in-creased radiosensitivity in both peripheral blood lymphocytes and skin fibroblasts[7,8]. The first hallmark is probably not a causative factor, but rather is secondary to liver dysfunction and hence will

not be discussed further. Enhanced radiosensitivity, on the other
hand, is almost certainly causally involved in some of the clinical
features of the disease; this trait has been the subject of much
research, and the findings at the cellular, cytogenetic, and biochemical (DNA repair) levels will now be summarized.

Cellular Response to DNA-Damaging Agents. AT fibroblast strains,
derived from skin biopsies of afflicted donors, are in general hypersensitive to the lethal action of ionizing radiation and radiomimetic
chemicals, such as bleomycin (BM) and N-methyl-N'-nitro-N-nitroso-
guanidine (MNNG), but respond like control strains from normal donors
to killing by far UV (254 nm) light and N-hydroxy-acetylaminofluorene
(N-hydroxy-AAF), a far UV-mimetic agent (see Table 1).

Numerous workers have measured the colony-forming ability of AT
fibroblasts after acute exposure to sparsely ionizing radiation (X-
or γ-rays)[10-15]. To date no less than 20 genealogically distinct AT
strains have been tested under oxia and/or hypoxia, and all 20 exhibit increased radiosensitivity, ranging from 2.6-3.5 fold at the
10% survival level. Studies of cell viability (assayed by dye ex-
clusion) indicate that cultured lymphocytes from AT patients are
as hypersensitive to radiation toxicity as their skin fibroblasts[23].
The striking consistency of these data, currently unprecedented among
hereditary disorders, documents enhanced radiosensitivity as an in-
variable *in vitro* marker of AT. This observation, aside from un-
covering a laboratory test for early (possibly prenatal) clinical
diagnosis, strongly implies that radiation hypersensitivity is
directly related to the primary defect in the disease.

AT fibroblast strains are also more sensitive than normal
strains to inactivation by densely ionizing radiation (14 MeV neutrons)[15,16]; however, the increase in sensitivity is only ∿1.5 times.
This difference for the two types of radiation is still compatible
with the presence of a DNA repair defect in AT cells, since a considerable fraction of the damage inflicted in DNA by neutrons (unlike
X- or γ-rays) goes unrepaired in normal cells, and thus a DNA repair
malfunction might be expected to be less critical for recovery from
densely compared to sparsely ionizing radiation. Likewise, anomalous
DNA repair can readily explain the deficiency exhibited by AT fibroblasts in recovering from "potentially lethal damage" induced by
X-rays. That is, holding cultures in a confluent (cell division-
arrested) state for a short period after X-ray exposure enhances
colony survival to a lesser extent in AT than in normal strains[24].

AT strains display much variation in their colony-forming
ability after treatment with near UV (313 nm) light or such radiomimetic chemicals as ethylnitrosourea, an alkylating agent noted for
its ability to induce tumors in the central nervous system (CNS) of
rodents[25]. These survival data for ionizing radiation and related
DNA-damaging agents are explicable in terms of a DNA repair abnormality in AT cells. On the one hand, radiation is known to produce

Table 1.   Sensitivity of AT fibroblast strains to cell killing by various DNA-damaging agents[a]

DNA-damaging agent	AT2BE[b]	AT3BI	AT4BI	AT5BI	Others[c]	References
Physical						
X- or γ-rays (Air,N₂)	+++	+++	+++	+++	+++ (16)	10-15
Neutrons (Air, N₂)	++	++	++	++		15, 16
254 nm (UV)	−	−	−	−	− (3)	17,19
313 nm (UV)	++	−	++	−		15,17,18
Chemical[d]						
MMS	+	−	+	++	−/++ (6)	15,17,20
MNNG	++	++	−	++	+++ (3)	8,19
ENU	+++	++	++			8
MMC					−/++ (4)	15,17,20
4NQO	++	−	+			8
BM		++	+++	++	+++ (1)	21,22
AD				++	(1)	20
N-hydroxy-AAF					− (?)	17

[a] −, normal (0.8 ≤ DRF ≤ 1.2); +, slightly sensitive (1.2 < DRF ≤ 1.5); ++, moderately sensitive (1.5 < DRF ≤ 2.0); +++, markedly sensitive (2.0 < DRF < 4.0) where DRF (dose reduction factor at 10% survival) = D_{10} (normal strain)/D_{10} (AT strain). Note: No entry, not tested.

[b] Also denoted as CRL 1343.

[c] Number in brackets, number of other strains tested.

[d] MMS, methylmethanesulfonate; MNNG, N-methyl-N′-nitro-N-nitrosoguanidine; ENU, ethyl-nitrosourea; MMC, mitomycin C; 4NQO, 4-nitroquinoline 1-oxide; BM, bleomycin; AD, actinomycin D; N-hydroxy-AAF, N-hydroxy-acetylaminofluorene.

a host of chemically diverse radioproducts in cellular DNA[3]; therefore a normal radioresponse by a given strain presumably demands
that a number of different repair pathways all be fully functional,
and, by the same token, a malfunction in any one pathway might be
expected to result in increased radiation cytotoxicity, as consistently found for AT strains. On the other hand, each of the
chemicals (and 313 nm light) produces a relatively narrow and
characteristic spectrum of reaction products in DNA[1,3]; hence if different AT strains are deficient in enzymes involved in different
repair pathways--a prognosis for which there is biochemical evidence
(see below)--it follows that such mutant strains should display, as
observed, different degrees of sensitivity to a given radiomimetic
chemical.

In short, the combined fibroblast survival and lymphocyte viability studies characterize AT as an ionizing radiation analogue of
XP, the inherited disorder linking sunlight-induced skin cancer with
UV hypersensitivity due to abnormal repair of DNA photoproducts (see
accompanying article by Friedberg). That is, AT and XP strains are
as a rule each hypersensitive to a specific and mutually exclusive
class of DNA-damaging agents. Near UV light and 4-nitroquinoline
1-oxide (4NQO) have both far UV- and radiomimetic properties[8,16], and
it should therefore come as no surprise that both XP and (some) AT
strains show elevated sensitivity to each agent. The elevation in
near UV sensitivity provides a rational explanation for the tendency
of some AT patients to develop telangiectasis over sunlight-exposed
regions of the skin.

The use of viruses as probes of cellular function has complicated
our insight into the phenotypic anomalies in AT fibroblasts. In
contradistinction to the cell survival and viability results,
limited host-cell reactivation studies show that AT cells possess
a normal capacity to promote the reproduction of X-ray- or MNNG-
inactivated adenovirus but are deficient in the reactivation of far
UV-damaged virus, as judged by plaque-forming ability and viral
(V antigen) protein synthesis[26]. An explanation for this paradox
may be that the agents produce different lesion spectra in viral
and cellular DNA, arising from differing complexities in the structural organization of the two DNAs.

An intriguing trait of AT fibroblasts is their apparent hypomutability by ionizing radiation[27]. The lack of correlation between
radiation-induced cell lethality and mutability in AT may result
from the mutant cells harboring a defect in an error-prone repair
pathway. This hypothesis fits with the proposal that the putative
repair process mediating recovery from potentially lethal damage
induced by radiation is relatively error-prone and consequently
mutagenic[28]; as stated earlier, the process is abnormal in AT cells
[24], thereby presumably enhancing cell death but decreasing mutation

induction[28]. Alternatively, the radiogenic genetic alterations
(e.g. deletions) which lead to mutations in normal cells may tend
to be lethal lesions in AT cells[27]. Be that as it may, it is noteworthy that the apparent hypomutability of AT cells by X-rays lends
no support to the somatic mutation theory of cancer.

Cytogenetic Properties. Although the modal karyotype is normal
in cultured AT cells, the chromosomes in both lymphocytes and fibroblasts tend to undergo breakage and rearrangement spontaneously.
This trait has led German to classify AT as one of the chromosome
breakage syndromes, the others being Fanconi anemia (FA), BS, and
XP[29]. Many types of gross chromosome aberrations (e.g. breaks, gaps,
fragments, dicentrics) occur excessively in a random fashion over
the entire genome[30-32]. The frequency of these chromosome alteration
varies greatly not only among but also within the same patients (as
a function of donor age and cell type), possibly reflecting the complex interplay of such factors as differences in the immune status
of the donors and fluctuations in their levels of exposure to
undetermined clastogenic agents[7,8,30].

A striking nonrandom cytogenetic aberration distinctive to AT
is the specific structural rearrangement in which band 14q12 is the
preferred exchange site leading to the transfer of the distal portion
of chromosome 14 to the terminal region of either the second 14 or,
on occasion, to 6, 7, or X[30,33]. Translocation clones displaying
decreased chromosome fragility may increase due to some proliferative
advantage to become the major type of peripheral blood lymphocyte[30].
The presence of these pseudodiploid clones may herald the development
of cancer; in one AT patient with chronic lymphocytic leukemia, the
neoplastic lymphocytes appeared to descend directly from a premalig-
nant clone marked by a 14q translocation[33].

AT lymphocytes also consistently exhibit unusual chromosome
behaviour after X-ray treatment[32]. Irradiation in G_0/early G_1 induces, in the following mitosis, numerous types of chromatid-type
aberrations (gaps, breaks, and notably triradials), a class of
damage rarely observed in irradiated normal cells. The same AT
lymphocytes also contain excess levels of chromosome-type aberrations after G_0/early G_1 exposure and chromatid-type aberrations
after G_2 exposure. These two sets of data suggest that both single-
and double-strand openings remain in the DNA of cultured AT lymphocytes for protracted times and act as foci for the observed exchange
events. The results further imply that AT cells might be defective
in carrying out at least two repair processes, one acting on one-
strand radioproducts and a second operating on two-strand radioproducts. The possibility that these putative strand openings arise
during faulty excision repair of one-strand lesions (e.g. base defects) and two-strand lesions (e.g. interstrand crosslinks), rather
than from inept rejoining of directly induced single- and double-
strand breaks, cannot be resolved by cytogenetic analysis.

The frequency of sister chromatid exchanges, both spontaneous[34] and induced by exposure to X-rays and three radiomimetic chemicals[35], is normal in cultured AT cells. The rate of cell transformation by simian virus 40 is also normal[36]. Thus the molecular defect giving rise to enhanced radiosensitivity and chromosome instability affects neither the mechanism mediating the exchange of sister chromatids nor the one promoting the insertion of viral DNA into the host cell genome.

DNA Repair Properties. As noted earlier, ionizing radiation induces a vast array of lesions in cellular DNA, including single- and double-strand breaks having chemically heterogeneous termini, numerous structural modifications in the four heterocyclic bases and the sugar moieties, and DNA-protein crosslinks[3]. Sites containing base, sugar or crosslink damage are thought to be restored to a normal configuration by an excision-repair process involving: (i) an incision near the modified residue; (ii) a second single-strand nick on the other side of the lesion, leading to the excision of the damaged segment; (iii) filling in of the resulting gap (repair synthesis); and (iv) ligation of preexisting and newly inserted material[3,5,6]. The processes mediating the rejoining of single- and double-strand breaks are poorly understood but presumably require in most cases some enzymatic surgery and subsequent repair synthesis to clean the "frayed" strand termini[5].

The DNA repair characteristics of AT fibroblast strains after insult by radiation or other DNA-damaging agents are summarized in Table 2. The 13 strains examined can be divided into two broad groups, denoted as exr⁻ and exr⁺, on the basis of their capacity, relative to that of normal strains, to execute in response to hypoxic γ-irradiation two conventional indices of DNA repair synthesis-- namely, DNA repair replication and unscheduled DNA synthesis (UDS). The diminished level of γ-ray-induced repair synthesis observed in the exr⁻ strains does not stem from a defect in strand-rejoining, because all AT strains restitute both single-strand (including alkali-labile lesions) and double-strand breaks with normal kinetics, as judged by several independent criteria[10,15,37,39,42-44]. (AT strains also display a normal ability to rejoin the mixture of clean breaks and alkali-labile defects produced in DNA by BM[21].) Instead the reduction in repair synthesis can be ascribed to a defective capacity to remove alkali-stable radioproducts which are detected as sites in DNA sensitive to the strand-incising activity of lesion-recognizing enzymes (endonucleases in concert with DNA glycosylases) present in crude protein extracts from *Micrococcus luteus*[37,39]. (See, however, conflicting results of van der Schans and coworkers elsewhere in this volume.) The site removal data imply that certain AT strains lack a fully functional enzyme (endonuclease or DNA glycosylase) or cofactor involved in the initial incision reaction in an excision-repair process (hence the exr⁻ denotation). The radioproduct(s) whose removal is presumed to be defective in exr⁻ AT strains

Table 2.   DNA repair properties of AT fibroblast strains

DNA-damaging agent	Repair hallmark	Strain[a]		References
		exr⁻	exr⁺	
Hypoxic γ-rays	DNA repair replication	D[b]	P[b]	37,38
	Unscheduled DNA synthesis	D	P	39
	M. luteus endonuclease-sensitive site removal	D	P	37,39
	Single-strand break rejoining:			
	Velocity sedimentation	P	P	15,37,39
Oxic γ-rays	Thymine glycol removal:			
	In cells	P	P	40
	By cell extracts	P	P	41
	Single-strand break rejoining:			
	Velocity sedimentation	P	P	10,15
	Alkaline elution	P	P	42
	Endonuclease S_1	P	P	43
	Double-strand break rejoining:			
	Velocity sedimentation	P	P	44
BM	Single- and double-strand break rejoining:			
	Velocity sedimentation	P	P	21
4NQO	Alkali-labile lesion removal	D,P	D	15
MNNG	DNA repair replication	D	P	19
MMS	DNA repair replication	P	P	19
Far UV light	UV endonuclease-sensitive site removal	P	P	37,45
	DNA repair replication	P	P	19,37
	Unscheduled DNA synthesis	P	P	15,45
	Postreplication repair	P	P	46
N-acetoxy AAF	DNA adduct removal	P		47
	DNA repair replication	P	P	45

[a] Exr⁻, deficient, and exr⁺, proficient in DNA repair replication induced by hypoxic γ-rays. Exr⁻ strains: AT1BE (CRL 1312), AT2BE, AT3BI, AT81CTO, AT82CTO, AT97CTO, AT194CTO[12,38]. Exr⁺ strains: AT3BE (CRL 1347), AT4BI, AT5BI, AT7BI[38].

[b] D, deficient; P, proficient; no entry, not tested.

has not as yet been identified; the whole spectrum of base, sugar, and crosslink lesions are all candidates with the exception of thymine glycols. AT strains, both exr⁻ and exr⁺, apparently repair this numerically important class of modified bases normally[40,41].

Laboratory corroboration for the clinical signs of genetic heterogeneity in the disorder derives in part from cell fusion analyses which have allocated three exr⁻ AT strains to two complementation groups: AT1BE (CRL 1312) and AT3BI to Group A and AT2BE (CRL 1343) to Group B[38]. Additional evidence that AT3BI and AT2BE strains carry different mutated loci comes from recent experiments on the repair of an "ionizing-radiation-like" reaction product formed in cellular DNA after 4NQO treatment[15]. DNA sites containing the reaction product (a guanyl adduct) are selectively rendered labile at alkaline pH, and hence their presence can be followed indirectly as single-strand breaks, using the standard alkaline sucrose-gradient method. Preliminary results from this experimental approach reveal that, unlike the normal removal capacity of AT3BI cells, AT2BE cells are markedly defective in their ability to remove the guanyl adduct. The rate of removal of this adduct is as protracted in AT4BI as in AT2BE fibroblasts; thus exr⁺ would appear to be a misnomer for this strain. Conceivably, all AT strains are exr⁻ but the deficiency simply went undetected in the earlier γ-ray experiments because of technical shortcomings[8]. The observation that cell extracts of both exr⁻ and exr⁺ strains have a reduced capacity to enhance the priming activity of irradiated DNA for purified DNA polymerase[49] is consistent with the notion that all AT strains are defective in incising radiation-damaged DNA.

The pattern of DNA repair replication for six AT strains when challenged with MNNG mimics exactly that observed after exposure to hypoxic γ-irradiation; four (exr⁻) exhibit diminished levels and two (exr⁺) exhibit normal levels[19]. Thus human cells possess at least one common pathway for the repair of γ-ray and MNNG damage in DNA.

Despite the enhanced sensitivity of some AT strains to the lethal effects of methylmethanesulfonate, all strains have an apparently normal capacity to repair damage caused by this alkylating agent, at least as reflected by DNA repair replication (Table 2). In agreement with cell survival results, all DNA repair hallmarks are normal in AT strains after exposure to either far UV light or N-acetoxy-AAF (Table 2).

Despite considerable success in elucidating the DNA repair deficiencies in AT strains, the precise enzymological defect responsible for the enhanced radiosensitivity observed at the cellular and cytogenetic levels remains unknown. Enzymological studies have provided little insight into the nature of the underlying defect;

cell extracts of both exr$^-$ and exr$^+$ strains contain normal levels of activity of the following putative repair enzymes: uracil-DNA glyco sylase[50], AP endonuclease[49,51,52] and a correxonuclease (J. Doniger, personal communication). An answer as to whether the supposed stra openings giving rise to the elevated incidence of chromosome aber- rations in AT cells result from faulty excision-repair events during postirradiation incubation[8] or failure to rejoin a biologically critical, albeit numerically minute [and hence biochemically unde- tectable (Table 2)], subclass of strand breaks[53,54] must clearly await further experimentation.

AT HETEROZYGOTES: A CANCER-PRONE AND RADIOSENSITIVE SUBGROUP?

Family medical records of AT patients reveal a recurrent his- tory of cancer, including leukemias, lymphomas, and ovarian, gastric biliary system, and pancreatic carcinomas[55,56]. In particular, a detailed medical follow-up study involving ∿600 blood relatives and ∿300 spouse controls from 27 AT kinships has permitted Swift and colleagues to estimate that heterozygous carriers of an AT gene run five times the normal risk of dying from malignancy before age 45[56]. Assuming that the Hardy-Weinberg equilibrium principle operates, AT gene carriers constitute approximately 1% of the entire population, an estimate that may be conservative in view of the likelihood of mo than one gene coding for the AT phenotype. These two estimates lead to the dramatic prediction that at least 5% of all cancer fatalities in early adulthood occur in AT gene carriers.

This prediction cannot be tested directly, because there is at present no physical sign or laboratory marker for identifying AT heterozygotous carriers. We were thus led to assay fibroblast strains from presumptive obligatory heterozygotes (i.e. parents of AT patients) for postirradiation colony-forming ability. Fig. 1 displays the hypoxic γ-ray sensitivity of 19 strains derived from nine afflicted children, nine parents and one unaffected sibling belonging to seven AT families. In five families, the survival capacity of the presumptive heterozygous strains is intermediate between that of the normal control and the AT homozygous strains, and this correlates in the three heterozygotes analyzed further with a moderately reduced ability to execute hypoxic γ-ray-induced DNA repair replication[12], implicating defective DNA repair as a causal factor. The intermediate radiosensitivity of strain ATH2NY suggests that its donor, an unaffected child in Family 7, is a gene carrier for AT. In the remaining two families, the radioresponse of the presumptive heterozygotes is not significantly different from normal. These and other data[13,23] raise the possibility that a meaningful portion of the human population may be predisposed to neoplasia due to, at least in part, an increased susceptibility to radiation and extrinsic radiomimetic chemicals. The verification of this prospect and its actual contribution to the total cancer

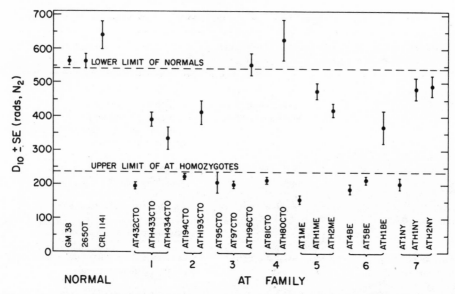

Fig. 1. Hypoxic γ-ray sensitivity of normal, AT homozygous and AT
heterozygous strains, expressed as D_{10} with associated SE
(vertical bars). Nomenclature for AT homozygotes, AT+number
+city of biopsy (e.g. AT432CTO≡AT432+Canada+TOronto); AT heterozygotes, ATH+number+city of biopsy (e.g. ATH433CTO).

load would seem to be a priority item in environmental carcinogenesis. The development of a simple and sensitive screening test for
detection of AT heterozygotes, possibly exploiting their apparent
moderate radiosensitivity, would go some distance toward the realization of this goal.

SEARCH FOR OTHER RADIOSENSITIVE DISORDERS

     The emergence of reduced radiation tolerance as an invariable
laboratory marker of AT has encouraged a hunt for additional radiosensitive disorders among the repertory of Mendelian inherited syndromes. In our selection of genetic traits for assay, special attention has been given to the following characteristics, all featured
in AT: (i) cancer proneness; (ii) impaired immunity; (iii) spontaneous chromosomal instability; (iv) oculodermal changes; (v)
neurological complications; and (vi) severe radiation response. Our
laboratory has surveyed 12 syndromes to date; the clinical description of each is summarized in Table 3, and the cellular radiosensitivity of fibroblast strains from patients afflicted with each disorder is given in Table 4.

Table 3. Heritable disorders sharing hallmarks with AT[a]

Disorder	Inheritance[b]	Associated neoplasms	Other relevant hallmarks
Fanconi anemia (FA)	AR	acute myelogenous leukemia; mucosal and hepatic tumors	chr. instab.[c]; pancytopenia; recurrent infection; dermal pigmentary changes; dwarfism; skeletal, cardiac and renal malformations
Bloom syndrome (BS)	AR	leukemias; intestinal tumors	chr. instab.; impaired immunity; facial telangiectasis
Chediak-Higashi syndrome (CHS)	AR	lymphomas	impaired immunity; recurrent infection; sun sensitivity; oculodermal pigmentary changes
Wiskott-Aldrich syndrome (WAS)	XR	lymphatic tissue tumors	pancytopenia; recurrent infection; bloody diarrhea; eczema
agammaglobulinemia (Bruton type) (AG)	XR	leukemias; lymphatic tissue tumors	chr. instab.; immature bone marrow; recurrent infection
dyskeratosis congenita (DC)	XR	epithelial tumors	pancytopenia; disturbed maturation of epithelium; dermal telangiectasis; mental retardation
incontinentia pigmenti (IP)	XD	acute myelogenous leukemia	chr. instab.; dermal pigmentary changes; skeletal, cardiac, ocular and dental malformations
Rothmund-Thomson syndrome (RTS)	AR	dermal tumors	sun sensitivity; dermal pigmentary changes; cataract formation; dwarfism; immature gonads
tuberous sclerosis (TS)	AD	brain, renal and cardiac tumors; facial and subungual fibromas	epilepsy; mental retardation; dermal pigmentary changes; skeletal malformations
neurofibromatosis (NF)	AD	dermal and CNS tumors	café-au-lait spots; mental retardation; skeletal and cardiac malformations; gastrointestinal bleeding
Huntington chorea (HC)	AD		motor(e.g. chorea, dysarthria) and mental (e.g. dementia) changes
Retinoblastoma, bilateral (RB)	AD	retinal tumors, osteosarcomas	

[a] Compiled from refs.16, 57, and 58.

[b] AR, autosomal recessive; AD, autosomal dominant; XR, X-linked recessive; XD, X-linked dominant.

[c] Chromosomal instability, spontaneous.

Table 4. γ-Ray survival of cell lines from patients with disorders sharing hallmarks with AT[a]

Strain	Donor			$D_{10} \pm SE$[b]	
	Disorder	Age (yr)	Sex[c]	Oxia	Hypoxia
"NORMAL"[d]		3-51	F(5),M(5)	385 ± 17	582 ± 29
CRL 1196	FA	5	F	364 ± 12	
GM 368	FA	8	M	389 ± 13	540 ± 51
GM 391	FA	6	F	360 ± 24	
GM 1309	FA	12	M	370 ± 18	
GM 2053	FA	12	F	329 ± 19*	368 ± 24*
GM 2361	FA	14	M	371 ± 9	575 ± 25
WG 71	FA	2	F	457 ± 17	
WG 331	FA	5	F	458 ± 13	
BL₁	BS		F	366 ± 23	
GM 2075	CHS	1	F	348 ± 26	
GM 1598	WAS	4	M		547 ± 40
GM 362	AG	21	M	400 ± 17	
GM 1774	DC	6	M	362 ± 20	
GM 492	IP	24	F	409 ± 20	
73-63	RTS	10	M	379 ± 19	482 ± 13*
78-34	RTS	11	F	406 ± 18	530 ± 18
WG 127	RTS	13	M		528 ± 33
WG 480	RTS	19	F	248 ± 36*	422 ± 17*
GM 1635	TS	17	M	276 ± 24*	382 ± 28*
GM 1643	(second biopsy)			311 ± 32*	345 ± 16*
GM 1644	TS	20	F	395 ± 15	534 ± 32
GM 2333	TS	23	M	308 ± 12*	303 ± 45*
GM 622	NF	8	M		609 ± 9
GM 1633	NF	61	M		566 ± 36
GM 1640	NF	19	F		478 ± 10*
GM 1136	HC	45	M	279 ± 24*	
GM 2079	HC	48	F	281 ± 22*	
GM 2147	HC	55	M	320 ± 21*	
GM 2173	HC	52	F	383 ± 31	
GM HCa	'at risk'[e]			317 ± 19*	
GM HCb	'at risk'[e]			254 ± 17*	
AG 1879	RB	11	F		453 ± 43*
AG 1880	RB	37	F		485 ± 17*
AG 1142	RB	2	F	338 ± 14*	
AG 3326	RB	30	F	363 ± 10	

[a] From refs. 15 and 16.

[b] Dose (rads) giving 10% survival ± standard error of mean, using linear regression analysis[12].

[c] F, female; M, male.

[d] Pooled data from 93 independent survival experiments (48, oxia; 45, hypoxia), using ten strains from unrelated normal donors whose ages span the first six decades.

[e] Offspring of HC patient and thus at risk for developing the disease.

* $P < 0.05$, comparing $D_{10} \pm SE$ of indicated strain to that of 'NORMAL', using standard error of the mean in the statistical test.

Cancer proneness *per se* does not appear to be a particularly useful indicator of inherent susceptibility to the cytotoxic action of γ-rays because fibroblasts from victims with any one of the first seven cancer-associated syndromes generally exhibit normal colony-forming ability after X-ray exposure. The first six disorders also share immune deficits (stemming from hemopoietic disturbances) as a common denominator, and four of the six have spontaneous chromosome instability in common; hence neither does immunodeficiency nor impaired chromosome stability correlate with elevated radiosensitivity *in vitro*.

One of the eight FA strains assayed may owe its moderate radiosensitive response to a decreased capacity to excise thymine glycols, as nuclear preparations of some other FA strains are faulty in the removal of these radioproducts[59]. The survey of Arlett and Harcourt[13] also suggests that radiosensitivity is only on occasion linked with FA. Table 4 provides evidence for radiation hypersensitivity in the last five disorders in Table 3. The biochemical basis for the increased sensitivity in two of the Rothmund-Thomson syndrome (RTS) strains, 73-63 and WG 480, may lie in the fact that both are deficient in the removal of *M. luteus* extract-sensitive sites induced by hypoxic γ-irradiation[15]. Some RTS patients are sunlight-sensitive, and the major oculodermal changes seen in the disease may arise from the radiomimetic action of the UV component of solar rays.

Tuberous sclerosis, neurofibromatosis, and Huntington chorea (HC) are all primarily neurological disorders. In each, the cardinal complication is progressive neurodegeneration, and there is a growing consensus that it is this clinical trait with which radiation hypersensitivity correlates best. Our HC cell survival data confirm earlier reports[60,61]; enhanced radiosensitivity has also been demonstrated in HC lymphocyte cell lines (J. Robbins, personal communication). Thus TS, NF, and HC join an expanding list of heritable disorders linking progressive neurodegeneration *in vivo* with radiation hypersensitivity *in vitro*; these include AT, Friedreich ataxia[62], olivopontocerebellar atrophy[63] and familial dysautonomia[63]. These findings provide evidence for the postulate of Robbins that heritable neuronal degenerative diseases are caused by premature cell death in the CNS due to defects in DNA repair or possibly other cellular processes, thereby rendering the victims' neurons hypersensitive to DNA-damaging agents[64].

Our survival results for strains from patients with bilateral retinoblastoma (RB) (Table 4) support the more elaborate studies of Weichselbaum and coworkers[14]. The elevated susceptibility of these strains to γ-ray killing is consistent with an untoward radioresponse clinically, manifested by a tendency to develop multiple radiogenic osteosarcomas subsequent to successful radiotherapeutic treatment of the retinal tumors[65]. Radiation hypersensitivity has also been associated with several other genetic disorders, such as progeria[13,14] and partial trisomy of chromosome 13[14], and several "cancer families"[14,16,39].

CONCLUDING REMARKS AND PROGNOSIS

In an increasing number of human heritable disorders, cell culture studies have revealed an association between cellular radiosensitivity *in vitro* and cancer proneness and especially progressive neurodegeneration *in vivo*. However, our understanding of the molecular basis for this relationship is only in its infancy. Substantial biochemical evidence for defects in the enzymatic repair of radiogenic DNA damage has only been obtained so far for AT, and, even here, it seems probable that faulty DNA repair may not be the primary causal factor in the development of many of the clinical features of this multifaceted disorder, including the predisposition to lymphoreticular neoplasia and leukemia. It should be emphasized, however, that, although the primary cause of cancer proneness in AT is probably an immunodeficient state, DNA repair anomalies could play a role. Chromosomal changes, occurring in a background of defective repair, could give rise to genetically altered cells having proliferative advantage and prone to neoplastic transformation, and thereby place increased stress on an already impaired immune apparatus. Disorders such as AT, TS, HC, and RB may serve as useful models for clarifying the role of elevated cellular radiosensitivity and imperfect DNA repair in the pathogenesis of progressive neurodegenerative processes and cancer. Abnormal radiosensitivity holds promise in genetic counselling as a diagnostic laboratory marker for presymptomatic detection of individuals at increased risk for any one of these radiosensitive disorders.

ACKNOWLEDGMENTS

Work conducted in the authors' laboratory was supported in part by US NCI Contract N01-CP-81002 with the Clinical Epidemiology Branch, NCI, Bethesda, Maryland. Drs. N.E. Gentner, D.K. Myers, and I.G. Walker graciously consented to review the manuscript and offered helpful suggestions for its improvement. Special thanks are extended to Colleen Walters for diligent secretarial assistance.

REFERENCES

1. J. J. Roberts, The repair of DNA modified by cytotoxic, mutagenic, and carcinogenic chemicals, Adv. Radiat. Biol. 7:211 (1978).
2. T. Lindahl, DNA repair enzymes acting on spontaneous lesions in DNA, in:"Cellular Senescence and Somatic Cell Genetics: DNA Repair Processes," W. W. Nichols and D. G. Murphy, eds.,

p. 225, Symposia Specialists, Miami (1977).

3. P. C. Hanawalt, E. C. Friedberg, and C. F. Fox, "DNA Repair
    Mechanisms," Academic Press, New York (1978).

4. C. F. Arlett and A. R. Lehmann, Human disorders showing in-
    creased sensitivity to the induction of genetic damage,
    Ann. Rev. Genet. 12:95 (1978).

5. M. C. Paterson, Environmental carcinogenesis and imperfect
    repair of damaged DNA in *Homo sapiens*: Causal relation
    revealed by rare hereditary disorders, in:"Carcinogens:
    Identification and Mechanisms of Action," A. C. Griffin and
    C. R. Shaw, eds., p. 251, Raven Press, New York (1979).

6. E. C. Friedberg, U. K. Ehmann, and J. I. Williams, Human dis-
    eases associated with DNA repair, Adv. Radiat. Biol. 8:85
    (1979).

7. K. H. Kraemer, Progressive degenerative diseases associated
    with defective DNA repair: Xeroderma pigmentosum and ataxia
    telangiectasia, See ref. 2, p. 37.

8. M. C. Paterson and P. J. Smith, Ataxia telangiectasia: An in-
    herited human disorder involving hypersensitivity to ionizing
    radiation and related DNA-damaging chemicals, Ann. Rev.
    Genet. 13:291 (1979).

9. T. A. Waldmann and K.R. McIntire, Serum-alpha-fetoprotein
    levels in patients with ataxia-telangiectasia, Lancet 2:1112
    (1972).

10. A. M. R. Taylor, D. G. Harnden, C. F. Arlett, S. A. Harcourt,
    A. R. Lehman, S. Stevens, and B. A. Bridges, Ataxia telan-
    giectasia: A human mutation with abnormal radiation sensi-
    tivity, Nature 258:427 (1975).

11. R. Cox, G. P. Hosking, and J. Wilson, Ataxia telangiectasia:
    Evaluation of radiosensitivity in cultured skin fibroblasts
    as a diagnostic test, Arch. Dis. Child. 53:386 (1978).

12. M. C. Paterson, A. K. Anderson, B. P. Smith, and P. J. Smith,
    Enhanced radiosensitivity of cultured fibroblasts from
    ataxia telangiectasia heterozygotes manifested by defective
    colony-forming ability and reduced DNA repair replication
    after hypoxic γ-irradiation, Cancer Res. 39:3725 (1979).

13. C. F. Arlett and S. A. Harcourt, Survey of radiosensitivity in
    a variety of human cell strains, Cancer Res. 40:926 (1980).

14. R. R. Weichselbaum, J. Nove, and J. B. Little, X-ray sensitivity
    of fifty-three human diploid fibroblast cell strains from
    patients with characterized genetic disorders, Cancer Res.
    40:920 (1980).

15. M. C. Paterson, P. J. Smith, N. T. Bech-Hansen, B. P. Smith,
    B. M. Sell, and M. V. Middlestadt, unpublished data.

16. M. C. Paterson, Environmental and genetic interactions in
    human cancer, in:"Proceedings of the First International
    Conference on Health Effects of Energy Production,"
    N. E. Gentner and P. Unrau, eds., p. 69, Atomic Energy of
    Canada Limited, Report AECL-6958, Chalk River (1980).

17. C. F. Arlett, Lethal response to DNA damaging agents in a

variety of human fibroblast cell strains, Mutat. Res. 46:106
(1977).

18. P. J. Smith and M. C. Paterson, Sensitivity to near UV light
     in cultured fibroblasts from various human syndromes, Proc.
     Ann. Meet. Am. Assoc. Cancer Res., New Orleans 20:88 (1979).

19. D. A. Scudiero, Decreased DNA repair synthesis and defective
     colony-forming ability of ataxia telangiectasia fibroblast
     cell strains treated with N-methyl-N'-nitro-N-nitrosoguan-
     idine, Cancer Res. 40:984 (1980).

20. D. I. Hoar and P. Sargent, Chemical mutagen hypersensitivity
     in ataxia telangiectasia, Nature 261:590 (1976).

21. A. R. Lehmann and S. Stevens, The response of ataxia telan-
     giectasia cells to bleomycin, Nucleic Acids Res. 6:1953
     (1979).

22. A. M. R. Taylor, C. M. Rosney, and J. B. Campbell, Unusual
     sensitivity of ataxia telangiectasia cells to bleomycin,
     Cancer Res. 39:1046 (1979).

23. P. C. Chen, M. F. Lavin, C. Kidson, and D. Moss, Identification
     of ataxia telangiectasia heterozygotes, a cancer prone po-
     pulation, Nature 274:484 (1978).

24. R. R. Weichselbaum, J. Nove, and J. B. Little, Deficient re-
     covery from potentially lethal radiation damage in ataxia
     telangiectasia and xeroderma pigmentosum, Nature 271:261
     (1978).

25. M. F. Rajewsky, R. Goth, O. D. Laerum, H. Biessmann, and
     D. F. Hulser, Molecular and cellular mechanisms in nervous
     system--specific carcinogenesis by N-ethyl-N-nitrosourea,
     in:"Fundamentals in Cancer Prevention," P. N. Magee,
     S. Takayama, T. Sugimura, and T. Matsushima, eds., p. 313,
     University of Tokyo Press, Tokyo, and University Park Press,
     Baltimore (1976).

26. R. S. Day, III, Viral probes for mammalian cell DNA repair:
     Results and prospects, see ref. 3, p. 531.

27. C. F. Arlett, Survival and mutation in gamma-irradiated human
     cell strains from normal or cancer-prone individuals, in:
     "Radiation Research: Proceedings of the Sixth International
     Congress of Radiation Research," S. Okada, M. Imamura,
     T. Terashima, and H. Yamaguichi, eds., p. 596, Toppan
     Printing Co., Tokyo (1979).

28. R. R. Weichselbaum and J. B. Little, Familial retinoblastoma
     and ataxia telangiectasia: Human models for the study of
     DNA damage and repair, Cancer 45:775 (1980).

29. J. German, Genes which increase chromosomal instability in
     somatic cells and predispose to cancer, Prog. Med. Genet.
     8:61 (1972).

30. F. Hecht and B. K. McCaw, Chromosome instability syndromes,
     in:"Genetics of Human Cancer," J. J. Mulvihill, R. W. Miller,
     and J. F. Fraumeni, Jr., eds., p. 105, Raven Press, New York
     (1977).

31. D. G. Harnden, Ataxia telangiectasia syndrome: Cytogenetic and cancer aspects, in:"Chromosomes and Cancer," J. German, ed., p. 619, Wiley Press, New York (1974).

32. A. M. R. Taylor, J. A. Metcalfe, J. M. Oxford, and D. G. Harnden, Is chromatid-type damage in ataxia telangiectasia after irradiation at G_0 a consequence of defective DNA repair? Nature 260:441 (1976).

33. B. K. McCaw, F. Hecht, D. G. Harnden, and R. L. Teplitz, Somatic rearrangement of chromosome 14 in human lymphocytes, Proc. Natl. Acad. Sci. USA 72:2071 (1975).

34. N. H. Hatcher, P. S. Brinson, and E. B. Hook, Sister chromatid exchanges in ataxia telangiectasia, Mutat. Res. 35:333 (1976).

35. S. M. Galloway, Ataxia telangiectasia: The effects of chemical mutagens and X-rays on sister chromatid exchanges in blood lymphocytes, Mutat. Res. 45:343 (1977).

36. T. Webb, D. G. Harnden, and M. Harding, The chromosome analysis and susceptibility to transformation by Simian virus 40 of fibroblasts from ataxia-telangiectasia, Cancer Res. 37:997 (1977).

37. M. C. Paterson, B. P. Smith, P. H. M. Lohman, A. K. Anderson, and L. Fishman, Defective excision repair of γ-ray-damaged DNA in human (ataxia telangiectasia) fibroblasts, Nature 260:444 (1976).

38. M. C. Paterson, B. P. Smith, P. A. Knight, and A. K. Anderson, Ataxia telangiectasia: An inherited human disease involving radiosensitivity, malignancy and defective DNA repair, in: "Research in Photobiology," A. Castellani, ed., p. 207, Plenum Press, New York (1977).

39. M. C. Paterson, P. J. Smith, N. T. Bech-Hansen, B. P. Smith, and B. M. Sell, γ-ray hypersensitivity and faulty DNA repair in cultured cells from humans exhibiting familial cancer proneness, see ref. 27, p. 484.

40. P. Cerutti, K. Shinohara, and J. Remsen, Repair of DNA damage induced by ionizing radiation and benzo(a)pyrene in mammalian cells, J. Toxicol. Envirn. Health 2:1375 (1977).

41. J. F. Remsen and P. A. Cerutti, Excision of gamma-ray induced thymine lesions by preparations from ataxia telangiectasia fibroblasts, Mutat. Res. 43:139 (1977).

42. P. V. Hariharan, S. Eleczko, B. P. Smith, and M. C. Paterson, unpublished data.

43. R. B. Sheridan, III, and P. C. Huang, Ataxia telangiectasia: Further considerations of the evidence for single strand break repair, Mutat. Res. 61:415 (1979).

44. A. R. Lehmann and S. Stevens, The production and repair of double strand breaks in cells from normal humans and from patients with ataxia telangiectasia, Biochim. Biophys. Acta 474:49 (1977).

45. F. E. Ahmed and R. B. Setlow, Excision repair in mammalian cells, see ref. 3, p. 333.

46. A. R. Lehmann, S. Kirk-Bell, C. F. Arlett, S. A. Harcourt,
     E. A. de Weerd-Kastelein, W. Keijzer, and P. Hall-Smith,
     Repair of ultraviolet light damage in a variety of human
     fibroblast cell strains, Cancer Res. 37:904 (1977).
47. D. E. Amacher and M. W. Lieberman, Removal of acetylaminofluorene
     from the DNA of control and repair-deficient human fibro-
     blasts, Biochem. Biophys. Res. Commun. 74:285 (1977).
48. M. Ikenaga, H. Takebe, and Y. Ishii, Excision repair of DNA base
     damage in human cells treated with the chemical carcinogen
     4-nitroquinoline 1-oxide, Mutat. Res. 43:415 (1977).
49. T. Inoue, K. Hirano, A. Yokoiyama, T. Kada, and H. Kato, DNA
     repair enzymes in ataxia telangiectasia and Bloom's syndrome
     fibroblasts, Biochim. Biophys. Acta 479:497 (1977).
50. U. Kuhnlein, B. Lee, and S. Linn, Human uracil DNA N-glycosi-
     dase: Studies in normal and repair defective cultured
     fibroblasts, Nucleic Acids Res. 5:117 (1978).
51. R. B. Sheridan, III, and P. C. Huang, Apurinic and/or apyrimi-
     dinic endonuclease activity in ataxia telangiectasia cell
     extracts, Mutat. Res. 52:129 (1978).
52. R. E. Moses and A. L. Beaudet, Apurinic DNA endonuclease
     activities in repair-deficient human cell lines, Nucleic
     Acids Res. 5:463 (1978).
53. A. R. Lehmann, Ataxia telangiectasia and the lethal lesion
     produced by ionizing radiation, see ref. 2, p. 167.
54. A. M. R. Taylor, Unrepaired DNA strand breaks in irradiated
     ataxia telangiectasia lymphocytes suggested from cytogenetic
     observations, Mutat. Res. 40:407 (1978).
55. W. B. Reed, W. L. Epstein, E. Boder, and R. Sedgwick, Cutaneous
     manifestations of ataxia-telangiectasia, J. Amer. Med. Assoc.
     195:746 (1966).
56. M. Swift, L. Sholman, M. Perry, and C. Chase, Malignant neo-
     plasms in the families of patients with ataxia-telangiectasia,
     Cancer Res. 36:209 (1976).
57. D. Bergsma, "Birth Defects Compendium (Second Ed.)," Alan R.
     Liss (for The National Foundation--March of Dimes), New
     York (1979).
58. V. A. McKusick, "Mendelian Inheritance in Man: Catalogs of
     Autosomal Dominant, Autosomal Recessive, and X-linked
     Phenotypes (Fifth Ed.)," The Johns Hopkins University Press,
     Baltimore (1978).
59. J. F. Remsen and P. A. Cerutti, Deficiency of gamma-ray exci-
     sion repair in skin fibroblasts from patients with Fanconi's

anemia, <u>Proc</u>. <u>Natl</u>. <u>Acad</u>. <u>Sci</u>. USA 73:2419 (1976).

60. C. F. Arlett, Presymptomatic diagnosis of Huntington's disease? <u>Lancet</u> 1:540 (1980).

61. A. N. Moshell, R. E. Tarone, S. F. Barrett, and J. H. Robbins, Radiosensitivity in Huntington's disease: Implications for pathogenesis and presymptomatic diagnosis, <u>Lancet</u> 1:9 (1980).

62. P. D. Lewis, J. B. Corr, C. F. Arlett, and S. A. Harcourt, Increased sensitivity to gamma irradiation of skin fibroblasts in Friedreich's ataxia, <u>Lancet</u> 2:474 (1979).

63. J. H. Robbins, A. N. Moshell, R. G. Scarpinato, and R. E. Tarone, Cells from patients with olivopontocerebellar atrophy and familial dysautonomia are hypersensitive to ionizing radiation, <u>Clin</u>. <u>Res</u>. 28:290A (1980).

64. J. H. Robbins, Workshop summary: Xeroderma pigmentosum, see ref. 3, p. 603.

65. R. H. Sagerman, J. R. Cassady, P. Tretter, and R. M. Ellsworth, Radiation induced neoplasia following external beam therapy for children with retinoblastoma, <u>Am</u>. <u>J</u>. <u>Roentgenol</u>. 105:529 (1969).

STUDIES ON THE REPAIR DEFECTS OF ATAXIA TELANGIECTASIA CELLS

G.P. van der Schans, H.B. Centen and P.H.M. Lohman

Medical Biological Laboratory TNO
Rijswijk,
The Netherlands

ABSTRACT

Patients suffering from ataxia telangiectasia (AT), a rare hereditary disease, show an elevated radiosensitivity extending from the clinical to the cellular level. Ataxia cells have been claimed to be deficient in repair of damage in DNA. Paterson et al.[1] have developed a method to measure the number of γ-ray induced lesions in DNA of mammalian cells, by treatment with damage specific endonucleases (or glycosylases) present in extracts of Micrococcus luteus. With this method we studied the removal of these lesions in normal and AT3BI cells, but no difference in the rate of repair was observed.

We also studied the repair of γ-rays induced single-strand breaks. After moderate to high doses (2-15 krad), a rapid and slow repair could be distinguished, representing two different types of lesions. The induction of the rapidly repaired single-strand breaks could be prevented by the presence of cysteamine during irradiation. Both types of single-strand breaks were also observed in irradiated AT3BI cells; they were repaired at the same rates as in normal cells.

After low irradiation doses ($\leq$ 400 rad) in normal cells more than 50 per cent of the single-strand breaks was repaired within 2 minutes after irradiation, whereas preliminary results with AT3BI cells indicate that here some 3-6 minutes are needed to reach this percentage.

INTRODUCTION

Primary fibroblasts derived from AT-patients are abnormally sensitive to ionizing radiation, which has been tentatively ascribed to a deficiency in repair of damage in DNA[1,2]. Paterson et al.[1] found some AT-strains (ex[-]) partially defective in repair replication after anoxic γ-irradiation, other AT-cells were normal. Their ex[-]-cells were shown to have a reduced capacity with respect to excision of γ-radiation induced lesions, recognized by specific endonucleases (or glycosylases) present in crude extracts of <u>Micrococcus luteus</u>. Taylor et al.[2] reported that AT-cells show an increased level of spontaneously occurring or radiation-induced chromosome- and chromatide aberrations. They suggest that these effects could be due to insufficient repair of single- or double-strand breaks. Further evident differences in radiosensitivity and repair capacity between AT-cells and normal cells have not been reported in literature.

We have attempted to contribute to the elucidation of the problems around the molecular basis for the hypersensitivity of AT-cells. To this end we have compared irradiated AT- and normal human fibroblasts with respect to repair of γ-endonuclease susceptible sites, to repair replication as detected by equilibrium centrifugation in alkaline NaI gradients and to the repair of single-strand breaks, measured by alkaline sucrose gradient centrifugation and by the alkaline elution method.

RESULTS

1.   γ-endotest

Efforts to reproduce the experiments of Paterson and co-workers[1] did not succeed. In our experiments the ex[-]-cell strain AT3BI, which was also used by Paterson et al. appeared to remove γ-endonuclease susceptible sites at the same rate and to the same extent as normal human fibroblasts. Attempting to reproduce the published results we varied the normal cell strain, growth media, dose of γ-rays, degree of anoxia during irradiation and the procedure of DNA extraction. None of the variations led to a difference in the rate of site removal between normal and AT3BI cells. Neither did the use of <u>M. luteus</u> extracts prepared in other ways, nor that of an extract sent to us by Paterson, lead to a distinct difference in removal of γ-endonuclease susceptible sites.

2.   Repair replication

Repair replication was determined by measuring the incorporation of (^{3}H) thymidine in the DNA of cells as the consequence of irradiation. After exposure to 50 krad of ^{60}Co-γ-rays (under N_2) AT3BI showed some 35% less repair synthesis of DNA than normal human fibroblasts (AH and RD). After UV-irradiation (7 J/m^2),

however, no difference was observed. These results are in agreement with those reported by Paterson and coworkers.

3. Repair of single-strand breaks

Besides an enhanced radiosensitivity with respect to cell-
killing, AT-cells show an enhanced induction of chromosome aberrations by X-rays, which may indicate a too slow rate of repair
of single-strand and double-strand breaks or other types of
damage, or that a small amount of damage is not repaired at
all[2,3]. In order to test these hypotheses we studied the induction
and repair of DNA breaks in various ways, including a technique
that allows the detection of very small numbers of breaks (low
radiation doses). The methods applied were sedimentation in high
salt alkaline sucrose gradients after direct lysis of the cells
on top of the gradients, and alkaline elution of single-stranded
DNA from polycarbonate membrane filters, according to the method
of Fornace and Kohn and coworkers[4,5].

3.1 Detection of single-strand breaks in high salt sucrose gradients

With this method a variety of mammalian cells (CHO, AH, RD,
AT3BI), labelled with (^{3}H)-thymidine, was studied for the repair
of single-strand breaks after irradiation with a moderate to high
dose of ^{60}Co-γ-rays, either under N_2 (5 and 15 krad) or in air (2
krad). Disappearance of the breaks, in cells incubated at 37 °C,
showed biphasic kinetics; about 60 per cent was repaired at a
rapid rate, with a half-life of approx. 5 min, the remaining 40
per cent disappeared more slowly, with a $t_{\frac{1}{2}}$ of about 15 min.

Cysteamine appeared to protect predominantly against the
induction of those single-strand breaks that are repaired rapidly.
There was no difference detectable between AT3BI-cells and normal
primary fibroblasts, neither with regard to the ratio between the
amount of the two types of breaks, nor with respect to the rates
of which they were repaired.

3.2 Alkaline elution

This sensitive method is very suitable to determine the
amount of single-strand breaks after irradiation of cells with
low radiation doses. Doses of X-rays of 50 to 400 rad (in air)
were given. Even the lowest dose resulted in a distinct effect on
the elution of the DNA from the membrane filter. After these low
doses the repair of single-strand breaks is remarkably rapid: in
AH, RD and CHO cells 50 per cent are repaired within 2 min of
incubation of the cells in medium at 37 °C (irradiation was at 0
°C), whereas AT3BI-cells needed 3-6 min to repair 50 per cent of
the radiation induced single-strand breaks.

The DNA was eluted from the filter according to a first
order process.

When this sensitive method is used, damages caused by the
labelling of DNA with radioactive precursors might become significant: cells exposed for 3 days to 0.05 µCi (^{3}H)-thymidine/ml
(20,000 dpm ^{3}H was incorporated in the DNA of 10^5 cells),
contained a higher amount of not repaired single-strand breaks
than cells labelled simultaneously with 0.01 µCi (^{14}C)-thymi-
dine/ml; the difference was equivalent to the effect of 25-35 rad
of X-rays (in air).

DISCUSSION

Although we could confirm the observation that AT3BI-cells
are abnormally sensitive to ionizing radiation and that these
cells show a decreased rate of repair-replication, we could not
detect a deficiency in the removal of γ-endonuclease susceptible
sites. So, our results do not support the conclusion of Paterson
et al.[1] that a defect in the removal of these sites in AT3BI-
cells is responsible for the hyperradiosensitivity.

We also tried to find arguments for the hypothesis of Taylor[2]
and Lehman[3] that AT-cells might be abnormal because radiation
damage is repaired too slowly or because some lesions are not
repaired at all. With regard to single-strand breaks our results
give some support to this explanation, as the data obtained with
the alkaline elution method indicate that, after low doses,
repair of these lesions proceeds 1.5-3 times less rapidly in
AT3BI-cells than in normal fibroblasts, however, repair is still
very fast, and it is hard to imagine that this difference in
repair is the cause of the enormous increase in the amounts of
chromosome-aberrations.

The observation we made that repair of single-strand breaks
after exposure to moderate to high doses of γ-rays (2-15 krad)
occurs at two distinct rates, suggests that two types of single-
strand breaks are induced. It is tempting to assign the slow
repair to single-strand breaks in the core-DNA of the nucleosomes
and to assume that the more rapidly repaired single-strand breaks
are those in the linker-DNA. If this explanation is correct, then
the ratio between the two types of breaks indicates that single-
strand breaks are induced more easily in the (relatively short)
pieces of the linker-DNA than in the much larger core-DNA. This
might mean that a large proportion of the damages induced in the
DNA of cells by ionizing radiation results from indirect action.
This would explain why cysteamine protects mainly against the
induction of the rapidly repairable single-strand breaks in the
linker-DNA. The linker-DNA is more accessible for nucleolytic
enzymes, and also - in this concept - for radiation induced
water-radicals and secondary radicals in the cell. It will be of
interest to know whether single-strand breaks induced by bleomycin,
known to produce predominantly damage in the linker-DNA, are
indeed repaired at a high rate.

It is realised, however, that other explanations have not yet been ruled out, such as the possibility that the different rates correspond to different phases in the cell-cyclus during irradiation of the a-synchronous cultures, or that the difference is due to different kinds of single-strand breaks (clean and dirty breaks).

The observation that the elution of the DNA from the filter occurred according to a first order process[6,7] does not agree with current theories about the unwinding of DNA. In these theories it is assumed that the speed of unwinding is inversely proportional to the length of the already unwound DNA pieces. The double-stranded part is thought not to rotate. When it is accepted that the rate at which DNA appears in the filtrate is governed by the unwinding, these assumptions cannot be maintained. So the unwinding seems to be more a kind of "breathing" of the DNA molecule, with starting points possibly in AT-rich parts of the DNA molecule, independent of the presence of single-strand breaks. It is also possible, however, that not the unwinding but the filtration of long single-stranded molecules through the membrane is the rate limiting step.

REFERENCES

1. M.P. Paterson, B.P. Smith, P.H.M. Lohman, A.K. Anderson and L. Fishman, Nature 260: 444 (1976).
2. A.M.R. Taylor, Mutation Res. 50: 407 (1978).
3. A.R. Lehmann and S. Stevens, Nucl. Acids Res. 6: 1953 (1979).
4. A.J. Fornace, K.W. Kohn, and H.E. Kan Jr., Proc. Nat. Acad. Sci. USA 73: 39 (1976).
5. K.W. Kohn, L.C. Erickson, R.A.G. Ewig, and C.A. Friedman, Biochemistry 15: 4629 (1976).
6. B. Rydberg, Radiation Res. 61: 274 (1975).
7. B. Rydberg, Radiation Res. 81: 492 (1980).

ATAXIA-TELANGIECTASIA: STUDIES ON DNA REPAIR SYNTHESIS IN

FIBROBLAST STRAINS

Yosef Shiloh[*], Maimon M. Cohen[§] and Yechiel Becker[+]

Departments of Human Genetics[*] and Molecular Virology[+],
The Hebrew University-Hadassah Medical School, Jerusalem,
Israel, and Department of Pediatrics[§], The Children's
Memorial Hospital, Chicago, Illinois 60614, USA

ABSTRACT

DNA repair synthesis was studied in human skin fibroblast
strains derived from six patients with ataxia telangiectasia (AT)
and four normal donors. The cells were treated with gamma irradiation under aerobic conditions and with the carcinogens methyl
methane sulfonate (MMS), N-methyl-N'nitro-N-nitrosoguanidine (MNNG)
and mitomycin C (MMC). The patterns of dose response and the variations in "DNA repair capacity" were similar in both the AT and normal cell strains, except in one of the normal strains which showed
a significantly reduced dose response towards gamma-irradiation
and MMS.

INTRODUCTION

Ataxia-telangiectasia (AT) is a multisystem autosomal recessive disease characterized mainly by cerebellar degeneration,
immunodeficiencies, cancer proneness, extreme sensitivity to
ionizing radiation and "spontaneous" chromosomal breakage[1]. It has
been suggested that a defect in certain DNA repair mechanisms is
involved in this syndrome, since cell lines derived from AT patients show increased sensitivity towards ionizing radiation and
various chemical carcinogens such as MMS, MNNG, MMC and actinomycin
D[2,3]. After hypoxic gamma irradiation, about half of the AT strains
examined so far showed a reduction in DNA repair synthesis and in
the rate of removal of certain lesions recognized by an endonuclease from M. luteus[3,4,5]. The DNA repair synthesis of these
strains after treatment with MNNG was also reduced, but was normal

361

after treatment with MMS[6]. The response of these and other AT fibroblast strains to UV irradiation was normal[2,3]. Phenotypic, as well
as genetic, heterogeneity seems to be characteristic of AT cells
from different patients with respect to their response to treatment
with either irradiation or carcinogens[2]. This heterogeneity, as
well as the existence of different "complementation groups" among
strains of the same phenotype[3] makes it essential to obtain additional data by examining AT cells from patients belonging to different "gene pools" who are genetically unrelated to the patients
examined so far. The AT patients from whom the skin biopsies were
taken for this study were of Jewish North African and of Middle
Eastern origin, and the data obtained from their cells add information to the existing knowledge on DNA repair mechanisms in skin
fibroblasts from AT patients.

METHODS AND RESULTS

    Fibroblast cell strains were established from skin biopsies
obtained from six AT patients and four normal donors (Table 1),
were maintained in Dulbecco's modified Eagle's medium (Grand Island
Biological Company, N.Y.) and were examined at passage levels 7-12.

Table 1. Fibroblast strains used in this study

	Designation	Age of Donor	Sex	Country of Origin	Chromosome breakage (breaks/cell)
Normal[*]	F-106	25	F	Israel	background level – 0.02-0.03
	F-107	28	M	Israel	
	F-187	24	M	Israel	
	F-196	11	M	Israel	
AT	F-131	9	F	Morocco	0.46
	F-144	9	M	Iran	0.48
	F-169	3	F	Israel (Arab)	0.61
	F-182	14	M	Morocco	0.95
	F-191	10	F	Saudi Arabia	0.13
	F-238[**]	10	F	Turkey	not examined

[*] All normal donors are of Eastern European descent.
[**] The biopsy was a gift of Dr. A.I. Berkel, Institute of Child
Health, Hacettepe University, Ankara, Turkey.

Treatment regimens included irradiation in a ^{60}Co source gamma-
cell 220 (Atomic Energy Commission of Canada, Ltd.) under aerobic
conditions, treatment for 60 minutes with MMS, for 120 minutes with
MNNG, and for 120 minutes with MMC. All treatments were done in the
presence of 2 mM hydroxyurea and 10 µCi/ml of ^{3}H-thymidine (specific
activity 36.5 Ci/mmol, Nuclear Research Center, Negev, Israel).

Chromatography of the cellular DNA on benzoylated, naphtoy-
lated DEAE (BND)-cellulose was used to determine the extent of DNA
repair synthesis. The amount of ^{3}H-thymidine incorporated into
double stranded DNA, which elutes from BND-cellulose columns with
1.0 M NaCl, has been established as a reliable measure for DNA
repair synthesis[7]. The growing point areas which contain single-
stranded regions of DNA adhere to the column and do not elute with
1.0 M NaCl, with the result that the residual background of semi-
conservatively synthesized DNA is removed. Dose response curves obtained by this method have been found to be much the same as those
obtained by the other traditional methods[8]. DNA extracted from the
AT and normal cells was sheared and eluted from BND-cellulose
columns with 1.0 M NaCl. Both the amount of DNA and the radioactivity incorporated into the DNA in the eluate were determined, and
corrections were made for differences in the thymidine pool size
among the different cell strains tested[7]. All AT strains were
examined at least twice at different passage levels and the controls at least four times.

The results are summarized in Figs. 1 and 2. Typical dose
response curves were obtained for each treatment, the highest
response being after gamma irradiation (Fig. 1A) and the lowest
after treatment with MMC (Fig. 1D). Certain variations in "DNA
repair capacity" existed among the different cell strains (Fig. 1)
and between cells at different passage levels of the same strain
(Fig. 2). There was no significant difference between the AT
strains and the normal strains, F-106, F-107 and F-187, after all
four treatments, and the dose response curves of all these strains
fell in the same range. However, the control strain F-196 derived
from a healthy boy aged 11 showed a significantly lower dose
response curve after treatment with gamme irradiation and MMS
(Fig. 1A and B).

DISCUSSION

The reason for choosing the particular DNA damaging agents
used in this study is that most of the AT cell strains examined so
far were found to be hypersensitive to at least one of them[2].
Furthermore, they induce different types of damage repaired by
different pathways[9]. In some of the AT strains examined by other
investigators, reduction in DNA repair synthesis was found after
gamma irradiation and treatment with MNNG, while MMS induced a
normal response in strains hypersensitive to this agent[2,3].

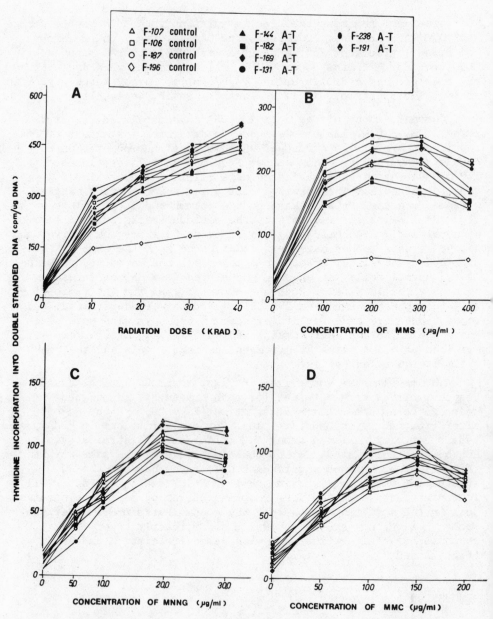

Fig. 1. DNA repair synthesis induced in six AT and four control fibroblast strains by gamma irradiation (A), methyl methane sulfonate (B), N-methyl-N'-nitro-N-nitrosoguanidine (C), and mitomycin C (D). Double stranded DNA was eluted from BND-cellulose columns with 1.0 M NaCl. Both the amount of DNA and the radioactivity incorporated were determined

Our results showed that, apart from strain F-196, derived from a healthy boy, all the AT and non-AT strains showed the same basic dose response in all four types of treatments used. The curve to curve variation obtained with the six AT cell strains tested probably falls in the same range as most cell strains from healthy donors. In this respect, our strains do not differ from some of the other AT strains tested which were reported to show normal DNA repair synthesis after gamma irradiation and treatment with MMS and MNNG[2,5].

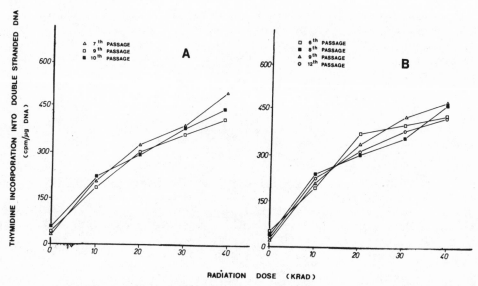

Fig. 2.   DNA repair synthesis as a function of gamma irradiation dose at different passage levels. A: strain F-144 (AT). B: strain F-107 (control).

Sensitivity, as measured by the survival curves of the cells, is another biological parameter for DNA damage and repair, and the sensitivity of our strains to the damage-inducing agents used in this study is at present being investigated.

It should be noted, however, that in many of the AT cell strains described recently, there has not been absolute correlation between hypersensitivity to a DNA damaging agent and a reduced DNA repair synthesis. Only about half of the AT cell strains hypersensitive to hypoxic gamma irradiation had a reduced amount of DNA repair synthesis after this treatment[3], while other strains which were hypersensitive to MMS and MNNG showed normal DNA repair synthesis following treatment with these carcinogens[2,3,6]. All three agents induce a large spectrum of DNA lesions, and Paterson & Smith[3] suggested that AT cell strains with different responses to the same agent were defective in removal of different lesions in their DNA.

It might be possible that different AT strains may indeed have a reduced ability to remove certain DNA lesions produced by alkylation or irradiation. These lesions may have a critical influence on cell survival, but may also constitute only a minor fraction of the large spectrum of the induced lesions. As a result of this, the possible small reduction in DNA repair synthesis disappears within the natural variation among strains and is not manifested.

The exceptionally low response of strain F-196 with regard to DNA repair synthesis after gamma irradiation and MMS is of interest since the donor is apparently a healthy individual. Other parameters of DNA repair in this particular strain are now under investigation.

REFERENCES

1. K.H. Kraemer, in: "DNA Repair Processes", W.W. Nichols, D.G. Murphy, eds., Symposia Specialists Inc., Miami, pp. 37-71 (1977).
2. E.C. Friedberg, U.K. Ehmann and J.I. Williams, Adv. Radiat. Biol. 8:86 (1979)
3. M.C. Paterson, and P.J. Smith, Ann. Rev. Genet. 13:291 (1979)
4. M.C. Paterson, B.P. Smith, P.H. Lohman, A.K. Anderson, and L. Fishman, Nature 260:444 (1976)
5. M.C. Paterson, A.K. Anderson, B.P. Smith, and P.J. Smith, Cancer Research 39:3725 (1979)
6. D.A. Scudiero, in "DNA Repair Mechanisms", P.C. Hanawalt, E.C. Friedberg, and C.F. Fox, eds., Academic Press, New York, pp. 655-658 (1978)
7. D. Scudiero, E. Henderson, A. Norin, and B. Strauss, Mutation Res. 29:473 (1975)
8. D. Scudiero, and B. Strauss, Mutation Res. 35:311 (1976)
9. P.C. Hanawalt, P.K. Cooper, A.K. Ganesan, and C.A. Smith, Ann. Rev. Biochem. 48:783 (1979)

Supported by a grant from the United States - Israel Binition Science Foundation.

THE RESPONSE OF COCKAYNE SYNDROME CELLS

TO UV-IRRADIATION

Alan R. Lehmann and Lynne Mayne

MRC Cell Mutation Unit
University of Sussex
Falmer, BRIGHTON BN1 9QG, England

Cockayne syndrome (CS) is a rare autosomal recessive disorder. The principal clinical features are dwarfism with a small head and senile appearance, skeletal abnormalities, loss of adipose tissue, severe mental retardation, retinal degeneration and severe photosensitivity. The onset occurs after several months of apparently normal development. The retinal, neurological and mental symptoms are progressive, as is the loss of adipose tissue[1]. There are no reports in the literature of elevated frequencies of cancer, as found for example in xeroderma pigmentosum. With one or two exceptions the karyotype is normal.

In cultured cells, a number of reports have demonstrated pronounced sensitivity of fibroblasts to the lethal effects of UV-irradiation[2-5], and also to a number of chemical mutagens[5]. The ability to reactivate UV-irradiated adenovirus-5 was reduced in a number of CS strains studied by Day and Ziolkowski[6] but normal in several strains investigated by Hoar and Davis[7]. UV-induced sister-chromatid exchanges were enhanced in four CS fibroblast strains[4], but not in two lymphoblastoid cell lines[8].

At the molecular level, after UV-irradiation, no defect has been detected in the rate of excision of pyrimidine dimer sites[9,10], in unscheduled DNA synthesis[2,3,5,9] or in repair replication[9]. These results suggest that excision-repair is normal in these cells. The molecular weight changes of newly-synthesized DNA in UV-irradiated Cockayne cells were also similar to those in normal cells, suggesting that there was no pronounced defect in daughter-strand (postreplication) repair[10].

The reduced rate of replicative DNA synthesis (measured by thymidine incorporation) seen immediately after UV-irradiation is similar in normal and CS cells.   Between 3 and 8 hours after exposur of normal cells to 4 J m^{-2} of UV-irradiation, there is a dramatic recovery to normal or greater than normal rates of DNA synthesis. This recovery does not occur in CS cells[10] nor in excision-defective XP cells of complementation group A[10,11].   These results showed that recovery of normal rates of DNA synthesis was probably dependent on excision-repair[11], and in addition on the action of the CS gene product.

In order to ascertain whether the CS gene product was involved solely in the recovery of DNA synthesis, and whether the failure to bring about this recovery was the primary defect in CS cells, the effects of UV irradiation on non-growing cells were analysed.   Human fibroblasts cease dividing when the serum concentration is reduced to 0.5%[12].   After 7 days in serum-deprived medium cells were UV irradiated.   A few days later, when dead cells had detached from the plates, the adhering cells were counted.   In this way the lethal effects of UV could be measured in cells which never subsequently synthesized DNA.   The results shown in Fig. 1 demonstrate that, as with XP cells[12], two CS cell strains were sensitive to UV

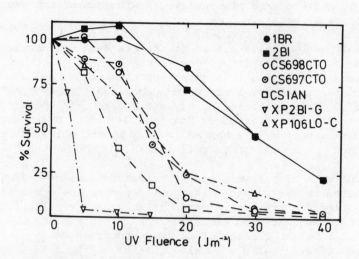

Fig. 1.   Lethal effects of UV on non-growing normal (1BR,2BI) Cockayne (CS698CTO,CS697CTO,CS1AN) and XP (XP2BI - group G,XP106LO - group C) cells.

even when they synthesized no DNA after UV-irradiation.   These
results show unambiguously that the primary defect in CS cells is
not solely connected with recovery of DNA synthesis, but rather that
the latter is a secondary manifestation of the primary defect.

    In order to investigate this further we have begun a detailed
study of RNA synthesis in normal, CS and XP cells after UV-
irradiation, both in growing and in stationary phase cells.   We
were surprised to find that in growing cells RNA synthesis (as
measured by incorporation of ^{3}H uridine) and DNA synthesis were
inhibited by UV to similar extents.   RNA synthesis, however,
recovered to normal rates more quickly than did DNA synthesis.   As
with DNA synthesis, this recovery did not occur either in CS cells
or in an XP cell strain of complementation group G with no detectable
excision-repair (Fig.2).   There was already a significant difference
between normal and defective cells one hour after a UV-irradiation
fluence of 4 J m^{-2}.   Similar results were obtained with cells in
stationary phase, except that RNA synthesis did eventually recover
in the CS cells after 4 J m^{-2}, as might be anticipated from the
survival data shown in Fig. 1.

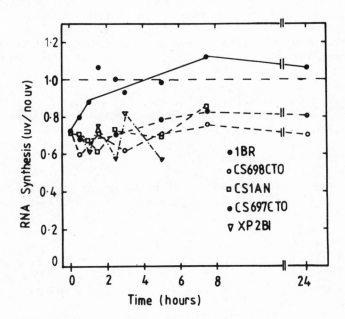

Fig. 2.   RNA synthesis in UV-irradiated growing cells.   Results are
         presented as the ratio of ^{3}H-uridine incorporated into cells
         exposed to 4 J m^{-2} to that in unirradiated cells.   Cells
         pulse labelled for 15, 30 or 60 min at different times after
         irradiation.

DISCUSSION

In earlier work we showed that unlike normal cells, CS cells failed to recover normal rates of DNA synthesis 5 h after UV-irradiation[10]. We have now shown that even earlier, CS cells fail to recover normal rates of RNA synthesis, so that a significant difference between normal and defective cells can be observed within 1 h of irradiation. A similar failure to recover normal rates of RNA synthesis is seen in an excision-defective XP cell strain (complementation group G).

One possible explanation for these observations is that immediately following irradiation, there is a rapid excision of pyrimidine dimers from the DNA of transcribing regions in normal cells. This specific excision brings about the restoration of RNA synthesis and is defective in CS cells. Quantitatively it is a negligible proportion of total excision-repair and its absence is therefore undetected. In XP cells both this specific excision and general excision repair are defective. There is at present no evidence to support this hypothesis, but it may be possible to test it in the future. Whether or not it proves correct we may consider the following scheme as a more general working hypothesis. Damage in DNA directly inhibits transcription into RNA. A restorative process present in normal but lacking in CS (and XP) cells comes into play, permitting the synthesis of new protein(s) which in turn bring about restoration of normal rates of DNA synthesis.

In the absence of these recovery processes, the interaction of the damage with DNA replication leads to cell death after low UV fluences in growing cells. In stationary cells the absence of DNA synthesis allows enough time for RNA synthesis to recover more slowly, so that the defective cells can still tolerate a UV fluence of 4 J m^{-2}. After higher fluences RNA synthesis does not recover even in stationary cells, and the cells die.

REFERENCES

1. F. Guzzetta, in "Handbook of Clinical Neurology", P. J. Vinken and G. W. Bruyn, Vol.13, pp.431-440, Amsterdam:North Holland Publishing Co. (1972).
2. R. D. Schmickel, E. H. Y. Chu, J. E. Trosko and C. C. Chang, Paediatrics 60: 135-139 (1977).
3. A. D. Andrews, S. F. Barrett, F. W. Yoder and J. H. Robbins, J. Invest. Dermatol. 70: 237-239 (1978).
4. R. R. Marshall, C. F. Arlett, S. A. Harcourt and B. C. Broughton, Mutation Res., 69: 107-112 (1980).
5. M. H. Wade and E. H. Y. Chu, Mutation Res. 59: 49-60 (1979).
6. R. S. Day and C. Ziolkowski, in "DNA Repair Mechanisms" P. C. Hanawalt, E. C. Friedberg and C. F. Fox, pp.535-539, New York, Academic Press (1978).

7. D. I. Hoar and F. Davis, Mutation Res., 62: 401-405 (1979).
8. W.-S. Cheng, R. E. Tarone, A. D. Andrews, J. S. Whang-Peng and
     J. H. Robbins, Cancer Res. 38: 1601-1609 (1978).
9. F. E. Ahmed and R. B. Setlow, Biochim. Biophys. Acta 521:
     805-817 (1978).
10. A. R. Lehmann, S. Kirk-Bell and L. Mayne, Cancer Res., 39: 4237-
     4241 (1979).
11. J. M. Rudé and E. C. Friedberg, Mutation Res., 42: 433-442 (1977).
12. G. J. Kantor, C. Warner and D. R. Hull, Photochem. Photobiol.
     25: 483-489 (1977).

CHROMOSOMAL RADIOSENSITIVITY IN FANCONI'S ANAEMIA

G. Duckworth and A.M.R. Taylor

Department of Cancer Studies
University of Birmingham
Birmingham, U.K.

INTRODUCTION

Fanconi's anaemia (FA) is an inherited condition in which congenital abnormalities are associated with bone marrow failure. Along with Ataxia telangiectasia and Bloom's syndrome, Fanconi's anaemia is one of the chromosome breakage syndromes displaying spontaneous chromosome instability. Although these diseases are rare the fact that patients carry an elevated risk of developing cancer makes them important and this cancer risk may be related to defective ability to repair certain kinds of damage to their DNA.

Spontaneous chromosomal abnormalities occurring in the lymphocytes of FA patients were first described by Schroeder et al[1]. These were mainly chromatid-type aberrations consisting of chromatid gaps and breaks and non-homologous interchanges.

In 1973 Sasaki & Tonomura[2] reported chromosomal sensitivity of FA lymphocytes to agents which cause DNA interstrand cross-linking, including mitomycin C, nitrogen mustards and psoralens with UV light. These agents induce high levels of chromosomal aberrations in lymphocytes from FA patients and the colony forming ability of cultured fibroblasts from FA patients is extremely sensitive to mitomycin C. This sensitivity of FA cells to cross-linking agents suggests that they are defective in the repair of interstrand cross-links.

373

There have been a number of reports that cells from FA patients are deficient in the ability to repair various types of damage to their DNA induced by different types of radiation including UV light, X-rays and γ-rays. In 1971 Higurashi & Conen[3] reported that the chromosomes of cultured lymphocytes and fibroblasts from FA patients were significantly more radiosensitive than those from normal controls, after treating cells with γ-rays at the G_0 stage of the cell cycle. In a further publication in 1973[4] they reported that the number of chromosome breaks per cell per rad in cultured lymphocytes from FA patients was greater than that in Ataxia telangiectasia lymphocytes. In contrast Sasaki & Tonomura in 1973[2] reported that the chromosomes of lymphocytes from FA patients responded normally to irradiation in the G_0 and S stages of the cell cycle. Finkelberg et al[5] reported that skin fibroblasts from FA patients did not show an enhanced sensitivity to cell killing by γ-rays, while Huang[6] reported that one out of four FA fibroblast cultures showed significant chromosomal radiosensitivity. A partial deficiency for the excision of γ-ray induced thymine damage from exogenous DNA in cell homogenates of two out of four FA fibroblast strains was reported by Remsen and Cerutti[7]. One of the strains found to have normal excision by Remsen and Cerutti was studied by Rainbow and Howes[8] using a host cell reactivation technique. They reported a deficiency in the repair of DNA damaged by either UV light or γ-rays in this strain. Using the micronucleus technique Heddle et al[9] reported a slight increase in chromosomal breakage in FA cells as compared with normal cells, after X-irradiation. Bigelow et al[10] reported that both the lymphocytes and fibroblasts obtained from six FA patients were twice as sensitive to the induction of chromatid type chromosomal aberrations by X-rays administered in G_2 as cells from normal controls. Recently Arlett and Harcourt[11] and Weichselbaum et al[12] have published reports on the radiosensitivity of a range of human cell strains. Arlett and Harcourt reported that one FA fibroblast strain was sensitive to cell-killing by γ-rays. Weichselbaum et al, however, found normal sensitivity to cell-killing by X-rays in the same strain and a second FA cell strain.

There have been many conflicting reports about the sensitivity of cells from FA patients to ionising radiation and although a large amount of work has been done in this area the question of whether radiosensitivity occurs in these cells is not clear.

We have studied chromosomal X-ray sensitivity in lymphocytes from four FA patients over a range of doses in G_2 and at one dose in G_0. These patients were all clinically diagnosed as having FA and we confirmed this diagnosis by testing the chromosomal sensitivity of the lymphocytes to mitomycin C. A further patient (NW) had no clinical symptoms of FA and was not diagnosed as such. However, lymphocytes from this patient showed high levels of spontaneous chromosome instability and chromosome damage after mitomycin C treatment.

MATERIALS AND METHODS

10 ml of heparinized whole blood was obtained from each of 4 FA patients (FA1BI, FA2BI, FA3BI, FA5BI) and 5 normal controls (C1, C2, C3, C4, C7). Blood was received from patient FA5BI on two separate occasions. Untreated lymphocyte cultures were set up for each blood sample by adding 0.5 ml blood to 4.5 ml Hams F10 medium containing 10% bovine serum, 1% phytohaemagglutinin, penicillin (100 IU ml^{-1}) and streptomycin (100 µg/ml). Untreated lymphocytes were harvested after 48h at 37°C following 1h in colchicine. Mitomycin C treatment was carried out at G$_0$ by adding 0.5 ml blood to 4.5 ml Hams F10 medium containing 10% bovine serum, penicillin and streptomycin and either 0.1 µg/ml or 0.5 µg/ml mitomycin C (Sigma). The cells were incubated in the dark at 37°C for 30 min and then washed twice in Hams F10 medium. They were cultured with phytohaemagglutinin in the usual way and colchicine was added after 47h. The cells were harvested after 48h. For irradiation experiments a Pantak X-ray machine was used (245 kV, 12mA, HVL 1.0 mmCu, 101 rad min^{-1} in water). Whole blood was irradiated with 400 rad at G$_0$ (immediately before being put into culture) and the lymphocytes were harvested at 52h following 1h in colchicine. G$_2$ irradiations were carried out at 50, 100, 150 and 200 rad 4h before harvesting the cultures at 48 h. The cells were treated for 10 min with 0.075M KCl at 37°C and then fixed in 3:1 methanol : glacial acetic acid. Cell suspensions were then dropped onto wet slides and air-dried. After aceto-orcein staining the preparations were scored for aberrations. Whenever possible 50 metaphases were examined for each treatment and each patient.

RESULTS

The figures of spontaneous aberrations occurring in the lymphocyte chromosomes of FA patients and controls illustrate the high spontaneous breakage frequency in FA cells (Table 1). All of the FA patients together with patient NW show higher aberration frequencies than any of the controls. Most of the spontaneous aberrations occurring in the FA cells are chromatid gaps and breaks and these occur at a greatly elevated level compared with controls. The frequency of spontaneous aberrations in the control lymphocytes is perhaps a little low. The frequency of spontaneously occurring chromosomal aberrations varies quite widely between different FA patients and also between samples from the same patient cultured at different times.

DISCUSSION

Data presented here confirm the abnormal chromosomal sensitivity of cells from FA patients to mitomycin C. However, the data also suggests that there is no obvious increased chromosomal

Table 1.   Chromosomal Aberrations in Untreated Lymphocytes
           from Fanconi's Anaemia and Normal Controls

Patient	Exp. No.	No. cells analysed	No. r	No. dic	No. f	No. ctg	No. ctb	No. csg	No. csb	No. tri	No. tr	No. qr
FA1BI		50	0	0	2	4	4	0	1	0	0	0
FA2BI		100	1	1	4	22	2	4	0	1	0	2
FA3BI		100	0	0	3	17	2	4	1	0	1	2
FA5BI	I	50	1	0	0	5	9	3	1	0	2	2
	II	50	0	1	5	18	12	8	0	0	3	2
NW		50	0	0	3	25	5	1	2	0	0	1
C1		50	0	0	0	1	0	1	0	0	0	0
C2		50	0	0	0	1	0	0	0	0	0	0
C3		50	0	0	0	0	0	0	0	0	0	0
C4		50	0	0	0	2	1	0	1	0	0	0

r = rings; dic = dicentrics; f = fragments; ctg = chromatid gaps; ctb = chromatid
breaks; csg = chromosome gaps; csb = chromosome breaks; tri = tricentrics;
tr = triradials; qr = quadriradials

Table 2 shows the frequency of chromosomal aberrations in
lymphocytes from FA patients and controls, after treatment with
mitomycin C at 0.1 and 0.5 µg/ml.   In cells from FA patients the
frequency of most types of chromosomal aberrations was increased
at both doses. This induced damage consisted mainly of chromatid
gaps and breaks. The frequency of triradials and quadriradials
was also elevated. This was also the case in cells from patient
NW. Cells from normal controls showed very little induction of
chromosome aberrations following mitomycin C treatment at either

Table 2.   Chromosomal Aberrations in Lymphocytes from Fanconi's
           Anaemia Patients and Normal Controls after Mitomycin C
           Treatment

Patient	Mitomycin C Dose µg/ml	No. Cells Analysed	No. r	No. dic	No. f	No. ctg	No. ctb	No. csg	No. csb	No. tri	No. tr	No. qr
FA2BI	0.1	50	3	6	8	22	11	7	0	0	4	5
	0.5	12	0	4	12	19	10	5	0	0	15	2
FA3BI	0.1	50	0	2	6	26	21	0	0	0	10	10
	0.5	50	1	3	3	61	89	3	1	0	39	35
FA5BI	0.1	50	0	0	6	16	19	0	0	0	0	1
NW	0.1	100	0	1	2	114	56	8	5	0	3	6
C1	0.1	50	0	0	0	0	0	0	0	0	0	0
	0.5	30	0	0	0	0	0	0	0	0	0	0
C2	0.1	50	0	0	0	0	0	0	0	0	0	0
	0.5	50	0	0	0	4	2	0	0	0	0	0
C3	0.1	50	0	0	0	1	0	0	0	0	0	0
	0.5	50	0	0	0	1	3	0	0	0	0	0
C4	0.1	100	0	0	0	10	0	1	0	0	0	0

r = rings; dic = dicentrics; f = fragments; ctg = chromatid gaps; ctb = chromatid breaks;
csg = chromosome gaps; csb = chromosome breaks; tri = tricentrics; tr = triradials;
qr = quadriradials

dose. Table 3 shows the frequency of chromosomal aberrations in lymphocytes from patients and controls after X-irradiation at various doses. These figures suggest that low levels of aberration were induced in the cells from both FA patients and controls by G_2 irradiation at various doses and that the main types of aberration induced were chromatid gaps and breaks. After accounting for the spontaneous aberration frequency there appears to be no clear difference between the levels of aberrations induced by X-rays in cells from FA patients and in cells from normal controls.

G_0 irradiation induced chromosome-type damage (rings, dicentrics and fragments) in cells from a single FA patient, NW and 2 normal controls. These were not seen to occur spontaneously to any extent in either patient or controls. The X-ray induced levels of aberration were very similar in both groups.

Following either G_0 or G_2 irradiation there was no obvious increased chromosomal radiosensitivity in FA patients under the conditions used.

Table 3. Chromosomal Aberrations in Lymphocytes from Fanconi's Anaemia Patients and Normal Controls After X-irradiation

Patient	X-ray Dose (rad)	No. Cells Analysed	No. r	No. dic	No. f	No. ctg	No. ctb	No. csg	No. csb	No. tri	No. tr	No. qr
G_2 Irradiation												
FA2BI	50	60	0	2	0	22	8	6	0	0	2	2
FA3BI	50	50	0	0	1	9	7	2	1	0	1	3
	150	34	0	0	0	10	3	1	0	0	1	3
FA5BI	100	50	0	3	5	12	8	4	0	0	0	0
	200	35	0	1	6	34	21	5	0	0	0	0
C1	50	27	0	0	0	1	2	0	0	0	0	0
	100	50	0	0	2	2	1	0	0	0	0	0
C2	50	50	0	0	1	1	4	0	0	0	0	0
	100	50	0	0	0	7	8	0	0	0	0	0
C3	50	21	0	0	0	0	0	0	0	0	0	0
G_0 Irradiation												
FA1BI	400	50	7	40	71	8	6	1	4	3	0	2
NW	400	50	20	82	127	18	10	3	5	1	2	0
C4	400	50	11	49	77	2	0	0	0	1	0	0
C7	400	50	12	55	96	5	0	1	0	0	0	0

r = rings; dic = dicentrics; f = fragments; ctg = chromatid gaps; ctb = chromatid breaks; csg = chromosome gaps; csb = chromosome breaks; tri = tricentrics; tr = triradials; q = quadriradials

radiosensitivity in the lymphocytes of FA patients as compared
with the lymphocytes of normal controls. This is true following
irradiation at either the G_0 or the G_2 stage of the cell cycle.
This conclusion is both supported[2] and contradicted[3,4,10] by
previously published work. Approaches to investigating the
increased radiosensitivity of cells from FA patients other than
by cytogenetic methods have not given unequivocal results.[5,6,7,8,9]

It can be concluded that if increased radiosensitivity does
occur in the cells of FA patients it must be fairly marginal and
not of the order that is seen in the cells of patients who are
clearly radiosensitive such as Ataxia telangiectasia patients.
To detect a small increase in radiosensitivity in FA patients
above the level in normal controls, cells from a large number of
patients will have to be studied over a range of radiation doses
and at different stages in the cell cycle. It may be useful to
compare the radiosensitivity of cells from FA patients with those
from say Down's syndrome patients which have been shown to have a
measurable but small increase in radiosensitivity as compared with
normal cells. Such a positive control may help to define the level
of radiosensitivity occurring in the cells of FA patients.

ACKNOWLEDGEMENTS

We thank Dr. Maj Hulten and Dr. F. Hill for their help in
obtaining blood samples, Mr. I. Cross for technical assistance and
the staff of the Birmingham Radiation Centre for the use of their
facilities. This work was supported by the Cancer Research
Campaign.

REFERENCES

1. T. M. Schroeder, F. Anschutz and A. Knopp, Spontane
    Chromosomen Aberrationen bei familiarer Panmyelopathie,
    Humangenetik, 1:194 (1964).
2. M. S. Sasaki and A. Tonomura, A High Susceptibility of
    Fanconi's Anaemia to Chromosome Breakage by DNA Cross-
    linking Agents, Cancer Research, 33:1829 (1973).
3. M. Higurashi and P. E. Conen, In Vitro Chromosomal Radio-
    sensitivity in Fanconi's Anaemia, Blood, 38:336 (1971).
4. M. Higurashi and P. E. Conen, In Vitro Chromosomal Radio-
    sensitivity in "Chromosomal Breakage Syndromes", Cancer,
    32:380 (1973).
5. R. Finkelberg, M. W. Thompson and L. Siminovitch, Survival
    after Treatment with EMS, γ-rays and mitomycin C of Skin
    Fibroblasts From Patients with Fanconi's Anaemia,
    Am. J. Human Genetics, 26:A30 (1975).

6. C. C. Huang, Sensitivity to Radiation and Chemicals of Cell
    Cultures Derived from Patients with Down's Syndrome,
    Fanconi's Anaemia and Normal Persons, Am. Ass. Cancer Res.,
    17:20 (1976).
7. J. Remsen and P. A. Cerutti, Deficiency of Gamma-ray Excision
    Repair in Skin Fibroblasts from Patients with Fanconi's
    Anaemia, Proc. Natl. Acad. Sci. USA, 73 no7:2419 (1976).
8. A. J. Rainbow and M. Howes, Defective Repair of Ultraviolet
    and Gamma-ray damaged DNA in Fanconi's Anaemia, Int J.
    Radiat. Biol, 31 no2:191 (1977).
9. J. A. Heddle, C. B. Lue, E. F. Saunders,and R. D. Benz,
    Sensitivity to Five Mutagens in Fanconi's Anaemia as
    Measured by the Micronucleus Methods, Cancer Research,
    38:2983 (1978).
10. S. B. Bigelow, J. M. Rary and M. A. Bender, G_2 Chromosomal
    Radiosensitivity in Fanconi's Anaemia, Mutation Research,
    63:189 (1979).
11. C. F. Arlett and S. A. Harcourt, Survey of Radiosensitivity in
    a Variety of Human Cell Strains, Cancer Research, 40:926
    (1980).
12. R. R. Weichselbaum, J. Nove and J. B. Little, X-ray Sensitivity
    of Fifty-three Human Diploid Fibroblast Cell Strains from
    Patients with Characterized Genetic Disorders, Cancer
    Research, 40:920 (1980).

DNA REPLICATION IN NORMAL AND DAMAGED CELLS

DNA REPLICATION IN MAMMALIAN CELLS DAMAGED BY MUTAGENS

Alan R. Lehmann

MRC Cell Mutation Unit
University of Sussex
Falmer, Brighton BN1 9QG, England

When growing mammalian cells are damaged by mutagenic treatment DNA replication is disturbed, presumably by the presence of damage in the DNA strands used as templates for replication.   In UV-irradiated cells, which have been investigated in great detail, five different phenomena have been observed:-

(1)   Rapid inhibition of the overall rate of DNA synthesis in each cell, as measured by incorporation of ^{3}H-thymidine.   This occurs to a similar extent in nearly all cell types.

(2)   Subsequent recovery of normal rates of DNA synthesis.   This recovery occurs at very different rates in different cell types[1-3], and is not solely dependent on the capability of the cell type to carry out excision-repair[3].

(3)   Reduction in the size of DNA labelled with a pulse of radio-active thymidine shortly after irradiation.   This is again very dependent on cell type (e.g. refs. 4-5).   It could result from (a) blocking of the progression of the growing forks by damage in the parental DNA, (b) the production of gaps in the daughter strands in the region of the damage (or possibly elsewhere) without the progression of the forks being halted.

(4)   Subsequent increase in size of the initially small DNA labelled soon after irradiation.   This implies that the hypothetical blocks are not permanent or that the gaps are sealed.

(5)   Recovery of the ability to synthesize normal-sized DNA as observed by pulse labelling with ^{3}H-thymidine at later times after irradiation.

THE FATE OF A REPLICATION FORK WHICH ENCOUNTERS UV-INDUCED DAMAGE

The above phenomena are manifestations of a complex interplay of a
variety of different processes, each one presumably under delicate
control.   Despite extensive study there continues to be controversy
about the mechanisms of these processes.   The reductions in rates
of DNA synthesis and in the size of newly-synthesized DNA could
result from (a) inhibition of the initiation of DNA synthesis in
individual replicons or replicon clusters, (b) complete blocks to
the progress of the replication forks, (c) indirect inactivation of
replicating replicons or replicon clusters, (d) delays in fork
progression followed by synthesis past the damage, (e) formation of
gaps in the daughter strands in the region of the damage (or possibly
elsewhere), with or without a delay in the progress of the replicatio
forks, (f) any combination of these.

Cleaver, Park and associates have recently carried out a series
of extensive studies on the detailed kinetics of the changes in rates
of DNA synthesis and of changes in the sizes of newly-synthesized
DNA after UV-irradiation of different human cell types[2,6,7].   Based
on their findings they have put forward a model[6] suggesting that,
on encountering a lesion in the parental strand, the replication
fork had only two alternatives, either it was blocked, which occurred
with a certain probability (P(block)), or it continued past the
damaged site without interruption.   They considered that it was
not necessary to invoke the existence of gaps in daughter strand DNA
to explain their data and that processes termed postreplication or
daughter strand repair may not in fact exist.   They suggested that
results previously interpreted as evidence for gaps in daughter-
strands near damaged sites could be explained as stretches of DNA
between forks blocked by damaged sites in adjacent replicons (see
Fig. 4 in ref. 6).

The model of Park and Cleaver is critically dependent on the
number of pyrimidine dimers per replicon.   In order to assess the
validity of their model it is therefore important to have an accurate
estimate of the sizes of the replicons in the cells studied.   The
value (17 µ) determined for untransformed human fibroblasts by Kapp
and Painter[8] and used by Cleaver, Thomas and Park[7] is unfortunately
some four-fold lower than that determined by fibre autoradiography
-- 60-70 µm[9,10], and the validity of values calculated by Kapp and
Painter has been questioned[8].   Until this discrepancy is resolved,
interpretation of data from UV-irradiated cells becomes extremely
difficult.   It is my contention, however, that the postulate of an
all-or-nothing block to replication is not easily reconciled with
a number of phenomena reported by other workers.   It is rather my
feeling that there are three (or more) possible fates for a replication fork which encounters damage, namely (1) a block, (2) an
interruption of synthesis leaving a gap but allowing fork progression

to continue and (3) synthesis continuing without interruption.
Depending on the cell type and UV fluence used any of these three
might predominate.   Below I list the evidence not readily recon-
cilable with an all-or-nothing block to synthesis:-

(1) Doniger[11], using fibre autoradiography and sucrose gradient
analysis, showed that in V79 hamster cells, after low UV fluences
(<10 J m^{-2}) fork progression (measured autoradiographically) was
hardly delayed and the length of labelled segments hardly affected,
yet the size of newly-synthesized DNA measured on alkaline sucrose
gradients was less than that in unirradiated cells.   This is only
consistent with the daughter strands containing small discon-
tinuities (which would not be detected autoradiographically).
This work also suggested that the inhibition of thymidine incorporation was the result of replicon inactivation.   These results
are to be contrasted with earlier work of Edenberg[12] who showed,
using similar techniques, that relatively high UV fluences blocked
fork progression in HeLa cells.

(2) Meneghini and co-workers[13] demonstrated that in excision-
defective XP cells irradiated with 20 J m^{-2} the reduction in
thymidine incorporation did not result from an accumulation of
growing forks (i.e. from blocks).   In cells density-labelled after
UV-irradiation such that the same amount of label was incorporated
into irradiated and unirradiated cells (i.e. the UV-irradiated cells
were labelled for longer times) the distribution of intermediate
density DNA (DNA in growing points) was similar in irradiated and
unirradiated cells.   Furthermore, under these conditions, sites
susceptible to single-strand nucleases were detected in the recently
replicated parental strands.   These sites were interpreted as being
opposite gaps in the daughter strands.   Earlier work[14,15] had
however shown that these sites were not sensitive to the T4 endo-
nuclease-V, as would be predicted if the gaps were opposite dimers.
This latter observation awaits a satisfactory explanation.   The data
of Meneghini and co-workers were consistent with DNA damage slowing
the progress of the replication forks, with gaps being formed in the
lagging strands, whereas the leading strands were synthesized continuously[13].

(3) My earlier findings of filled-in segments in daughter-strand
DNA[16] could not in my view be interpreted as spaces between blocked
forks in adjacent replicons, as suggested by Park and Cleaver[6].

(4) An increase in the value of P(block) should have two effects,
namely a decreased size of newly-synthesized DNA, and a lower overall
rate of DNA synthesis.   There should thus be some correlation between
the inhibition of thymidine incorporation and the reduction in size
of newly-synthesized DNA, (although blocking of fork progression
may not be the only cause of the inhibition of DNA synthesis).   In
fact, numerous studies have shown that there is no such correlation,
e.g. (a) in some established cell lines the inhibition of thymidine

incorporation by UV is very similar, whereas the sizes of newly-synthesized DNA in irradiated cells are quite different[4];   (b) in a human neuroblastoma line, thymidine incorporation is not inhibited by UV, although the newly-synthesized DNA is reduced in size[17]; (c) in a number of systems, if caffeine is added after irradiation, there is a greater reduction in molecular weight of newly-synthesized DNA and an inhibition in its subsequent increase in size, but there is either no effect on thymidine incorporation or a small increase, not a decrease[18,19].   Similar observations have been made after treatment with chemical mutagens[20];   (d) as shown very clearly by Park and Cleaver[2,7], the ability to synthesize normal-sized DNA recovers much more rapidly than the ability to synthesize DNA at normal rates.   Indeed in some instances the synthesis of normal-sized DNA has completely recovered at times when thymidine incorporation is still at a minimum level (e.g. see Fig.1 in ref.2); (e) in cells from patients with Cockayne syndrome, the reduced rate of thymidine incorporation persists[3] whereas the ability to synthesize normal-sized DNA recovers.   These observations are not easily explained by an all-or-nothing block mechanism.   On the other hand a model envisaging gaps as a possible option does not require any correlation between rates of synthesis and size of newly-synthesized DNA.   The latter is dependent upon how frequently gaps are produced and how rapidly they are sealed, whereas the former may result from blocks, delays, inactivation of replicons etc.

Although none of the above experiments presents definite proof of any particular mechanism, they are difficult to reconcile without invoking the existence of daughter-strand gaps.   This is not to suggest that gaps are left near every damaged site, merely that *this is a possible option* when a replication fork encounters damage. Indeed the evidence suggests that only in the most extreme cases (e.g. XP variants) does a discontinuity result close to every pyrimidine dimer[21].   Conversely, in other systems, such as mitochondrial DNA[22], and HeLa cells exposed to high UV fluences[12], the evidence is more consistent with damage blocking progression of the replication fork.

A single scheme, at present entirely speculative, could give rise to gaps, blocks, delays and direct synthesis past damage.   Let us consider two enzyme systems, X, which carries out normal DNA synthesis, and Y, a special system necessary for effecting synthesis past damage by some unknown mechanism.   Damage in the DNA dislodges system X.   If the local concentration of X ((X)) is low, fork progression will be blocked either before (low(Y)) or after (high (Y)) the damaged site.   If (Y) is low but (X) is high, fork progression will continue but a gap will be formed.   If both (X) and (Y) are high, synthesis can continue without a gap.   It should be emphasized that the crucial parameters are the local concentrations in the vicinity of damaged sites.   These could be different in different cells, after varying UV fluences etc., and could therefore give rise to the apparently contradictory results obtained using different experimental systems.   Chang has suggested that DNA polymerase β may be able to synthesize past DNA damage (ref. 23 and this volume), and it could therefore be a component of Y, whereas the replicative DNA polymerase α would be part of X.

Whatever the cause of the various perturbations of DNA synthesis, all of them can be overcome eventually, by mechanisms which are still completely obscure.   The variations in the responses of cells with different genotypes, such as XP variants[5,7,19] and Cockayne syndrome[3,24], and under different physiological conditions, suggest that special mechanisms are called into play and these can with validity be termed post-replication, or daughter-strand repair.

EFFECTS ON THE INITIATION OF DNA SYNTHESIS

Recently it has been shown that a number of mutagens inhibit initiation of DNA synthesis in clusters of replicons.   This is particularly marked with ionizing radiation (e.g. see ref.25), whereas chemical mutagens affect both initiation and chain elongation in different ways[26].   Recent studies suggest that low UV fluences may also inhibit replicon initiation[27].   In no case, however, has complete inhibition of initiation been observed, even after high doses of mutagen.   This suggests that some replicon clusters may be resistant to such inhibition.   It has been suggested that inhibition of initiation of DNA synthesis in replicon clusters results from strand-breaks, either directly produced by mutagens like ionizing radiation[25], or enzymatically produced during excision repair, after UV[2,27].   This seems, however, to be an oversimplification.   After ionizing radiation, the rejoining of strand breaks is very rapid, and the rates of DNA synthesis are slowest after most of the strand-breaks have been rejoined[28].   After UV-irradiation, the inhibition of DNA synthesis in excision-defective XP cells (in which no enzymatic breaks are formed) is at least as great as that in normal cells[2].   These observations cannot therefore be interpreted in a simple way in terms of interactions of strand breaks with replicon clusters.

CONCLUSIONS

The confusion concerning the response of DNA replication to DNA damage will not be resolved in the near future.   As pointed out by several workers, the processes involved will depend not only on the nature and frequency of the damaged sites, but also on the organization of DNA in the mammalian chromosome, and the relationship between replicons, replicon clusters, nucleosomes and other chromosomal subunits.   The purpose of the above discussion has been to demonstrate that each situation must be analysed in its own right and that caution should be exercised when generalising from results obtained under a particular set of experimental conditions.

REFERENCES

1. J. M. Rudé and E. C. Friedberg, Mutation Res. 42, 433-442 (1977)
2. S. D. Park and J. E. Cleaver, Nucleic Acids Res. 6, 1151-1159
    (1979).
3. A. R. Lehmann, S. Kirk-Bell and L. Mayne, Cancer Res. 39, 4237-
    4241 (1979).
4. A. M. Rauth, M. Tammemagi and G. Hunter, Biophys. J. 14, 209-
    220 (1974).
5. A. R. Lehmann, S. Kirk-Bell, C.F. Arlett, S. A. Harcourt, E. A.
    de Weerd-Kastelein, W. Keijzer and P. Hall-Smith, Cancer Res.
    37, 904-910 (1977).
6. S. D. Park and J. E. Cleaver, Proc. Nat. Acad. Sci., U.S. 76,
    3927-3931 (1979).
7. J. E. Cleaver, G. H. Thomas and S. D. Park, Biochim. Biophys.
    Acta 564, 122-131 (1979).
8. L. N. Kapp and R. B. Painter, Biophys. J. 24, 739-748 (1978).
9. R. Hand, Hum. Genet., 37, 55-64 (1977).
10. C. H. Ockey, J. Cell Sci. 40, 125-144 (1979).
11. J. Doniger, J. Mol. Biol. 120, 433-446 (1978).
12. H. J. Edenberg, Biophys. J. 16, 849-860 (1976).
13. M. Cordeiro-Stone, R. I. Schumacher and R. Meneghini, Biophys.
    J. 27, 287-300 (1979).
14. R. Meneghini and P. Hanawalt, Biochim. Biophys. Acta 425, 428-
    437 (1976).
15. J. M. Clarkson and R. R. Hewitt, Biophys. J. 16, 1155-1164 (1976).
16. A. R. Lehmann, J. Mol. Biol. 66, 319-337 (1972).
17. M. R. James, M.Sc. Thesis, University of Queensland (1979).
18. A. R. Lehmann and S. Kirk-Bell, Mutation Res. 26, 73-82 (1974).
19. A. R. Lehmann, S. Kirk-Bell, C. F. Arlett, M. C. Paterson,
    P. H. M. Lohman, E. A. de Weerd-Kastelein and D. Bootsma,
    Proc. Nat. Acad. Sci., U.S. 72, 219-223 (1975).
20. J. J. Roberts, Advances Radiat. Biol. 7, 211-436 (1978).
21. A. R. Lehmann, Nucleic Acids Res. 7, 1901-1912 (1979).
22. D. A. Clayton, J. N. Doda and E. C. Friedberg, Proc. Nat. Acad.
    Sci. U.S. 71, 2777-2781 (1974).
23. L. M. S. Chang and F. J. Bollum, J. Biol. Chem. 248, 3398-3404
    (1973).
24. A. R. Lehmann and L. Mayne.   This volume.
25. R. B. Painter and B. R. Young, Biochim. Biophys. Acta 418, 146-
    153 (1976).
26. R. B. Painter, Cancer Res. 38, 4445-4449 (1978).
27. W. K. Kaufmann, J. E. Cleaver and R. B. Painter, Biochim.
    Biophys. Acta 608 (1980).
28. L. J. Tolmach and R. W. Jones, Radiation Res., 69, 117-133 (1977).

REPLICATION PROCESSES IN CHINESE HAMSTER CELLS TREATED WITH

MUTAGENIC AGENTS: EFFECTS OF SPLIT DOSE TREATMENTS

Gunnar Ahnström and Edward Hefner

Wallenberg Laboratory
University of Stockholm
S-106 91 Stockholm

INTRODUCTION

Postreplication repair in UV-irradiated hamster cells seems to be enhanced if the cells have been preirradiated with low doses of UV radiation compared to cells irradiated only once. This points to the existence of inducible repair processes also in mammalian cells (D'Ambrosio and Setlow, 1976).

Alkylating chemicals have also been found to enhance postreplication repair in UV-irradiated cells (Johansson and Ahnström, 1980). The combination of alkylating chemicals and UV light in split dose experiments has been further studied in this investigation.

MATERIAL AND METHODS

Chinese hamster cells were grown in multi-well plastic trays. They were seeded in the evening in numbers to have a maximum in dividing cells next morning. UV-irradiations were done on ice with the cells covered by a thin layer of buffer solution. Mutagenic chemicals were dissolved in medium and the cells treated for 15 minutes at 37° C followed by a recovery in fresh medium for 30 minutes. Cells were pulse-labelled for 30 minutes in ^{3}H-thymidine and then chased in medium for various periods of time.

Strand interruptions in newly synthesized DNA were determined by a DNA-unwinding technique (Johansson and Ahnström, 1980) based on the principle that DNA strand separation in weak alkali only occurs from free end points and strand breaks (Ahnström and Erixon, 1973).

Cells were cooled down to ice temperature and washed with ice-cold
buffer. 0.5 ml of 0.03 M NaOH solution was added and the samples
stored in the dark at $0°$ C for 30 minutes. After neutralization with
1.1 ml 0.02 M NaH_2PO_4 solution, sonication and addition of SDS,
single and double stranded DNA was separated by hydroxyapatite chromatography (Ahnström and Erixon, 1981).

RESULTS

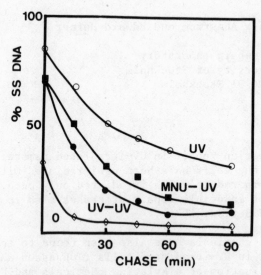

Fig. 1.   Postreplication repair in UV-irradiated hamster cells.
UV: Cells were irradiated with 15 J/m^2, allowed to recover
in medium for 30 minutes and the labelled in ^{3}H-thymidine
for 30 minutes.
MNU-UV: 15 minutes treatment in 0.25 mM MNU - 3 hours in
medium - 15 J/m^2 - 30 minutes in medium - 30 minutes in
^{3}H-yhymidine.
UV-UV: 2.4 J/m^2 - 3 hours in medium - 15 J/m^2 - 30 minutes
in medium - 30 minutes in ^{3}H-thymidine.
0: No treatment - 30 minutes in ^{3}H-thymidine.
All samples were chased in medium after the thymidine pulse
for times indicated in the figure.

Fig. 1. shows the percentage of ^{3}H-labelled DNA-segments containing strand interruptions or replication forks as a function of

the chase time. The UV-induced strand interruptions decay at a higher rate if the cells previously were exposed to UV radiation or methyl nitrosourea (MNU). Methylmethanesulfonate and X-rays were also found to enhance postreplication repair in UV-irradiated cells (data not shown). This in contrast to MNU-treated cells where none of the agents mentioned above were found to have any stimulating influence on postreplication repair. Fig. 2. illustrates the effect of a pretreatment with UV radiation.

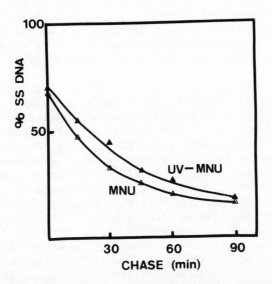

Fig. 2.   Postreplication repair in MNU-treated hamster cells. UV-MNU: 2.4 J/m^2 UV – 3 hours in medium – 15 minutes in 2 mM MNU – 30 minutes in medium – 30 minutes in ^{3}H-thymidine MNU: 15 minutes in 2 mM MNU – 30 minutes in medium – 30 minutes in ^{3}H-thymidine. The pulse-labelled cells were chased in medium for times indicated in the figure.

DISCUSSION

To explain the apparent enhancement in postreplication repair in cells preexposed to small doses of UV radiation it was suggested that an inducible repair component is involved (D´Ambrosio and Setlow, 1976). An other explanation was put forward by Painter (1978) who pointed out that the first irradiation with UV had resulted in an abnormal distribution of sizes of growing replicons at the time when the second dose was given. This could fully account for the observed enhancement effects after split dose treatments.

D´Ambrosio and Setlow concluded that there are at least two postreplication pathways in chinese hamster cells. The constitutive pathway slowly fills in gaps left behind oppsite dimers and this pathway is resistent to cycloheximide. Details of the pathway which required protein synthesis were not discussed.

Compared to UV radiation, MNU was found to induce strand interruptions in newly synthesized DNA in high numbers without seriously inhibiting the ^{3}H-thymidine incorporation. It was therefore assumed that the main effect of MNU on DNA-replication was to induce gaps. If the gapfilling process as suggested by D´Ambrosio and Setlow is constitutive, this would explain why there is no enhancement of postreplication repair in MNU-treated cells by pretreatments.

We now hypothesize that dimers could either block the replication or induce gaps depending on circumstances. The rate of DNA polymerization is lower in mammalian cells compared to bacteria. Mammalian chromosomes have a more complex structure and more work is therefore required to prepare the DNA for replication. We assume that a lesion in the DNA specificly can inhibit the polymerization and that other processes responsible for preparing the DNA-protein complex are uncoupled and may continue along the chromosome on their own hands. This would permit a higher rate of polymerization behind the lesion and might also enhance the bypass of the next lesions. In the case of pyrimidine dimers we propose that the first dimer encountered in a replicon blocks the replication and uncouples the processes which are preparing the chromatin for replication. When the polymerase reaches the next dimer the DNA is already prepared for replication and the dimer will be bypassed leaving a gap in the new strand. Pretreated cells now have replicons which have overcome the first dimer and replicate faster than replicons encountering their first dimer, i.e. in cells irradiated only once.

ACKNOWLEDGEMENTS

    This investigation was supported by funds from the Swedish Natural Science Research Council.

REFERENCES

    Ahnström, G., and Erixon, K., 1973, Radiation induced
        strand breakage in DNA from mammalian cells:
        Strand separation in alkaline solution, Int. J.
        Radiat. Biol., 23:285.

Ahnström, G., and Erixon, K., 1981, The measurement of
        strand breaks by DNA unwinding in alkali and
        hydoxyapatite chromatography, in: "DNA Repair:
        A Laboratory Manual of Research Procedures,"
        E. C. Friedberg and P. C. Hanawalt, eds.,
        Marcel Dekker, New York.
D῾Ambrosio, S. M., and Setlow, R. B., 1976, Enhancement
        of postreplication repair in Chinese hamster cells,
        Proc. Natl. Acad. Sci. USA., 73:2396.
Johansson, B., and Ahnström, G., 1980, Replication and
        repair processes in Chinese hamster cells: The
        effects of ultraviolet light, alkylating chemicals
        and X-rays studied by DNA unwinding in alkali,
        in: "DNA: Recombination, Interactions and Repair,"
        S. Zadrazil, and J. Sponar, eds., Pergamon Press,
        Oxford, 1980.
Painter, R. B., 1978, Does ultraviolet light enhance
        postreplication repair in mammalian cells?,
        Nature, 275:243.

ROLE OF MAMMALIAN DNA POLYMERASES IN REPLICATION AND REPAIR

Hans Krokan

Institute of Medical Biology
University of Tromsø
9001 Tromsø, Norway

INTRODUCTION

Replication of eukaryotic nuclear DNA is discontinuous at two levels; it takes place within numerous replicons, and within each replicon at least one of the
two nascent DNA strands is made discontinuously by formation of small (average 135 nucleotides) primary pieces
which are subsequently ligated (reviewed in ref. 1-6).
Although there is no definite proof for the participation in DNA replication of any known enzyme activity,
this model suggests that a minimum of six proteins are
involved. These are an unwinding protein (DNA dependent
ATPase?), an RNA polymerase ("primase"? RNA pol I?),
a DNA polymerase (DNA pol α?), an RNAse (RNAse H?), a
gap-filling DNA polymerase (DNA pol α?) and a DNA ligase
(DNA ligase I? DNA ligase II?), (Fig. 1). The enzymology of mammalian DNA repair is very complex, and even less
well understood than that of DNA replication.

ANIMAL DNA POLYMERASES

Animal cells contain at least three distinct DNA
polymerase activities (α, β and γ) which are generally
identified by their characteristic chromatographic properties, responses to inhibitors, primer-template pre-
ferences, salt requirements and molecular weights. The
functional roles of these enzymes remain unclear, but
enough data has accumulated to allow some tentative conclusions (Table 1).

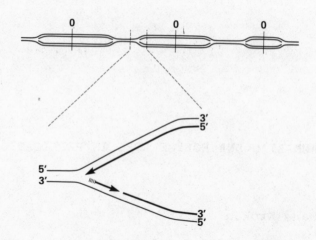

Fig. 1. Discontinuous replication of eukaryotic DNA
within replicons. Schematic representation
of replicons with replication origins (0) and
a replication fork with an RNA primer (ⅿⅿ)
and nascent DNA (heavy lines).

DNA polymerase α

DNA polymerase α represents the major DNA polymerase activity in proliferating cells (2,7), but may be
completely absent in some differentiated cells (8).
The α-polymerase class is defined operationally as a
high molecular weight (> 100.000) enzyme which is sensitive to N-ethylmaleimide (NEM) and which has a rather
distinct pattern of primer-template utilization and ab-

Table 1. Mammalian DNA polymerases

DNA pol	Mol. weight	NEM[*]	d_2TTP	Aphidi-colin	Function
α	>100.000	sens	insens	sens	DNA repli-cation DNA repair
β	45.000	insens	sens	insens	DNA repair
γ	>100.000	sens	sens	insens	Strand dis-placement synthesis

[*]N-ethylmaleimide

sence of 3' to 5' exonuclease activity. The lack of
exonuclease activity is also characteristic for DNA polymerases β and γ, but was described in one form of DNA
polymerase α from mouse myeloma cells (9) and in "DNA
polymerase δ" which in some respects resembles DNA polymerase α (10).

Complete purification has for a long time been very
difficult, partially due to an apparent heterogeneity
(2,7,11,12). However, several laboratories have recently succeeded in obtaining near homogeneous forms of DNA
polymerase α, although reports on the size, number of
subunits and other properties vary (9,12,13). This may
be due to species differences, but may also be a reflection of the complexity of the enzyme. It is possible
that some subunits are only loosely associated with the
"core enzyme" and therefore easily lost during purification. DNA polymerase α is the only mammalian DNA
polymerase known that can start synthesis on an RNA
primer (7). It does not appear to recognize specific
sequences (2), although it has a very high affinity for
poly(dT) over other homopolymers (14). It binds specifically to single-strand regions (12).

Recently, it was shown that the unusual dinucleotide
diadenosine 5',5'''-P^1, P^4-tetraphosphate (Ap$_4$A) has a
high affinity for one of the subunits in DNA polymerase
α (13). Furthermore, Ap$_4$A is found in higher concentrations in proliferating cells than in resting ones (15),
and at least in vitro stimulate G$_1$-cells to enter S-
phase (16).

During cell fractionation in aqueous media, most
(70-95%) of the DNA polymerase α activity is usually
found in the cytoplasma and the remainder in the nucleus
(2,7,17). This may be an artefact resulting from redistribution during cell fractionation, since most of the
activity was found in the nucleus after fractionation
in nonaquous glycerol, or in very low ionic-strength
buffers containing Ca^{2+}, or when "minicells" were prepared using the cytochalasin B enucleation method (18-
20). Therefore, it has been accepted that DNA polymerase α normally resides inside the nucleus. However,
these conclusions may be premature. In fact, it has
been shown that the low salt/ Ca^{2+} method may cause an
preferential inactivation of the cytoplasmic DNA polymerase α, thus creating a false impression of intracellular distribution (21).

Several lines of evidence have implicated DNA poly-

merase α in nuclear DNA replication. First, a 3-10 fold
increase in DNA polymerase α activity consistently
appeared concomitant with the onset of DNA synthesis,
while DNA polymerase β activity remained relatively constant (2,7). Second, DNA polymerase α cosedimented with
replicating papova-virus chromosomes, as did DNA polymerase γ (2). Third, DNA polymerase α is the only animal
DNA polymerase that can start synthesis on an RNA primer
(2,7). Finally, the combined results from the use of
several selective DNA polymerase inhibitors have suggested strongly that DNA polymerase α is involved in nuclear
DNA replication.

Association of DNA polymerase α with replicating SV40 nucleoprotein complexes

When nuclei from SV40-infected monkey cells are incubated at 0°C in a hypotonic buffer, SV40 nucleoprotein
complexes leak out and can be separated from nuclei by
low speed centrifugation (22). Mature chromosomes can be
partially separated from replicating chromosomes which
are enriched in DNA polymerase α and γ (2,23 and Fig. 2).
However, less than 5% of the α-polymerase in the extract
cosedimented with SV40 DNA. DNA polymerase γ sedimented

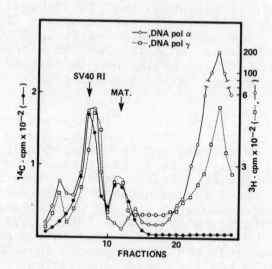

Fig. 2. DNA polymerase activities associated with SV40
        chromatin. SV40 chromatin prepared from [14]C-
        thymidine-labelled CV-1 cells infected with SV40
        was separated into replicating (SV40 RI) and ma-
        ture (mat.) chromatin by sedimentation in neutral
        sucrose, and DNA polymerase activity analysed in
        each fraction.

slightly behind replicating SV40 nucleoprotein-complexes (Fig. 2). The DNA polymerase that cosedimented with re-plicating SV40 nucleoprotein complexes was DNA polyme-rase α as judged from its sensitivity to N-ethylmaleimide and aphidicolin, its resistance to d_2TTP, and its elu-tion from DEAE-cellulose (peak at 0.19 M KCl at pH 7.5).

Selective inhibitors of DNA polymerases

Recently, several compounds have been shown to have selective effects on one or two of the three mammalian DNA polymerases. Aphidicolin and araCTP, which preferen-tially inhibit DNA polymerase α, and 2'3'-dideoxythymi-dine 5'-triphosphate (d_2TTP), which preferentially in-hibits DNA polymerase β and γ, have been particularly useful (2,24-28) (Table 2). Recently, it was shown that butylanilino-uracil, a dGTP analog, is a selective inhibitor of DNA polymerase α and of cellular DNA repli-cation (29). The clear picture that emerges is that DNA polymerase α is the only DNA polymerase required for cellular nuclear DNA synthesis as well as for SV40 DNA synthesis.

Table 2.   Involvement of DNA pol α in DNA replication[*]

Inhibitor	Subcellular system (nuclei or nuclear ex- tract)	DNA pol α	β	γ	Conclusion
Aphidicolin (μg/ml)	CV-1, 0.47	· 1.8	>125	>125	DNA pol α(26)
"	CV-1/SV40, 2.4				DNA pol α(26)
"	HeLaS$_3$,0.1-0.5	0.1-0.5	>30	>30	DNA pol α(**)
d_2TTP/dTTP	CV-1,>20	20	0.2	0.08	DNA pol α(26)
"	CV-1/SV40,>20				DNA pol α(26)
"	HeLaS$_3$, 10	5	0.2	<0.01	DNA pol α(27)
araCTP/dCTP	CV-1, 0.6	0.8	23	14	DNA pol α(26)
"	CV-1/SV40,0.7-1.0				DNA pol α(26)
"	HeLaS$_3$, 0.3	1.25	>10	>10	DNA pol α(27)
Butylanilino uracil (mM)	HeLaS$_3$, 0.05 (intact cells)	0.05	> 1	> 1	DNA pol α(29)

[*] Inhibitor concentration that gives 50% inhibition of DNA synthesis.
[**]Ref. 25 and Krokan and Wist, unpublished.

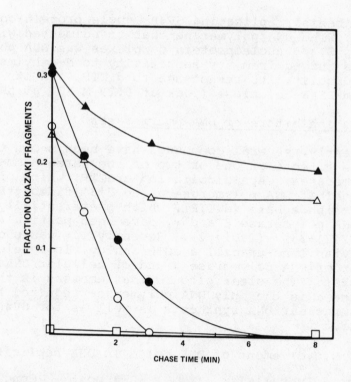

Fig. 3. Effect of aphidicolin on joining of Okazaki frag-
         ments. Replicating SV40 DNA in isolated nuclei
         (26) was labelled with [α-^{32}P]dNTP for 30 sec
         after which aphidicolin (25 μg/ml) or buffer was
         added, and the incubation continued. At 2 min
         cytosol (26) and a 200-fold excess dNTP were added
         and the incubation continued for 0-8 min after
         which SV40 DNA was purified. The fraction of
         radioactivity in Okazaki-fragments was then cal-
         culated after sedimentation in alkaline sucrose
         gradients either without further treatment of
         purified SV40 DNA (—●—, no aphidicolin, —▲—,
         aphidicolin), or after treatment of purified SV40
         DNA with E.coli ligase (—o—, no aphidicolin,
         —△—, aphidicolin) or E.coli pol I plus E.coli
         ligase (—□—, aphidicolin and no aphidicolin).
         (Weaver, Krokan & DePamphilis, unpublished).

     In a separate series of experiments it was shown
that DNA polymerase α is also the enzyme that fills in
gaps between primary pieces (Okazaki fragments). It was
found that d2TTP had no effect on the joining of Okazaki
fragments (30), whereas aphidicolin severely inhibited this
process (31, 32 and Fig. 3). Interestingly, aphidicolin

also inhibited joining of those fragments that were separated from the longer strand only by a nick, indicating
that DNA polymerase α may interact with DNA ligase (Fig.3)
Aphidicolin has no effect on partially purified mammalian
DNA ligase (31). A cytosol fraction devoid of DNA polymerase and DNA ligase appears to cooperate with DNA polymerase α in the gap filling process (31). The nature of
this protein(s) is entirely unknown. While the experiments reported above, as well as several other reports
indicate that DNA polymerase α is involved in DNA replication, it is still unclear whether there is one subspe-
cies which is the active form.

DNA polymerase γ

DNA polymerase γ has been found both in the nucleus
and in the cytoplasm and is similar, if not identical,
to mitochondrial DNA polymerase (33,34). It represents
only a minor part (1-2%) of the total polymerase activity
in growing cells, although it was reported to be the major DNA polymerase activity detected in human placenta
(35). Estimates of molecular weight range from 110.000
to 330.000 (17) although the highest molecular weights
probably are due to aggregation of the enzyme (36). The
HeLa cell enzyme has been purified to a specific activity
of 25.000 units per mg protein, representing a 60.000-
fold purification (36). The γ-polymerase can copy several synthetic ribohomopolymers primed with a complementary deoxyribo-oligonucleotide, although it has a strong
preference for poly(rA) oligo(dT). It can use activated
DNA at an efficiency of 5-30% of that seen with poly (rA)
oligo (dT) (17,36). Unlike reverse transcriptase, which
can also copy synthetic ribohomopolymers, it cannot copy
natural RNA (17). DNA polymerase γ is also unable to
copy poly(Cm) oligo(dG), which can be copied by all reverse transcriptases tested, although quite inefficiently
by some (37).

Role of DNA polymerase γ in strand displacement synthesis

Although the exact mechanism varies in each case,
strand displacement synthesis is seen in mitochondrial
DNA replication, adenovirus replication and in DNA synthesis by the rolling circle model (1-3).

Since DNA polymerase γ is the only DNA polymerase
present in isolated mitochondria (34), it seems very
likely that this enzyme is involved in replication of
mtDNA. This was also suggested by the sensitivity of
mtDNA synthesis in isolated, permeabilized neuronal syn-

Table 3. DNA synthesis by the strand displacement model[*]

Inhibitor	Subcellular system (nuclei)	DNA pol α	β	γ	Conclusion
d$_2$TTP/dTTP	HeLa/Ad2, 0.02	10	0.2	0.01	DNA pol γ (26)
Aphidicolin (μg/ml)	-	1.8	>125	>125	DNA pol α (26)
"	HeLaM/Ad2, 3.0	1.0	>125	>125	DNA pol α (26)
"	HeLaS$_3$/Ad2, 1	1	> 30	> 30	DNA pol α (41)
"	HeLaS$_3$/Ad2, 0.8	1	in-sens	in-sens	Complex of DNA pol α and γ? (39)
d$_2$TTP/dTTP	synaptosomal mito-chondria	0.2	DNA pol γ: <0.2 (only γ present)		DNA pol γ (8)

[*]Inhibitor concentration that gives 50% inhibition of
DNA synthesis.

aptosomes to d$_2$TTP (8). However, it is not ruled out that
DNA polymerase α could participate in mitochondrial DNA
replication in vivo. Several groups have now shown that
adenovirus DNA synthesis in isolated nuclei or nuclear
extracts is extremely sensitive to d$_2$TTP (26,38,39).
This together with the observation that DNA polymerase γ
is the major DNA polymerase present in isolated replicating adenovirus complexes (40), suggests strongly that
DNA polymerase γ is involved in adenovirus DNA replication. In addition, adenovirus DNA replication also
appears to require DNA polymerase α (26,39,41),(Table 3).
DNA synthesis by the rolling circle model has not been
established in mammalian cells, but has been detected as
an alternative mode of replication of polyoma virus (42).
It is unknown which DNA polymerase is responsible for
this synthesis. However, elongation of purified RI SV40
DNA in vitro appeared to take place according to the
rolling circle model and was insensitive to aphidicolin,
indicating that DNA polymerase γ may be involved (Krokan,
unpublished). Finally it should be mentioned that single-
stranded non-nascent DNA both of low (43) and high (44)
molecular weight has recently been detected in animal
cells. How these fragments are generated and their
possible function are completely obscure, but they might
be generated by strand displacement synthesis.

DNA polymerase β

DNA polymerase β is present predominantly in the cell
nucleus (2,17). The homogenous protein comprises a single
peptide chain of about 45.000 daltons (7). Its activity
is essentially equal in proliferating and differentiated
cells (7,17) and it is thought to be involved mainly in
repair of DNA, although there are reports suggesting that
it participates in DNA replication (45,46). It is most
active with activated DNA as template-primer, although
it, like DNA polymerase γ, can copy poly(rA) oligo(dT)(7,
17). DNA polymerase β is absent in some lower eukaryotes
(17).

FUNCTION OF DNA POLYMERASES IN DNA REPAIR

Generally, a DNA polymerase is needed to fill in
gaps resulting from a repair process initiated by several

Table 4.   DNA polymerases involved in DNA repair.

Treatment	System	Inhibitor		Conclusion
		d_2TTP	Aphidi-colin	
UV-irrad.	H.fibrobl. (lysate or cells)	-	+++	DNA pol α (49)
"	HeLa cells (G_2-nuclei)		+++	DNA pol α (50)
"	H.lymphocyt. (permeabil.)		+++	DNA pol α (51)
"	Rat neurons (nuclei)	+++		DNA pol β (8)
Bleomycin	Mouse sarc. (permeabil.) and rat (nuclei)	+++		DNA pol β (53)
N-methyl-N'-nitroso-urea	H.fibrobl. (intact or lysate)		+++	DNA pol α (49)
N-methyl-N'-nitro-N-nitroso-guanidine	H-lymphocyt. (permeabil.)		++	DNA pol α (51)

other repair enzymes (47). There is evidence for the
participation of both DNA polymerase α and β in DNA repair. Involvement of β-polymerase was suggested by correlation with repair activity (48). Furthermore, rat neuronal brain nuclei, which lack DNA polymerase α and have
a low level of DNA polymerase γ, can still repair UV-
irradiated DNA (8). However, this does not exclude the
possibility that DNA polymerase α, when present, can
participate in DNA repair. In fact, inhibitor studies
have strongly suggested the involvement of DNA polymerase
α in the repair of UV-lesions and alkylations in several
different cell types (Table 4). DNA polymerase β also
seems to be involved, although the selectivity of some
of the inhibitors used is questionable (49-54).

In several respects it is more complicated to study
the involvement of DNA polymerases in repair than in replication. Firstly, the level of repair activity is
usually low compared to that of replication, so that it
is sometimes necessary to induce massive lesions in order
to reach a satisfactory level of repair activity. It is
possible that repair in massively damaged cells may differ
qualitatively from the repair processes in lightly damaged
cells. Secondly, it is possible that different types of
lesions require different polymerases, and finally, it is
possible that DNA polymerases may substitute for each
other in DNA repair depending on the relative abundancy
of the various polymerases.

REFERENCES

1. Kornberg, A. (1980) In "DNA Replication", W.H. Freeman
2. DePamphilis, M.L. & Wassarman, P.M. (1980) Ann. Rev.
   Biochem. 49:627
3. Weissbach, A. (1979) Arch. Biochem. Biophys. 198:386.
4. Cremisi, C. (1979) Microbiol. Rev. 43:297.
5. Seale, R.L. (1978) In "The Cell Nucleus", 4, (Bush,
   H., Ed.) pp. 155-172 Academic Press, Inc. N.Y., London
   and San Francisco.
6. Sheinin, R. & Humbert, J. (1978) Ann. Rev. Biochem.
   47:277.
7. Bollum, F. (1975) In "Progr. Nucl. Acid. Res. and
   Mol. Biol.", 15:109. Ed. W.E. Cohn, Academic Press,
   N.Y., London and San Francisco.
8. Hübscher, U., Kuenzle, C.C. & Spadari, S. (1979)
   Proc. Natl. Acad. Sci. 76:2316.
9. Chen, Y.C., Bohn, E.W., Planck, S.R. & Wilson, S.H.
   (1979) J. Biol. Chem. 254:11678.

10. Byrnes, J.J. & Black, V.L. (1978) Biochemistry, 17:4226.
11. Hesslewood, I.P., Holmes, A.M., Wakeling, W.F. & Johnston, I.R. (1978) Eur. J. Biochem. 84:123.
12. Fisher, P.A. & Korn, D. (1979) J. Biol. Chem. 254: 11033.
13. Grummt, F., Waltl, G., Jantzen, H.M., Hamprecht, K., Huebscher, V. & Kuenzle, C. (1979) Proc. Natl. Acad. Sci. USA, 76:6081.
14. Wilson, S.H., Matuskage, A., Bohn, E.W., Chen, Y.C. & Sivarajan, M. (1977) Nucl. Acids Res. 4:3981.
15. Rapaport, E. & Samecnik, P. (1976) Proc. Natl. Acad. Sci. USA, 73:3984.
16. Grummt, F. (1978) Proc. Natl. Acad. Sci. USA, 75:371.
17. Weissbach, A., (1977) Ann. Rev. Biochem. 46:25.
18. Foster, D.N. & Gurney, T. (1976) J. Biol. Chem. 251: 7893.
19. Lynch, W.E., Surrey, S. & Lieberman, I. (1975) J. Biol. Chem. 250:8179.
20. Herrick, G., Spear, B. and Veomett, G. (1976) Proc. Natl. Acad. Sci. USA, 73:1136.
21. Korn, D., Eichler, D.C., Fisher, P.A. & Wang, T.S.F. (1977) In "DNA Synthesis, Present and Future", pp. 517-558, ed. Molineux, I. & Kohiyama, M. Plenum Press, N.Y. & London.
22. Su, R.T. & DePamphilis, M.L. (1976) Proc. Natl. Acad. Sci., USA 73:3466.
23. Edenberg, H.J., Anderson, S. & DePamphilis, M.L. J. Biol. Chem. 253:3273.
24. Ikegami, S., Taguchi, T., Ohashi, M., Oguro, M., Nagano, H. & Mano, Y. (1978) Nature, 275:458.
25. Wist, E. & Prydz, H. (1979) Nucl. Acids. Res. 4:1583.
26. Krokan, H., Schaffer, P. & DePamphilis, M.L. (1979) Biochemistry, 18:4431.
27. Wist, E. (1979) Biochim. Biophys. Acta 562:62.
28. Oguro, M., Suzuki-Hori, C., Nagano, H., Mano, Y. & Ikegami, S. (1979) Eur. J. Biochem. 97:603.
29. Wright, G.E., Baril, E.F. & Brown, N.C. (1980) Nucl. Acids. Res. 8:99.
30. DePamphilis, M.L., Anderson, S., Bar-Shavit, R., Collins, E., Edenberg, H., Herman, T., Karas, B., Kaufmann, G. Krokan, H., Shelton, E., Su, R., Tapper, D. and Wassarman, P.M. (1978) Cold Spring. Harb. Symp. Quant. Biol. 43:679.
31. Weaver, D., Krokan, H. & DePamphilis, M.L. (1980) J. Supramol. Str., Suppl. 4, Abs. 895:333.
32. DePamphilis, M.L., Anderson, S., Cusick, M., Hay, R., Herman, T., Krokan, H., Shelton, E., Tack, L., Tapper, D., Weaver, D. & Wassarman, P.M. (1980) In "Mechanistic Studies of DNA Replication and Ge-

netic Recombination ICN-UCLA Symposia on Molecular and Cellular Biology" 19, in press.

33. Hübscher, U., Kuenzle, C.C. & Spadari, S. (1977) Eur. J. Biochem. 81:249.

34. Bertazzoni, U., Scovassi, A.I. & Brun, G. (1977) Eur. J. Biochem. 81:237.

35. Krauss, S.W. & Linn, S. (1980) Biochemistry 19:220.

36. Knopf, K.-W., Yamada, M. & Weissbach, A. (1980) Biochemistry, 15:4540.

37. Robert-Guroff, M., Schrecker, A.W., Brinkmann, B.J. & Gallo, R.C. (1977) Biochemistry, 16:2866.

38. van der Vliet, P.C. & Kwant, M.M. (1978) Nature, 276:532.

39. Habara, A., Kano, K., Nagano, H., Mano, Y., Ikegami, S. & Yamashita, T. (1980) Biochem. Biophys. Res. Comm. 92:8.

40. Green, M., Arens, M.Q., Yamashita, T., Wold, W.S.M. & Brackmann, K.H. (1978) Cold Spring Harb. Symp. Quant. Biol. 43:755.

41. Longiaru, M., Ikeda, J.-E., Jarkovsky, Z., Horwitz, M. (1979) Nucl. Acids Res. 6:3369.

42. Bjursell, G. (1978) J. Virol. 26:136.

43. Denhardt, D.T., Kowalski, J. & Miyamoto, C. (1978) Cold Spring Harb. Symp. Quant. Biol. 43:243.

44. Bjursell, G., Gussander, E. & Lindahl, T. (1979) Nature, 280:420.

45. Butt, T.R., Wood, W.M., McKay, E.L. & Adams, R.L.P. (1978) Biochem. J. 173:309.

46. Hobart, P.M. & Infante, A.A. (1978) J. Biol. Chem. 253:8229.

47. In "DNA Repair Mechanisms",(1978) (Hanawalt, P.C., Friedberg, E.C. & Fox, C.F. Eds.) Academic Press, N.Y., San Francisco, London.

48. Bertazzoni, U., Stefanini, M., Pedrali-Noy, G., Giullotto, E., Nuzzo, F., Falaschi, A. & Spadari, S. (1976) Proc. Natl. Acad. Sci. USA, 73:7888.

49. Ciarrocchi, G., Jose, J.G. & Linn, S. (1979) Nucl. Acids. Res. 7:1205.

50. Hanaoka, F., Kato, H., Ikegami, S., Ohashi, M. & Yamada, M. (1979) Biochem. Biophys. Res. Comm. 87:575

51. Berger, N., Kusohara, K.K., Petzold, S.J. & Sikurski, G.W. (1979) Biochem. Biophys. Res. Comm. 89:218.

52. Wawra, E. & Dolejs, I. (1979) Nucl. Acids. Res. 7:1675.

53. Seki, S. & Oda, T. (1980) Biochim. Biophys. Acta 606:246.

54. Castellot, J.J., Miller, M.R., Lehtomaki, D.M. & Pardee, A.B. (1979) J. Biol. Chem. 254:6904.

55. Coetzee, M.L., Chou, R. & Ove, P. (1978) Cancer Res. 38:3621.

EVIDENCE FOR A ROLE OF DNA HELICASE II IN DNA REPLICATION

Mo-Quen Klinkert and Mahmoud Abdel-Monem

Max-Planck-Institut für medizinische Forschung
Abteilung Molekulare Biologie
Jahnstr. 29, 6900 Heidelberg, West Germany

ABSTRACT   Studies using antibodies raised against DNA helicase II
of E.coli and applied to subcellular E.coli DNA replicating systems
suggest that DNA helicase II participates in the unwinding of DNA
molecules, which initiate replication as closed circles. A similar
role for the rep protein has been implicated in the rolling circle
model of replication of the double-stranded forms of fd and øX174.

INTRODUCTION

At least four ATP-dependent DNA unwinding enzymes have been
isolated from Escherichia coli. DNA helicase I (M_r 180.000) (1)
(present in 500-700 copies/cell) and DNA helicase II (75.000) (2)
(5000-8000 copies) are capable of separating double-stranded DNA
without the help of other proteins (3). rep Protein (75.000) (4)
(50 copies) requires a stoichiometric amount of the (single-
stranded) DNA binding protein I (DBPI) (5), presumably to keep the
separated strands from renaturing. DNA helicase III (2x20.000) (6),
most recently isolated, is also capable of unwinding DNA and does
so more efficiently when DBPI is present.

Mutants for the rep protein have been isolated, the most well-
known being rep3 (7). This mutant is not able to replicate the
double-stranded forms (RF) of phages øX174 (8) and fd (9). rep
Mutants however are viable; replication synthesis in rep3 cells
is reduced by approximately 40-50% (7). Most recently Kohiyama
and coworkers have found E.coli mutants which may be temperature-
sensitive in DNA helicase II (pers. comm.).

In order to look for functions of DNA helicase I and DNA helicase II in DNA replication, antibodies raised against these enzymes were applied to subcellular E.coli DNA replicating systems. In this way a role of DNA helicase II in DNA replication has been elucidated. It appears from our findings that unwinding during replication involves DNA helicase II - or alternatively the rep protein - depending on whether replication is initiated as a closed circle or as a rolling circle. A function for DNA helicase I remains unknown.

METHODS

Replication of E.coli chromosome

Elongation DNA synthesis was carried out using cells lysed on a cellophane disc (10) in the presence or absence of purified antibodies. Incubation was carried out for 20 min at 22°C.

Replication of λ DNA

Elongation synthesis of λ DNA was measured on cellophane discs using lysates prepared from cells infected with the wild type phage (11). In another experiment synthesis of λ phage DNA was initiated in vitro on cellophane discs by complementing the lysates of cells infected with λ P mutant phage with the viral P protein necessary for the initiation of viral replication (12). Incubation was for 20 min at 28°C.

Replication of ColE1 DNA

ColE1 DNA synthesis was initiated in vitro by adding the DNA to crude E.coli extracts (13) in the presence or absence of antibodies. Incubation continued for 80 min at 30°C.

Replication of fd RF

fd RF synthesis was measured by incubating fd RF in an E.coli extract supplemented with the phage-specific initiating endonuclease (9) with or without antibodies (20 min 30°C).

RESULTS AND DISCUSSION

Our results have shown that the replication of the E.coli chromosome (Fig. 1) and also that of phage λ and ColE1 plasmid (Fig. 2) is inhibited by antibody against DNA helicase II. Further

experiments have shown that both λ and ColE1 multiply in rep mutant
cells and additionally that they are replicated in lysates of rep
mutant cells. These results suggest that the replication of λ DNA
and ColE1 DNA, like that of the E.coli DNA requires the activity of
DNA helicase II but not that of the rep protein. On the other hand
we have found that fd RF replication is not inhibited by anti-DHII
antibody; this is expected for a rep-dependent process.

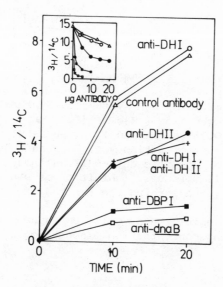

Fig. 1.   Kinetics of E.coli chromosomal elongation synthesis.
E.coli cells grown in the presence of [^{14}C]thymine (2 µg/
ml 600 mCi/mole), were spread and lysed on cellophane
discs in the presence of 10 µg anti-DHI or 10 µg anti-
DHII or 5 µg anti-DBPI or 2 µg anti-dnaB or 10 µg control
antibody. Where both anti-DHI and anti-DHII were present,
5 µg of each antibody were added. The lysate was incubated at 22°C on a DNA synthesis mixture containing [^{3}H]
dTTP (500 Ci/mole). Approximately 3000 ^{3}H-counts were
calculated to be incorporated per cell in 20 min (∿ 6 x
10^5 deoxynucleotides), comparable to values given in the
literature. The ^{3}H-value was corrected for variation in
the number of cells on the disc by the ^{14}C-value.
Insert: Inhibition of E.coli DNA synthesis as a function
of antibody concentration. Incubation was carried out for
20 min.

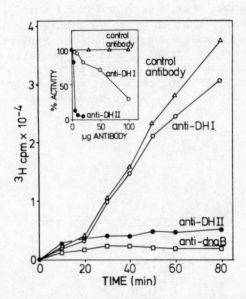

Fig. 2. Kinetics of ColE1 DNA synthesis. Synthesis was measured
         in the presence of [³H]dTTP (500 Ci/mole) in 18 μl E.coli
         extract, 0.25 μg DNA and 10 μg anti-DHI or 10 μg anti-DHII
         or 30 μg control antibody or 3 μg anti-dnaB at 30°C. One
         molecule of specific anti-DHII antibody inhibited one DNA
         helicase II molecule. Inhibition of DNA synthesis was re-
         versed by preadsorbing the antibody to purified DNA heli-
         case II.
         Insert: Inhibition of ColE1 DNA synthesis as a function
         of antibody concentration. Incubation was for 30 min.

         In interpreting these results we consider the two types of
replication of these molecules. Like the E.coli chromosome (14)
λ DNA and ColE1 DNA initiate replication as closed circles (15,
16) while the replication of fd RF (9) (as observed for φX174 RF)
(8) occurs in open (rolling) circles. Our results would therefore
suggest that the initiation of replication in closed circles involves the activity of DNA helicase II while the initiation in
open circles the rep protein. λ DNA replication during the late
phase of viral development appears to be an exception here: it
replicates during this phase as a rolling circle. However the
break in this molecule serves to convert the early closed circle
intermediate into a late rolling circle (15), rather than to initiate replication as in fd RF. λ replication is therefore uniformly initiated as closed circles.

The role of DNA helicase II and <u>rep</u> in the initiation process
is discussed as follows: DNA helicase II needs a single-stranded
region of 12 nucleotides to initiate unwinding (3). DNA polymerase
I-catalysed synthesis of a "DNA leader fragment" known to initiate
the replication of the superhelical ColE1 DNA (16) presumably leads
to the displacement of a single-stranded loop and thus would provide a site for DNA helicase II. With <u>E.coli</u> DNA and λ DNA one of
the processes of "transcriptional activation" preceding their replication might serve to expose an initiating site for the enzyme.

In contrast, <u>rep</u> protein can initiate unwinding starting at a
nick - provided the enzyme is acting in conjunction with the site-
specific endonucleases of phages ⌀X174 and fd (4,9). Such a mode of
action would explain why fd RF does not replicate in <u>rep</u> cells, i.e.
why <u>rep</u> cannot be replaced by DNA helicase II. However there is no
obvious reason why <u>rep</u> cannot unwind the DNA for DNA helicase II,
when the latter is inactivated by antibody.

To explain this we consider the fact that DNA helicase II un-
winds DNA in a 5' to 3' direction of the enzyme-bound strand (3).
It should therefore be found on the strand lagging in DNA synthesis.
<u>rep</u> Protein unwinds in the opposite direction (3' to 5') (17), and
is thus located on the leading strand. This difference in their location at the replication fork could conceivably be the reason why
the two enzymes are not easily exchangeable. (Fig. 3).

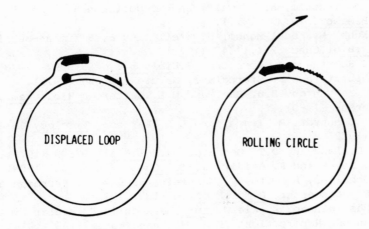

Fig. 3.   Schemes of DNA replication. The hook represents the 5' DNA
          terminus, the black dot DNA polymerase III. The thickened
          arrow represents the position and direction of action of
          DNA helicase II (in the displaced loop) or <u>rep</u> (in the
          rolling circle).

In addition to our general conclusion that DNA helicase II is active in replicative unwinding of the DNA, the quantitative aspect of the action of antibody against DNA helicase II at replication forks remains to be discussed. We observe strong inhibition of replication by anti-DHII antibody only when the replication is initiated in vitro, as can be done with λ DNA (using the λ P mutant) and ColE1 DNA. On the other hand replication is only partially inhibited (50%) when studied as elongation of in vivo initiated E.coli or λ DNA. We consider the possibility that part of this residual elongation synthesis (which can be completely inhibited by anti-dnaB antibody) occurs on in vivo unwound DNA. On the other hand, because of the relatively high level of residual synthesis measured (∿ 3 x 10^5 deoxynucleotides per cell in 20 min), other unwinding factors might well be active at the fork. These are probably not DNA helicase I or rep since inactivation of these enzymes by antibody or rep mutation does not increase the inhibition achieved with anti-DHII antibody.

In summary we conclude that while DNA helicase II plays an important role in the replicative unwinding of the DNA in closed circle replication, the enzyme alone is probably not responsible for this process. What the additional factors functioning in replicative unwinding are remains to be seen.

REFERENCES

1. M. Abdel-Monem, H. Dürwald, and H. Hoffmann-Berling, Eur.J. Biochem. 65:441 (1976).
2. M. Abdel-Monem, H. Dürwald, and H. Hoffmann-Berling, Eur.J. Biochem. 79:39 (1977).
3. B. Kuhn, M. Abdel-Monem, H. Krell, and H. Hoffmann-Berling, J.Biol.Chem. 254:11343 (1979).
4. J.F. Scott and A. Kornberg, J.Biol.Chem. 253:3292 (1978).
5. B. Alberts and R. Sternglanz, Nature 269:655 (1977).
6. G.T. Yarranton, R.H. Das, and M.L. Gefter, J.Biol.Chem., in press (1980).
7. D.T. Denhardt, M. Iwaya, and L.L. Larison, Virology 49:486 (1972).
8. H.E.D. Lane and D.T. Denhardt, J.Mol.Biol. 97:99 (1975).
9. T.F. Meyer and K. Geider, J.Biol.Chem. 254:12636 (1979).
1Q. H. Schaller, B. Otto, V. Nüsslein, J. Huf, R. Herrmann and F. Bonhoeffer, J.Mol.Biol. 63:183 (1972).
11. V. Nüsslein and A. Klein, in: "Methods in Molecular Biology on DNA Replication", R.B. Wickner, ed., 7:163 (1974).
12. A. Klein, B. Bremer, H. Kluding, and P. Symmons, Eur.J.Biochem. 83:59 (1978).
13. W.L. Staudenbauer, Molec.Gen.Genet. 145:273 (1976).
14. J. Cairns, Cold Spring Harbor Symp.Quant.Biol. 28:44 (1963).
15. A.M. Skalka, Curr.Topics Microbiol.Immunol. 78:201 (1977).

16. W.L. Staudenbauer, <u>Curr. Topics Microbiol.Immunol.</u> 83:93
        (1978).
17. G.T. Yarranton and M.L. Gefter, <u>Proc.Natl.Acad.Sci.USA</u> 76:1658
        (1979).

DNA Replication in Cells with Polytene Chromosomes

Ulf Lönn

Department of Histology, Karolinska Institutet

S-104 01 Stockholm , Sweden

INTRODUCTION

To investigate whether the structural features of the chromosome may determine the size of the replication units the polytene chromosomes of Diptera offer exceptional experimental possibilities. A polytene chromosome consists of a large number of homologous chromatids located side by side and is characterized by a specific and constant band- interband structure (1). This band- interband pattern corresponds on the chromatid level to a pattern of repeated units consisting of a region of highly compacted chromatin (the chromomere) followed (or preceded) by a non-compact region (the interchromomere). Up to 95% of the total DNA is located in the chromomere.

The size of the replication units are most commenly determined using fiber autoradiographic methods. However, a novel aproach is reported by Taylor and co-workers who have shown that it is possible to release a double-stranded DNA fragment from the active replication units of the chromosome (2,3). The average size of these DNA fragments is half the size of the average replication unit. We describe here the release of double-stranded DNA fragments from the polytene chromosomes and discuss the results in relation to an earlier proposal that each band in the polytene chromosome may function as a separate replication unit.

MATERIALS AND METHODS

Larvae of Chironomus tentans were raised in the laboratory as described earlier (4). Late fourth instar larvae 8 week old and weighing about 25 mg were used. For labelling of the salivary gland DNA,

the animals were injected into the blood with 20 uCi of tritiated
thymidine (27 Ci/mmol) dissolved in 1ul of 0.67% NaCl, 0.04%
KCl. The excised salivary glands of one animal were immeadatly covered with 75 ul of a solution containing 0.05M Tris-HCl (pH 7.4),
0.002M Na$_2$EDTA, 1% SDS. After 3 hours at +25^{0}C the labelled DNA was
analyzed either by agarose gel electrophoresis or CsCl-gradient centrifugations. For further details see (5).

RESULTS AND DISCUSSION

   To investigate whether nascent DNA can be released from the polytene chromosomes during cell lysis the following experiments were
performed. The animals were injected with tritiated thymidine and
then sacrificed 30 min later. The salivary gland DNA was extracted
at +25^{0}C with the neutral non-denaturing buffer as described in Materials and Methods and then separated by gel electrophoresis in a
0.75% agarose flat bed gel. The results showed a heterogeneous population of labelled DNA molecules located between slices 33 - 47
(Fig 1a). There is no or little labelled material at slices 4 - 6
which is the location of stable and/or steady-state labelled
DNA. Four hours after precursor injection the electrophoretic profile has changed mainly insofar that apart from the labelled material located at slices 33 - 47, there is also labelled material close
to the trough at slices 4 - 6 (Fig 1b). Finally forty hours after
precursor injection the main part of the radioactivity is located
at slices 4 - 6 (Fig 1c). Since the labelling procedure used here
has the character of a pulse (5) the results indicate that the DNA
fragments located at slices 33 - 47 after a time-lag are joined together to produce a high-molecular weight DNA. Moreover, CsCl-gradient centrifugations were used to establish that the DNA fragments
located at slices 33 - 47 are a double-stranded DNA (Fig. 1d). The
release of the fragments from the polytene chromosome is probably
enzymatic since it can be prevented by lysing the cells at 0^{0}C
instead of at +25^{0}C (See 5).
   The size of the double-stranded DNA fragments range between
3.75 - 6x10^6 daltons with an average value of 4.75 x10^6 daltons(5).
This corresponds to a mean replication unit size of 28 kilobases.
This value is the same as reported by Blumenthal et al for Drosophila non-polytene chromosomes using fiber autoradiographic methods (6).
   Early autoradiographic investigations revealed a positive correlation between DNA content in a band and the duration of DNA replication in the same band (7). From this finding it was inferred that
each band may function as a separate replication unit. However,
later autoradiographic investigations showed that in certain situations the time required for DNA synthesis in a band may be less
than that predicted from the DNA content (8). In Chironomus the
average chromomere contains about 100 kilobases of DNA (9) suggest-

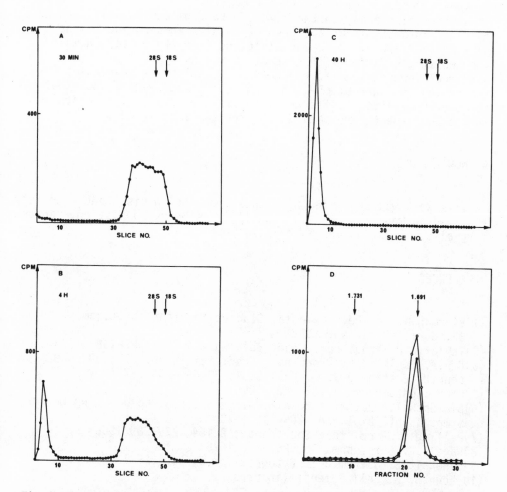

Fig.1 Electrophoretic separation of labelled salivary gland DNA. Larvae were injected with 20 uCi tritiated thymidine and sacrificed 30min (a), 4h (b) or 40h (c) later. The salivary gland DNA was extracted at +25°C and then analyzed in 0.75% agarose flat-bed gels. Arrows indicate the position of marker Chironomus 28 S and 18 S ribosomal RNA. (d) CsCl-gradient centrifugation. The animals were sacrificed either 30 min or 40h after precursor injection. The arrow to the left in-indicate the position of Micrococcus DNA (1.731 g/cm³). The labelled Chironomus DNA banded in the same position in the gradient (1.691g/cm³) in the two samples. -o- 30 min, -•- 40 h.

ing that several replication units of the size described here can
be located side by side in the chromomere.

We have also analyzed the size of the replication units in the
polytene chromosomes at two developmental stages during the development
of the larvae (red-head stage and late 4th instar animals).
The two stages differ with respect to intensity of DNA replication
with much lower values at red-head stage. Also the level of poly-
tenization is lower at the red-head stage. The results show that ther
is no or very little differences in size of active replication units
in the polytene chromosomes during the two stages (10).

ACKNOWLEDGEMENTS

This work was supported by grants from the Swedish Cancer Society,
the Swedish Medical Research Council,Åke Wibergs Foundation, Magnus
Bergwalls Foundation and Karolinska Institutet.

REFERENCES

(1)Beerman,W. In: Developmental Studies in. giant chromosomes. pp 1-
    33. Springer Verlag. 1972
(2)Taylor,J.H. Proc. nat. Acad. Sci (Wash.) 70, 1083-1087 1973.
(3)Guy,A.,Taylor,J.H. Proc. nat. Acad. Sci (Wash.) 75,6088.6092 1978
(4)Lönn,U.,Edström,J-E. J. Cell Biol. 73,696-704 1977.
(5)LönnU. Chromosoma 77, 29-40 1980.
(6)Blumenthal,A.B.,Kriegstein,H.J.,Hogness,D.S. Cold Spring Harbor
    Symp. quant. Biol. 38, 205-223 1973
(7)Pelling,C. Proc. roy. Soc.(Lond.) B 164, 279-289 1966
(8)Hägele,K. Chromosoma 55, 253-258 1976.
(9)Daneholt,B.Edström,J-E. Cytogenetics 6, 350-356 1967
(10)Lönn,U. J. Cell Science (in press).

ACCURACY OF DNA REPLICATION: EFFECTS OF LESIONS IN THE TEMPLATE

ACCURACY AND AGEING

Robin Holliday

National Institute for Medical Research

The Ridgeway, Mill Hill, London, NW7 1AA

INTRODUCTION

It is taken for granted that present day organisms have a variety of devices to ensure that macromolecules are made accurately and also that, in the case of DNA, structural integrity is maintained by elaborate repair mechanisms. Yet this cannot always have been the case, since primitive organisms must have been far less accurate. Increased accuracy and repair would confer a selective advantage, leading eventually by evolution to the situation we now observe. However, the avoidance or elimination of errors requires the consumption of metabolic energy. This is well known in the case of DNA synthesis, where errors can be removed by a proofreading exonuclease mechanism, and each wrong nucleoside triphosphate is converted to a monophosphate. The synthesis of repair enzymes is also energy-consuming. Less is known about the mechanisms which ensure accurate protein synthesis, but it is likely that ribosomes and associated factors not only allow the assembly of polypeptide chains, but also effectively discriminate against the incorporation of wrong amino acids, perhaps by a kinetic proof reading mechanism or its equivalent (1,2). Another way of increasing accuracy is to reduce the rate of synthesis of macromolecules, since this allows more time for the dissociation of incorrect substrate-enzyme complexes (3). There is evidence that mutations in ribosomal proteins which increase accuracy also reduce the rate of translation (4). There is also much evidence that when defective proteins are synthesised, they are preferentially degraded (5,6).

The general principle emerges that organisms have to strike a balance between accuracy and energy expended, or accuracy and speed

of synthesis.   It is clear that in the evolution of mechanisms to
eliminate errors, the law of diminishing returns applies.   There
must come a point where it is no longer advantageous to improve the
fidelity of information transfer between macromolecules and the
conclusion can therefore be drawn that there must always be some
mistakes in the synthesis of these molecules.   With regard to DNA
synthesis, it is also worth pointing out that complete accuracy
would eliminate mutations and therefore make it impossible for any
such organism to evolve further.

ERROR PROPAGATION

     Since some errors in macromolecules will always be present, it
is very important to consider their possible effects on cells or
organisms.   Orgel (7) first pointed out that a proportion of
errors will be in proteins which are themselves essential for
fidelity in information transfer.   If these errors lead to a loss
of specificity, then they may introduce further errors in macromolecular synthesis, and particularly in proteins.   Thus, organisms
are potentially unstable, since error feed-back could increase the
level of mistakes with time or in successive cell generations,
until a lethal "error catastrophe" occurs.   The important parameter
is the feed-back factor (8).   If this is less than the intrinsic
or spontaneous error level (i.e. the number of errors introduced
after one cell generation, starting with a hypothetical error-free
cell), then a steady state level of errors is reached.   If the
feed-back factor is greater than the intrinsic error level, then
the errors will increase exponentially and eventually lead to cell
death.   An increase in errors through feed-back is referred to as
error propagation.   Kirkwood (9) has recently reviewed various
theoretical models which consider the possible significance of
error propagation in biological systems.

     Orgel's theory is a simplification, since it assumes that the
feed-back factor is independent of the intrinsic error level.   It
is more realistic to suppose that there is a critical intrinsic
error level, below which cells can grow and survive indefinitely,
but above which they are subject to escalating error propagation
and will eventually die.   This is illustrated in Figure 1.
Experiments with Neurospora crassa (10) and Escherichia coli (11,12)
strongly suggest that increasing the normal error level in protein
synthesis lead, after many cell generations, to a lethal error
catastrophe.

THE PROCESS OF AGEING

     It is clear that most micro-organisms must have a steady level
of errors, since they can be grown indefinitely.   This is also
likely to be true for germ line cells in sexually reproducing
organisms, since otherwise any such species could not survive.

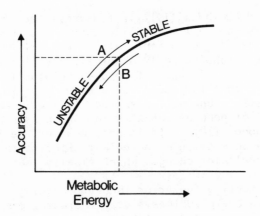

Fig.1.The likely relationship between the accuracy of synthesis
       of macromolecules (or their maintenance by repair) and
       the energy expended.   Below the point indicated by the
       dashed lines, organisms may be unstable owing to error
       feed-back (7).   Primitive organisms presumably evolved
       by natural selection (arrow A) to produce non-ageing
       species.   Somatic cells of higher organisms, which age,
       may have returned (arrow B), or have a given probability
       of returning, to the unstable situation (13,14).

However, the situation may be quite different for somatic cells.
As we have seen, organisms have to strike a balance between accuracy
and expenditure of energy.   The biological function of the soma
is, in effect, to ensure the propagation of germ cells, or more
specifically, the DNA of the organism in question (15).   In
natural environments individuals have a fairly short life expect-
ancy, since they usually die from predation, disease or starvation,
rather than "old age" (16).   It is, therefore, most advantageous to
develop and reproduce quickly and of little importance to maintain
the soma indefinitely.   Therefore, we might expect accuracy to be
relaxed in somatic cells, in order to ensure maximum reproductive
capacity.   With regard to the process of ageing, the crucial
question is whether accuracy falls below the critical level,
where error feed-back becomes significant.   If it does, then
somatic cells or tissues will necessarily have a finite life span.
Ageing would then be attributed to a build up of errors in essential
macromolecules (7).   Whereas primitive organisms must have evolved
by natural selection from the unstable to the stable state (Fig.1),
the "disposable soma" theory of the evolution of ageing proposes
that the somatic cells of higher organisms have reverted from the
stable to the unstable state (13,14).

FIDELITY OF DNA POLYMERASE DURING AGEING

Hayflick and Moorhead (17) first demonstrated that diploid
human fibroblasts have a limited life span in culture. After a
long period of normal growth, the rate of cell division slows down,
the cells become morphologically abnormal, and finally the cultures
die out after a total of about 60 population doublings. The exact
relationship between this process of ageing in vitro and the ageing
of whole organisms is unclear, although it is well known that the
growth potential of populations of skin fibroblasts decreases with
the age of the donor (18). It is not, however, seriously in doubt
that these cells do provide a convenient experimental system to
study intrinsic cellular changes which finally lead to cell death.

One way of testing the error theory of ageing is to measure
the accuracy of protein synthesis at different times during the
life span. It is very difficult to do this directly, because
methods currently available for detecting the random misincorporation
of amino acids are not very sensitive (19). Several indirect
methods have been used, some of which have produced results which
support the error theory; however, these will not be reviewed here.
Another approach is to examine the specificity of an enzyme which
is essential for one of the pathways of information transfer. DNA
polymerase is an obvious candidate, since methods are well established for measuring the fidelity of replication of defined templates
in vitro. Previously, Linn et al (20) showed that DNA polymerase
α from young cultures of human foetal lung fibroblast strain MRC-5
was 2-10 times more accurate than the same enzyme from late passage
senescent cultures. In these experiments the cells were disrupted
by sonication and the enzyme recovered had a lower sedimentation
coefficient (5S) than expected for DNA polymerase α. Also, in
most of the measurements the divalent cation Mn^{2+} was used, since
this amplifies error levels.

Further experiments have now been carried out by Murray (21),
in which cells were gently disrupted with a homogeniser to prevent
the breakage of nuclei and the DNA polymerase α recovered had the
expected 7S sedimentation coefficient. (It is likely that the
active 5S form is produced by proteolytic cleavage.) In these
experiments, the enzyme was also examined at various intervals
throughout the life span of MRC-5, and, in addition, a procedure
for greatly reducing background counts made it possible to use the
physiological cation, Mg^{2+}. With poly[d(A-T)] as template, normal
synthesis with [^{3}H]dTTP and dATP is compared with the misincorporation assay, in which the two complementary substrates are unlabelled and the non-complementary [^{3}H]dGTP is added. Using Mg^{2+}
as cation, there was a highly significant increase in enzyme error
levels throughout the life span, the overall increase being 2-5
fold in different experiments. A similar rate of increase was
found with Mn^{2+} as cation, but the error frequency was about five

times higher throughout.   With another template, poly[d(I-C)],
using Mg^{2+}, the misincorporation of [^{3}H]dTTP was measured.   The
error levels were less than with poly[d(A-T)], but again there was
a significant increase throughout the life span.   In each series
of experiments with enzyme from cells of different age, the results
strongly indicate that the increase in error level is exponential
rather than linear.

A number of essential controls have been carried out, using
the poly[d(A-T)]/dGTP/Mg^{2+} system.   For example, it was shown that
there was no substrate contamination by using DNA polymerase 1 from
E. coli, which is intrinsically more accurate in vitro than DNA
polymerase α.   The addition of nucleotides from terminal transferase activity was ruled out.   The time course of the polymerisation reaction was followed and it was shown that the misincorporation level remained constant.   This was also the case when the
enzyme concentration was varied.   Nearest neighbour analysis
demonstrated that the incorrect base, guanine, was incorporated
adjacent to thymine, as expected if it substitutes for adenine
during polymerisation.   Mixing experiments with enzyme from young
and old cells gave the expected intermediate misincorporation
values.   Finally, preliminary experiments with DNA polymerase γ ,
using poly (A) oligo (dT) as template, showed that this enzyme
also lost fidelity during fibroblast ageing.

It is known that mutations in bacteriophage T4 DNA polymerase
can reduce fidelity (22), and it would be expected that random
amino acid substitution in DNA polymerase from transcriptional or
translational errors would, overall, reduce the specificity of DNA
polymerase α.   An exponential increase in misincorporation frequency
is predicted, if error propagation continues throughout the life
span.   An alternative possibility is that an altered error-prone
form of the enzyme is being produced in increasing amounts as the
cells age.   It is hard to see why this should happen, and it also
seems very unlikely that the specificity of two polymerases would
be altered in the same way.

INCREASE IN MUTATIONS DURING AGEING

A decline in the accuracy of DNA polymerase would also result
in an increase in the frequency of gene mutations.   A previous
study provided evidence that this is indeed the case with MRC-5,
since the number of cell variants containing elevated levels of
the enzyme glucose-6-phosphate dehydrogenase increased exponentially
throughout the culture life span (23,24).   New experiments have now
been carried out by Morley et al (25) in which the frequency of
mutations in T lymphocytes has been measured throughout the human
life span.

Lymphocytes from 10 ml blood samples were isolated, stimulated with phytohaemaglutinin and treated with the purine analogue, 6-thioguanine (6TG) for 40 hours.   Normal cells incorporate 6TG and are killed, but occasional mutant cells lack the enzyme hypoxanthine guanine phosphoribosyl transferase and cannot incorporate the analogue into nucleic acids.   Their de novo synthesis of purines allows them to complete a round of DNA synthesis, and the mutants can therefore be scored by the incorporation of ^{3}H thymidine into DNA, using autoradiography.   In all, 37 measurements were made from donors between 9 and 95 years of age.   The mutation frequency increased from 4×10^{-6} to 7×10^{-5} over the 86 year range.   (The frequency of mutations in young individuals is considerably lower than that reported by Strauss and Albertini (26), using a similar method).   The data give a better fit to an exponential increase than a linear one, with a 3% increase in mutation frequency per year.   This result is expected if there is an accumulation of errors in macromolecules by a feed-back mechanism.   Simple accumulation of DNA damage would be expected to give a linear increase in mutations.

THE SOMATIC MUTATION THEORY OF AGEING

The somatic mutation theory predates the general error theory of ageing.   It proposes that genetic damage accumulates with time until the load of mutations becomes so high that cells can no longer survive.   The first mathematical formation of the theory was by Szilard (27), who assumed that "hits" on genetic targets were recessive and that cell viability was lost when hits occurred on the same target in homologous chromosomes.   For man, it turned out that it was only possible to match the theoretical and observed distributions of life span, and to explain the small number of inherited hits, if the number of targets was quite small.   Szilard was therefore forced to conclude that targets are either whole chromosomes or large parts of them.

It was also suggested by Hayflick (28) that the ageing of human fibroblasts might be due to a multi-hit process, but no quantitative treatment of this hypothesis was developed.   We have, therefore, constructed a modified version of Szilard's theory to see whether it is compatible with the ageing of fibro-blast cultures.   The following assumptions are made: (i) targets for mutational hits are single essential genes (i.e. indispensible for growth) and we assume the number of these is in the range $10^3 - 10^4$; (ii) 95% of these genes are on the autosomes, so they exist in duplicate copies, while the remaining 5% are on the X-chromosome and exist only as single active copies; (iii) auto-somal hits are recessive; (iv) the hit rate per target per cell generation is constant in all cells in the culture and at all times; (v) all viable cells divide at the same rate.   We have calculated the probability that a cell is rendered non-viable by a lethal hit in any one generation (29).   The main conclusions are as follows:

If there are 10^4 essential genes, then to explain the dying out of
fibroblast populations after 50-60 population doublings, the
mutation rate per gene per generation must be as high as $3-4 \times 10^{-4}$.
This is considerably higher than the estimated rate of mutation in
the human germ line (30) or the observed frequency in lymphocytes
(25) or in cultured human fibroblasts (31).   (The calculated mutation rate is even higher if there are only 10^3 essential genes.)
The theory also predicts that cell viability will be as low as 50-
60% in early passage cultures.   Several studies demonstrate that
this is not the case (32,33,34).   An additional strong argument
against the mutation theory comes from experiments in which the
longevity of diploid and tetraploid fibroblast cells is compared.
Two independent investigations showed that the life span of the two
types of cell is similar (35,36), whereas the mutation theory of
ageing would predict a considerably increased longevity for
tetraploid cells.

CONCLUSIONS

From these results it can be concluded that ageing, at least
in fibroblasts, is unlikely to be primarily due to a simple
accumulation of mutations in DNA, linearly with time or during
successive cell divisions.   There is probably a fundamental
distinction between the intrinsic cellular changes which bring about
ageing and the final causes of cell death.   It is certainly quite
possible that the latter are due to deleterious gene mutations.
An exponential increase in mutations would initially have very
little effect on cell populations, but eventually all the cells
may well lose at least one essential gene function.   It is then
necessary to explain why mutations should increase in this way.

The general error theory of ageing takes into account all
pathways of information transfer between macromolecules (37,38),
and the basic postulate is that errors are themselves responsible
for a significant increment of errors in successive generations of
these molecules.   The theory proposes that cells which do not age
have adjusted to a steady state level of errors, whereas those that
do have entered an irreversible downward path, which leads eventually
to a lethal error catastrophe.   The steady state could, however,
be metastable, with a given probability that any cell would cross
a critical threshold and begin to propagate errors (14,39 and Fig.
1).   Since it is probable that random amino acid substitutions
in proteins are much more common than base changes in DNA (perhaps
by a factor of about 10^4), it may well be that protein error feed-
back constitutes the primary cause of ageing, as originally suggested
by Orgel (7).   This will affect an increasing number of DNA polymerase molecules or other proteins required for accurate replication
or repair.   As a consequence, there will be an escalating rate of
increase in defects in DNA, which could either contribute to or be
the major cause of cell death.

REFERENCES

1. J.J. Hopfield, Proc. Nat. Acad. Sci. U.S.A. 71: 4135 (1974)
2. A.B. Caplan and J.R. Menninger, J. Mol. Biol. 134: 621 (1979)
3. J. Ninio, Biochemie 57: 587 (1975)
4. D.J. Galas and E.W. Branscomb, Nature, London. 262: 617 (1976)
5. A.L. Goldberg and J.F. Dice, Biochem. Rev.43: 835 (1974)
6. A.L. Goldberg and A.C. St. John, Biochem. Rev. 45: 747 (1976)
7. L.E. Orgel, Proc. Nat. Acad. Sci. U.S.A. 49: 517 (1963)
8. L.E. Orgel, Proc. Nat. Acad. Sci. U.S.A. 67: 1476 (1973)
9. T.B.L. Kirkwood, J. Theoret. Biol. 82: 363 (1980)
10. C.M. Lewis and R. Holliday, Nature, London. 228: 877 (1970)
11. E.W. Branscomb and D.J. Galas, Nature, London. 254: 161 (1975)
12. R.F. Rosenberger, G. Foskett and R. Holliday, Mech. Age. Devel
        (in press)
13. T.B.L. Kirkwood, Nature, London. 270: 301 (1977)
14. T.B.L. Kirkwood and R. Holliday, Proc. Roy. Soc., London B.
        205: 531 (1979)
15. R. Dawkins in The Selfish Gene, Oxford University Press (1976)
16. P.B. Medawar, in An Unsolved Problem in Biology, H.K. Lewis,
        London. (1952)
17. L. Hayflick and P.S. Moorhead, Exp. Cell Res. 25: 585 (1961)
18. G.M. Martin, C.A. Sprague and C.J. Epstein, Lab. Invest. 23:
        86 (1970)
19. J.D. Buchanan and A. Stevens, Mech. Age. Devel. 7: 321 (1978)
20. S. Linn, M.V. Kairis and R. Holliday, Proc. Nat. Acad. Sci.,
        U.S.A. 73: 2818 (1976)
21. V. Murray, Changes in DNA polymerase during ageing, PhD. thesis
        Council for National Academic Awards, U.K. (1979) and in
        preparation.
22. J.F. Speyer, J.D. Karam and A.B. Lenny, Cold Spr. Harb. Symp.
        Quant. Biol. 31: 693 (1966)
23. S. Fulder and R. Holliday, Cell 6: 67 (1975)
24. S. Fulder, Mech. Age. Devel. 10: 101 (1979)
25. A. Morley, S.A. Cox and R. Holliday (submitted for publication)
26. G.H. Strauss and R.J. Albertini, Mut. Res. 61: 353 (1979)
27. L. Szilard, Proc. Nat. Acad. Sci., U.S.A. 45: 30 (1959)
28. L. Hayflick, Exp. Cell Res. 37: 614 (1965)
29. T.B.L. Kirkwood and R. Holliday in Structural Pathology of
        DNA and the Biology of Ageing, L. Schoeller, ed., Freiburg
        Deutsche Forschungsgemeinschaft (in press)
30. L.L. Cavalli-Sforza and W.F. Bodmer, in The Genetics of Human
        Populations, W.H. Freeman, San Francisco (1971)
31. R. DeMars, Radiation Res. 24: 335 (1974)
32. V.J. Cristofalo and B.B. Sharf, Exp. Cell Res. 76: 419 (1973)
33. R.A. Vincent and P.C. Huang, Exp. Cell Res. 102: 31 (1976)
34. R. Holliday, L.I. Huschtscha and T.B.L. Kirkwood (in prepara-
        tion)
35. K.V.A. Thompson and R. Holliday, Exp. Cell Res. 112: 281 (1978

36.   H. Hoehn, E.M. Bryant, P. Johnston, T.H. Norwood and G.M.
            Martin, Nature, London. 258: 608 (1975)
37.   L.E. Orgel, Nature, London. 243: 441 (1973)
38.   R. Holliday and G.M. Tarrant, Nature, London. 238: 26 (1972)
39.   T.B.L. Kirkwood and R. Holliday, J. Mol. Biol. 97: 257 (1975)

HIGH FIDELITY DEVICES IN THE REPRODUCTION OF DNA

M. Radman, C. Dohet, M.-F. Bourguignon, O.P. Doubleday
and P. Lecomte

Département de Biologie Moléculaire
Université Libre de Bruxelles, rue des Chevaux 67
B - 1640 Rhode-St-Genèse, Belgium

INTRODUCTORY COMMENTS ON GENETIC STABILITY

The basic characteristics of genetic material that ensure the
long term population fitness and biological evolution are its
stability and variability in the course of reproduction. This
apparent paradox ceases to be the paradox as soon as we start asking
the questions when, how much and what kind of genetic variability is
required for fitness. It seems trivial to speculate that a maximal
genetic stability would be required in populations perfectly adapted
to their environment, whereas an appreciable genetic variability
would be required for adaptation to new selective growth conditions.
As to the kind of genetic variability, both theoretical considerations (Leigh, 1973) and a few experimental approaches suggested that
random mutagenesis is the main route in promoting fitness under
selective conditions for vegetatively reproducing haploid organisms,
such as bacteria (Cox and Gibson, 1974), whereas gene rearrangements
through recombinations, but not random mutagenesis, is the effective
mechanism of fitness-increasing variability in sexually reproducing
diploid organisms, such as Drosophila (Ayala, 1967 and refs.
therein).

We have learned, in the last couple of decades, that genetic
stability and variability are under genetic controls, since numerous
mutants have been isolated which are less or more genetically stable
than the wild type organism, e.g. mutators, antimutators, recombination deficient and hyper-recombination mutants, and mutants that
are refractory to the mutagenic effects of various powerful mutagenic agents (reviews, Cox 1976, Radman et al. 1978).

This paper is a review, seasoned with two original experimental contributions, dealing with the molecular mechanisms involved in the accuracy of DNA replication. It will become evident that in addition to the hypothesized kinetic proofreading mechanisms, such as branched energy-driven reactions (Hopfield 1974), delayed reactions (Ninio 1975) and the hydrolytic destruction of the incorrect intermediary product (Brutlag and Kornberg 1972), even the final replication product (double stranded DNA) is still subject to highly specific and efficient error-correction mechanisms (Radman et al. 1978, 1980).

Table 1 lists our current knowledge and thinking about the mechanisms of replicational fidelity in E.coli, and the following text presents comments to Table 1.

NON-ENZYMATIC MECHANISMS

When the difference in the free energy of binding of complementary and non-complementary nucleotides is translated into error frequency in non-enzymatic copying of DNA ($-\Delta G = RT \ln$ correct/incorrect), one to ten percent error is expected (review, Loeb et al 1978) and indeed found in some experiments (Orgel and Lohrmann 1974). Replication models which assume incorporation of pre-paired nucleotide pairs, for which there is still little experimental evidence (A.R. Morgan, this volume), imply a higher contribution of non-enzymatic nucleotide selectivity due to a double base pair checking.

REPLICATIVE PROOFREADING MECHANISMS

DNA Polymerases

Base selection. It has been demonstrated, both genetically and biochemically, that DNA polymerases actively contribute to the selection of correct nucleotides. At least one mutator mutant in phage T4 DNA polymerase gene affects nucleotide selection but not the 3' to 5' exonuclease (Hershfield 1973). DNA polymerase bound to its template-primer substrate shows increased affinity for the correct nucleotide triphosphate (Gillin and Nossal 1976). The efficiency of base selection can be affected by (1) the alterations of the relative concentrations of dNTP pools (Fersht 1979, Hibner and Alberts 1980, review Loeb et al, 1978), (2) the presence of divalent cations competing with Mg^{++}, such as Mn^{++} (Loeb et al 1978, Hibner and Alberts 1980) and Be^{++} (Loeb et al 1978, Radman et al 1977) and (3) by the polymerisation rate which can be affected e.g. by changes in the relative concentrations of dNTPs and PPi. The latter point was discussed by Ninio (1975) in his time-delay scheme "slow is accurate" which however was not supported by an analysis of 2-aminopurine misinsertion by T4 DNA polymerase (Clayton et al. 1979). Our results (Figure 2) show that the decrease in the

Table 1. Replicational and postreplicational fidelity mechanisms in E.coli

Mechanism involved	Cumulative error frequency	Affected by	Respective mutators	Epigenetic effects
A) Non-enzymatic	~10^{-2}-10^{-1}	base modification	-	-
B) Replicative proofreading by (1) DNA polymerase a) base selection		(a) -dNTPs pool ratio -cations Mn^{++},Be^{++} etc. -speed (dNTP, PPi conc.)	T4L88(*) dnaE (polC) dnaQ mutT mutD	SOS induction modifies some of the B mechanisms to produce mutator effect (***)
b) 3'→5' exonuclease	~10^{-6}-10^{-5}	(b) -exo/pol ratio -dNMP conc. -speed	T4L56(*) L98 T4L42,L141 CB120(**)	
(2) Accessory proteins (ssb)	~10^{-7}	?	(?)ssb,lexC T4 gene 32(*)	
C) Postreplicative mismatch correction	~10^{-10}	-Methylation of the 5'GA^{me}TC 3' sequence -mismatch chemistry(Py-Pu, frameshifts)	dam mutH mutL mutS mutU	No evidence (***)
		-mismatch concentration (saturation effects)	mutD(?)	

References to A are Orgel and Lohrmann (1974) and reviews by Topal and Fresco (1976) and Loeb et al.(1978) ; to B1(a) are Hershfield (1973), Gillin and Nossal (1976), Clayton et al.(1979), reviews Loeb et al.(1978), Radman et al.(1978), Bernardi and Ninio (1978) and to B1(b) are Clayton et al.(1979) and references therein, Fersht (1979), Hibner and Alberts (1980) and Bernardi et al.(1979), to B2 Kunkel et al.(1979) and ref. therein and to C Nevers and

Spatz (1975), Radman et al.(1978), Radman et al.(1980), Rydberg
(1978), Glickman and Radman (1980), Bale et al.(1979). The references
to mutator mutants : review Cox (1976), Rydberg (1978), Radman et al.
(1978), Glickman and Radman (1980).

(*) Phage T4 mutants mapping in the gene 43 are biochemically well
characterized : L56 and L98 have a decreased exo/pol ratio (Clayton
et al. 1979 and ref. therein).

(**) These T4 antimutator mutants have an increased exo/pol ratio,
either because of an increased exonuclease activity (L42 and L141)
or a decreased polymerase activity (CB120) as referenced in Clayton
et al.(1979).

(***) SOS induction (Radman 1974) occurs following damage to DNA
and triggers, among other phenotypes, a transitory mutator effect
(review, Witkin 1976). The best evidence, although still indirect,
that SOS induction modifies some of the B fidelity mechanisms is by
Caillet-Fauquet et al.(1977) and review by Radman et al.(1978).
C fidelity mechanisms do not appear to be rate limiting in the SOS
induced mutagenesis in phage λ (G.Michel-Maenhaut and P.Caillet-
Fauquet, unpubl. observations).

polymerisation rate by an exonuclease-free DNA polymerase α is
directly proportional to the increase in its fidelity. This effect
however, applies only to specific DNA polymerases and specific base
pair mismatches (Lecomte, Doubleday and Radman, in prep. and below).

   3' to 5' exonuclease proofreading activity. An impressive
combination of genetic and biochemical studies of mutator and antimutator mutants affecting T4 DNA polymerase, shapes our current
understanding of the role of polymerase-associated 3' to 5' exonuclease in replicational fidelity (reviews, Alberts and Sternglanz
1977, Bernardi and Ninio 1978). The principal conclusion is that
the ratio of exonuclease/polymerase activities determines the
fidelity of DNA synthesis (Bessman et al. 1974, Clayton et al.1979).
Mutator and antimutator mutations merely decrease and increase
respectively the relative exonuclease/polymerase rates and do not
affect exonuclease specificity towards the terminal mismatch.
Therefore, the mutator and antimutator effects can be mimicked in
vitro either by manipulating the polymerisation rates (Clayton et
al. 1979) or by inhibiting the exonuclease activity (Que et al.1978).

   The proofreading function of the replicative E.coli enzyme,
DNA polymerase III, has been demonstrated by Fersht (1979) using
an assay analogous to the one used to define this activity in DNA
polymerase I (Brutlag and Kornberg, 1972): excision of the 3'OH
terminal mismatch as the prerequisite for subsequent chain elongation. However, the best assay to demonstrate the proofreading exonuclease activity in the course of replication is the DNA dependent
turnover of dNTPs to free dNMPs ("abortive incorporation")(Hershfield
and Nossal, 1972).

Two recent papers present a comprehensive analysis of replicational fidelity of the reconstituted E.coli (Fersht 1979) and T4
phage (Hibner and Alberts 1980) DNA replication complexes using the
phage ØX174 am ⟶ am+ reversion system detecting A→G transitions.
Compilation of their data suggests that the base substitution rate
at a given site is (1) directly proportional to the [incorrect]^2/
[correct] ratio of dNTPs up to the maximal polymerisation rate after
which there is a linear increase in error rates which is now
proportional to the [incorrect] / [correct] dNTPs and (2) directly
proportional to the next-to-misinserted complementary nucleoside
triphosphate corresponding to the "pulling effect" preventing the
efficient competitive exonucleolytic removal of the misincorporated
nucleotide.

Pyrophosphorolysis. This genuine reverse polymerisation reaction
removing the 3'OH terminal residue from the primer end in the form
of the triphosphate, (unlike 3' to 5' exonucleolytic removal of the
nucleoside monophosphate) was hypothesized as a possible proofreading mechanism (Bernardi et al. 1979). In spite of the lack of
experimental evidence, theoretical virtues of pyrophosphorolysis
as a proofreading mechanism will be discussed below.

Accessory Proteins

Weak mutator mutants mapping in the phage T4 gene 32, coding
for a DNA binding protein (Bernstein et al. 1972, Koch et al. 1976)
as well as the decrease in error rates during in vitro DNA synthesis
by the addition of E.coli single strand DNA binding protein (ssb)
(Kunkel et al. 1979), suggest that ssb, and probably other components
of the DNA replication machinery, contribute to the replicational
fidelity. This effect could be direct, e.g. alignement of the
template and prevention of base rotation in the template (see Lomant
and Fresco, 1975) by the phosphate-bound ssb proteins, or indirect,
by affecting the architecture of the multienzyme "replisome" complex.
It is therefore of considerable interest to know what kind of
mutations (base substitutions versus frameshifts) are affected by
ssb and other accessory proteins.

POST-REPLICATIVE MISMATCH CORRECTION SYSTEM

There is now good evidence that a post-replicative excision
repair, excising mismatched nucleotides exclusively from the newly
synthesized strand and copying the gap using parental strand as the
template, contributes by three orders of magnitude to the final
accuracy of DNA replication (Radman et al. 1978, 1980, Glickman and
Radman 1980). The discrimination between the old parental strand
and the new strand (carrying replicational errors) seems to be based
upon the transitory undermethylation of the newly synthesized strand,
as initially suggested by Wagner and Meselson (1976), and in
particular upon the undermethylation of the 5' GATC 3' sequences

(Radman et al. 1980). The latter sequence which occurs in E.coli
DNA roughly once per thousand nucleotides accounts for the vast
majority of adenine methylation which is dependent upon an intact
dam+ gene (Marinus 1973, Lacks and Greenberg 1977, Geier and Modrich
1979). The mutator phenotype of all dam⁻mutants (Bale et al. 1979)
and of mismatch repair deficient mutants, mutH, mutS, mutL and mutU
(uvrE) (Nevers and Spatz 1975, Rydberg 1978, Radman et al. 1978,
Glickman and Radman 1980), as well as their similar mutator
potencies and specificities, are all consistent with a mismatch
repair system correcting transition and frameshift type of replicational errors (see Glickman and Radman 1980).

    Direct support for this methylation-instructed error-correction
model came from experiments which have shown that the undermethyla-
ted DNA strand is preferentially mismatch repaired. The substrate
for mismatch correction in these experiments was phage λ heteroduplex
DNA, with strands varying both in adenine methylation and in heterozygous genetic markers (mismatches). These were transfected into
E.coli hosts and the progeny phage analysed for the frequency of
parental markers carried initially by the methylated (me+) and non-
methylated (me⁻) strands (Radman et al. 1980, Rykowsky, Pukkila,
Radman, Wagner and Meselson, in prep.; Dohet, Bourguignon and Radman,
in prep.). Transfection of the wild type or dam⁻ E.coli gave always
a bias in favour of the me+ strand genotype. Such bias was abolished
in mismatch repair deficient mutants (Radman et al. 1980).

    Table 2 shows the parental genome output from heteroduplex
transfections of all potent E.coli mutator mutants with me+/me⁻
heteroduplexes. The observed bias in favour of the initial me+
strand is an underestimated effect since (1) me+ strands of λ DNA
unlike E.coli DNA, are roughly only half methylated (Dreiseikelmann
et al. 1979 and unpublished experiments from Meselson's and our
laboratories), (2) there is a significant contamination of the
l-strand by its complementary r-strand, and vice versa, in the
course of strand separation, and (3) we can detect only mismatch
correction events that were completed before two competitive
processes occured : DNA methylation and replication. Note that
(1) heteroduplexes A and B have the same mismatch chemistry (i.e.
l(c+ P3)/r(cP+)) which is different from that of C and D (l(cP+)/
r(c+P3)), e.g. an A:C mismatch in A and B would correspond to a
G:T mismatch at that site in C and D, everything else being identical
and (2) A and D have the same methylation pattern (r(me+)/l(me⁻))
as opposed to B and C (r(me⁻)/l(me+)). We draw the following conclusions from the Table 2 :
(a) When me+/me⁻ and me⁻/me+ (the order refering to the methylation
of r and l strand respectively) λ DNA heteroduplexes are compared,
methylation always favours the representation of that genotype in
the progeny from transfected wild type E.coli (exp. 1 and 5).

Table 2.   Influence of DNA methylation and of mismatch chemistry
           upon mismatch correction in various mutator strains :
           λ heteroduplex transfections.

Exp.	Transfected E.coli strain	PERCENT PARENTAL GENOME OUTPUT			
		c^+P3	$c\ P^+$	c^+P3	$c\ P^+$
		A: heteroduplex $1\xrightarrow{+\ \ 3}me^-$ $r\xrightarrow[c\ \ +]{}me^+$		B: heteroduplex $1\xrightarrow{+\ \ 3}me^+$ $r\underset{c\ \ +}{\leadsto}me^-$	
1	NR3752, wild	2.4	94.8	96.5	2.0
2	NR3773, mutH101	15	75.6	10.4	64.8
3	NR3774, mutL101	36.4	59.9	45.8	49.3
4	NR3775, mutS101	29.3	63.0	47.0	50.5
		C: heteroduplex $1\xrightarrow{c\ \ +}me^+$ $r\underset{+\ \ 3}{\leadsto}me^-$		D: heteroduplex $1\underset{+\ \ 3}{\overset{c\ \ +}{\leadsto}}me^-$ $r\xrightarrow[+\ \ 3]{}me^+$	
5	NR3752, wild	0.6	97.0	78.7	16.6
6	GM33, dam 3	-	100	31.6	51.9
7	AB1157 uvrE502 (mutU)	0.3	21.3 ($c^+P^+=78.4$)	19.4	65.8
8	C600, mutT1	-	96.3	61.3	28.8
9	KD1079, mutD5	-	100	-	97.3

The methodology of strand separation and heteroduplex transfection
has been described elsewhere (Wagner and Meselson 1976, Radman et
al. 1980). Strains have been described by Glickman and Radman (1980)
and by Cox (1976).

(b) Since such a bias is absent in mutH, mutL and mutS mutator
mutants, which are deficient in mismatch repair (Rydberg 1978,
Radman et al. 1978, Glickman and Radman 1980), we assume that the
bias is provoked by mismatch repair.
(c) Mismatch chemistry determines the efficiency of correction
(compare het A, exp.1 and het D, exp.5) ; furthermore, the combination of the methylation pattern and of mismatch chemistry influences
correction efficiency (compare het C and het D).
(d) Mutators mutH, mutL and mutS are similarly deficient in repair
of both c/c^+ and $P3/P^+$ mismatches, whereas mutU (uvrE502) efficiently
corrects $1\ P^+(me^+)/r\ P3(me^-)$ mismatch but not $1\ c(me^+)/r\ c^+(me^-)$
mismatch, hence yielding an excess recombinant type c^+P^+ in C

heteroduplex transfections (exp.7).

(e) Dam⁻ and mutT-1 (Treffers et al. 1954) mutators are fully
proficient in mismatch repair (het C, exp.6 and 8).

(f) MutD5, the most potent mutator known (Fowler et al. 1974),
yields only the progeny of the l-strand, irrespective of methylation.
This complex phenomenon will be commented elsewhere by Dohet and
Radman. The lack of marker rescue by mismatch repair of the me⁻
strand before strand loss (presumably by replication) is evident
(het D, exp.9), suggesting the possibility that mismatch correction
may be inefficient in mutD5 mutant, perhaps as the result of the
saturation of mismatch repair system by a high amount of mismatches
in the replicating mutD5 DNA (experiments in progress).

Other Post-Replicative Correction Mechanisms. There are also
error avoidance mechanisms to prevent uracil entering or remaining
in DNA since dut (sof) mutants (deficient in deoxyuridinetriphospha-
tase) (Tye et al. 1977, Hochhauser and Weiss 1978) and ung mutants
(deficient in uracil-DNA-glycosidase) (Duncan et al. 1978) exhibit
a weak mutator effect.

Mutator Mutants and Mutation Rates

Figure 1 shows a current map of E.coli mutator genes. With the
exception of mutator effects of some dnaE (Kondrad 1978, Sevastopoulos and Glaser 1977) and polA mutants (Vaccaro and Siegel 1975)
as well as ung and dut mutants (already mentioned), for which there
is no reason to expect mismatch correction deficiency, only the
mutT1 mutant seems to be fully proficient in the methylation-
instructed mismatch correction (tif-1 and mutG mutators have not
been tested). As indicated in Table 1, DNA polymerase mutants and
the mutT and mutD mutators cause an increased misincorporation rate
during DNA synthesis. If mutD mutator effect, which causes all base
changes and frameshifts (review Cox, 1976), indeed saturates the
mismatch correction system, the failure of the mutT to do so could
be explained by the fact that mutT causes only the AT→ CG transversions (review Cox, 1976), which are not corrected by the
mismatch repair discussed above. When both replicative and post-
replicative error-correction mechanisms are considered, in vivo
mutation rates can be accounted for (Table 1).

Slow Replication may be a High Fidelity Device in Mammalian Cells

The average replication fork movement in mammalian cells was
estimated to be about one hundred times slower than in bacterium
E.coli (review, Edemberg and Huberman 1975), which incorporates
roughly a thousand nucleotides per second. What is the "reason"
to evolve a slowing-down of the replicational speed when the amount
of total cellular DNA has increased by a thousand-fold ? We propose
that one reason might be the achievement of the replicational
fidelity at no energy cost.

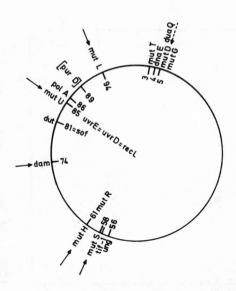

Figure 1. E.coli mutator gene map.

Slow processes are expected to be more accurate (Ninio 1975) one example being the correlation between the translational speed and infidelity of ribosomal mutants (Gallant et al.1980, Hopfield et al. 1980, Kurland 1980). Phage T4 DNA polymerase is more accurate at low polymerisation rates than at high polymerisation rates, however the increase in fidelity (measured by 2-aminopurine incorporation) seems to be due exclusively to the increased exonuclease/polymerase ratio at low polymerisation rates (Clayton et al. 1979).

We show in Figure 2 that the slowing-down of DNA synthesis catalysed by DNA polymerase ɑ from Drosophila embryos (Banks et al. 1979) leads to a remarkable increase in fidelity, although the enzyme is devoid of any detectable exonuclease activity (G. Villani, personal commun.).

The fidelity of DNA synthesis was measured as the misincorporation frequency (A versus G) from a constant incorrect/correct substrate ratio, on a poly(dC):oligo(dG) homopolymer (see Legend to Figure 2). The successful detection of the proportionality between error and polymerisation rates was perhaps due to the known high tautomerisation rate of cytosine to its imino form (pairing with adenine), since neither misincorporation of dTTP on poly(dC) nor the use of a poly(dT) homopolymer did show this effect. This observation encourages us to search for the mechanism by which DNA polymerase ɑ (but not β, results not shown) proofreads during slow synthesis. This proofreading is not likely to use any energy, which makes it an attractive alternative to the 3' to 5' exonuclease

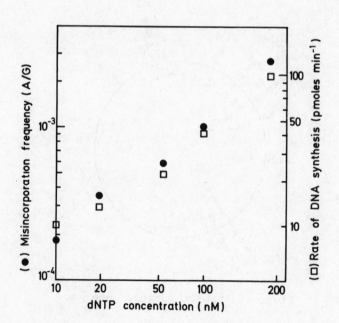

Figure 2. The reaction mixture (100 μl) contained 50 mM Tris pH 8.5
1 mM dithiothreitol, 10 mM MgCl$_2$, 0.3 mg/ml bovine serum albumin,
[^{3}H] dGTP (50 cpm/pmol) and α - [^{32}P] dATP (20,000 cpm/pmol) were in
a ratio 5:1, 60 μM poly(dC):oligo(dG) (hybridisation ratio 2:1)
"cleaned" by ura-N-glycosylase (kind gift of Dr. P. Karren, University of Sussex) to remove uracil residues (Lecomte, Doubleday and
Radman, in prep.) and 1 unit DNA polymerase α from Drosophila
embryos (kind gift of Dr. G. Villani, Stanford University) described
by Banks et al.(1979). Reaction was carried out at 37°C for 30 min
and the misincorporation frequency (A/G) in the polymer determined
as described (Villani et al. 1978).

and to mismatch correction mechanisms whose amplification could
become prohibitive due to excessive energy cost at high rates.

Accuracy versus Energy Cost Considerations in Proofreading and
DNA Repair Processes for Small and Large Genomes.

    Figure 3 is a generalised qualitative representation of the
accuracy (or genetic stability) versus energy cost, inspired by the
work on T4 DNA polymerase and its mutator and antimutator variants
affected in the exonucleolytic turnover of dNTPs to free dNMPs
(Clayton et al. 1979).

Since T4 DNA polymerase antimutator mutants can be obtained in the laobratory, the question arises why were such mutants not selected for in nature ? An answer is suggested by Figure 2 : the energy cost to achieve high replicational accuracy by an increase in 3' to 5' exonuclease becomes prohibitive, in addition to its increasing antagonism with the polymerisation process itself. For example, one antimutator T4 DNA polymerase can hydrolyse about 30 percent of all inserted dNMPs (Clayton et al. 1979). Such a situation would cause in mammalian cells a waste of a billion dNTPs per DNA replication. Any energy-driven kinetic proofreading mechanism (Hopfield 1974) is likely to behave similarly to the illustration in Figure 2. Even the mismatch correction process might be quite costly considering that the mismatch excision tract was estimated to amount to about two thousand nucleotides (plus the same number of dNTPs consumed by DNA repair synthesis) in E.coli (Wagner and Meselson 1976), which is not essentially different from similar estimates of gene conversion tract in eukaryots (review, Radding 1978).

Table 1 indicates that the E.coli replicative machinery makes roughly only one mistake per genome replication, hence only one mismatch to be corrected per generation. There is some experimental evidence that mismatch repair takes place also in mammalian cells (Wilson 1977), but its role in mutation avoidance has not yet been tested.

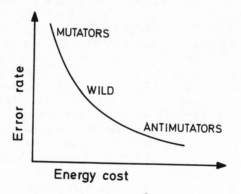

Figure 3. Accuracy (or genetic stability) versus energy cost according to Clayton et al.(1979).

These considerations lead us to think that the problem of
energy cost for proofreading and DNA repair processes becomes
orders of magnitude more acute as the amount of genomic DNA
increases by orders of magnitude. Did mammalian cells evolve new
less costly devices to maintain genetic stability ? Are there
polymerase-associated proofreading mechanisms less wasteful than
3' to 5' exonuclease ? Pyrophosphorolysis could formally fulfill
such a role (Bernardi et al. 1979, Ninio, pers. commun.) by
removing the misinserted or misincorporated terminal 3' nucleotide,
in the form of dNTP, when polymerisation ceases, e.g. due to its
own mistake, i.e. a 3'OH terminal mismatch. However, a terminal
mismatch seems to block pyrophosphorolysis by various DNA polymerases (Chang and Bollum 1973). Still pyrophosphorolysis seems to be
correlated with a kinetic proofreading mechanism of DNA polymerase
(Lecomte, Doubleday and Radman, in prep.). Another energy-saving
device would be the existance of a mismatch correction process with
a much smaller excision repair tract. Slow replication is yet
another fidelity device discussed above.

Another speculation to be tested is invited when we consider
the cost of repair of mismatches and of DNA lesions which, at a
given flux of DNA damaging treatment, will be a thousand times more
abundant in a mammalian cell than in a bacterium. Since the vast
majority of mammalian DNA is apparently non-functional (see Orgel
and Crick 1980), its repair might be a serious energy waste.
Perhaps in relation to the latter point, an interesting evolution
in DNA repair processes seems to have taken place : unlike in
bacteria where very few, if any, bulky DNA lesions (e.g. pyrimidine
dimers) are copied in the course of replication, such lesion-bypass
synthesis is highly efficient in differentiated mammalian cells
(review, Radman 1980). On the other hand, the excision repair of
pyrimidine dimers is quite inefficient leaving often as many as
20 percent dimers in the DNA of surviving mammalian cells (Cornelis,
1978). The extinction of specific excision repair pathways in the
course of differentiation has been demonstrated (Rajewsky et al.
1977).

These observations suggest a strategy adopted by differentiated
mammalian cells : rather than repairing all DNA lesions by removal,
a lesion-bypass synthesis, which is probably mutagenic, has evolved.
This strategy sounds reasonable in the light of the disposable soma
theory (Dawkins 1976), yet we would expect that the epithelial stem
cells, and certainly germ line cells, may not use such a mutagenic
lesion-bypass replication. The lack of the bypass replication should
lead to an increased sensitivity of stem and germ line cells to
DNA damaging agents, unless DNA repair by lesion removal compensates
the loss of bypass in survival. Extreme radiation sensitivity of the
stem cells (Potten 1977) is consistent with the loss of some repair
function(s). Therefore, it would be interesting to test whether
there is more excision repair and less lesion-bypass "post-

replicative repair" in germ and stem cells.

Mutagenic lesion-bypass synthesis could, in the light of this discussion, be an evolutionary product of cells with large amounts of DNA, hence a "recent" rather than a primitive form of DNA replication. Although the preliminary experiments on erroneous pyrimidine dimer bypass synthesis with DNA polymerases α and β are negative (results not shown), the suggestion of Dr. Lucy M.S. Chang, at this meeting, that β-polymerase is a good candidate for the bypass polymerase is intriguing since (1) β-polymerase is found only from metazoa (Chang 1976), (2) it can extend synthesis upon unremoved 3'OH mismatched templates (Chang 1973) and (3) it can perform limited terminal addition at the non-templated primer terminus (Chang 1973).

ACKNOWLEDGEMENTS

We thank Dr. Penny Jeggo for discussions and manuscript reading. This research was supported by Euratom Contract 156-76-1 BIOB, 224-76-1 BIOB and an agreement between the Belgian Government and the Université Libre de Bruxelles "Actions de Recherches Concertées" to M.R. and to Prof. M. Errera.

REFERENCES

Alberts, B.M. and Sternglanz, R., 1977, Nature 269:655-661.
Ayala, F., 1967, Proc. Natl. Acad. Sci.USA 58:1919-1924.
Bale, A., d'Alarcao, M. and Marinus, M.G., 1979, Mutat.Res. 59:157-165.
Banks, G.R., Boczi, J.A. and Lehman, I.R., 1979, J. Biol. Chem.254:9886-9892.
Bernardi, F. and Ninio, J., 1978, Biochimie 60:1083-1095.
Bernardi, F., Saghi, M., Dorizzi, M. and Ninio, J., 1979, J. Mol. Biol. 129:93-112.
Bernstein, C., Bernstein, H., Mufti, S. and Strom, B., 1972, Mutat. Res. 16:113-119.
Bessman, M.J., Muzycska, N., Goodman, M.F. and Schnaar, R.L., 1974, J. Mol. Biol. 88:409-421.
Brutlag, D. and Kornberg, A., 1972, J. Biol. Chem. 247:241-248.
Caillet-Fauquet, P., Defais, M. and Radman, M., 1977, J. Mol. Biol. 117:95-112.
Chang, L.M.S., 1973, J. Biol. Chem. 248:6983-6992.
Chang, L.M.S. and Bollum, F.J., 1973, J. Biol. Chem. 248:3398-3404.
Chang, L.M.S., 1976, Science 191:1183-1185.
Clayton, L.K., Goodman, M.F., Branscomb, E.W. and Galas, D.J., 1979, J. Biol. Chem. 254:1902-1912.
Cornelis, J.J., 1978, Biochem. Biophys. Acta 521:134-143.
Cox, E.C. and Gibson, T.C., 1974, Genetics 77:169-184.
Cox, E.C., 1976, Annu. Rev. Genet. 10, 135-156.
Dawkins, R., 1976, The Selfish Gene, Oxford University Press.

Dreiseikelmann, B., Eichenlaub, R. and Wackernagel, W., 1979, Biochem. Biophys. Acta 562:418-428.

Duncan, B.K., Rockstroh, P.A. and Warner, H.R., 1978, J. Bact. 134:1039-1045.

Edemberg, H.J. and Huberman, J.A., 1975, Ann. Rev. Genet. 9:245-284.

Fersht, A., 1979, Proc. Natl. Acad. Sci. USA 76:4946-4950.

Fowler, R.G., Degnen, G.E. and Cox, E.C., 1974, Mol. Gen. Genet. 133:179-191.

Gallant, J. and Foley, D., 1980, in:"Ribosomes", Chambliss, Craven, Davies, Davis, Kahan and Nomura Ed., University Park Press, Baltimore, pp.615-638.

Geier, E. and Modrich, P., 1979, J. Biol. Chem. 254:1408-1413.

Gillin, F.D. and Nossal, N.G., 1976, J. Biol. Chem. 251:5219-5224.

Glickman, B.W. and Radman, M., 1980, Proc. Natl. Acad. Sci. USA 77:1063-1067.

Hershfield, M.S. and Nossal, N.G., 1972, J. Biol. Chem. 247:3393-3404.

Hershfield, M.S., 1973, J. Biol. Chem. 248:1417-1423.

Hibner, U. and Alberts, B.M., 1980, Nature 285:300-305.

Hochhauser, S.J. and Weiss, B., 1978, J. Bact. 134:157-166.

Hopfield, J.J., 1974, Proc. Natl. Acad. Sci. USA 71:4135-4139.

Hopfield, J.J. and Yamane, T., 1980, in:"Ribosomes", Chambliss, Craven, Davies, Davis, Kahan and Nomura Ed., University Park Press, Baltimore, pp. 585-596.

Koch, R.E., McGraw, M.K. and Drake, J.W., 1976, J. Virol. 19:490-494.

Kondrad, B.E., 1978, J. Bact. 133:1197-1202.

Kunkel, T.A., Meyer, R.R. and Loeb, L.A., 1979, Proc. Natl. Acad. Sci. USA 76:6331-6335.

Kurland, C.G., 1980, in:"Ribosomes", Chambliss, Craven, Davies, Davis, Kahan and Nomura Ed., University Park Press, Baltimore, pp. 597-614.

Lacks, S. and Greenberg, B., 1977, J. Mol. Biol. 114:163-168.

Leigh, E.G., 1973, Genetics 73(suppl.):1-18.

Loeb, L.A., Weymouth, L.A., Kunkel, T.A., Gopinathan, K.P., Beckman, R.A. and Dube, D.K., 1978, Cold Spring Harbor Symp. Quant. Biol. 43:921-927.

Lomant, A.J. and Fresco, J.R., 1975, Progr. Nuc. Acid. Res. Molec. Biol. 15:185-218.

Marinus, M.G., 1973, Mol. Gen. Genet. 127:47-55.

Nevers, P. and Spatz, H., 1975, Mol. Gen. Genet. 139:233-243.

Ninio, J., 1975, Biochimie 57: 587-595.

Orgel, L.E. and Lohrmann, R., 1974, Accounts of Chem. Res. 7:368-375.

Orgel, L.E. and Crick, F.H.C., 1980, Nature 284:604-607.

Potten, C.S., 1977, Nature 269:518.

Que, B.G., Downey, K.M. and So, A.G., 1978, Biochemistry 17:1603-1606.

Radding, C.C., 1978, Annu. Rev. Biochem. 47:847-880.

Radman, M., 1974, Molecular and Environmental Aspects of Mutagenesis, Ed. Prakash, Sherman, Miller, Lawrence, Taber, Springfield, Ill., pp.128-142.

Radman, M., Villani, G., Boiteux, S., Defais, M., Caillet-Fauquet,P. and Spadari, S., 1977, in: Origins of Human Cancer, Hiatt,H., Watson, J.D. and Winsten, J.A., Eds., Cold Spring Harbor, pp.903-922.

Radman, M., Villani, G., Boiteux, S., Kinsella, A.R., Glickman, B.W., and Spadari, S., 1978, Cold Spring Harbor Symp. Quant. Biol. 43: 937-945.

Radman, M., Wagner, R.E., Glickman, B.W. and Meselson, M., 1980, in:"Progress in Environmental Mutagenesis", M. Alacevic Ed., Elsevier/North Holland, Amsterdam, pp.121-129.

Radman, M., 1980, Photochem. Photobiol. 32:823-830.

Rajewsky, M.F., Augerslicht, L.H., Bressmann, H., Goth, R., Hülser, D.F., Lacrum, O.D. and Lomakina, L.Ya, 1977, in:"Origins of Human Cancer" Book B, Hiatt, Watson and Winsten Eds., Cold Spring Harbor Conferences on Cell Proliferation, pp.709-729.

Rydberg, B., 1978, Mutat. Res. 52:11-29.

Sevastopoulos, G.G. and Glaser, D.A., 1977, Proc. Natl. Acad. Sci. (Wash.) 74:3947-3950.

Topal, M.D. and Fresco, J.R., 1976, Nature 263:285.

Treffers, H.P., Spinelli, V. and Belser, N.O., 1954, Proc. Natl. Acad. Sci. USA 40:1064-1071.

Tye, B.K., Nyman, P.O., Lehman, I.R., Hochhauser, S. and Weiss, B., 1977, Proc. Natl. Acad. Sci. USA 74:154-157.

Vaccaro, K.K. and Siegel, E.C., 1975, Mol. Gen. Genet. 141:251-262.

Wagner, R.E.Jr. and Meselson, M., 1976, Proc. Natl. Acad. Sci. USA 73:4135-4139.

Wilson, J.H., 1977, Proc. Nat. Acad. Sci. USA 74:3503-3507.

Witkin, E.M., 1976, Bacteriological Reviews 40:869-907.

INHIBITION OR ABSENCE OF DNA PROOFREADING EXONUCLEASE IS NOT SUFFICIENT TO ALLOW COPYING OF PYRIMIDINE DIMERS

O.P. Doubleday, G. Michel-Maenhaut, A. Brandenburger,
P. Lecomte and M. Radman

Département de Biologie Moléculaire
Université Libre de Bruxelles
B - 1640 Rhode-St-Genèse (Belgium)

SUMMARY

Speculation about the role of the 3' to 5' proofreading
exonuclease function of DNA polymerases during UV mutagenesis has
led us to examine the extents of DNA synthesis performed by a
variety of different DNA polymerases, upon intact and UV-irradiated
$\emptyset$X174 DNA. The relative inhibition of DNA synthesis after UV-
irradiation in the presence of the heavy metal ions Ag^+ and Hg^{++},
and use of the T4 pyrimidine dimer specific endonuclease, confirms
that the majority of the UV-induced inhibition of DNA synthesis
is due to the presence of pyrimidine dimers in the template DNA.
Experiments with DNA polymerases lacking 3' to 5' exonuclease
activity, or with inhibited or denatured 3' to 5' exonuclease
activity, revealed that pyrimidine dimers constitute an absolute
block for DNA synthesis, irrespective of 3' to 5' exonuclease
proofreading activity. It is therefore concluded that, although it
might be required, inhibition or absence of 3' to 5' proofreading
activity is not sufficient to allow efficient transdimer synthesis.

INTRODUCTION

The photoreversibility of UV-induced mutagenesis (Witkin,1966)
and studies of UV-induced mutations at the level of DNA sequence
specificity (Coulondre & Miller,1977) suggest strongly that

pyrimidine dimers are the principal premutagenic lesions in E.coli.
Although much is known about the genetic control of UV-mutagenesis
the underlying molecular mechanism is not clear, since it is known
that none of the three constitutive DNA polymerases of E.coli are
capable of fully replicating UV-irradiated templates, presumably
as the result of the presence of non-coding lesions, such as
pyrimidine dimers.

UV-irradiated single-stranded ØX174 DNA has been shown to be
replicated to a greater extent under conditions of (SOS-induced)
error-prone repair (Caillet-Fauquet et al, 1977). However, it is
still not clear whether this is due to transdimer DNA synthesis
(i.e. with pyrimidine dimers persisting in the DNA), or to the
removal of pyrimidine dimers in a manner not affecting the
integrity of the DNA sugar-phosphate backbone (e.g. by a glycosylase) which would facilitate an erroneous copying of the damaged
site. The finding that DNA polymerase III is essential for UV-
induced mutagenesis of E.coli (Bridges et al, 1976) is possibly
more consistent with the first of these two alternatives, since
it is likely that a modification of the template DNA would
facilitate the replication of UV-irradiated DNA by all the cellular
DNA polymerases.

The 3' to 5' proofreading function of DNA polymerases is one
of the determining factors in the fidelity of DNA replication
and spontaneous mutagenesis (for example see Clayton et al, 1979).
Several observations have led to speculation as to the role of
DNA proofreading function in radiation-induced mutagenesis.
Diminished proofreading activity has been correlated with increased
misincorporation on intact templates (Hall & Lehman 1968, Byrnes
et al 1977, Clayton et al 1979) and on templates containing
miscoding DNA lesions (Lecomte, Doubleday and Radman, in prep.).
Villani et al (1978) found that the blockage of DNA synthesis on
UV-irradiated templates is accompanied by the generation of
deoxynucleoside monophosphates. They proposed that this is due to
polymerase "idling", whereby the DNA polymerase incorporates
bases opposite pyrimidine dimers (non-coding lesions), but
subsequently, by proofreading exonuclease, excises these mismatched
bases. This raised the question, is the blockage of DNA synthesis
on UV-irradiated templates caused by this proofreading activity
"idling" at pyrimidine dimers ? The observation that the extent
of overall DNA synthesis upon UV-irradiated templates was increased
when replication was performed by a polymerase lacking 3' to 5'
proofreading function offered indirect support for this hypothesis
(Villani et al 1978, Radman et al 1977).

In this paper we report the results of experiments designed
to measure directly the influence of DNA polymerase proofreading
function upon the elongation of new DNA chains during the
replication of UV-irradiated templates.

MATERIALS AND METHODS

DNA Templates. 5 µg of ØX174 RF 1 DNA (covalently closed, double-stranded replicative form), prepared by the method of Godson and Vapnek (1973), was labelled with ^{32}P-thymidine by a nick-translation reaction : 50 mM Tris pH 8.0, 10 mM MgCl$_2$, 1 mM DTT, 1 mM dATP, dCTP and dGTP, 200 µCi ^{32}P-TTP (400 Ci/mmol, Amersham) and 3 units of E.coli DNA polymerase I (Boehringer, Mannheim) in 100 µl at 14°C. DNAse activity endogenous to the DNA polymerase was sufficient to allow the reaction to proceed, and no further DNAse was added. When approximately 50 % of the label had been incorporated into TCA-insoluble material the DNA was extracted twice with phenol, three times with ether, passed through a 1 ml G-50 Sephadex column and digested with the restriction enzyme Hpa 1 (kind gift of Dr. Micke Lupker, Leiden) or Hind II (Boehringer, Mannheim). The fragments were separated on a 1 % or 2 % agarose gel respectively (SeaKem ME), using 5 mM sodium acetate, 1 mM EDTA and 40 mM Tris pH 8.0 as buffer, and recovered by electroelution (McDonell et al 1977). Specific DNA fragments were annealed to single-stranded ØX174 DNA in the proportion 1:3 by heating to 100°C for 2 minutes and then incubating at 70°C for 15 minutes in 10 mM Tris pH 7.4, 0.3 M NaCl and 1 mM EDTA. The DNA was alcohol precipitated and resuspended in 25 mM Tris pH 7.4, 1 mM EDTA, 5 mM NaCl.

Analysis of the extent of DNA synthesis. The DNA polymerase reaction was terminated by the addition of 1/10 th volume Stop solution, containing 0.2 % Bromophenol Blue, 50 % Sucrose, 5 % SDS, 25 mM EDTA, heating to 90°C for 2 minutes, chilling in ice, centrifuging (12,000 g for 2 minutes) and adding 1/10 volume 2 N NaOH solution. The denatured DNA, and appropriate size markers, were electrophoresed on agarose gels, placed on Whatman 3 mm chromatography paper, covered with Saran wrap, dried (Hoefer Scientific gel drier) and autoradiographed at - 70°C, using Kodak RX-1 film and Dupont Cronex Screens.

DNA polymerases and synthesis conditions. DNA synthesis by the large fragment of E.coli DNA polymerase I (Boehringer Mannheim) was in a 30 µl reaction containing template DNA, 50 mM Tris HCl pH 8.0, 10 mM MgCl$_2$, 1 mM dithiothreitol (DTT), 25 µM each of dATP, dTTP, dGTP and dCTP (dNTPs). One unit of enzyme activity incorporates 10 nmol of dNTP into acid-insoluble material during 30 min at 37°C using the conditions of Richardson et al (1964).

Cell-free extracts containing E.coli DNA polymerase III activity were prepared by the method of Conrad and Campbell (1979). These preparations were judged to have a high proportion of polymerase III dependent DNA synthesis activity on the basis of their sensitivity to 150 mM KCl (Kornberg 1974) and their resistance to polymerase I antibody (a kind gift from Dr. L. Loeb).

DNA synthesis was performed in a 30 μl reaction containing template
DNA, 50 mM Tris HCl pH 7.4, 10 mM MgCl$_2$, 10 mM DTT, 50 μM dNTPs,
5 mM ATP (in some assays 5 mM spermidine.HCl was added, without
apparent effect).

Avian myeloblastosis virus (AMV) DNA polymerase (lot No.
G-1178, greater than 95 % pure) was a kind gift from Dr. J.Beard.
DNA synthesis was performed in a 30 μl volume containing template
DNA, 50 mM Tris HCl pH 8.3, 40 mM KCl, 6 mM MgCl$_2$, 5 mM DTT,
50 μM dNTPs, 3 μg BSA. One unit of enzyme activity incorporates
1 nmol of dTTP into acid-insoluble material in 10 min at 37°C
using poly rA:dT$_{12-18}$ as template.(Kacian and Spiegelman 1973).

DNA polymerase from Drosophila, fraction VII according to
the scheme of Banks et al (1979) was the kind gift of G. Villani.
DNA synthesis was performed in a 30 μl reaction containing template
DNA, 50 mM Tris HCl pH 8.5, 20 mM (NH$_4$)$_2$SO$_4$, 10 mM MgCl$_2$, 5 mM DTT,
250 μM dNTPs and 10 μg BSA. One unit of enzyme activity incorporates
1 nmol of dNTP into acid-insoluble material in 60 min at 37°C
(Banks et al 1979).

Measurement of 3' to 5' exonuclease activity of DNA polymerases. The 3' to 5' exonuclease activity of DNA polymerases was
measured by determining the kinetics of excision of the ^{3}H-dAMP
from the 3' terminus of oligo (dG)$_{12-18}$ (^{3}H-dA)$_{0.5}$ annealed to
poly(dC) under the conditions used for DNA synthesis (Lecomte et
al, in prep.).

Ultraviolet irradiation. The UV source was a Mineralight with
a maximum energy output at 2537 A. DNA solutions (50-250 μl),
placed on hydrophobic plastic petri dishes were gently agitated
during exposure to UV at a dose rate of 1-3 Jm^{-2}s^{-1}. Doses were
measured with a Latarjet dosimeter.

For irradiation in the presence of heavy metal ions the DNA
template was passed through a 1 ml Sephadex G-50 column and
recovered in water, since the presence of excess anions such as
Cl$^-$ inhibits the effects of Hg^{++} (Rahn et al 1970). After the
addition of equimolar amounts of AgNO$_3$ or HgCl$_2$ solutions the
DNA was UV-irradiated. NaCl was added to a final concentration
of 1 M and the solutions kept on ice for 1 hour before being passed
through a G-50 column to remove the metal ions. The DNA was
recovered in water, lyophilised to reduce the volume and resuspended in 25 mM Tris HCl pH 7.4, 1 mM EDTA and 5 mM NaCl.

RESULTS

We have developed a system to investigate the effects of
lesions in template DNA upon the extent of DNA replication and in

particular upon the elongation of newly synthesised strands. By
using a radioactively labelled DNA primer and performing DNA
synthesis with non-labelled dNTPs, we are able to analyse electro-
phoretically in DNA denaturing gels the extent of continuous DNA
synthesis primed by this specific fragment. Our use of natural
DNAs as template (ØX174 viral strand) and primers (restriction
fragments of ØX174 RF 1 DNA) eliminates some of the problems
inherent in the use of homopolymer templates (see, for example,
Loeb et al 1978). We experimented with various methods of labelling
primer fragments, but found that only nick-translation with
$32P$-dNTP was able to generate the high specific activity DNA
required for this assay. The disadvantage of this method is that
some of the primer fragments are of less than unit size (due either
to the nicks introduced by the nick translation reaction, or to
decay of $32P$ giving rise to DNA strand breaks). However, since we
are interested in the maximum size of molecules formed, the
presence of smaller molecules does not interfere with the interpretation of our gels. Furthermore, this method was found to be
capable of detecting sites of inhibition of DNA synthesis by T4
DNA polymerase and DNA polymerase ∝ from Drosophila (results not
shown), presumably resulting from secondary structure effects
(Challberg and Englund 1979, Huang and Hearst 1980). Primers in
which the 5' terminus were labelled by polynucleotide kinase were
found to be too sensitive to 5' → 3' exonuclease or phosphatase
activity to be of general use.

DNA Synthesis on Intact ØX174 Templates Gives Rise to Fully Replicated Molecules.

The extent of DNA synthesis upon intact and UV-irradiated
ØX174 DNA after 15 and 60 min incubation with the large fragment
of E.coli DNA polymerase I is shown in figure 1. On the basis of
the relationship between electrophoretic mobility and DNA size
(Maniatis et al 1975, McDonell et al 1977), it is clear that, in
the presence of sufficient DNA polymerase, and given sufficient
time of incubation, fully replicated ØX174 molecules (5386 bases,
Sanger et al 1978) are formed. The kinetics of DNA synthesis by
the large fragment of DNA polymerase I were found to be linear
until the DNA template was fully replicated (data not shown), an
observation consistent with those found measuring the kinetics of
DNA synthesis by the uptake of labelled dNTP into TCA-insoluble
material (see for example Villani et al 1978).

UV-irradiation of ØX174 templates blocks DNA synthesis.

DNA synthesis on UV-irradiated templates is blocked, even
when incubation is carried out for times longer than those necessary
for full replication of intact ØX174 templates (figures 1, 2, 3 and

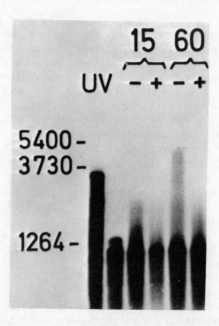

Figure 1. <u>Kinetics of DNA synthesis on intact and UV-irradiated</u>
<u>DNA template</u>. Intact and UV-irradiated (200 Jm^{-2}) template DNA,
made by hybridising 2 nanomoles of ØX174 viral DNA with an excess
of ^{32}P-labelled Hpa I fragment 2, was replicated by 0.6 Units of
DNA polymerase I large fragment for 15 and 60 minutes at 32°C. The
denatured DNA products were electrophoresed in a 1 % agarose gel
with Hpa I fragments 2 (1264 bases) and 1 (3730 bases) as standards.
The predicted position of a fully replicated ØX174 molecule (5400
bases) is shown.

4). The extent of residual DNA synthesis is inversely proportional
to the dose of UV used to irradiate the template (figure 2 and
unpublished data). These results, which merely confirm the well
known phenomenon of inhibition of DNA synthesis by UV irradiation
(Setlow et al 1963) are, nevertheless, a reassuring characterisation of our system since they show that we do not have apparent
'transdimer' synthesis arising, for example, by primer slippage.

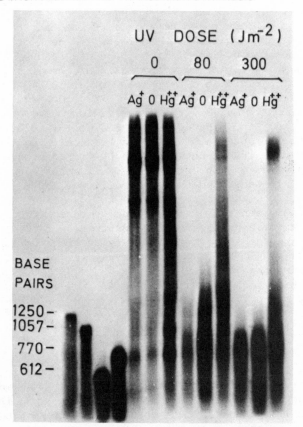

Figure 2. The extent of DNA synthesised upon DNA templates UV-
irradiated in the absence or presence of Ag^+ or Hg^{++}.
The Hind II fragment 2 of ØX174 was annealed to an excess of viral
DNA and UV-irradiated in the absence of metal (O) or in the
presence of $AgNO_3$ (Ag^+) or $HgCl_2$ (Hg^{++}). 200-250 picomoles of
primed ØX174 viral DNA were replicated by 8 μl of DNA polymerase
III fraction (approximately 100 nanograms of protein) for 20
minutes at 32°C, denatured, and electrophoresed in a 2 % agarose
gel with the size markers as indicated.

The UV-induced blockage of DNA synthesis is caused by pyrimidine
dimers.

     The demonstration by Moore and Strauss (1979) that when UV-
irradiation of ØX174 templates blocks DNA synthesis it does so one
(or more) bases before any pair of pyrimidine bases suggests
strongly that pyrimidine dimers are responsible for the majority
of the blockage of DNA synthesis by UV. In an attempt to corroborate

this conclusion we have taken advantage of the fact that the heavy
metal ions Ag[+] has been shown to increase the rate of formation of
pyrimidine dimers by UV-irradiation (Rahn and Landry 1973) whereas
Hg[++] has been shown to decrease the ratio of formation, by UV-
irradiation, of pyrimidine dimers (Rahn et al 1970), but not of
other lethal lesions (Rahn et al 1973). Electrophoretic analysis
of ØX174 viral DNA treated with T4 pyrimidine dimer specific
endonuclease (the kind gift of Dr. P. Seawell) after UV-irradiation
in the presence or absence of Hg[++] and Ag[+] confirmed these
observations (results not shown). When DNA irradiated under these
conditions was used as a template for DNA synthesis by E.coli DNA
polymerase III it was found that synthesis was more inhibited
upon DNA irradiated in the presence of Ag[+], and was less inhibited
when the DNA had been irradiated in the presence of Hg[++] (figure 2).
Similar effects of sensitization and protection to UV were found
when ØX174 DNA was assayed for viability by transfection
(Brandenburger et al, in prep.). These results agree with other
studies (Setlow et al 1963, Villani et al 1977, Moore and Strauss
1979), indicating that the majority of UV-induced lesions blocking
DNA synthesis are pyrimidine dimers.

Inhibition or lack of 3' to 5' proofreading exonuclease is not sufficient to allow transdimer DNA synthesis.

Complete inhibition of the 3' to 5' exonuclease activity of
E.coli DNA polymerase I by the presence of 5 mM dNMPs (data not
shown) did not influence the extent of synthesis upon irradiated
DNA (figure 3), although effects upon the fidelity of DNA synthesis
may be seen under these conditions (Lecomte, Doubleday and Radman,
in prep.) or under equivalent conditions (Byrnes et al 1977).
DNA polymerase I whose 3' to 5' exonuclease activity was irreversi-
bly denatured (Lecomte and Doubleday, in prep.) was also found
to be incapable of replicating UV-irradiated DNA (data not shown).

Two eukaryotic DNA polymerases known to lack 3' to 5' exonuclease activity, AMV polymerase (Battula and Loeb, 1975) and
polymerase from Drosophila (Banks et al 1979) were also tested for
their ability to replicate UV-irradiated templates (figures 3 & 4).
Neither enzyme was capable of copying UV-irradiated DNA. Both
polymerases had relatively low activity upon our templates, almost
certainly as the result of template secondary structure effects.
In the case of α polymerase we observed very strong inhibition of
synthesis around base 3980, corresponding to a site of inhibition
of vaccinia DNA polymerase which Challberg and Englund (1979) have
correlated with a hairpin structure in the template.

Experiments were also performed with T4 DNA polymerase, whose
proofreading activity was inhibited by dNMPs (data not shown) and
E.coli DNA polymerase III (due to contaminating exonuclease activity
in our polymerase III preparation we were unable to measure the

3' to 5' exonuclease activity of this enzyme, consequently we tested
the effects of increasing concentrations of dNMPs, up to those
inhibitory for synthesis, upon the replication of UV-irradiated
templates). No effect of proofreading activity was found in these
experiments, nor was any effect observed when the speed of replication, which influences proofreading activity (Ninio 1975, Clayton
et al 1979, Radman et al, this volume) was altered by changing the
concentrations of dNTPs in the reactions. We confirmed previous
observations (Challberg and Englund, 1979, Huang and Hearst, 1980),
as to the sensitivity of T4 DNA polymerase to the effects of template
secondary structure (data not shown).

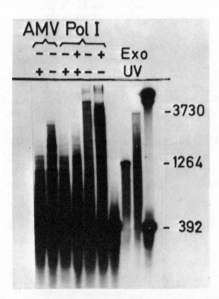

Figure 3. The lack, or inhibition, of 3' to 5' exonuclease proofreading activity does not influence the extent of replication of
UV-irradiated ØX174 template by E.coli DNA polymerase I and AMV
DNA polymerase. 200-250 picomoles of primed intact or UV-irradiated
(200 Jm^{-2}) template DNA, made by hybridising Hpa I fragment 3
(392 bases) to an excess of the viral strand, were replicated by
3 Units of AMV polymerase, or by 0.5 Units of E.coli DNA polymerase
I large fragment by incubation at 32°C for 40 minutes. The proofreading activity of polymerase I was inhibited by the presence of
1.25 mM of each dNMP during the reaction as indicated (exo -). The
denatured DNA products were electrophoresed in a 2 % agarose gel,
with the Hpa I fragments 1 (3730 bases) and 2 (1264 bases) and 3
(392 bases) as size markers.

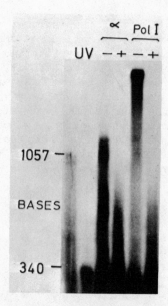

Figure 4. <u>DNA polymerase α from Drosophila, which lacks 3' to 5'
exonuclease activity, is not capable of replicating UV-irradiated
DNA</u>.   100-150 picomoles of intact or UV-irradiated (500 Jm^{-2})
template DNA, prepared by hybridising Hind II fragments 6 (345,
341, 335 bases, see Sanger et al 1977) to an excess of viral DNA,
were replicated by 0.1 Units of <u>E.coli</u> DNA polymerase I, or 1.0
Units of DNA polymerase α . After incubation at 32°C for 3 hours,
the DNA was denatured and electrophoresed in a 2.2 % agarose gel,
with Hind II fragments 1 (1057 bases) and 6 (approx. 340 bases)
as size markers.

DISCUSSION

     On the basis of their mechanism of action, mutagens can be
divided into two classes : <u>direct mutagens</u>, which cause subtle
modification of DNA and are mutagenic by virtue of their direct
mispairing properties when replicated, and <u>indirect mutagens</u>, which
result in bulky DNA lesions, thereby inhibiting DNA synthesis
(Radman et al 1977). Direct mutagens do not greatly inhibit DNA
synthesis,for example 5 bromouracil substituted DNA is a reasonable
substrate for DNA polymerase I (Trautner et al 1962), as is
partially deaminated poly dC, which we have found (unpublished) may
be efficiently replicated in the presence or absence of dATP (the
formation of uracil constitutes a mechanism of direct mutagenesis

in vivo, especially if the deaminated residue was previously in the form of 5-methyl cytosine, which deaminates to thymine (see Coulondre et al. 1978, Bird 1980, for a discussion of the biological implications). In contrast to direct mutagens, indirect mutagens block DNA chain elongation and their mutagenic effects are dependent upon the expression of recA and lex genes in E. coli (reviews, Witkin 1976, Radman et al. 1977).

The observation by Villani et al (1977) that the blockage of DNA synthesis on UV-irradiated templates is accompanied by the generation of dNMPs ("polymerase idling") led to the question : is the blockage of DNA synthesis by pyrimidine dimers due to polymerase proofreading activity excising the (mismatched) primer nucleotide from which the polymerase might otherwise have extended, albeit inefficiently, over the non-coding DNA lesion ? Thus arose the hypothesis that SOS-repair may be due to inhibition of DNA polymerase proofreading function (Radman et al, 1977, Villani et al 1978, Bridges 1978), a hypothesis fully consistent with the phenomenon of untargeted mutagenesis. Recently Schroeder (1979) has postulated that mutagenesis may result from a reversal of the DNA proofreading function, which might be expected to give rise to a terminal transferase-like activity. Such an activity, with no requirement for a DNA template, might allow synthesis to proceed beyond lesions which normally block DNA synthesis. The experiments that we report here discourage the hypotheses that either inhibition or reversal (at least in its simplest form, driven by dNMPs) of the 3' to 5' proofreading activity of DNA polymerases is sufficient to give rise to transdimer synthesis.

How might transdimer DNA synthesis occur ? It is conceivable that diminished nucleotide-pair discrimination by a DNA polymerase, such as has been observed for certain T4 mutator mutants (Hershfield 1973) might allow DNA synthesis to occur over non-coding DNA lesions. However, it is difficult to imagine that such synthesis would not be inhibited by 3' to 5' proofreading activity, since it would involve elongation from a mismatched 3' terminus, a substrate par excellence for proofreading exonuclease. Thus, while our results show that inhibition, or lack, of proofreading activity is not alone sufficient to allow transdimer synthesis, its involvment in a more complex system cannot be excluded.

ACKNOWLEDGEMENTS

We thank Dr. Penny Jeggo for discussions and manuscript reading. This research was supported by Euratom Contract 156-76-1 BIOB, 224-76-1 BIOB and by an agreement between the Belgian Government and the Université Libre de Bruxelles "Actions de Recherches Concertées" to M.R. and to Prof. M. Errera.

REFERENCES

Banks, G.R., Boezy, J.A. and Lehman, I.R., 1979, J. Biol. Chem.
     254: 9886-9892.
Battula, N. and Loeb, L.A., 1975, J. Biol. Chem. 251:982-986.
Bird, A.P., 1980, Nucl. Acids Res. 8:1499-1504.
Bridges, B.A., Mothershead, R.P. and Sedgwick, S.G., 1976,
     Molec. gen. Genet. 144:53-58.
Bridges, B.A., 1978, Nature 275:591-592.
Byrnes,J.J., Downey, K.M., Que, B.G., Lee, M.Y.W., Black, V.L. and
     So, A.G., 1977, Biochemistry 16:3740-3746.
Caillet-Fauquet, P., Defais, M. and Radman, M., 1977, J. Mol. Biol.
     117:95-112.
Challberg, M.D. and Englund, P.T., 1979, J. Biol. Chem. 254:7820-
     7826.
Clayton, L.K., Goodman, M.F., Branscomb, E.W. and Galas, D.J., 1979
     J. Biol. Chem. 254: 1902-1912.
Conrad, S.E. and Campbell, J.L., 1979, Nucleic Acids Res. 6:3289-
     3303.
Coulondre, C. and Miller, J.H., 1977, J. Mol. Biol. 117:577-606.
Coulondre, C., Miller, J.H., Farabaugh, Ph.J. and Gilbert, W., 1978,
     Nature 274:775-780.
Godson, G.N. and Vapnek, D., 1973, Biochem. Biophys. Acta 299:516-
     520.
Hall, Z.W. and Lehman, I.R., 1968, J. Mol. Biol. 36:321-333.
Hershfield, M.S., 1973, J. Biol. Chem. 248:1417-1423.
Huang, C-C. and Hearst, J.E., 1980, Anal. Biochem. 103:127-139.
Kacian, D.L. and Spiegelman, S., 1973, in:"Methods in Enzymology"
     Vol 29E, L. Grossman and K. Moldave, ed., Academic Press New
     York.
Kornberg, A., 1974, DNA Synthesis, Freeman, San Francisco CA.
Loeb, L.A., Weymouth, L.A., Kunkel, T.A., Gopinathan, K.P.,
     Bechman, R.A. and Dube, D.K., 1978, On the fidelity of DNA
     Replication, Cold Spring Harbor Symposia on Quantitative
     Biology XLIII, 921-927.
McDonell, M.W., Simon, M.N. and Studier, F.W., 1977, J. Mol. Biol.
     110:119-146.
Maniatis, T., Jeffrey, A. and van de Sande, H., 1975, Biochemistry
     14:3787-3794.
Moore, P. and Strauss, B.S., 1979, Nature 278:664-666.
Ninio, J., 1975, Biochimie 57:587-595.
Radman, M., Villani, G., Boiteux, S., Defais, M., Caillet-Fauquet,P.
     & Spadari, S., 1977, in "Origins of Human Cancer, H. Hiatt et
     al. ed., Cold Spring Harbor, New York.
Rahn, R.O., Battista, M.D.C. and Landry, L.C., 1970, Proc. Natl.
     Acad. Sci. USA 67:1390-1397.
Rahn, R.O. and Landry, L.C., 1973, Photochem. Photobiol. 18:29-38.
Rahn, R.O., Setlow, J.K. and Landry, L.C., 1973, Photochem.
     Photobiol. 18:39-41.

Richardson, C.C., Schildkraut, C.L., Aposhian, H.V. and Kornberg, A. 1964, J. Biol. Chem. 239:222-232.

Sanger, F., Air, G.M., Barrell, B.G., Brown, N.L., Coulson, A.R., Fiddles, J.C., Hutchison, C.A., Slocombe, P.M. and Smith, M. 1977, Nature 265:687-695.

Sanger, F., Coulson, A.R., Friedman, T., Air, G.M., Barrell, B.G., Brown, N.L., Fiddles, J.C., Hutchison, C.A., Slocombe, P.M. and Smith, M., 1978, J. Mol. Biol. 125:225-246.

Schroeder, C., 1979, Eur. J. Biochem. 102:291-296.

Setlow, R.B., Swenson, P.A. and Carrier, W.L., 1963, Science 142:1464-1466.

Trautner, T.A., Swartz, M.N. and Kornberg, A., 1962, Proc. Natl. Acad. Sci. USA 48:449-455.

Villani, G., Boiteux, S. and Radman, M., 1978, Proc. Natl. Acad. Sci. USA 75:3037-3041.

Witkin, E.M., 1966, Science 152:1345-1353.

ON THE POSSIBLE ROLE OF THE MISCODING DNA-LESIONS, 1,$\underline{N}^6$-ETHENO-ADENINE AND 3,$\underline{N}^4$-ETHENOCYTOSINE, IN VINYL CHLORIDE-INDUCED MUTAGENESIS AND CARCINOGENESIS[1]

Alain Barbin*, Helmut Bartsch*, Philippe Lecomte** and
Miroslav Radman**

*   International Agency for Research on Cancer, 150
    cours Albert-Thomas, F-69372 Lyon Cédex 2, France
** Free University of Brussels, B-1640 Rhode-St-Genèse,
    Belgium

ABSTRACT

Chloroacetaldehyde (CAA) and chloroethylene oxide (CEO), two
reactive metabolites of vinyl chloride, were used to introduce
increasing amounts of 1,$\underline{N}^6$-ethenoadenine (εA) and 3,$\underline{N}^4$-etheno
cytosine (εC) residues in poly(dA) and poly(dC), respectively.
The modified polynucleotides were assayed with $\underline{E.\ coli}$ DNA polymerase I for their template activity and for misincorporation.
The miscoding properties of εA and εC that we observed may explain
the mutagenic effects reported for vinyl chloride and its metabolites;
these lesions may also represent one of the initial steps in vinyl
chloride or CEO-induced carcinogenesis.

INTRODUCTION

The biological effects of vinyl chloride, a recognized
carcinogen in animals and humans[1,2], appear to depend on its
conversion by microsomal cytochrome P-450-dependent monooxygenases[3]
into CEO[4], a mutagenic[5-8] and carcinogenic[9] metabolite that can
rearrange to CAA[10]. εA and εC are two major fluorescent adducts
formed by the reaction of adenine and cytosine derivatives with
metabolically activated vinyl chloride[4,11,12] or with CEO and
CAA[4,13]. They have also been isolated from DNA and RNA of rats

---

[1] This work was supported in part by a contract, no. N01 CP-55630,
from the National Cancer Institute, USPHS.

treated with vinyl chloride[11,12,14]. We therefore investigated whether εA and εC are promutagenic DNA lesions, using Escherichia coli DNA polymerase I and synthetic templates prepared by reaction of the homopolymers poly(dA) and poly(dC) with CEO or CAA. Similar experiments previously revealed that O^6-methylguanine miscodes for DNA and RNA polymerases[15,16]. There is accumulating evidence that O^6-methylguanine is involved in the induction of mutagenesis and carcinogenesis by alkylating agents[17]. Our data report for the first time that a human carcinogen, vinyl chloride, can form miscoding DNA adducts.

METHODS

Poly(dA) and poly(dC) dissolved in 50 mM sodium cacodylate buffer pH 7.0 were treated at 37°C with increasing concentrations of CAA (for 20 hrs) or CEO (for 20 mins). The molar ratios of CAA or CEO to nucleotide phosphorus [as poly(dA) or poly(dC)] concentrations in the reaction mixture are given in Table 1, first column. Treated and untreated polynucleotides were then purified by chromatography on a Sephadex G10 column with 50 mM sodium cacodylate buffer pH 7.0 as eluent. CAA- or CEO-treated poly(dA), referred to as poly(dA,dεA), showed the characteristic fluorescence spectrum of $1,N^6$-ethenoadenosine. CAA- or CEO-treated poly(dC), referred to as poly(dC,dεC), as expected, were not fluorescent.

Treated and untreated poly(dA) [or poly(dC)] were annealed at a 1:1 nucleotide phosphorus ratio with oligo(dT)$_{12-18}$ [or oligo(dG)$_{12-18}$]. DNA synthesis was assayed on the primed templates using E. coli DNA polymerase I. Each reaction mixture (150μl) contained 2.5 nmoles homo- or copolymer (nucleotide phosphorus), 4 nmoles ^{3}H-dTTP (0.67 μCi) or ^{3}H-dGTP (3.4 μCi), 0.2 units polymerase, 10 mM Mg Cl$_2$ and 50 mM Tris-HCl pH 8.5. After 1 hr incubation at 30°C the solutions were pipetted onto glass microfibre filters, and DNA was precipitated in cold 5% trichloroacetic acid. Following extensive rinsing, the radioactivity present was determined by liquid scintillation counting. Misincorporation assays were performed using a double-labelling technique. The reaction mixtures were as those described previously, except for the addition of a ^{32}P-labelled non-complementary nucleotide (0.8 nmoles; 32 μCi).

RESULTS AND DISCUSSION

The introduction of etheno groupings in poly(dA) and poly(dC) templates resulted in up to a 100-fold inhibition of DNA synthesis and in up to a several hundred-fold increase in incorporation of non-complementary bases (Table 1). Both effects were dependent on

Table 1. Incorporation of complementary and non-complementary deoxyribonucleotides using CAA- or CEO-treated poly(dA) and poly(dC) templates.

Molar ratio: CAA(CEO) to nucleotide phosphorus	CAA-treated poly(dA)			CEO-treated poly(dA)		
	dTMP incorporation (pmoles)[a]	error rate x 10^5 [b] dCMP/dTMP	dGMP/dTMP	dTMP incorporation (pmoles)[a]	error rate x 10^5 [b] dCMP/dTMP	dGMP/dTMP
0.0	2000	29	27	2000	3	2
1.7	1500	28	19	2000	3	2
17.0	600	21	46	1000	4	5
170.0	30	43	3600	200	4	26
1700.0	30	40	1100	20	16	320

Molar ratio: CAA(CEO) to nucleotide phosphorus	CAA-treated poly(dC)			CEO-treated poly(dC)		
	dGMP incorporation (pmoles)[a]	error rate x 10^5 [b] dAMP/dGMP	dTMP/dGMP	dGMP incorporation (pmoles)[a]	error rate x 10^5 [b] dAMP/dGMP	dTMP/dGMP
0.0	1000	200	10	1200	200	9
0.6	1000	170	13	1200	180	6
6.3	1000	170	12	1200	180	6
63.0	400	150	28	1000	180	11
630.0	60	100	78	120	430	140

[a] The template activity is expressed as the number of pmoles of the complementary nucleotide (dTMP or dGMP) incorporated into the newly synthesized DNA.

[b] The misincorporation or error rate is expressed as the molar ratio of non-complementary to complementary nucleotide incorporated on the template.

the extents of reaction, i.e., the molar ratio of CAA or CEO to homopolymer nucleotide phosphorus.

Misincorporation of dGMP into newly-synthesized DNA was observed when CAA- or CEO-treated poly(dA) was used as a template (Table 1). That misincorporation of dGMP was mainly a consequence of depurination of poly(dA,dεA) template during the assay could be excluded, since the predicted dCMP incorporation which, on partially depurinated poly(dA), should be about four times higher than the incorporation of dGMP[18], was not observed. However, incubation of the poly(dA) templates for 20 hrs probably results in a low depurination: this idea is supported by the differential incorporation of dCMP on the two unreacted templates when either incubated for 20 min or for 20 hrs. The high misincorporation rate of dGMP observed strongly infers miscoding properties for εA. εA may thus represent a promutagenic lesion which can be expected to lead to A-T→C-G transversions.

Misincorporation of dTMP was observed on poly(dC,dεC) templates (Table 1). Untreated poly(dC) incorporated 20 times more dAMP than dTMP; this difference was shown previously for heat-denatured poly(dC) templates[19]. Therefore, our findings may be explained by deamination of cytosine residues prior to the assay to give uracil, which codes for adenine. As indicated by a decrease in the misincorporation of dAMP, the rate of deamination in the template was reduced after CAA or CEO treatment because of reaction with the N^4 amino group of cytosine. Only when poly(dC) was treated with CEO at the highest concentration (Table 1) was an exception noticed: the error rate for dAMP increased drastically. Further experiments are required to investigate the possible role as a miscoding base, of 2,3-dihydro-2-hydroxy-imidazo[1,2-c]-pyrimidin-5(6H)-one, which is the precursor of εC in the reaction of CAA with cytosine residues[20]. Our data clearly demonstrate that εC and/or its precursor miscode for T and thus possibly induce C-G→A-T transversions.

In conclusion, our data provide evidence for the miscoding properties of εA and εC, major adducts that are formed when electrophilic vinyl chloride metabolites, CEO and CAA, react with adenine and cytosine residues in DNA. The induction of A-T→C-G and C-G→A-T transversions resulting from such promutagenic lesions may explain the fact that metabolically activated vinyl chloride, CAA or CEO specifically induce base-pair substitution mutations in Salmonella typhimurium[7,21,22]. The miscoding properties of εA and εC may be responsible for the mutagenic effects of CEO found in all the genetic indicator organisms tested so far[2]. The formation of such miscoding lesions in DNA may well represent a critical step in vinyl chloride- or CEO-induced carcinogenesis[9]. Akin to alkylated oxygen atoms of DNA bases[16,23,24], we have shown that new types of modified DNA bases (i.e., adenine and cytosine with an additional imidazole ring between the exo nitrogen and the adjacent endo nitrogen) also miscode during in vitro DNA replication. This type of lesion may have relevance for other carcinogens and mutagens, like vinyl bromide, which produce εA and εC residues in RNA[25] or haloethylnitrosoureas, which give rise to structurally-related ethano derivatives of nucleic acid bases[26,27].

REFERENCES

1. J.L. Creech and M.N. Johnson, Angiosarcoma of the liver in the manufacture of polyvinyl chloride, J. Occup. Med., 16:150-151 (1974).

2. IARC Monographs on the Evaluation of the Carcinogenic Risk of Chemicals to Humans, Vol. 19, International Agency for Research on Cancer, Lyon, France (1979).

3. H. Bartsch and R. Montesano, Mutagenic and carcinogenic effects of vinyl chloride, Mutat. Res., 32:93-114 (1975).

4. A. Barbin, H. Brésil, A. Croisy, P. Jacquignon, C. Malaveille, R. Montesano, and H. Bartsch, Liver-microsome-mediated formation of alkylating agents from vinyl bromide and vinyl chloride, Biochem. Biophys. Res. Commun., 67:596-603 (1975).

5. E. Huberman, H. Bartsch, and L. Sachs, Mutation induction in Chinese hamster V79 cells by two vinyl chloride metabolites, chloroethylene oxide and 2-chloroacetaldehyde, Int. J. Cancer, 16:639-644 (1975).

6. N. Loprieno, R. Barale, S. Baroncelli, H. Bartsch, G. Bronzetti, A. Cammellini, C. Corsi, D. Frezza, R. Nieri, C. Leporini, D. Rosellini, and A.M. Rossi, Induction of gene mutations and gene conversions by vinyl chloride metabolites in yeast, Cancer Res., 36:253-257 (1977).

7. C. Malaveille, H. Bartsch, A. Barbin, A.M. Camus, R. Montesano, A. Croisy, and P. Jacquignon, Mutagenicity of vinyl chloride, chloroethyleneoxide, chloroacetaldehyde and chloroethanol, Biochem. Biophys. Res. Commun., 63:363-370 (1975).

8. U. Rannug, R. Göthe, and C.A. Wachtmeister, The mutagenicity of chloroethylene oxide, chloroacetaldehyde, 2-chloroethanol and chloroacetic acid, conceivable metabolites of vinyl chloride, Chem.-biol. Interactions, 12:251-263 (1976).

9. F. Zajdela, A. Croisy, A. Barbin, C. Malaveille, L. Tomatis, and H. Bartsch, Carcinogenicity of chloroethylene oxide, an ultimate reactive metabolite of vinyl chloride, and bis(chloromethyl)ether after subcutaneous administration and in initiation-promotion experiments in mice, Cancer Res., 40:352-356 (1980).

10. G. Bonse, T. Urban, D. Reichert, and D. Henschler, Chemical

reactivity, metabolic oxirane formation and biological
reactivity of chlorinated ethylenes in the isolated perfused
rat liver preparation, Biochem. Pharmacol., 24:1829-1834
(1975).

11. R.J. Laib and H.M. Bolt, Alkylation of RNA by vinyl chloride
metabolites in vitro and in vivo: formation of $1,N^6$-etheno-
adenosine, Toxicology, 8:185-195 (1977).

12. R.J. Laib and H.M. Bolt, Formation of $3,N^4$-ethenocytidine
moieties in RNA by vinyl chloride metabolites in vitro and
in vivo, Arch. Toxicol., 39:235-240 (1978).

13. J.R. Barrio, J.A. Secrist III, and N.J. Leonard, Fluorescent
adenosine and cytidine derivatives, Biochem. Biophys. Res.
Commun., 46:597-604 (1972).

14. T. Green and D.E. Hathway, Interactions of vinyl chloride
with rat-liver DNA in vivo, Chem.-biol. Interactions,
22:211-224 (1978).

15. L.L. Gerchman and D.B. Ludlum, The properties of O^6-methyl-
guanine in templates for RNA polymerase, Biochim. Biophys.
Acta, 308:310-316 (1973).

16. P.J. Abbott and R. Saffhill, DNA synthesis with methylated
poly(dC-dG) templates. Evidence for a competitive nature
to miscoding by O^6-methylguanine, Biochim. Biophys. Acta,
562:51-61 (1979).

17. A.E. Pegg, Formation and metabolism of alkylated nucleosides:
possible role in carcinogenesis by nitroso compounds and
alkylating agents, Adv. Cancer Res., 25:195-269 (1977).

18. C.W. Shearman and L.A. Loeb, Effects of depurination on the
fidelity of DNA synthesis, J. Mol. Biol., 128:197-218 (1979).

19. S. Boiteux, G. Villani, and M. Radman, Replicational fidelity
of E. coli DNA polymerase I and III and AMV reverse
transcriptase on intact, UV irradiated and heat deaminated
poly(dC) template, Biochim. Biophys. Acta, in press (1980).

20. J. Biernat, J. Ciesiolka, P. Gornicki, R.W. Adamiak, W.J.
Krzyżosiak, and M. Wiewiórowski, New observations concerning
the chloroacetaldehyde reaction with some tRNA constituents.
Stable intermediates, kinetics and selectivity of the
reaction, Nucleic Acids Res., 5:789-804 (1978).

21. J. McCann, V. Simmon, D. Streitwieser, and B.N. Ames,
Mutagenicity of chloroacetaldehyde, a possible metabolic
product of 1,2-dichloroethane (ethylene dichloride), chloroethanol (ethylene chlorohydrin), vinyl chloride, and cyclophosphamide, Proc. Natl. Acad. Sci. USA, 72:3190-3193 (1975).

22. U. Rannug, A. Johansson, C. Ramel, and C. Wachtmeister,
Mutagenicity of vinyl chloride after metabolic activation,
Ambio, 3:194-197 (1974).

23. P.J. Abbott and R. Saffhill, DNA synthesis with methylated
poly(dA-dT) templates: possible role of O^4-methylthymine as
a pro-mutagenic base, Nucleic Acids Res., 4:761-769 (1977).

24. B. Singer, N-Nitroso alkylating agents: formation and persistence
of alkyl derivatives in mammalian nucleic acids as

contributing factors in carcinogenesis, J. Natl. Cancer
Inst., 62:1329-1339 (1979).

25. H. Ottenwälder, R.J. Laib, and H.M. Bolt, Alkylation of RNA by
    vinyl bromide metabolites in vitro and in vivo, Arch.
    Toxicol., 41:279-286 (1979).
26. D.B. Ludlum, B.S. Kramer, J. Wang, and C. Fenselau, Reaction of
    1,3-bis(2-chloroethyl)-1 nitrosourea with synthetic poly-
    nucleotides, Biochemistry, 14:5480-5485 (1975).
27. W.P. Tong and D.B. Ludlum, Mechanism of action of the nitroso-
    ureas-III. Reaction of bis-chloroethyl nitrosourea and
    bis-fluoroethyl nitrosourea with adenosine, Biochem.
    Pharmacol., 28:1175-1179 (1979).

THE FUNCTION OF METHYLATED BASES IN DNA OF Escherichia coli

M.G. Marinus

Department of Pharmacology, University of Massachusetts
Medical School, Worcester MA   01605   U.S.A.

The DNA of E.coli K-12 contains 6-methyladenine (6-meA)
and 5-methylcytosine (5-meC). Approximately two percent of all
adenines are methylated and one percent of all cytosines
(1,2,3). These methylated bases are formed by DNA methylases
which transfer methyl groups from S-adenosyl-L-methionine to
specific base sequences in newly synthesized daughter strand DNA
(4,5). Three distinct DNA methylases have been detected in
E.coli K-12. The hsd DNA adenine methylase, the dam DNA adenine
methylase and the dcm DNA cytosine methylase. The hsd DNA
adenine methylase is involved in classical restriction and
modification and produces very few methylated adenines in DNA.
For further information regarding this enzyme, the reader is
referred to recent reviews (6,7).

The dcm DNA cytosine methylase recognises
5'-CC(A/T)GG-3' sequences in DNA and methylates the internal
cytosine residue (8). A mutant strain in which the dcm gene is
deleted has been isolated and characterized (9) and shows no
gross abnormality with respect to growth, mutagenesis, or
survival after ultra-violet (UV) irradiation (10). Since this
mutant strain is viable, the DNA cytosine methylase is not
required for viability.

Plasmids propagated in dcm mutants are cleaved to
completion by R.EcoRII whereas plasmids propagated in wild type
strains are resistant to cleavage (11). This finding can be
interpreted to mean that a function of the dcm gene product is
to protect host DNA from endonucleases produced by plasmids.

A role for 5-meC residues in spontaneous mutagenesis
has been proposed (12) and is based on the finding that hotspots
in the lacI gene occur in sequences containing 5-meC.
Spontaneous deamination of 5-meC yields thymine which, if

unrepaired, after the next round of DNA replication should
result in a G-C to A-T base pair change. dcm⁻ strains, as
expected, do not yield hotspots (12). Creation of a new
5'-CCAGG-3' sequence in DNA yields a hot spot at this sequence
(12). Although the balance of experimental data favors a role
for 5-meC in spontaneous mutagenesis, it remains to be
explained, however, that of four 5'-CCAGG-3' sequences in the
lacI gene, only two promote hotspot formation.

The dam adenine methylase recognizes the sequence
5'-GATC-3' in double stranded DNA (13,14,15). The DNA of dam-3
bacteria has no detectable methylation of adenine in these
sequences (13), nor any detectable enzyme activity in vivo or in
vitro (10) suggesting that the dam adenine methylase may be
dispensable for viability.

Mutation in the dam gene results in a variety of
phenotypes designated Dam, AP, UV, Vrm, Flh, Sli and Smf (16).
The Dam phenotype refers to the amount of 6-meA in DNA; AP^s to
inhibition of colony formation in the presence of 2-aminopurine;
UV^s to a slight increase in sensitivity to UV light; Vrm⁻ to
the inviability of dam⁻ recA⁻, dam⁻ recB⁻, dam⁻
recC⁻ and dam⁻ lexA⁻ strains; Flh to the "hyper-rec"
phenotype (17); Sli^H to an increased number of free lambda
phages in lysogenic cultures (18) and Smf to an increased rate
of spontaneous mutation (19,20). The reversion of missense,
nonsense (21) and + or − frameshift mutations (Table 1) can
occur in dam mutants.

Table 1

Effect of dam-3 on reversion of frameshift mutations as
measured by revertant frequency

Frameshift mutation	Revertant Frequency dam⁺	dam-3	$\dfrac{\text{dam-3}}{\text{dam}^+}$
lacZ (ICR36)	4.4×10^{-10}	1.3×10^{-8}	30
lacZ (ICR48)	3.9×10^{-8}	3.1×10^{-7}	7.9
trpA21 (−)	2×10^{-11}	5.3×10^{-9}	250
trpA9813 (−)	4.3×10^{-9}	5.4×10^{-8}	13
trpE9777 (+)	2.5×10^{-8}	1.6×10^{-6}	64.

The phenotypes listed above are not present in dam⁺
revertants isolated from dam-3 strains (16). Similarly, dam-3
bacteria harboring the plasmid pMQ3 do not show these
phenotypes. pMQ3 is a recombinant plasmid containing the dam
gene of E.coli (Arraj and Marinus, unpublished data).

Some of the phenotypes of dam mutants are suppressed by
mutation in other genes (16). These include mutS, mutL, uvrE,
recL and sin. Mutation in mutH can also suppress the AP
phenotype of dam mutants (22).

To explain the early observations with dam mutants Marc

Shulman in a letter to the author in 1974 proposed that DNA "replication per se is not sufficiently accurate. The bacterium recognizes that it made a replication error by perceiving a base pair mismatch. Perhaps at this point in time, just after the error has been made, only the parental, the correct, strand is methylated, thus affording the bacterium a criterion for choosing the correct strand as the master strand when it corrects the mismatch."

The idea that methylation is involved in repair of mismatched bases was also suggested by Wagner and Meselson (23) and subsequent experimental data has shown that 6-meA residues determine strand specificity during repair (Ryokowski, Pukkila, Radman, Wagner and Meselson, unpublished data). In addition, Nevers and Spatz (24) and Rydberg (25,26) have shown that the uvrD, uvrE, mutL, mutR and mutS gene products are required for efficient removal of mismatched base pairs from DNA. From studies on the AP phenotype of dam mutants, Glickman and Radman found that double mutants of the genotype dam⁻ mutL⁻, dam⁻ mutR⁻ and dam⁻ mutS⁻ are resistant to 2-aminopurine, from which they formulated a model (22,27) similar to that of Shulman.

A model for the role of 6-meA in mismatch repair during E.coli DNA replication supposses that newly synthesized daughter strand DNA is not methylated immediately after synthesis. Instead there is a delay during which mismatched base pairs are corrected using the methylated parental strand as template.

In dam⁻ mut⁺ mutants, neither strand is methylated and since the mismatch repair system is operative, mistakes occur because the repair system cannot discriminate between parental and daughter strands. The phenotypes of dam mutants presumably are consequences of repair in the parental strand and should not be observed in the absence of such repair. The phenotypes of dam⁻ mutL⁻ and dam⁻ mutS⁻ are those of the wild type except for a high rate of spontaneous mutagenesis (16) suggesting that mismatch repair in such strains is inefficient and that mismatched base pairs persist in DNA leading to mutations after DNA replication.

The above model can account for the Smf, Sli, Flh and AP phenotypes of dam mutants. The Vrm and UV phenotypes, however, are not readily explained by the model. The model also does not explain the failure of dam mutants to survive a challenge with a high dose of methyl-nitro-nitrosoguanidine (MNNG) after adaption with a low dose of MNNG (28). dam⁻ mutL⁻ or dam⁻ mutS⁻ strains also cannot survive a challenge dose of MNNG after adaption with MNNG (Marinus, unpublished data). The simplest explanation for the phenotypes which do not fit the model would be that DNA adenine methylation has another role in cellular metabolism apart from that in mismatch repair or that there may be multiple mismatch repair pathways.

Another function of 6-meA may involve DNA replication.

An interesting feature of the E.coli K-12 replication origin is
that the-GATC-sequence is eight times more abundant than
expected (34,35). The preponderance of these sequences at the
origin may be to insure a high level of mismatch correction, or
it may have some other function for the initiation of DNA
replication. The observation by Lark (29) that unmethylated DNA
cannot serve as a template for further DNA synthesis remains
unexplained. Gomez-Eichelman and Lark (30) have found that the
ends of Okazaki pieces are non-randomly distributed with respect
to 6-meA. Lark (31) has shown that lysates prepared from E.coli
cells grown in the presence of ethionine show a marked decrease
in ability to support DNA synthesis. Restoration of synthesis
by DNA polymerase I can be achieved by the addition of
S-adenosyl-L-methionine and ATP.

A further area of interest is the interaction of the
dam function of E.coli and the mom function of bacteriophage
Mu. Resistance to restriction of Mu is controlled by a
modification function designated mom. For efficient restriction
of mom⁻ Mu phage, it is necessary to propagate it in a dam
mutant (32). mom function does not alter the sequence
specificity of the dam DNA methylase but the interaction leads
to the formation of a new modified base in a DNA sequence
distinct from that for dam (33).

In conclusion, the roles of 5-meC and 6-meA in
spontaneous mutagenesis appear to be established. The role of
6-meA in mismatch repair also appears established, at least for
bacteriophage lambda. It is also clear that there remain
several unresolved problems that may lead to the discovery of
other biological functions for methylated bases.

Acknowledgements

This work was supported by USPHS grant GM22055 and a
Faculty Research Award (FRA-149) from the American Cancer
Society.

References

(1)   Srinivasan, P.B. and E. Borek (1964) Science 145:548-553
(2)   Dunn, D.B. and J.D. Smith (1955) Nature, 175:336-337
(3)   Doskocil, J. and Z. Sormova (1965) Biochim. Biophys.
        Acta 95:513-515
(4)   Borek, E., and P.R. Srinivasan (1966) Ann. Rev.
        Biochem. 35:275-297
(5)   Billen, D. (1968) J. Mol. Biol 31:477-486
(6)   Meselson, M., R. Yuan and J. Heywood (1972) Ann. Rev.
        Biochem. 41:447-466
(7)   Arber, W. (1974) Prog. Nucleic Acid Res. Mol. Biol.
        14:1-50
(8)   May, M.S. and S. Hattman (1975) J. Bacteriol.
        123:768-770

(9)   Kondoh, H. and H. Ozeki (1977) Genetics 84:403-421

(10)   Bale, A., M. d'Alarcao and M.G. Marinus (1979) Mutation Res. 59:157-165

(11)   Schlagman, S., S. Hattman, M.S. May and L. Berger (1976) J. Bacteriol. 126:990-996

(12)   Coulondre C., J.M. Miller, P.J. Farrabaugh and W. Gilbert (1978) Nature 274:775-780

(13)   Lacks, S. and B. Greenberg (1977) J. Mol. Biol. 114:153-168

(14)   Hattman, S., J.E. Brooks and M. Masurekar (1978) J. Mol. Biol 126:367-380

(15)   Geier, G.E. and P. Modrich (1979) J. Biol. Chem. 254:1408-1413

(16)   McGraw, B.R. and M.G. Marinus (1980) Mol. Gen. Genet. 178:309-315

(17)   Marinus, M.G. and E.B. Konrad (1976) Mol. Gen. Genet. 149:273-277

(18)   Marinus, M.G. (1980) J. Bacteriol 141:223-226

(19)   Marinus, M.G. and N.R. Morris (1974) J. Mol. Biol. 85:309-322

(20)   Glickman, B.W. (1979) Mutation Res. 61:153-162

(21)   Marinus, M.G. and N.R. Morris (1975) Mutat. Res. 28:15-26

(22)   Glickman, B.W. and M. Radman (1980) Proc. Natl. Acad. Sci., U.S.A. 77:1063-1067

(23)   Wagner, R.W. and M. Meselson (1976) Proc. Natl. Acad. Sci., U.S.A. 73:4135-4139

(24)   Nevers, P. and H.C. Spatz (1975) Mol. Gen. Genet. 139:233-248

(25)   Rydberg, B. (1978) Mutat. Res. 52:11-24

(26)   Rydberg, B. (1977) Mol. Gen. Genet. 152:19-28

(27)   Radman, M., G. Villani, S. Boiteux, A.R. Kinsella, B.W. Glickman, and S. Spadari Cold Spring Harbor Symp. Quant. Biol. 43:937-946

(28)   Jeggo, P., M. Defais, L. Samson and P. Schendel (1978) Molec. Gen. Genet. 162:299-205

(29)   Lark, C. (1968) ᵀ. Mol. Biol. 31:401-414

(30)   Gomez-Eichelmann, M.C. and K.G. Lark (1977) J. Mol. Biol. 117:621-635

(31)   Lark, C. (1979) J. Bacteriol. 137:44-50

(32)   Toussaint, A. (1977) J. Virol. 23:825-826

(33)   Hattman, S. (1979) J. Bacteriol. 32:468-475

(34)   Sugimoto, K., A. Oka, H. Sugisaki, M. Takarami, A. Nishimura, Y. Yasuda and Y. Hirota (1979) Proc. Natl. Acad. Sci., U.S.A. 76:575

(35)   Meijer, M., E. Beck, F.C. Hansen, H.E.N. Bergmans, W. Messer, K. von Meyenberg and H. Schaller (1979) Proc. Natl. Acad. Sci., U.S.A. 76:580

ROLE OF ADP-RIBOSE IN DNA REPAIR

$(ADP\text{-}RIBOSE)_n$, A NEW COMPONENT IN DNA REPAIR

Sydney Shall, Barbara Durkacz, Diane Ellis, Judy Irwin,
Pamela Lewis and Mohan Perera
Biochemistry Laboratory
University of Sussex
Brighton, Sussex BN1 9QG, England

It has been known for about 25 years that DNA-damaging chemicals and radiation will lower cellular NAD levels. We provide a biochemical explanation for these observations which leads us into the hypothesis that NAD metabolism participates in DNA repair.

The conventional view of NAD is of an oxidation-reduction cofactor. However, NAD is also the substrate of transferase enzymes that transfer the ADP-ribose fraction of NAD (Fig.1) to proteins or to other ADP-ribose molecules. This reaction has been extensively reviewed (1-4). These enzymes catalyse the formation of mono-(ADP-ribose)-protein conjugates and of poly(ADP-ribose). The structure of these $(ADP\text{-}ribose)_n$ compounds is shown in Fig.1; the diagnostic feature is the unusual ribosyl (1"-2')-riboside bonds. The $(ADP\text{-}ribose)_n$ are covalently attached to chromatin proteins. In undamaged cells about 90% of the (ADP-ribose) is present as monomers: the small percentage present as polymer may contain up to 70 (ADP-ribose) units in a chain. However, the total amount of (ADP-ribose) present in undamaged cells is very small; it is estimated that there is about 1 (ADP-ribose) residue for every 5000 DNA nucleotides.

$(ADP\text{-}ribose)_n$ biosynthesis is totally dependent on the presence of DNA; furthermore, closed, circular DNA does not activate the enzyme; but the ability of this DNA to activate the enzyme is proportional to the number of nicks in the DNA. Various experiments indicate that fragmentation of DNA increases its ability to activate the enzyme (4,5).

In nucleated cells all NAD synthesis is chromosomal; the vast majority of NAD formed is used for the DNA-dependent biosyn-

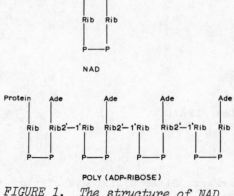

FIGURE 1. *The structure of NAD and of ADP-ribose.*

thesis of $(ADP\text{-}ribose)_n$. Both the synthesis and the degradation of $(ADP\text{-}ribose)_n$ are confined to chromatin. The requirement of (ADP-ribose) transferase is for double-stranded DNA. (ADP-ribose) is hydrolysed by a specific, chromatin-bound glycohydrolase which is inhibited by single-stranded DNA. The product of the reaction is ADP-ribose, which is presumably hydrolysed to AMP and ribose-phosphate, both of which enter the nucleotide pool very rapidly.

Because DNA fragmentation activates the enzyme, reliable enzyme assays may be done best in cells made permeable to nucleotides rather than with isolated nuclei or chromatin, in which some DNA single-strand breaks are inevitable (5). Therefore, the enzyme activity in permeabilised cells approximates the physiological activity in intact cells. It turns out that there is in addition, a large potential pool of enzyme activity in permeabilised cells - this can be revealed by fragmenting the DNA with DNAases. The expression of this potential activity does not require protein synthesis.

A variety of compounds inhibit $(ADP\text{-}ribose)_n$ biosynthesis, with Ki values between 1 and 50 µM. Nicotinamide, a product of the reaction, is a good inhibitor, but it is also an autocatalytic substrate for the biosynthesis of NAD. Therefore, we use 5-methyl-nicotinamide which is a powerful inhibitor of $(ADP\text{-}ribose)_n$ biosynthesis, but does not stimulate NAD synthesis. Secondly, methyl-xanthines such as theobromine, theophylline, IBMX and caffeine are inhibitors. While the first three methylxanthines are good inhibitors, caffeine is a very poor inhibitor indeed. Thirdly, thymine and thymidine are good inhibitors, but neither TMP nor any of the other nucleosides or bases are inhibitory. Fourthly, benzamides are very powerful inhibitors, with a Ki in permeable cells of around 4 µM. We use 3-aminobenzamide which is non-toxic up to 10 mM and does not inhibit incorporation of thymidine, uridine or leucine in exponentially-growing cells. These observations show that $(ADP\text{-}ribose)_n$ biosynthesis is not needed in undamaged, steady-state, exponential cultures for growth, for DNA, RNA or for protein biosynthesis. Whilst nicotinamide and amino-benzamide are good inhibitors, nicotinic acid and aminobenzoic acid are not inhibitory at all.

Finally, the direct hydrolysis of NAD by the non-nuclear enzyme, NAD glycohydrolase is not inhibited by theobromine,

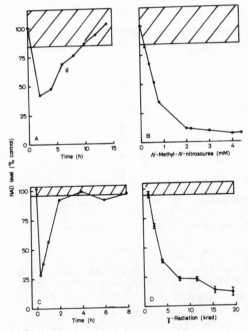

FIGURE 2. Effect of N-methyl-N-nitrosourea and γ-radiation on NAD levels in mouse leukaemia (L1210) cells). The hatched area shows the control levels ± 1 S.D. Top left: Time course of the effect of 800 μM MNU.
Top right: Dose response of the drop in cell NAD 2 hours after treatment with MNU.
Lower left: Time course of the effect of 12 krad of γ-radiation.
Lower right: Dose response of the drop in cell NAD 15 minutes after γ-radiation. (Reproduced from ref. 7).

thymidine or 3-aminobenzamide. Conversely, isonicotinic acid hydrazide inhibits the direct hydrolysis of NAD, but does not inhibit the biosynthesis of $(ADP\text{-}ribose)_n$.

We have confirmed that a wide variety of DNA-damaging agents will lower cellular NAD. The biochemical explanation for this result was suggested when we showed in 1975 that an alkylating agent both lowered the NAD level and activated (ADP-ribose)$_n$ transferase activity (6). The drop in cellular NAD is demonstrated (Fig.2) for both MNU and for γ-radiation (7,8). This decrease is caused also by streptozotocin, DMS (Fig.3, left), MMS, EMS, BCNU, MeCCNU, BMBA and neocarcinostatin. This decrease in NAD level can be totally prevented by the enzyme inhibitors, 5-methylnicotinamide, theobromine, theophylline, IBMX, thymidine or by 3-aminobenzamide (Fig.3, centre), but not by 3-aminobenzoic acid.

The specific activity of (ADP-ribose)$_n$ transferase in permeabilised cells is increased by the above DNA-damaging agents in a dose-dependent way. Moreover, the kinetics with which the NAD content decreases is different for each agent; it is very fast for γ-radiation and much slower for MNU (Fig.2). The kinetics of activation of the transferase enzyme correspond; there is a very rapid rise and fall after γ-radiation and a lower rise and fall with MNU or with DMS (Fig.3, right).

The conclusion from these observations is that DNA-damaging agents cause the utilisation of cellular NAD; and that this is mediated by the chromatin-bound, DNA-dependent enzyme, (ADP-ribose)$_n$ transferase.

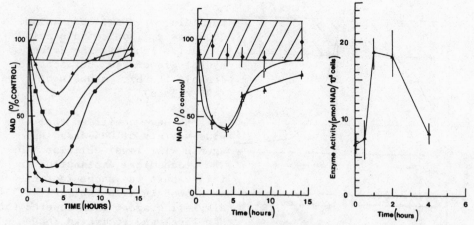

FIGURE 3. *The effect of DMS and of (ADP-ribose) transferase inhibitors on the NAD content of L1210 cells.*
Left: Cells were treated with increasing doses of DMS, and samples were removed at intervals for NAD assay. ▲, *10.5 μM;* ■, *52.6 μM;* ●, *105.3 μM;* ◆, *210.6 μM. Middle: L1210 cells were treated with 53 μM DMS either alone (Δ) or in the presence of either 2mM 3-aminobenzamide (◆) or of 2mM 3-aminobenzoic acid (Δ). Samples were removed at intervals for NAD estimation. Right: Cells were treated with 1mM DMS and the ADP-ribose transferase activity was determined at the indicated times in permeabilised cells (8, **reprinted with permission from Macmillan Journals Ltd.**).*

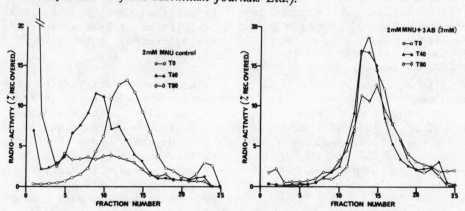

FIGURE 4. *The effect of 3-aminobenzamide on excision repair in L1210 cells after MNU. The cells were exposed to 2mM MNU for 20 minutes. Samples were taken at the end of this period (T0), 40 minutes later and 80 minutes later, and analysed on alkaline sucrose gradients. Left (4a): MNU treatment only. Right (b): The cells were exposed for 20 minutes to MNU, which was then washed out, and 3mM 3-aminobenzamide was added to the medium for the 80 minute recovery period. The bottom of the gradient is to the left, in this and in all subsequent gradients.*

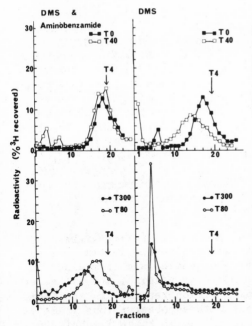

FIGURE 5. The effect of 3-amino-benzamide on excision repair in L1210 cells after DMS. Right hand panels: The cells were exposed to 100 µM DMS for 20 minutes and then resuspended in medium for 300 minutes. Left hand panels: The cells were exposed for 20 minutes to DMS, which was then washed out, and 3mM 3-aminobenzamide was added to the medium for the 300 minute recovery period. The DNA was then analysed on alkaline sucrose gradients (8) (T4), is bacterio-phage T4 DNA, single strand M.W. 55×10^6.(8, reprinted with permission from Macmillan Journals Ltd.).

If DNA-damaging agents induce the biosynthesis of (ADP-ribose)$_n$, then we may infer that perhaps (ADP-ribose)$_n$ biosynthesis is part of the cellular response to DNA-damage and that it is required for efficient repair. This hypothesis has been tested in three quite independent ways and in each case the observations support the hypothesis.

We have examined with alkaline sucrose gradients the resealing of single-strand breaks. This procedure includes breaks produced by the alkaline condi-tions of the gradient. After exposure to 2 mM MNU for 20 minutes the DNA sediments as a broad peak in the middle of the gradient (Fig.4a). After 40 minutes the DNA has increased in size and by 80 minutes, most of the DNA is found at the bottom of the tube (Fig.4a). By contrast, when 3 mM 3-aminobenzamide is present during the recovery period there is essentially no increase in the size of the DNA in 80 minutes (Fig.4b). Similarly, when cells are treated with 100 µM DMS, the DNA is fragmented (Fig.5, top right), and during the subsequent 80 minutes the DNA increases in size so that it sediments to the bottom of the tube (Fig.5, bottom right). However, when the enzyme inhibitor 3-aminobenzamide (3 mM) is added to the cells following removal of DMS, the DNA increases in size very slowly indeed. There is no observable increase in molecular weight of the DNA for 40 minutes after the removal of the DMS (Fig.5, top left). After 80 minutes, the DNA is increased slightly in size, but still sediments in the middle of the gradient (Fig.5, bottom left). Even after 5 hours the bulk of the DNA has not yet returned to its original size (Fig.5, bottom left). Addition of 3 mM 3-aminobenzoic acid does not retard excision repair following DMS treatment (8).

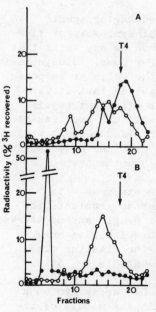

FIGURE 6. *The effect of NAD depletion on excision repair after DMS. L1210 cells were grown for 4 generations in nicotinamide-free medium (NAD was 10 to 20% of control). The culture was divided into two aliquots and nicotinamide was added to one to 9 µg/ml. 4 hours later the cells were exposed to 100 µM DMS for 20 minutes and then their ability to carry out excision repair was tested on alkaline sucrose gradients. Time: 0, 0 minutes; ●, 80 minutes; A, NAD depleted cells. B, Nicotinamide added back; the NAD level had returned to normal* **(8, reprinted with permission from Macmillan Journals Ltd.)**

Other inhibitors of the transferase also inhibit excision repair in L1210 cells. Both 5-methylnicotinamide and thymidine (in the presence of deoxycytidine) are effective inhibitors of excision repair. In addition, we have repeated these observations in mouse fibroblast 3T3 cells, in human lymphoid cells and in human, normal glial cells. Thus, we have demonstrated that a range of inhibitors of the enzyme, (ADP-ribose)$_n$ transferase, inhibit excision repair in a variety of cells from both mouse and humans.

The preceding evidence that (ADP-ribose)$_n$ biosynthesis participates in DNA repair rests heavily on experiments with enzyme inhibitors. Although we have shown that all four types of enzyme inhibitors are effective (8), the possibility remains that all these polymerase inhibitors might also coincidentally inhibit some aspect of DNA repair directly. Therefore, we have tested the ability of cells to carry out excision repair when the cellular NAD content is lowered by nutritional deprivation. The NAD levels of cells may be depleted by growing them in a nicotinamide-free medium. Even with quite low NAD content (10 to 20% of normal) the L1210 cells continue to grow and divide at the normal rate. The rate of incorporation of ^{3}H-thymidine by these NAD-depleted cells is the same as that of cells grown either in complete medium or in depleted medium supplemented with nicotinamide for 4 hours. The rate of incorporation of both uridine and leucine is the same in depleted and normal cells. The NAD content of L1210 cells is 750 pmol per 10^6 cells; the NAD concentration is therefore about 1000 µM. The K_m value for the DNAase-activated transferase in permeabilised cells is about 300 µM. Thus, lowering the NAD content of these cells should decrease the enzyme activity, and thereby inhibit DNA repair.

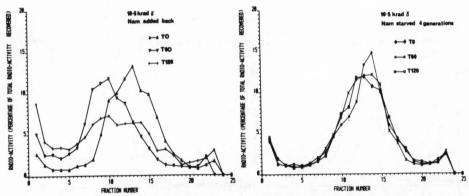

FIGURE 7. The effect of NAD depletion on strand rejoining in L1210 cells after γ-radiation. The experiment was as described in Figure 6, except that the cells were exposed to 10.5 krad of γ-radiation in air instead of to DMS.

After exposure to 100 μM DMS, the DNA in NAD-depleted L1210 cells becomes fragmented (Fig.6a), and not only is strand-rejoining totally inhibited, but the DNA is further decreased in size during the subsequent 80 min (Fig.6a). The ability to rejoin the strand breaks is recovered if, before exposure to DMS, the cells are preincubated for 4 hours in the presence of nicotinamide which returns the NAD to normal levels (Fig.6b).

Similarly, after 10.5 krad of γ-radiation, NAD-depleted cells cannot rejoin the single-strand breaks (Fig.7,right), although cells with a normal NAD level can do so (Fig.7, left). This has been shown in mouse fibroblasts, too. When mouse fibroblast, 3T3 cells, are depleted of NAD, they are unable to rejoin strand breaks induced by 4mM MNU (Fig.8, right), although when re-fed with nicotinamide, they are able to increase the size of the fragmented DNA (Fig.8, left).

We show by these experiments that cells with a low NAD level cannot rejoin strand breaks induced by chemicals or by radiation.

There has been some discussion of the function of NAD glyco-hydrolase, which directly hydrolyses NAD to nicotinamide and ADP-ribose. Is this enzyme involved in DNA repair? We have shown that theobromine, thymidine and 3-aminobenzamide, which do not inhibit the NAD glycohydrolase, are (1) able to prevent the lowering of cellular NAD by chemicals and by radiation, (2) do inhibit excision repair, and (3) do potentiate the cytotoxicity of these agents (see below). It would seem that NAD glycohydro-lase is not involved in DNA repair. More direct evidence for this conclusion comes from examining the effect of isonicotinic acid hydrazide on strand rejoining. We observe that 3 mM isonicotinic acid hydrazide does not at all inhibit single-strand

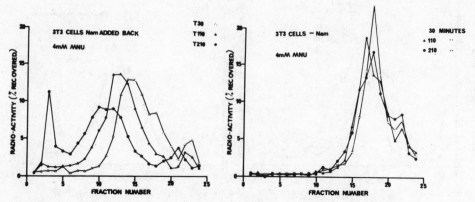

FIGURE 8.   The effect of NAD depletion on excision repair in mouse
fibroblast, 3T3, cells after MNU.   The experiment was as described
in Figure 6, except that the cells were exposed to 4mM MNU instead
of to DMS.

rejoining after exposure to 100 μM DMS.

    If (ADP-ribose)$_n$ biosynthesis participates in the cellular
recovery from DNA damage, then inhibition of (ADP-ribose)$_n$ biosynthesis would be expected to increase the lethality (and perhaps
also the mutagenicity) of DNA-damaging agents.

    We have tested this prediction with a variety of DNA-damaging
agents and all four types of enzyme inhibitors. Our test is the
ability of Ll210 cells to grow and form colonies in soft agar.
Undamaged cells have a plating efficiency of 90 ± 10%. We express
all our results as relative plating efficiencies, comparing the
survival of treated cells to that of untreated cells. The cells
were treated with the DNA-damaging agent, washed and plated in
soft agar. The soft agar contained the enzyme inhibitor and
therefore the cells were exposed to the inhibitor during all the
time subsequent to DNA damage. Furthermore, we chose concentrations of enzyme inhibitors which are by themselves non-toxic, as
shown in separate plating experiments. If the enzyme inhibitors
cause any increase in cell killing then this is clearly a synergistic effect with the DNA-damaging agent.

    In Fig.9 we display the synergistic potentiation of cell
killing by MNU and (ADP-ribose)$_n$ transferase inhibitors (9). A
dramatic, synergistic enhancement of MNU cytotoxicity by the enzyme
inhibitors is observed in Ll210 cells. When 2 mM 5-methylnicotina-
mide is present during a 1 hour exposure to MNU in addition to
being present during the subsequent cloning assay, the lethality
of the MNU is considerably increased (Fig.9A). At the 10% and 50%
survival levels there are, for MNU, dose enhancement factors of
14 induced by 2 mM 5-methylnicotinamide. The enhancement of MNU

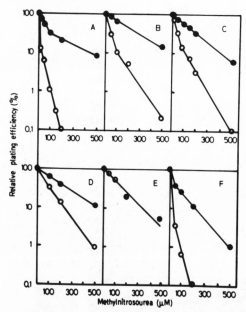

FIGURE 9. Synergistic potentiation of MNU toxicity by inhibitors of ADP-ribose transferase. (A-F) (●) Methyl-nitroso-urea; O, (A) plus 2mM 5-methylnicotinamide; (B) plus 250 μM theobromine; (C) plus 250 μM theophylline; (D) plus 2mM thymidine and 3mM deoxycytidine; (E) plus 2mM nicotinate; (F) plus 250 μM caffeine. *(Reproduced from ref. 9).*

toxicity by non-toxic doses of the enzyme inhibitors is evident also with the methylxanthines. 250 μM theobromine (Fig.9B) and 250 μM theophylline (Fig.9C) give a marked increase in toxicity. Caffeine, despite being a much weaker inhibitor of (ADP-ribose) transferase, gives an effect similar to the other methylxanthines (Fig.9F) possibly related to inhibition of post-replication repair. Thymidine, in the presence of deoxycytidine also potentiates the toxicity of MNU (Fig.9D). 3 mM deoxycytidine alone does not affect MNU toxicity. In contrast to the enzyme inhibitors, 2 mM sodium nicotinate, which does not inhibit the enzyme, does not enhance the toxicity of MNU (Fig. 9E).

Similarly, the enzyme inhibitors enhance the lethality of DMS. Exposure of L1210 cells to DMS for 1 hour results in a dose-dependent cytotoxicity (Fig. 10, left) (6). 90 μM DMS gives 10% survival. However, the toxicity of DMS is greatly increased when 2 mM 3-aminobenzamide is present during both exposure of the cells to DMS and in the subsequent cloning assay. In these conditions 22 μM DMS is sufficient to give 10% survival; this represents a four-fold dose enhancement. 3-aminobenzoic acid (2mM) has no effect on the cytotoxicity of DMS (Fig.10, left). In addition, 5-methylnicotinamide, thymidine/deoxycytidine (Fig.10, right) and theobromine potentiate the cytotoxicity of DMS. This potentiation of cytotoxicity is observed also with several other monofunctional alkylating agents including MMS and EMS, and with representatives of each class of enzyme inhibitor.

This synergistic increase in toxicity is evident also with γ-radiation and with neocarcinostatin, but with these agents the dose enhancement factor is very small. Only very large, supralethal doses of γ-radiation cause a lowering of cellular NAD. In addition, it has not yet been possible to demonstrate inhibition of single-strand rejoining by the enzyme inhibitors on alkaline gradients

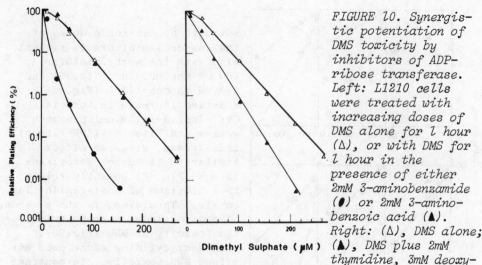

FIGURE 10. Synergistic potentiation of DMS toxicity by inhibitors of ADP-ribose transferase. Left: L1210 cells were treated with increasing doses of DMS alone for 1 hour (Δ), or with DMS for 1 hour in the presence of either 2mM 3-aminobenzamide (●) or 2mM 3-aminobenzoic acid (▲). Right: (Δ), DMS alone; (▲), DMS plus 2mM thymidine, 3mM deoxycytidine. Cells were plated for colony counting in 0.5% (w/v) agar in the presence or absence of inhibitors.

after γ-radiation. In marked contrast, in nicotinamide-deprived cultures strand-rejoining is completely inhibited (Fig.7) and NAD-deprived cultures show increased cell killing after γ-radiation compared to re-fed cultures. This difference between the responses to γ-radiation and to simple, monofunctional, alkylating agents implies that, of the characteristic set of lesions induced by each agent, one subset involves (ADP-ribose) and another subset does not. The magnitude of the (ADP-ribose) response depends, presumably on the proportion of these two subsets.

DNA repair after alkylation damage was severely inhibited in mouse lymphoid, L1210 cells both by the enzyme inhibitors and by nicotinamide deprivation. However, in mouse fibroblast (3T3) and in human glial cells inhibition of strand-rejoining by the enzyme inhibitors was much less marked, although nicotinamide deprivation totally inhibited strand-rejoining. The reason for the discrepancy between the enzyme inhibitors and NAD deprivation is unknown; perhaps the two procedures inhibit repair by the same mechanism but the NAD deprivation does so much more effectively; alternatively, it may cause other, incidental metabolic disturbances which prevent efficient repair.

The evidence that (ADP-ribose)$_n$ is involved in DNA repair may be summarised. Both alkylating agents and radiation lower cellular NAD contents, and they both elevate (ADP-ribose)$_n$ transferase activity. The kinetics of NAD disappearance corresponds with the kinetics of (ADP-ribose) transferase activation. The drop in NAD content and the rejoining of single-strand breaks are inhibited by

the specific inhibitors of (ADP-ribose)$_n$ transferase, all of which
synergistically potentiate the cytotoxicity of DNA-damaging agents.
In addition, cells that are nutritionally deprived of NAD cannot
efficiently execute excision repair. N-methyl-N-nitro-N-nitroso-
guanidine lowers cellular NAD and increases poly(ADP-ribose) in
intact cells (10).

We can safely conclude that NAD metabolism, and specifically
(ADP-ribose)$_n$ biosynthesis participates in DNA excision repair.

The work was supported by the Medical Research Council, the
Science Research Council and the Cancer Research Campaign.

1. T. Sugimura. Prog.Nucleic Acid Res. Mol.Bio. 13:127-151 (1973).
2. H. Hilz and P.R. Stone. Rev.Physiol.Biochem.Pharmacol. 76:1-58
   (1976).
3. O. Hayaishi and K. Ueda. Ann.Rev.Biochem. 46:95-116 (1977).
4. M.R. Purnell, P.R. Stone and W.J.D. Whish. Biochem.Soc.Trans. 8:
   215-227 (1980).
5. H. Halldorsson, D.A. Gray and S. Shall. FEBS Letters 85:349-352
   (1978).
6. W.J.D. Whish, M.I. Davies and S. Shall. Biochem.Biophys.Res.
   Commun. 65:722-730 (1975).
7. C.J. Skidmore, M.I. Davies, P.M. Goodwin, H. Halldorsson, P.
   Lewis, S. Shall and A-A. Zia'ee. Eur.J.Biochem. 101:135-142.
   (1979).
8. B.W. Durkacz, O. Omidiji, D.A. Gray and S. Shall. Nature 283:
   593-596 (1980).
9. N. Nduka, C.J. Skidmore and S. Shall. Eur.J.Biochem. 105:525-
   530 (1980).
10. H. Juarez-Salinas, J.L. Sims and M.K. Jacobson. Nature 282:740-
    741 (1979).

ADDITION OF UV ENDONUCLEASE TO PERMEABLE XP CELLS TO ANALYZE

THE RELATION OF POLY(ADPR) SYNTHESIS TO DNA REPAIR

Nathan A. Berger and Georgina W. Sikorski

The Hematology/Oncology Division of the Department of
Medicine, Washington University School of Medicine at
The Jewish Hospital of St. Louis, 216 South Kingshighway
St. Louis, Missouri 63110

Poly(adenosine diphosphoribose) is synthesized from NAD[+] by
poly(ADPR) polymerase, which is a tightly bound chromosomal enzyme
whose activity is stimulated when cells are treated with various DNA-
damaging agents.[1-4] The rapid synthesis and degradation of ADP ribose
polymers has the potential for causing drastic but reversible alterations in chromatin conformation. Since poly(ADPR) synthesis increases
in response to DNA damage, it was proposed that poly(ADPR) might be
involved in the DNA repair process, altering chromatin structure so as
to make DNA-damaged regions more readily accessible to the enzymes of
DNA repair.[2-5] Part of this proposal has been confirmed by the demonstration that inhibitors of poly(ADPR) synthesis interfere with the
ability of cells to recover and proliferate following DNA damage.[6]
In addition, cells made NAD[+]-deficient by nicotinamide starvation are
unable to carry out unscheduled DNA synthesis after treatment with
N-methyl-N'-nitro-N-nitroso-guanidine (MNNG); they are also unable to
reseal DNA strand breaks after treatment with dimethyl sulfate.[6,7]

We have shown that cells from normal human donors develop an increase in poly(ADPR) synthesis after treatment with various DNA-damaging agents including MNNG, N-acetoxy-acetylaminofluorine, Bleomycin,
and UV irradiation.[3] In cells from patients with xeroderma pigmentosum, poly(ADPR) synthesis increases in response to DNA damage induced
by MNNG but not in response to damage from UV irradiation.[8] We proposed that XP cells failed to increase their poly(ADPR) synthesis in
response to UV irradiation because of their defective ability to perform the initial incision at the sites of UV-induced DNA damage.[8,9]
In the present study, we supplied UV-irradiated, XP cells with M.
luteus UV endonuclease to determine how this would affect the synthe-

sis of poly(ADPR) and to investigate the relation of poly(ADPR) synthesis to the DNA repair process.

These studies were conducted with XP3BE lymphoblasts from a donor with xeroderma pigmentosum, Group C.[10] DNA and poly(ADPR) synthesis were measured in cells made permeable to exogenously supplied components and incubated at 37° C in a reaction system containing ATP, dATP dGTP, dCTP, [3H]dTTP, [14C]NAD, Mg^{++}, 2-mercaptoethanol and Triton.[11] M. luteus UV endonuclease, purified through the Sephadex G-75 stage,[1] was a gift from Dr. Lawrence Grossman and was included in the indicated reactions.

Table 1 shows the effects of UV irradiation and UV endonuclease on DNA and poly(ADPR) synthesis in XP3BE cells. The levels of DNA and poly(ADPR) synthesis measured in the control unirradiated cells were typical for cells in mid log phase growth. Upon addition of M. luteus UV endonuclease, unirradiated cells showed no change in the levels of DNA or poly(ADPR) synthesis. UV irradiation resulted in about a 50% decrease in the rate of DNA synthesis and essentially no change in the level of poly(ADPR) synthesis. When the UV-irradiated cells were treated with the UV endonuclease, DNA synthesis increased to a level slightly greater than that present in the unirradiated control cells. Concomitantly, the level of poly(ADPR) synthesis increased to about three times that of the control cells. The demonstration that the UV endonuclease caused DNA and poly(ADPR) synthesis to increase in UV-irradiated cells but had no effect in control cells indicates that such increases depend on the enzyme carrying out its specific function, which is to introduce DNA strand breaks at the sites of pyrimidine dimers.

One approach to determining the relation of poly(ADPR) to DNA repair is to examine the time course of their synthesis in UV-irradiated cells in the presence and absence of UV endonuclease. The first panel in Figure 1 shows that in permeabilized mid log phase cells, the synthesis of DNA and poly(ADPR) continued for 30 minutes and then leveled off. The second panel shows that UV irradiation resulted in a marked depression of DNA synthesis. However, even at the reduced rate, both DNA and poly(ADPR) continued to be synthesized during the first 30 minutes before leveling off. The third panel shows that when the UV endonuclease was added to the UV-irradiated cells, there was a rapid increase in poly(ADPR) synthesis which appeared to precede the increase in DNA synthesis.

Endonuclease-dependent DNA synthesis is determined by subtracting the DNA synthesis values for cells treated with UV irradiation alone from the values for cells treated with UV irradiation and UV endonuclease. Similarly, the endonuclease-dependent poly(ADPR) polymerase activity is determined by subtracting the values for UV-irradiated cells from those for UV-irradiated cells treated with UV endonuclease.

Table 1. Effect of <u>M. Luteus</u> UV Endonuclease on Synthesis of
          DNA and Poly(ADPR) in XP3BE (Group C) Lymphoblasts.

Treatment	DNA Synthesis	Poly(ADPR) Synthesis
	DPM/10^6 Cells	
Control	98,300	2,000
Control + UV Endonuclease	97,800	2,000
UV, 50 J/M^2	56,900	2,300
UV, 50 J/M^2 + UV Endonuclease	102,000	6,600

XP3BE lymphoblasts were grown to mid log phase. UV-irradiated
cells received 50 J/M^2, principal irradiation at 2537 Å; control cells were not irradiated. Immediately after irradiation,
cells were permeabilized as described[11] and 1 x 10^6 permeable
cells were incubated with a reaction mix containing ATP, all
4 deoxynucleoside triphosphates, [^{3}H]dTTP and [^{14}C]NAD for 30
minutes at 37° C. The indicated reactions contained 10 units
of <u>M. luteus</u> UV endonuclease. Reactions were terminated with
cold 20% TCA, 1% pyrophosphate and prepared for scintillation
counting on GF/C discs.[11]

The values derived from these calculations are plotted in the fourth
panel of Figure 1. Clearly, the addition of the UV endonuclease to
UV-irradiated XP cells resulted in an abrupt increase in poly(ADPR)
synthesis which preceded the increment in DNA synthesis. The rapid
increase in poly(ADPR) synthesis leveled off after 15 minutes; then
the amount of radioactive poly(ADPR) in the system decreased, suggesting that it was undergoing degradation. The amount of DNA synthesized
leveled off after 30 minutes incubation and remained constant thereafter.

    To investigate further the relation of poly(ADPR) synthesis to
DNA synthesis in the UV endonuclease-dependent reaction, we examined
the effects of inhibitors which were selected for their abilities to
inhibit selectively the synthesis of either DNA or poly(ADPR).
5-methylnicotinamide, 3-aminobenzamide, and theophylline were selected
for their ability to inhibit poly(ADPR) polymerase.[11,13] Cytosine
arabinoside triphosphate (araCTP), dideoxythymidine triphosphate
(ddTTP), phosphonoacetic acid, and aphidicolin are all DNA polymerase
inhibitors.[14] As shown in Table 2, 3-aminobenzamide, 5-methylnico-
tinamide, and theophylline all inhibited the UV endonuclease-dependent
stimulation of poly(ADPR) synthesis. In addition, all three agents
caused a partial inhibition of the endonuclease-dependent DNA synthesis. When araCTP was added to inhibit DNA synthesis, there was also a

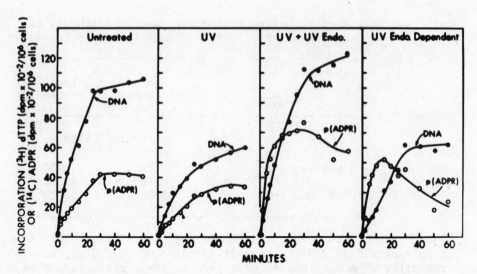

Fig. 1.   Time course of DNA (●—●) and poly(ADPR) (○—○) synthesis in
          XP3BE lymphoblasts. Cells were permeabilized and the reac-
          tions performed as described in Table 1. First panel on
          left: cells in mid log phase growth. Second panel: cells
          irradiated with 50 J/M^2 incubated without UV endonuclease.
          Third panel: cells irradiated with 50 J/M^2 and incubated with
          10 units of UV endonuclease in each reaction. Fourth panel:
          UV endonuclease-dependent reaction, obtained by subtracting
          results in second panel from those in the third.

partial inhibition of poly(ADPR) synthesis. DdTTP, phosphonoacetic
acid and aphidicolin all inhibited the endonuclease-dependent DNA synthesis, but had no effect on poly(ADPR) synthesis. These studies demonstrate that inhibition of poly(ADPR) synthesis interferes with the
DNA repair process that occurs after UV-irradiated XP cells are treated with M. luteus UV endonuclease. In contrast, inhibition of the UV
endonuclease-dependent DNA synthesis had no effect on poly(ADPR) synthesis. Thus, the UV endonuclease-dependent DNA synthesis appears to
depend, in part, on the prior synthesis of poly(ADP-Ribose), but not
vice-versa.

    In summary, we have taken advantage of the failure of XP cells to
repair UV-induced DNA damage to reconstitute these cells with the M.
luteus UV endonuclease, thus providing a system in which to examine
the relation between poly(ADPR) synthesis and the DNA repair process.
These studies confirm our proposal that poly(ADPR) is not synthesized
in response to the DNA damage created by UV irradiation but rather in
response to the specific action of the UV endonuclease on the UV-irra-

Table 2. Effect of Inhibitors on UV Endonuclease-
Dependent Synthesis of DNA and Poly(ADPR)
in UV-irradiated XP3BE Lymphoblasts.

Inhibitor	DNA Synthesis	Poly(ADPR) Synthesis
	(Percent of Control)	
Control	100	100
5 mM 3-aminobenzamide	43	5
5 mM 5-methylnicotinamide	66	32
5 mM theophylline	65	24
1 mM araCTP	26	67
1 mM ddTTTP	7	101
1 mM phosphonoacetic acid	19	107
1 µg/ml aphidicolin	14	103

Conditions for UV irradiation, cell permeabilization and
reaction conditions were the same as described in Table 1.
The effect of each agent on synthesis of DNA and poly(ADPR)
was determined in the presence and absence of 10 units of
UV endonuclease. The specific effect of each agent on
the UV endonuclease-dependent reaction was determined by
subtracting the values obtained in the absence of endonuclease from those obtained in the presence of endonuclease. Percentages were then calculated relative
to the amount of UV endonuclease-dependent synthesis
in cells incubated without any inhibitors.

diated DNA. The rapid increase in poly(ADPR) synthesis that occurs in
UV-irradiated cells treated with UV endonuclease precedes the increase
in DNA synthesis. Inhibition of this poly(ADPR) synthesis partially
prevents the endonuclease-dependent increase in DNA synthesis.

The results of our experiments are consistent with the pathway
for repair of UV-induced DNA damage outlined in Fig. 2. According to
this proposal, UV irradiation induces DNA damage which is recognized
by an endonuclease system whose function includes nicking the DNA
phosphodiester backbone at or near the site of the damage. Poly(ADPR)
polymerase, which is tightly bound and widely distributed in reserve
quantities in the chromatin,[15] is stimulated by the DNA strand breaks
to synthesize poly(ADPR) which in turn alters the chromatin structure
so as to permit the enzymes of DNA repair to gain access and repair
the regions of DNA damage. After the damage is repaired and the DNA

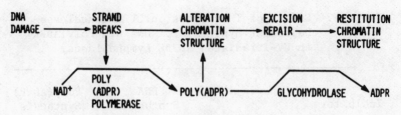

Fig. 2. Proposed pathway for repair of UV-induced DNA damage.

strand breaks are ligated, the stimulus for poly(ADPR) synthesis is
removed. Poly(ADPR) glycohydrolase is suppressed by single-stranded
DNA but is fully active in the presence of double-stranded DNA.[16]
When the DNA strand breaks are repaired, the glycohydrolase is free
to degrade the poly(ADPR) and the chromatin can return to its predam-
aged conformation.

ACKNOWLEDGEMENTS

     We thank Dr. A. Andrews for the XP3BE cells and Dr. L. Grossman
for the M. luteus UV endonuclease. This work was supported by funds
from American Cancer Society Grant CH-134, NIH grants CA23986 and
GM26463. N.A.B. is a Leukemia Society of America Scholar.

REFERENCES

1. O. Hayaishi and K. Ueda, Poly (ADP-Ribose) and ADP ribosylation of
     proteins, Ann. Rev. Biochem. 46:95 (1977).
2. M.I. Davies, S. Shall, and C.J. Skidmore, Poly (adenosine diphos-
     phate ribose) polymerase and deoxyribonucleic acid damage, Bio-
     chem. Soc. Trans. 5:949 (1977).
3. N.A. Berger, G.W. Sikorski, S.J. Petzold, and K.K. Kurohara, Asso-
     ciation of poly (adenosine diphosphoribose) synthesis with DNA
     damage and repair in normal human lymphocytes, J. Clin. Invest.
     63:1164 (1979).
4. E.G. Miller, Stimulation of nuclear poly (adenosine diphosphate
     ribose) polymerase-activity from Hala-cells by endonuclease,
     Biochim. Biophys. Acta 395:191 (1975).
5. J.E. Cleaver, Xeroderma pigmentosum, in: "The Metabolic Basis of
     Inherited Disease," J.B. Stanbury, J.B. Wyngaarden, and D.S.
     Fredrickson, eds., McGraw-Hill, New York (1978).
6. B.W. Durkacz, O. Omidiji, D.A. Gray, and S. Shall, (ADP-ribose)
     participates in DNA excision repair, Nature 283:593 (1980).
7. E.L. Jacobson, D. Juarez, and J.S. Sims, Absence of DNA repair

and poly(ADP-ribose) synthesis in NAD-depleted 3T3 cells,
Fed. Proc. 39:1739 (1980).

8. N.A. Berger, G.W. Sikorski, S.J. Petzold, and K.K. Kurohara, Defective poly(adenosine diphosphoribose) synthesis in xeroderma pigmentosum, Biochemistry 19:289 (1980).

9. A.J. Fornace, Jr., K.W. Kohn, and H.E. Kann, Jr., DNA single-strand breaks during repair of UV damage in human fibroblasts and abnormalities of repair in xeroderma pigmentosum, Proc. Natl. Acad. Sci. 73:39 (1976).

10. A.D. Andrews, J.H. Robbins, K.H. Kraemer, and D.N. Buell, Xeroderma pigmentosum long-term lymphoid lines with increased ultraviolet sensitivity, J. Natl. Canc. Inst. 53:691 (1974).

11. N.A. Berger, G. Weber, and A.S. Kaichi, Characterization and comparison of poly(adenosine diphosphoribose) synthesis and DNA synthesis in nucleotide permeable cells, Biochim. Biophys. Acta 519:87 (1978).

12. S. Riazuddin and L. Grossman, Micrococcus luteus correndonucleases. I. Resolution and purification of two endonucleases specific for DNA containing pyrimidine dimers, J. Biol. Chem. 252:6280 (1977).

13. M.R. Purnell and W.J.D. Whish, Novel inhibitors of poly(ADP-ribose) synthetase, Biochem. J. 185:775 (1980).

14. N.A. Berger, K.K. Kurohara, S.J. Petzold, and G.W. Sikorski, Aphidicolin inhibits eukaryotic DNA replication and repair --implications for involvement of DNA polymerase in both processes. Biochem. Biophys. Res. Commun. 89:218 (1979).

15. N.A. Berger, A.S. Kaichi, P.G. Steward, R.R. Klevecz, G.L. Forrest, and S.D. Gross, Synthesis of poly(adenosine diphosphate ribose) in synchronized Chinese hamster cells, Exp. Cell Res. 117:127 (1978).

16. L.O. Burzio, P.T. Riquelme, E. Ohtsuka, and S.S. Koide, Evidence for two variants of poly(adenosine diphosphate ribose) glyco-hydrolase in rat testis, Arch. Biochem. Biophys. 173:306, 1976.

ALKYLATION OF EHRLICH ASCITES TUMOR CELLS ELICITS

A RESPONSE FROM POLY(ADP-RIBOSE) POLYMERASE

John Michael Dornish and Ingrid Smith-Kielland

Department of Biochemistry, University of Oslo

Oslo, Norway

ABSTRACT

The activity of poly(ADP-ribose) polymerase in Ehrlich ascites tumor cell nuclei is seen to increase in a dose-dependent manner when cells are alkylated 1 hr with 1,3-bis(2-chloroethyl)-1-nitrosourea or nitrogen mustard. Stimulation appeared at drug amounts far in excess of that required for total cell death. At biologically relevant levels of alkylation (LD_{50}) there is no increase in enzyme activity relative to controls seen until repair of DNA is allowed to occur. During repair, enzyme activity reaches a maximum two to four hrs after removal of the drug and returns to control levels after six hrs. Alkaline sucrose gradient centrifugation reveals fragmentation of DNA during the period concomitant with maximum poly(ADP-ribose) polymerase activity and a return to high molecular weight DNA six hrs after removal of the drug.

INTRODUCTION

Poly(ADP-ribose) polymerase, a chromatin-bound nuclear enzyme using nicotinamide adenine dinucleotide as substrate has been implicated in participating in cellular DNA repair processes (1,2). Enzyme activity appears to be influenced by the state of cellular DNA (3,4). The enzyme is known to respond to breaks in DNA caused by nuclease action (5) and irradiation (3). Some monoalkylating agents have also caused increases in enzyme activity (1,2). We were interested in examining how dialkylating agents such as 1,3-bis(2-chloroethyl)-1-nitrosourea (BCNU) and nitrogen

mustard influence the activity of poly(ADP-ribose) polymerase, at physiological as well as non-physiological
conditions. In addition we wished to investigate the
relationship between increases in enzyme activity and
repair of damaged DNA.

MATERIALS AND METHODS

Cell line and Nuclei Isolation

Ehrlich ascites tumor cells adapted to in vitro
suspension culture and kept in exponential growth by
daily dilution with Eagle's MEM supplemented with 10%
fetal calf serum were used. Nuclei from control or drug-
treated cells were isolated by gentle hypotonic homogenization, remaining cytoplasm was removed with detergent
(6). Alkaline sucrose gradient centrifugation showed
little difference in sedimentation pattern between whole
cell and isolated nuclei DNA.

Poly(ADP-ribose) Polymerase Assay

The enzyme activity was measured as the incorporatio
of radioactivity from nicotinamide [^{14}C]adenine dinucleotide. The reaction mixture contained 0.1 M Tris-HCl,
pH 8.0, 12 mM $MgCl_2$, 1 mM mercaptoethanol, 1 mM [^{14}C]NAD,
9.6% ethanol. Incubation with 1 x 10^6 nuclei was at 25°C
for 15 min. The reaction was stopped by adding 10% TCA
and the precipitate collected onto Whatman GF/C filters.
TCA-insoluble radioactivity remaining on the filters was
estimated in a toluene/PPO scintillant.

Alkaline Sucrose Gradient Centrifugation

Cells were prelabeled overnight with [methyl-^{3}H]
thymidine. After drug treatment, 1.5-2 x 10^4 cells were
pipetted onto a 2% SDS layer over 5-20% linear sucrose
gradients containing 0.1 M NaOH and 0.1 M NaCl. Centrifugation in a SW 50.1 rotor was at 25,000 rpm for 60 min.
Fractions of 10 drops were collected onto disks of Whatma
3MM and the TCA-insoluble radioactivity was estimated.

RESULTS

Figure 1A shows the sedimentation pattern in alkalin
sucrose gradients of BCNU-treated cells. One hr treatmen
with 1 ug/ml (50% cell survival) did not differ in
sedimentation pattern from a control culture while 25 ug/
and 50 ug/ml BCNU treatments caused increased fragmentati
of DNA. No cell viability was seen at these high dosage

levels as measured by lack of colony formation in soft
agar.

Poly(ADP-ribose) polymerase activity also increased
with higher doses of BCNU (Figure 1B), where 40 ug/ml
showed maximum stimulation of the enzyme. Percent
stimulation is relative to control cells incubated simultaneously with the treated culture. Increases over
40 ug/ml BCNU caused cell disintegration leading to a loss
of enzyme activity. Although data is not shown, treatment
with nitrogen mustard, also produced similar increases in
both enzyme activity and in fragmentation of cellular DNA.

Alkaline sucrose gradient sedimentation patterns for
LD_{50} doses of both drugs are seen in Figure 2. Here
samples were taken after cells were washed free of drug
(1 hr treatment) and repair of damage to DNA allowed to
occur (re-incubation in fresh medium). At 0 hr after
removal of the drug, there is no evident decrease in DNA
size (Figure 2, A and C). Four hrs after drug treatment,
DNA is fragmented in both BCNU- and nitrogen mustard-
treated cultures. Six hrs after drug removal, the
sedimentation pattern for DNA is identical with a non-
treated control indicating complete repair and a return
to high molecular weight DNA (Figure 2, B and D).

Figure 3 shows the poly(ADP-ribose) polymerase enzyme
profile during repair of LD_{50} alkylation damage in a
typical experiment. For both drugs there was maximum
stimulation two to four hrs after drug treatment and a
return to a basal control level after six hrs.

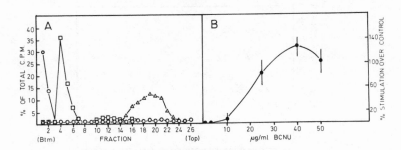

Fig. 1. The effect of BCNU on Ehrlich ascites cells.
         A. Alkaline sucrose gradient centrifugation.
         [^{3}H]TdR labeled cells treated 1 hr with 1 ug/ml
         (O), 25 ug/ml (□), or 50 ug/ml (Δ) BCNU.
         B. Poly(ADP-ribose) polymerase activity profile
         Percent stimulation measured as cpm/ug protein
         over control.

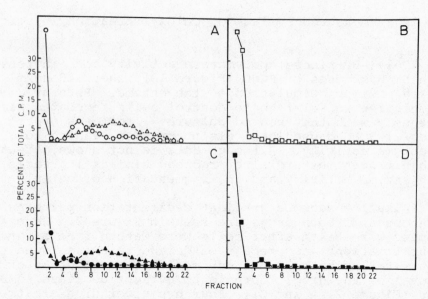

Fig. 2.   The effect of BCNU and nitrogen mustard on repair
          in Ehrlich ascites cells.   [³H]TdR labeled cells
          were treated 1 hr with BCNU or nitrogen mustard.
          Washed cells were incubated in fresh medium for
          0 hr (O,●), 4 hr (△,▲) or 6 hr (□,■) after drug
          removal.   Alkaline sucrose gradient centrifugation
          followed.   A and B: 1 ug/ml BCNU.
          C and D: 0.035 ug/ml nitrogen mustard.

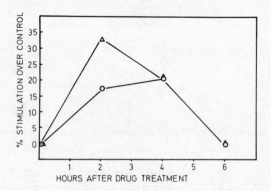

Fig. 3.   Poly(ADP-ribose) polymerase activity during repair
          of Ehrlich ascites cells.   % stimulation measured
          as cpm/ ug protein over control.   1 hr treatment
          with 1 ug/ml BCNU (O) or 0.035 ug/ml nitrogen
          mustard (△).   Washed cells re-incubated in fresh
          medium for time shown.

DISCUSSION

Experiments described in this study show that poly (ADP-ribose) polymerase activity is stimulated in a dose-dependent manner when Ehrlich ascites cells are alkylated with 1,3-bis(2-chloroethyl)-1-nitrosourea or nitrogen mustard. Alkaline sucrose gradient centrifugation shows that maximum fragmentation of cellular DNA due to alkylation damage is concurrent with maximum poly(ADP-ribose) polymerase activity after 1 hr treatment with drug. It appears that increasing fragmentation of DNA is associated with increasing damage due to alkylation, and this results in a stimulation of enzyme activity.

However, the increase in enzyme activity is seen at a dosage level far in excess of that which is biologically relevant and cannot be associated with repair due to cell death. No repair was seen in cells treated with 40 ug/ml BCNU (data not shown). During repair of alkylation damage due to physiologically tolerable doses of drugs, poly (ADP-ribose) polymerase responds to breaks in DNA. The percent stimulation over control during repair is lower than that seen in the dose-dependent experiments indicating that maximum enzyme stimulation is reached at high fragmentation of DNA while lesser increases in activity are associated with repairable damage. Proper interpretation of increases in poly(ADP-ribose) polymerase activity is thus dependent upon knowledge of the dose-response characteristics of a particular cell line for the drug or radiation treatment involved. As this study indicates, poly(ADP-ribose) polymerase activity will increase if DNA is fragmented by alkylating agent treatment, but stimulation of enzyme activity is not necessarily related to repair of damaged DNA.

REFERENCES

1. B. W. Durkacz, O. Omidiji, D. A. Gray and S. Shall, Nature 283:593 (1980).
2. M. E. Smulson, P. Schein, D. W. Mullins Jr. and S. Sudhakar, Cancer Res. 37:3006 (1977).
3. R. C. Benjamin and D. M. Gill, in "DNA Repair Mechanisms," P. C. Hanawalt, E. C. Freidberg and C. F. Fox, eds., pp. 337-340, Academic Press, New York, (1978).
4. N. A. Berger, S. J. Petzold and S. J. Berger, Biochim. Biophys. Acta 564:90 (1979).
5. E. G. Miller, Biochim. Biophys. Acta 395:191 (1975).
6. H. Krokan, E. Bjørklid and H. Prydz, Biochem. 14:4227 (1975).

LETHAL AND MUTAGENIC EFFECTS OF DNA DAMAGE

PROGRAMMED DEATH OF Escherichia coli

AFTER UV RADIATION

Željko Trgovčević, Erika Salaj-Šmic,
Mirjana Petranović, and Drago Petranović
Institute "Ruđer Bošković"
41001 Zagreb
Croatia, Yugoslavia

It is known that unicellular organisms are able to perpetuate themselves indefinitely. Under unfavorable conditions, however, they can meet their biological end-point. This end-point, i.e. cell death, is usually thought to be the result of passive "wearing out" processes. Senescence and death of human fibroblasts in vitro [1,2] and even biological aging of higher organisms are also ascribed to such processes. It is assumed that errors in protein and DNA synthesis passively accumulate to the point where, due to impaired cell functions, death ensues [3,4]. In marked contrast to this, our recent results [5,6] indicate that death of Escherichia coli cells after UV radiation might be an active "self-destruct" program.

The existence of a "self-destruct" program in E. coli was deduced from the results of experiments on the post-UV fate of the bacterial chromosome "labeled" with lambda cI857 ind prophage. (In general, the prophage DNA can be looked upon as a cluster of bacterial genes [7].) The lambda cI857 ind prophage genes are turned on by raising the temperature from 30°C to 42°C but are not induced by UV light [8]. Thus, the biological activity of the prophage part of the lysogenic chromosome can be measured by heat-inducing the lysogens at any time during post-UV incubation. At the beginning of post-UV growth, the prophage, due to its smaller size, is much more resistant than its host. At later times, however, its resistance decays rapidly and

505

reaches the survival level of the bacterial cell approximately 4 h
after UV. This rapid inactivation depends on the bacterial gene
recA[5,6].

If the fate of the prophage reflects the fate of its resident bacterial chromosome, then the above results may be interpreted to
mean that chromosomal inactivation and cell death are programm ed
into the recA-dependent ("SOS") sequence of post-UV events. We
have suggested that this inactivation takes place in all cells that
have not fully repaired their DNA[5].

One implication of the hypothesis of programmed death is that
the cells with inactivated chromosome will be incapable of synthesizing their genetic material, nam ely their DNA. The experiments
presented here were done to test this prediction.

DNA synthesis in UV-irradiated bacteria has been extensively
studied over the last two decades (for a review, see ref. 9). In
general, DNA synthesis is severely depressed after UV radiation;
but, after a dose-dependent lag, it eventually recovers. The results presented in Fig. 1 are in agreement with this general pattern: DNA synthesis is inhibited for about 90 min after exposing
the cells to UV light. Upon resuming, between 90 min and 4 h
of post-UV growth, DNA synthesis proceeds at the same rate as
in the unirradiated control. Two facts should be stressed in this
connection. First, only one locus for initiation of DNA replication
is activated on the chromosome of UV-exposed cells, preferentially at the chromosomal origin[10]. Second, the rate of DNA synthesis is about equal to that of an unirradiated control after all
UV fluences lower than a certain "critical dose"[9]; in our experiments this dose was about 110 J/m^2 (10% survivors). Together,
these data strongly suggest that, during the time when the normal
rate of DNA synthesis is observed, all irradiated cells are engaged in DNA synthesis (and possibly in postreplication repair).
Some of the cells will eventually re-establish the original structure of their DNA, while the others will die. Under our experimental conditions, these different fates of irradiated cells are
expected to be met within 4 h after UV exposure[5,6]. It is interesting to note that just after 4 h of post-UV incubation DNA synthesis abruptly slows down. The question arises whether, in accordance with the hypothesis of programmed death, this change
in the rate of DNA synthesis reflects the complete chromosomal

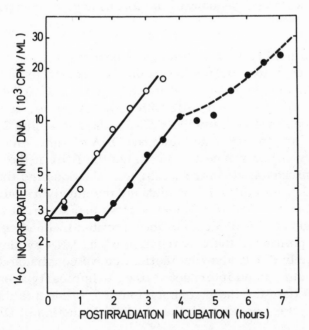

Fig. 1. DNA synthesis in repair-proficient <u>E. coli</u> AB1157 bacteria[11]
after exposure to UV light. The cells were grown in radioactive TP medium (TP medium[12] supplemented with 56 μg
and 2 μCi of ^{14}C adenine per ml). The overnight culture
was diluted 1/20 into 0.067 M phosphate buffer (pH 7.0),
exposed to 100 J/m^2 (23% survivors), centrifuged, resuspended into radioactive TP medium and incubated at 30°C.
The control culture was treated in the same way except
for radiation. For the determination of DNA synthesis
kinetics, samples from the control (o) and irradiated
culture (•) were taken at 30-min intervals. DNA synthesis
was measured according to the protocol of Smith and
O'Leary[13] except that, instead of radioactive thymine,
^{14}C adenine was used for labeling the DNA. ^{14}C RNA was
hydrolyzed with 0.5 N NaOH and then, following acidifi-
cation, ^{14}C DNA precipitated. The interrupted line represents the hypothetical kinetics of DNA synthesis subsequent
to the fourth hour of postirradiation incubation (for details,
see text).

failure in dead cells. To answer this question, we will consider DNA
synthesis kinetics subsequent to the fourth hour of postirradiation
incubation.

As shown in Fig. 1, 4 h after UV exposure, 1 ml of the irradiated culture contains 10,500 counts of ^{14}C adenine incorporated
into the DNA. Since the survival level is about 23%, 8,085 counts
(77%) are taken up by dead cells which, at this time, are expected
to have stopped synthesizing their DNA. The remaining 2,415
counts (23%) are incorporated into the DNA of surviving cells.
The survivors will continue to synthesize the DNA at the same
rate as an unirradiated control, i.e. with the doubling time of 70
min. Thus, 310 min after UV radiation, the survivors will contain
4,830 counts, after 380 min 9,660 counts, and so on. If these
values are added to 8,085 counts incorporated by dead cells, the
hypothetical points for the construction of the DNA synthesis curve
are obtained. In Fig 1, the hypothetical curve constructed in this
way is presented as an interrupted line. It is clearly seen that the
experimental data for the incorporation of ^{14}C adenine into cell
DNA (full circles) agree with the predicted kinetics of DNA synthesis.

Essentially the same results were obtained after several other
doses of UV light. Again, the analysis of the DNA synthesis curve
strongly suggests that, following the well-defined time point of
postirradiation incubation, E. coli cells with the damaged genomes
cease to replicate their DNA. Further support for this concept
was also obtained by experiments in which DNA was pulse-labeled
at various times after UV radiation and then analyzed on alkaline
sucrose gradients (Fig. 2). Between 90 min and 4 h of post-UV
growth, the pulse-labeled DNA sedimented more or less slowly
through alkaline sucrose. Normal-sized DNA was detected only
after the hypothetical time point at which the cells with incom-
pletely repaired chromosome had been stopped from synthesizing
their DNA. This is exactly what might be expected if the concept
of programmed death is correct.

Death inevitably occurs in all higher organisms but the mechanism and control of this general biological event is poorly understood. Here and in the previous reports[5,6] we have presented
the results which show that the controlled mechanism of dying
can be traced back to the simple prokaryotic organisms. We may

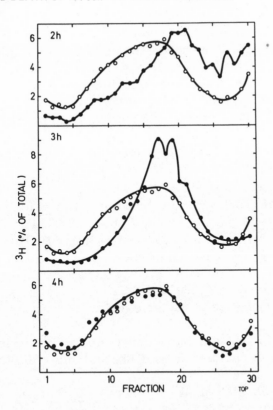

Fig. 2. Sedimentation in alkaline sucrose of ^{3}H labeled DNA from
       <u>E. coli</u> AB1157 bacteria. The overnight culture grown in TP
       medium was diluted 1/20 into 0.067 M phosphate buffer
       (pH 7.0), exposed to 100 J/m^2 (18% survivors), centrifuged,
       resuspended into fresh TP medium supplemented with 200
       μg of deoxyadenosine per ml and incubated at 30°C. Symbols:
       o, cells incubated for 60 min in non-radioactive medium
       and then for 15 min with 25 μCi of ^{3}H thymidine (specific
       activity: 20 Ci/mmol) per ml; ●, cells exposed to 100 J/m^2,
       incubated for 2 h or 3 h or 4 h in non-radioactive medium
       and then for 15 min with 25 μCi of ^{3}H thymidine (specific
       activity: 20 Ci/mmol) per ml. Sucrose gradient analysis
       was carried out as described earlier[14].

hope that by studying this mechanism in a bacterial cell we will be
able to learn more about death in higher organisms, including man.
In particular, it remains to be elucidated whether a "self-destruct"
biological program, which would readily explain clonal senescence,
exists in human fibroblasts.

REFERENCES

1. L. Hayflick and P.S. Moorhead, The serial cultivation of human diploid cell strains, Exp. Cell Res. 25:585 (1961).
2. L. Hayflick, The limited in vitro lifetime of human diploid cell strains, Exp. Cell Res. 37:614 (1965).
3. L.E. Orgel, Ageing of clones of mammalian cells, Nature 243: 441 (1973).
4. R. Holliday and G.M. Tarrant, Altered enzymes in ageing human fibroblasts, Nature 238:26 (1972).
5. M. Petranović, E. Salaj-Šmic, D. Petranović, and Ž. Trgovčević, Inactivation of prophage in uv-irradiated Escherichia coli: Dependence on recA gene activity, J. Bact. 140:848 (1979).
6. Ž. Trgovčević, M. Petranović, E. Salaj-Šmic, D. Petranović, and H. Qerimi, The cascade of "SOS" events, in: Progress in Environmental Mutagenesis, M. Alačević ed., Elsevier/North Holland, Amsterdam (1980).
7. W. Hayes, The Genetics of Bacteria and their Viruses, Blackwell, Oxford (1968).
8. R. Sussman and F. Jacob, Sur un système de répression thermosensible chez le bactériophage d'Escherichia coli, C.R.Acad. Sci. 254:1517 (1962).
9. P.A. Swenson, Physiological responses of Escherichia coli to far-ultraviolet radiation, in: Photochemical and Photobiological Reviews, vol. 1, K.C. Smith ed., Plenum, New York (1976).
10. D. Billen, Replication of the bacterial chromosome: Location of new initiation sites after irradiation, J. Bact. 97:1169 (1969).
11. B.J. Bachmann, Pedigrees of some mutant strains of Escherichia coli K-12, Bacteriol. Rev. 36:525 (1972).
12. M. Medić-Petranović, Ž. Trgovčević, Đ. Novak, and D. Petranović, Inactivation of the Escherichia coli chromosome during growth after uv-irradiation, Int. J. Rad. Biol. 32:103 (1977).
13. K.C. Smith and M.E. O'Leary, The pitfalls of measuring DNA synthesis kinetics as exemplified in ultraviolet radiation studies, Biochim. Biophys. Acta 169:430 (1968).
14. Ž. Trgovčević and W.D. Rupp, Lambda bacteriophage gene products and x-ray sensitivity of Escherichia coli: Comparison of red-dependent and gam-dependent radioresistance, J. Bact. 123:212 (1975).

EXCISION REPAIR AFFECTS THE SPECIFICITY OF UV MUTAGENESIS IN THE LACI[+]
GENE OF Escherichia coli: POSSIBLE INVOLVEMENT OF SECONDARY DNA
STRUCTURE.

Peter A. Todd and Barry W. Glickman[1]

Laboratory of Molecular Genetics,
State University of Leiden
Wassenaarseweg 64
2333 AL Leiden, The Netherlands

ABSTRACT

Using a system which allows the determination of 72 base pair
changes in the lacI[+] gene of Escherichia coli carried on an F' plasmid, it was found that the distribution of UV-induced mutations occurring in a ΔUvrB strain was only different at one particular site when
compared to the wild-type strain. It is proposed that secondary DNA
structure is important in the UV-induction of mutation, which is discussed with reference to the formation of modified tRNA genes: nonsense suppressors. Furthermore, the mutational spectrum obtained at
low UV fluence in the ΔUvrB strain was significantly different from
that at high UV fluence.

INTRODUCTION

Until the development of the lacI system (Coulondre and Miller,
1977a,b) few detailed studies were reported on the specificity of UV
mutagenesis in DNA repair proficient or deficient strains of E.coli.
Those studies which have been performed (Osborn et al., 1967; Person
and Osborn, 1968; Person et al.,1974) suffer from certain disadvantages. They rely on the reversion analysis of nonsense auxotrophic mutants, often mutagen-induced in the first place,which may preselect
certain sites or hotspots that behave atypically. In fact, the majority of UV-induced revertants are nonsense suppressors formed by mutation of the DNA coding for the anticodon of tRNA molecules (Bridges
et al., 1967). Several observations suggest that such nonsense sup-
pessors behave differently from true structural revertants.

---

[1] Laboratory of Molecular Genetics, National Institute of Health Sciences, P.O.B.12233, Research Triangle Park, NC27709, USA.

The lacI mutational specificity system overcomes the inherent
disadvantages of the earlier reversion systems and allows the detection of 72 forward base pair changes without selection bias. Each mutation can be assigned to a specific transition or transversion event
Since the whole DNA sequence is known insight may be gained into factors affecting mutagenesis such as the initial UV photoproduct formed
and the effect of neighbouring DNA sequences. Initially, we obtained
a UV spectrum at 1000 ergs/mm^2 in a wild-type strain to determine the
reproducibility of the system. The effect of excision repair capacity
was then determined by obtaining a spectrum at 50 ergs/mm^2 in a ΔUvrB
strain. In addition, a mutational spectrum for the ΔUvrB strain was
obtained at 5 ergs/mm^2.

RESULTS

The materials and methods for the lacI mutational specificity
system were basically those used by Coulondre and Miller (1977a,b).

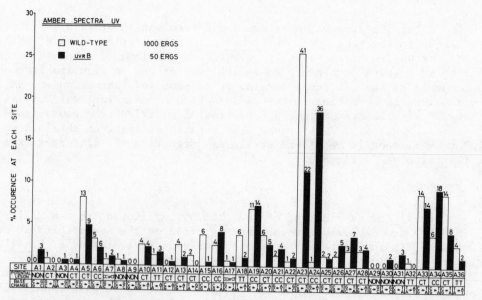

Fig. 1. Amber lacI mutant spectra. Wild-type (open bars) percentage
distribution of 160 mutants at 1000 ergs/mm^2. ΔUvrB (solid bars) percentage distribution of 193 mutants at 50 ergs/mm^2. Figures above the
histograms represent the number of independent mutational occurrences
at each site. Base change shows the relevant transition or transversion at that site and the orientation of the base pair target. NON
indicates when a pyrimidine dimer cannot be formed at that site and
CC, CT or TT indicate the respective dimer that may be formed.

Figs. 1 and 2 show the percentage distribution of lacI point mutants
between sites for the Uvr⁺ (1000 ergs/mm²) and ΔUvrB (50 ergs/mm²)
strains. There was no significant difference between the results obtained with the wild-type strain and those previously obtained by Coulondre and Miller (1977b). A striking difference between the Uvr⁺ and
ΔUvrB spectra is the appearance of an additional hotspot at site A24
in the excision repair deficient strain. The target lesion at this
site is a CC pyrimidine dimer. This site is relatively non-mutable in
the wild-type. It may be noted that all the hotspots occur in a relatively small region of the lacI⁺ gene (Fig.3). In fact, in this region
of approximately 60 base pairs, representing 6% of the gene, about 30%
of amber and 50% of ochre lacI mutants were detected with either
strain.

A UV-induced spectrum was also obtained at 5 ergs/mm² for the
ΔUvrB strain (results not shown). At this low level of induced mutagenesis the spontaneous spectrum influences the results. The A6, A15
and A34 deamination, DNA sequence specific hotspots appeared (Coulondre et al., 1978), but the A24 UV hotspot remained accounting for 20%

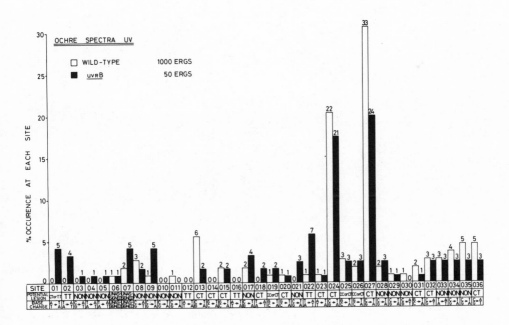

Fig. 2.   Ochre lacI mutant spectra. For details see Fig. 1. 108 and
120 ochre lacI mutants were analyzed for the Uvr⁺ and ΔUvrB strains
respectively. Sites 06 and 07 have not been assigned a definite position in the gene.

of mutagenesis. In contrast, mutation at sites A23, O24 and O27 was reduced from 20–30% to 8–10%, showing that site A24 behaves differently from the other UV hotspots.

In the wild-type strain, the frequency of G:C→A:T transitions is about one order of magnitude higher than transversion mutagenesis. Since CT pyrimidine dimers are the major UV lesion producing G:C→A:T transitions a comparison can be made of the frequency of this base change at different CT sites within the gene. The percentage occurrences for the amber sites are: A5, 8.1; A23, 25.6; A26, 2.2; A33, 8.7; A35, 8.7; and for the ochre sites: O13, 5.5; O24, 20.3 and O27, 30.5. Of these one class is poorly mutable, A26; another mutable, A5, A33, A35 and O13; and another highly mutable, i.e. the hotspots A23, O24 and O27.

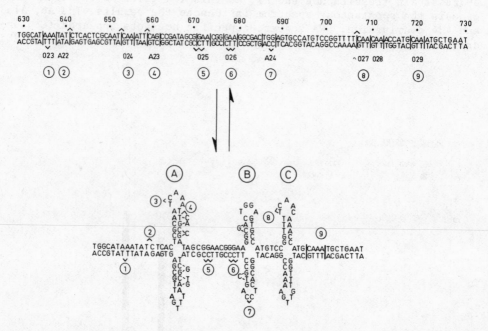

Fig.3.   Hypothetical secondary DNA structures for the 630–730 base sequence of the lacI$^+$ gene. (A), (B) and (C) are possible palindromic structures discussed in the text. The possible dimer positions at the relevant sites are indicated by ^.

DISCUSSION

A striking difference between the Uvr$^+$ and the ΔUvrB mutational spectra is the appearance of the A24 hotspot in the mutant. The similarity between the A24 CC dimer target position in a palindrome (Fig. 3) and the anticodon target of tRNA genes suggests that mutational frequency decline (MFD) may occur (Witkin, 1956). MFD was thought to be specific for suppressor mutations (Bridges et al., 1967) and is also dependent upon excision repair (Witkin, 1966). Since MFD is rapid, approximately 50% of mutants are lost in the first two minutes after UV irradiation (Bockrath and Palmer, 1977), it is likely that the absence of the A24 hotspot in the Uvr⁺ strain can be ascribed to MFD occurring at this site.

Since approximately 30% of amber and 50% of ochre mutants occur in 6% of the gene in the A24 area, we carefully examined the DNA sequence of this region (Fig.3). The hotspots A23, O24 and O27 may also involve potential dimer sites in unpaired structures. Using data obtained for RNA (Tinoco et al., 1973) we calculated ΔG values for these structures: (A) -0.8 kcal; (B) -7.4 kcal and (C) -3.6 kcal. It is possible that only structure (B) is stable enough to demonstrate MFD. The possibility that the other hotspots are in some way related to secondary DNA structure cannot be disregarded. Specific protein interactions with ssDNA during periods of replication or repair may temporarily stabilize the palindromic conformation. However, the CT target of site A35, also detecting a $G:C \to A:T$ transition, is located in the unpaired loop of a stable palindrome ($\Delta G = -12.4$ kcal) and yet is only moderately mutable. Since variation in the initial amount of damage (Grossman et al., 1978) is unlikely to fully account for the different mutation frequencies, it is possible that the secondary DNA structure may influence the fate of premutational lesions. The absence of a hotspot at the A35 site underlines the complexity of the situation.

Mutation by UV is thought to occur by error-prone DNA repair (Witkin, 1976). It has been suggested that a suppression of the 3'→5' exonucleolytic proofreading function of polymerase III allows transdimer synthesis and mutation occurs by the insertion of noncomplementary bases (Radman et al., 1979). If this base insertion were random it is difficult to explain the one order of magnitude higher frequency of transition to transversion mutagenesis seen in these experiments. Stabilization of the helical structure of ssDNA at gaps opposite dimers, perhaps by the ssb⁺ protein (Glassberg et al., 1979) might possibly result in the preferential insertion of purines rather than pyrimidines, consequently leading to mainly transition mutagenesis.

ACKNOWLEDGEMENTS

Appreciation is due to Nico Guijt for constructing the ΔUvrB strain, Corrie van Teylingen for her skillful technical assistance and Kees van Sluis for his constant encouragement and helpful discussions. This work was supported by grants from EURATOM, No. BION 194-76 and Koningin Wilhelmina Fonds, No. 79-68.

BIBLIOGRAPHY

Bockrath, R. C., and Palmer, J. E., 1977, Differential repair of pre-
    mutational UV-lesions at tRNA genes in E. coli, Mol. Gen. Genet.,
    156:133.
Bridges, B. A., Dennis, R. E., and Munson, R. J., 1967, Differential
    induction and repair of ultraviolet damage leading to true rever-
    sions and external suppressor mutations of an ochre codon in
    Escherichia coli B/r WP2, Genetics, 57:892.
Coulondre, C., and Miller, J. H., 1977a, Genetic studies of the lac
    repressor, III. Additional correlation of mutational sites with
    specific amino acid residues, J. Mol. Biol., 117:525.
Coulondre, C., and Miller, J. H., 1977b, Genetic studies of the lac
    repressor, IV. Mutagenic specificity of the lacI gene of Escher-
    ichia coli, J. Mol. Biol., 117:577.
Coulondre, C., Miller, J. H., Farabaugh, P. J., and Gilbert, W., 1978,
    Molecular basis of base substitution hot spots in Escherichia
    coli, Nature, 274:775.
Glassberg, J., Meyer, R. R., and Kornberg, A., 1979, Mutant single-
    strand binding protein of Escherichia coli: genetic and physio-
    logical characterization, J. Bacteriol., 140:14.
Grossman, L., Riazuddin, S., Haseltine, W. A., and Lindan, L., 1978,
    Nucleotide excision repair of damaged DNA, Cold Spr. Harb. Symp.
    Quant. Biol., 43:947.
Osborn, M., Person, S., Phillips, S., and Funk, F., 1967, A determin-
    ation of mutagenic specificity in bacteria using nonsense mutants
    of bacteriophage T4, J. Mol. Biol., 26:437.
Person, S., McCloskey, J. A., Snipes, W., and Bockrath, R. C., 1974,
    Ultraviolet mutagenesis and its repair in an Escherichia coli
    strain containing a nonsense codon, Genetics, 78:1035.
Person, S., and Osborn, M., 1968, The conversion of amber suppressors
    to ochre suppressors, Biochem., 60:1030.
Radman, M., Villani, G., Boiteux, S., Kinsella, A. R., Glickman, B. W.
    and Spadari, S., 1979, Replicational fidelity: mechanisms of mu-
    tational avoidance and mutation fixation, Cold Spr. Harb. Symp.
    Quant. Biol., 43:937.
Tinoco, I., Borer, P. N., Dengler, B., Levine, M. D., Uhlenbeck, O. C.
    Crothers, D. M., and Gralla, J., 1973, Improved estimation of
    secondary structure in ribonucleic acids, Nature New Biology,
    246:40.
Witkin, E. M., 1956, Time, temperature and protein synthesis: a study
    of ultraviolet-induced mutation in bacteria, Cold Spr. Harb.
    Symp. Quant. Biol., 21:123.
Witkin, E. M., 1966, Radiation induced mutations and their repair,
    Science, 152:1345.
Witkin, E. M., 1976, Ultraviolet mutagenesis and inducible DNA repair
    in Escherichia coli, Bacteriol. Rev., 40:869.

HYDRAZINE AND METHYLHYDRAZINE AS *recA*[+] INDEPENDENT

MUTAGENS IN *Escherichia coli*

Atte von Wright

Technical Research Centre of Finland
Food Research Laboratory
SF 02150 ESPOO 15, Finland

INTRODUCTION

Hydrazine (H_2N-NH_2) and methylhydrazine ($CH_3-HN-NH_2$) are both carcinogenic (IARC Monograph Vol.4, 1974). They react with pyrimidine bases in DNA causing ring opening and loss of pyrimidines, or more commonly, forming different adducts. Some of these adducts, especially N^4-aminocytosine are suspected to cause mutations by a base mispairing mechanism (see the review of Kimball, 1977). In this work the mutagenic action of hydrazine and methylhydrazine was studied using both bacterial and phage test systems. Both the bacterial test strains and the phage hosts used in the experiments differed in their repair capacities and in their expression of the postulated *recA+*-dependent, error-prone, inducible repair system ("SOS repair") (Radman, 1975). The aim of the work has been to find out whether "SOS repair" is involved in the mutagenic action of hydrazine.

MATERIALS AND METHODS

Bacterial tests

The tests were performed using the modified "spot test" and "treat and plate test" procedures described by von Wright and Tikkanen, 1980 to score reversions from tryptophan auxotrophy to prototrophy.

The near-isogenic tryptophan auxotroph test strains used were *Escherichia coli* B/r WP2 *trp*, WP2 *uvrA, trp* and CM871 *uvrA recA lexA trp*. References and more detailed descriptions of the strains are given in previous papers (von Wright and Tikkanen, 1980a; von Wright and Tikkanen, 1980b). In the course of experiments the mutator plasmid pKM101, which is supposed to cause constitutive expression of some of the "SOS" phenomena (Mortelmans and Stocker, 1976) was introduced into WP2 and CM871 by mating these strains with *Salmonella typhimurium* TA100 (McCann et al. 1975).

Phage tests

The test system to score clear plaque mutants of phage lambda described by von Wright and Bridges (1980) was used with the following modifications. For the tests the phages were incubated with various concentrations of the test agent in phosphate buffer at 37°C for 6 hours. Catalase was present in the incubation mixture at a concentration of 200 µg/ml to eliminate the H_2O_2 formed in aqueous hydrazine solutions (Freese et al. 1967). After incubation the phages were adsorbed at a multiplicity of infection of about 0.1. The hosts were *Escherichia coli* AB2480 *uvrA recA* (Howard-Flanders et al. 1969), AB1157 (Howard-Flanders et al. 1966) and JM 12 *tif* (Castellazi et al. 1972). The induction of *tif* phenomena, which are supposed to reflect con stitutive expression of "SOS repair", was performed as described by von Wright and Bridges (1980).

RESULTS

The results of the bacterial tests are given in Tables 1 and 2. As can be seen both hydrazine and methylhydrazine cause mutation in approximately equal frequencies in WP2, WP2 *uvrA* and CM871 *uvrA recA lexA* regardless of the *recA* genotype. However, the presence of the plasmid pKM101 enhances this mutagenic effect in WP2 but not in CM871.

Table 1. The results of the spot tests (means of three experiments)

Chemical	Amount/disc (µmole)	trp+ revertants/plate		
		WP2	WP2 *uvrA*	CM871 *uvrA recA lexA*
none	–	4	11	2
hydrazine[a]	0.5	54	76	64
	1.0	66	87	82
	2.0	98	140	201
methyl-hydrazine	0.5	28	48	37
	1.0	40	63	58
	2.0	62	98	86

[a] as hydrazine sulphate

Table 2.   Results of the "treat and plate" tests (means of two
           experiments)

Chemical	Concentration (mM)	WP2	CM871	WP2 + pKM101	CM871 + pKM101
none	–	1.5	2.0	2.8	2.5
hydrazine	0.5	183.0	102.0	522.0	125.0
	1.0	621.5	493.0	1215.5	555.5
methyl-	0.5	11.0	.5.5	186.0	6.5
hydrazine	1.0	49.5	21.0	316.5	22.0

In phage experiments (Table 3) all three hosts produced clear plaque
mutants in equal frequencies.   The *tif* expression did not affect
induced mutation rate although spontaneous phage mutation rate in
the *tif* host was higher than in the other two strains.

DISCUSSION

     The results show that hydrazine mutagenesis is independent of
recA genotype.   This indicates that the main mutagenic pathway in
this case does not results from "SOS" type of repair, base mispairing being probably the next most obvious alternative.

     At present there is no explanation of the enhancement of the
mutagenic action of hydrazine and methylhydrazine by the plasmid
pKM101 and of the lack of similar effect by *tif* expression.   The
reason for this discrepancy might be that pKM101-mediated phenomena
differ from those caused by *tif* expression or induction of "SOS
repair".   This has been in fact indicated recently by Goze and
Devoret (1979).

Table 3.   Results of the phage tests (means of two experiments)

Chemical	Concentration (mM)	frequency of clear plaque mutants[a]		
		AB2480	AB1157	JM12[a]
none	–	2.2×10^{-4}	1.4×10^{-4}	4.2×10^{-4}
hydrazine	10	13.2×10^{-4}	11.2×10^{-4}	16.0×10^{-4}
	20	12.0×10^{-4}	7.0×10^{-4}	16.5×10^{-4}
methyl-	10	4.2×10^{-4}	3.8×10^{-4}	6.0×10^{-4}
hydrazine	20	10.0×10^{-4}	8.5×10^{-4}	9.0×10^{-4}

[a]with *tif* expression induced

Castellazzi, M., George, J., and Buttin, G., 1972, Prophage induction
    and cell division in *Escherichia coli*. I. Further characteri-
    zation of the thermosensitive mutation *tif-l* whose expression
    mimics the effect of UV irradiation. Molec. gen. Genet. 119:
    139-152.

Freese, E. B., Gerson, J., Taber, H., Rhaese, H. J., and Freese, E.,
    1967, Inactivating DNA alterations induced by peroxides and
    peroxide producing agents, Mutation Res., 4:517-531.

Goze, A., and Devoret, R., 1979, Repair promoted by plasmid pKM101
    is different from SOS repair, Mutation Res., 61:163-179.

Howard-Flanders, P., Boyce, R. P., and Theriot, L., 1966, Three loci
    in *Escherichia coli* K-12 that control the excision of pyrimi-
    dine dimers and certain other mutagen products from DNA,
    Genetics, 53:1119-1136.

Howard-Flanders, P., Theriot, L., and Stedeford, J. B., 1969, Some
    properties of excision defective recombination deficient
    mutants of *Escherichia coli* K-12, J. Bacteriol., 97:1134.

IARC Monographs on the Evaluation of Carcinogenic Risk of Chemicals
    to Man, Some Aromatic Amines, Hydrazine and Related Substances,
    N-Nitroso Compounds and Miscellaneous Alkylating Agents,
    Vol.4, 1979, International Agency for Research on Cancer,
    Lyon.

Kimball, R. F., 1977, The mutagenicity of hydrazine and some of its
    derivatives, Mutation Res., 39:111-126.

Mortelmans, K. E., and Stocker, B. A. D., 1976, Ultraviolet light
    protection, enhancement of ultraviolet light mutagenesis and
    mutator effect of plasmid R46 in *Salmonella typhimurium*,
    J. Bacteriol., 128:271-282.

McCann, J., Spingarn, N. E., Kobori, J., and Ames, B. N., 1975,
    Detection of carcinogens and mutagens. Bacterial tester
    strains with R factor plasmids, Proc. Natl. Acad. Sci. (U.S.
    A.) 72:979-983.

Radman, M., 1975, SOS-repair hypothesis: Phenomenology of an
    inducible DNA repair which is accompanied by mutagenesis, in
    "Molecular Mechanisms for Repair of DNA", P. C. Hanawalt and
    R. B. Setlow, eds., Plenum Press, New York.

von Wright, A., and Tikkanen, L., 1980, The comparative mutageni-
    cities of hydrazine and its mono- and dimethylated deriva-
    tives in bacterial test systems, Mutation Res., 78:17-23.

von Wright, A., and Tikkanen, L., 1980, Hydrazine and methyl-
    hydrazine as *recA* independent mutagens in *Escherichia coli*
    Mutation Res., 71:269.

von Wright, A., and B. A. Bridges, 1980, Mutagenesis of lambda
    phage by *tif* expression or host irradiation function is
    largely independent of damage in phage DNA, Mutation Res.,
    in press.

EFFECTS OF PLATINUM COORDINATION COMPLEXES ON MUTANTS OF *Escherichia coli* DEFECTIVE IN DNA REPAIR AND GENETIC RECOMBINATION

Robert Alazard and Maryse Germanier

Laboratoire de Pharmacologie et de Toxicologie Fonda-
mentales du CNRS
205 route de Narbonne, 31078 Toulouse Cédex, France

ABSTRACT

The effect of three platinum coordination complexes on wild
type and repair deficient mutant of *Escherichia coli* has been
investigated. The antitumor drug *cis*-platinum (II) diamminodichloride (*cis*-PDD) produces an inhibition of DNA synthesis and reduces
the viability of excision repair and recombination defective mutants. The *trans* isomer (*trans*-PDD) has less effect than *cis*-PDD
on all of these mutants while [Pt(dien)Cl]Cl has no effect. Incorporation of radioactive precursors under non-replicating conditions
and density gradient centrifugation analysis of DNA indicate that
repair synthesis occurred after treatment of wild type cells with
cis-PDD or *trans*-PDD.

INTRODUCTION

Platinum coordination complexes were first noted as potent
anticancer agents by Rosenberg *et al.*[1]. The most widely investigated complex is cisplatin (*cis*-Platinum (II) diamminodichloride).
This compound is a mutagen which induces filamentous growth in
wild type and in certain repair deficient mutants of *Escherichia
coli* and inhibits DNA synthesis[2,3,4]. Subsequent studies on the
interaction of platinum compounds with macromolecules *in vitro*
and with cellular components *in vivo* have shown that their cytotoxic effect is mainly due to their ability to bind to DNA[5]. The
exact structure of the DNA adducts responsible for the biological
effect is still unknown. However physical chemical studies have
differentiated three classes of platinum binding to DNA : *cis*
bidentate, *trans* bidentate and monodentate[6]. We have investigated
the effect of three platinum compounds, one for each type of DNA

binding (Figure 1) on repair proficient and repair deficient
strains of *Escherichia coli*. This paper reports preliminary characterisation of the ability of these cells to repair the different platinum DNA adducts.

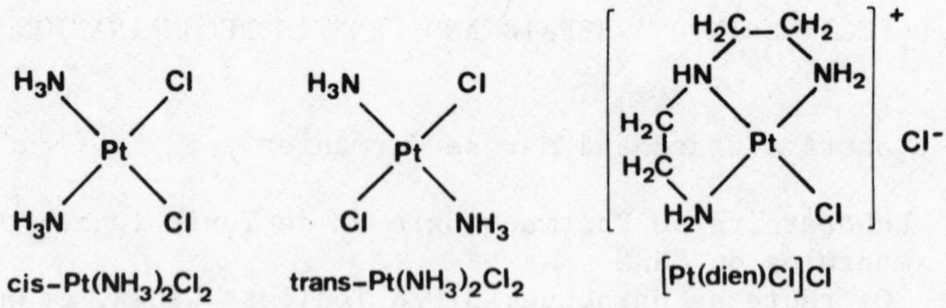

Figure 1 : Structure and formulae of the selected platinum compounds

RESULTS

Survival of *Escherichia coli* strains exposed to platinum compounds

The exposure reducing survival to 37 % (D37) for a series of
wild type and repair deficient *Escherichia coli* are summarized in
table 1. For *cis*-PDD, strains carrying rec B or pol A mutations
have a slightly increased sensitivity whereas uvr A, uvr B, lex A
and rec A mutants are 3 to 15 times more sensitive than their wild
type parent. The double mutant uvr A6 rec A13 is 2 times more sensitive than the rec A single mutant. *Trans*-PDD is toxic at 3 to
10 times higher doses than *cis*-PDD and the differences in sensitivity between wild type, uvr A and rec A cells respectively are less
apparent. [Pt(dien)Cl]Cl has no effect up to concentration greater
than 250 μM.

These data indicate that strains deficient in excision and
recombination repair have a reduced capacity to survive *cis*- and
trans-PDD treatment. The order of sensitivity observed for *cis*-
PDD is comparable to the results found for lesions induced by UV
light or "UV like" compounds[7,8,9]. This suggest that removal of
these chemical adducts is mediated by excision repair and the
rec A dependent repair pathways as already suggested by Beck *et
al*[4] for *cis*-PDD.

Table 1 : Concentrations of platinum compounds corresponding to
37 % survival of *Escherichia coli*. Bacteria were grown
in M63 medium to 4.10^7 cells/ml. Various platinum concentration were added and incubation was continued for
2 hours. Survival was measured on yeast extract-tryptone
plates. Numbers in brackets indicate the amount of platinum adducts bound per genome (10^7 bases) as measured
by atomic absorption spectroscopy ($\pm$ 35 %)

Strain	Markers relevant to DNA repair	Concentration corresponding to 37 % survival (colony forming ability) (µM)		
		cis-PDD	*trans*-PDD	dien
W 3110	+	26.5	93.3	
JG 112	pol A	13.3	33.3	
AB 1157	+	21.7 [800-1000]	73.3 [∿25000]	> 270
AB 1886	uvr A6	7 [300]	32.5 [5000]	> 270
AB 1885	uvr B5	7	26.7	
AB 2463	rec A13	1.5	16.7	80
AB 2470	rec B21	11.7	32.5	
AB 2494	lex A1	6.7	50	
AB 2480	uvr A6 rec A13	0.80	7.3	
MS 23	alk 1	29	71.7	> 135

Inhibition of DNA synthesis

Figure 2 shows the effect of *cis*-PDD on the incorporation of
^{3}H thymidine into acid precipitable material in the parent AB 1157
and its derivatives AB 1886 (uvr A6) and AB 2464 (rec A13). The
post treatment incorporation of ^{3}H thymidine in wild type cells is
almost normal for low doses. For example, treatment with 6.6 µM
cis-PDD corresponding to about 400 platinum adducts per genome
does not alter ^{3}H thymidine incorporation. Under these conditions
there is no change in ^{3}H leucine and ^{3}H uridine incorporation. A
decrease in the incorporation of ^{3}H thymidine is caused by higher
concentration of drug. This inhibition is enhanced for uvr A cells
whereas the mutant AB 2464 (rec A13) shows a severe inhibition of
incorporation over the whole period of incubation.

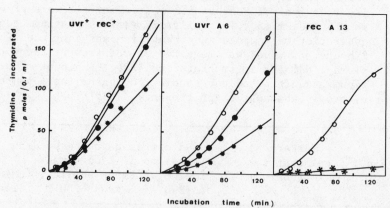

Figure 2 : Incorporation of ^{3}H thymidine into DNA of *Escherichia coli* treated with various amount of *cis*-PDD. Log-phase cultures were grown in M 63 medium and treated for 2 hours. Cells were then washed with M 63 saline buffer and resuspended in complete medium. [3]H thymidine (2 µg/ml, 6 µci/ml) and uridine (50 µg/ml) were added and the amount of radioactive material in the acid insoluble fraction was determined as a function of time.
(○) control ; (*) 6.6 µM ; (●) 13.3 µM ; (*) 26.6 µM.

The lack of inhibition of DNA synthesis in wild type cells at low doses suggests that most of these lesions are quickly excised from DNA or that DNA replication can occur in the presence of these adducts by some mechanism which circumvents unexcised lesions. Similar strain dependent inhibition of DNA synthesis occurs with *trans*-PDD using doses 2 to 5 times higher.

Repair replication *in vivo*

To determine whether bacteria treated by *cis*-PDD could perform repair replication, cells were synchronized by amino acid starvation, treated or not by this compound and post incubated in the presence of tritiated thymidine. Under these starvation conditions, no replication can be initiated. Figure 3A shows that cells which have been treated by *cis*-PDD can incorporate exogeneous thymidine more readily than control cells. This *de novo* incorporation of the precursor does not appear with a uvr A mutant (figure 3B) and might be related to repair synthesis. This interpretation is supported by density gradient centrifugation profiles of the DNA[10] from wild type bacteria treated with *cis*- or *trans*-PDD which show that after treatment with these compounds repair replication occurs while DNA synthesis is greatly reduced. (R. Alazard and M. Germanier, manuscript in preparation).

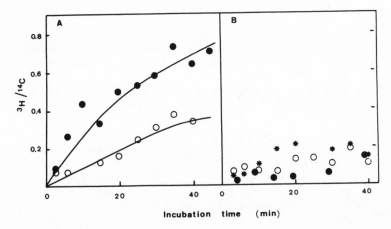

Figure 3 : Incorporation of ^{3}H thymidine into DNA by bacteria treated with *cis*-PDD in the absence of replication. Log phase cells (arg$^-$) pre-labelled for several doubling time with ^{14}C thymidine were resuspended in an arginine free medium and aerated during 45 minutes. Half of the culture was treated with *cis*-PDD during 2 hours and the other half was used as a control. Cells were then washed and resuspended in complete M 63 medium (without arginine). ^{3}H thymidine (2 µg/ml ; 6 µci/ml) and uridine (50 µg/ml) were added and the amount of acid insoluble radioactive material was determined. The ratio ^{3}H/^{14}C was calculated as a function of time for wild type AB 1157 (A) and its uvr A6 derivative (B). (o) control ; (*) 5 µM ; (•) 16.6 µM.

CONCLUSIONS

Our data indicate that removal of platinum adducts in *Escherichia coli* is controled by normal excision repair and recombination repair processes. Experiments in progress will determine whether the decreased activities of *trans*-PDD and [Pt(dien)Cl]Cl are a result of the repair response to these compounds.

ACKNOWLEDGMENTS

The authors wish to thank Dr. J.P. Macquet for making available the platinum compounds and Dr. N.P. Johnson for atomic absorption measurement and for his review of the manuscript. We are grateful to Dr. R. Devoret for providing most of the strains used in this work.

REFERENCES

1. B. Rosenberg, L. Van Camp, J.E. Trosko and V.H. Mansour, Platinum compounds : a new class of potent antitumor agents, Nature 222:385 (1969).
2. D.J. Beck and R.R. Brubaker, Mutagenic properties of cis-Platinum diaminodichloride in Escherichia coli, Mut. Res. 27:181 (1975).
3. B. Rosenberg, E. Renshaw, L. Van Camp, J. Hartwich and J. Drobnik, Platinum induced filamentous growth in Escherichia coli, J. Bacteriol. 93:716 (1967).
4. D. Beck and R.R. Brubaker, Effect of cis-Platinum diaminodichloride on wild type and DNA repair deficient mutants of Escherichia coli, J. Bacteriol. 116:1247 (1973).
5. J.J. Roberts and A.J. Thompson, The mechanism of action of antitumor platinum compounds, in Prog. Nucl. Acid Res. and Mol. Biol. 21:71 (1979).
6. J.P. Macquet and J.L. Butour, Modifications of the DNA secondary structure upon platinum binding : a proposed model, Biochimie, 60:901 (1978).
7. P. Howard-Flanders and R.P. Boyce, DNA repair and genetic recombination : studies on mutants of Escherichia coli defective in these processes, Radiat. Res. Suppl. 6:156 (1966).
8. M. Ikenaga, H. Ichikawa-Ryo and S. Kondo, The major cause of inactivation and mutation by 4-nitroquinoline 1-oxide in Escherichia coli : excisable 4 NQO-purine adducts, J. Mol. Biol. 92:341 (1975).
9. S.C. Cole, D. Levitan and R.R. Sinder, Removal of psoralen interstrand crosslink from DNA of Escherichia coli : Mechanism and Genetic control, J. Mol. Biol. 103:39 (1976).
10. P.C. Hanawalt and P.K. Cooper, Determination of repair replication in vivo in Methods in Enzymology, 21:221 (1971).

DNA REPLICATION, REPAIR, AND MUTAGENESIS IN PERMEABLE E. coli CELLS THAT ARE FULLY VIABLE

Erik Boye and Stein Alver

The Norwegian Radium Hospital

Montebello, Oslo 3, Norway

When suspended in an isotonic buffer E. coli cells repair radiation-induced DNA single-strand breaks by a very rapid, polA-dependent process (Town et al., 1971; Boye et al., 1974). Figure 1 shows the result of an experiment where intracellular, circular phage λ DNA molecules have been irradiated in wild type host cells suspended in different buffers. The cells were exposed to 8 krad of 4 MeV electrons and the fraction of broken circles was measured as a function of postirradiation incubation time at 37°C. As expected, repair was very rapid in the isotonic phosphate-buffered saline (PBS). However, rejoining is completely abolished in the hypotonic 40 mM Tris buffer. This loss of repair capacity depends on the tonicity of the buffer, since normal repair occurs in 40 mM Tris buffer containing 0.13 M NaCl or 0.2 M sucrose (data not shown). When the Tris buffer is supplemented with all four deoxyribonucleoside triphosphates (dNTP's), Mg^{2+}, and NAD, the rapid repair seems to be restored (Fig. 1). This suggests that factors necessary for repair leak out of the cells during treatment with hypotonic buffer. Also, the cells remain permeable so that the necessary factors, when supplied to the suspending buffer, can penetrate into the cells.

Strand break rejoining was absolutely dependent upon the presence of all four dNTP's and Mg^{2+}, while NAD had a stimulatory effect. Since NAD is known to be required for the final sealing of nicks by DNA ligase, the partial dependence of NAD observed in our experiments may indicate that a fraction of the ligase molecules are present in an activated form, and therefore has no requirement for the NAD cofactor.

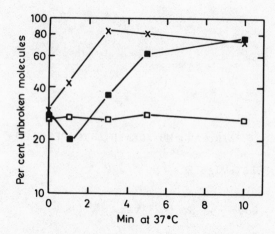

Fig. 1. Repair of radiation-induced DNA single-strand breaks in
         circular phage λ DNA molecules in host cells of strain
         AB 1157(λind⁻) treated with PBS (x) or Tris buffer (□,■).
         The superinfected complexes were exposed to 8 krad and
         strand breakage measured as before (Boye et al., 1974).
         The fraction of covalently closed λ circles was measured
         after irradiation in buffers without (x,□) or with (■)
         all four dNTP's (33 μM each), $MgCl_2$ (10 mM), and NAD
         (0.25 mM).

        There are several methods to make E. coli cells permeable to
small molecular-weight compounds. These include treatment with
toluene (Moses and Richardson, 1970), ether (Vosberg and Hoffmann-
Berling, 1971), or a high sucrose concentration (Wickner and
Hurwitz, 1972). In contrast to the above-mentioned methods the
present involving a hypotonic buffer treatment does not affect
cell viability. Cell survival was the same in 40 mM Tris as in the
isotonic PBS, suggesting that the present technique might be used
in studies of cell survival and mutagenesis. It was therefore of
interest to further characterize what happens to E. coli cells when
suspended in a hypotonic buffer.
        We have measured the effect of hypotonic buffers on the pool
of intracellular sulfhydryl compounds. A large fraction of these
compounds, which consist mainly of the tripeptide glutathione, was
found to leak out of the cells (Boye, 1980). Sulfhydryl compounds
are effective scavengers of the OH radical, and consequently a
good protector against ionizing radiation. In agreement with this
increased yields of radiation-induced DNA single-strand breaks
were found in hypotonic buffers (Boye, 1980).

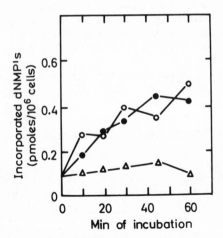

Fig. 2. DNA synthesis in different buffers. Cells of strain
W3110 (●,△) or p3478 <u>polAl</u> (o) were washed three times in
Tris buffer and resuspended in replication buffer (Boye,
1980) with (o, ●) or without (△) ATP.

Normally, dNTP's are not able to penetrate into <u>E. coli</u> cells.
However, if the cells are permeable exogenous dNTP's can enter the
cells and may be incorporated into the DNA. We have measured the
uptake of radioactively labeled dNTP's into TCA-insoluble material
in <u>E. coli</u> cells washed in 40 mM Tris buffer. The rate of incorporation is linear (Fig. 2) and comparable to the rates obtained
with toluene-, ether-, or sucrose-treated cells. The synthesis
measured probably represents semiconservative replication since it
is not affected by the presence of the <u>polAl</u> mutation and is
dependent upon ATP. This conclusion is further strengthened by
experiments showing that the synthesis is sensitive to novobiocin
and to non-permissive temperature in the replication mutant <u>dnaB</u>
(to be published elsewhere).

We have performed some introductory survival and mutagenesis
studies with cells permeabilized with 40 mM Tris buffer. Permeable
cells have been exposed to different doses of ionizing radiation
before incubation with all four dNTP's, Mg^{2+}, and NAD in the
presence or absence of ATP. In both cases normal repair of lesions
occurs, while the presence or absence of ATP determines whether
semiconservative replication is allowed or inhibited. Thus, if the

passing of the replication fork up to or past a lesion represents
a lethal or a mutagenic event we would expect to see a difference
in these two parameters with and without ATP. Preliminary results
indicate that cell survival is enhanced and mutagenesis is
decreased in the absence of ATP. One should be careful to draw any
conclusions from this experiment except that it shows that this
method has a great potential in studies of intracellular in vivo
processes in relation to cell survival and mutagenesis.

This work was supported by The Norwegian Cancer Society.

Boye, E., Johansen, I., and Brustad, T., 1974, Time scale for
    rejoining of bacteriophage λ deoxyribonucleic acid molecules
    in superinfected pol⁺ and polA1 strains of Escherichia coli
    after exposure to 4 MeV electrons, J. Bacteriol., 119:522.
Boye, E, 1980, DNA repair and replication in cells of Escherichia
    coli made permeable with hypotonic buffers, Radiat. Res.
    (in press).
Moses, R.E. and Richardson, C.C., 1970, Replication and repair of
    DNA in cells of Escherichia coli treated with toluene,
    Proc. Natl. Acad. Sci. U.S.A., 67:674.
Town, C.D., Smith, K.C., and Kaplan, H.S., 1971, DNA polymerase
    required for rapid repair of x-ray-induced DNA strand breaks
    in vivo, Science, 172:853.
Vosberg, H.-P. and Hoffmann-Berling, H., 1971, DNA synthesis in
    nucleotide-permeable Escherichia coli cells. I. Preparation
    and properties of ether-treated cells, J. Mol. Biol., 58:739.
Wickner, R.B. and Hurwitz, J., 1972, DNA replication in Escherichia
    coli made permeable by treatment with high sucrose, Biochem.
    Biophys. Res. Comm., 47:202.

INVOLVEMENT OF ATP IN REPAIR AND

MUTAGENESIS OF X-IRRADIATED MAMMALIAN CELLS

Françoise Laval

Groupe de Recherches de Radiologie Clinique

Institut Gustave-Roussy, 94800 VILLEJUIF, FRANCE

In mammalian cells, the existence of an error-prone repair process has been reported after UV-irradiation[1-2]. However, in X-irradiated cells, the existence of a similar process is only suggested by the suppression of malignant transformation by protease inhibitors[3]. We have previously shown that treatment with an uncoupler of oxidative phosphorylation increased the survival of X-irradiated cells[4].Therefore experiments were designed to determine whether this survival increase was related to the cellular ATP concentration and could be explained by the existence of an error-prone repair system.

RESULTS AND DISCUSSION

When H_4 cells (derived from a rat hepatoma[5]) are incubated with an uncoupler of oxidative phosphorylation for 30 min prior to and during X-irradiation, their survival is increased. Table 1 shows the Do measured when plateau cells are irradiated in MEM or treated with the uncouplers prior to irradiation. The drugs were dinitrophenol (DNP), m-chloro-carbonylcyanide phenylhydrazone (CCCP) and carbonyl cyanide p-trifluoromethoxyphenylhydrazone (TFCP).

The survival increase[3] (i) is more important in the case of plateau cells than in the case of exponential cells ; (ii) is related to the potency of the uncoupler ; (iii) is observed when the cells are incubated with the drugs immediatly prior to or immediatly after irradiation ;(iv) has been observed with different cells lines from rodent or human origin.

Table 1 : Survival curve parameters for H_4 plateau cells
         incubated for 30 min with uncouplers prior to
         X-irradiation.

UNCOUPLER	PLATING EFFICIENCY	Do (rad)
0	75 - 85	$152,8 \pm 1,3$
DNP (10^{-4}M)	72 - 84	$225,5 \pm 4,6$
CCCP (10^{-5}M)	73 - 84	$269,0 \pm 12,8$
TFCP ($2,5 \times 10^{-6}$M)	73 - 85	$334,2 \pm 6,4$

The influence of uncouplers on X-ray induced mutation frequency was studied by the appearance of 8-azaguanine resistant mutants[6]. The maximum number of mutants was measured after an
expression time of three doublings for cells irradiated in MEM or
irradiated after TFCP ($2,5 \times 10^{-6}$M) treatment. The results showed
that incubation with the uncoupler markedly increases the mutation
frequency in irradiated cells[7]. For example, when 10^6 cells are
irradiated with a dose of 300 rads in MEM, the total number of
mutants is 20 ; when the cells are incubated with TFCP then irradiated, this number is 210.

In order to check if the survival increase observed after
uncoupler treatment needs protein synthesis, the cells were incubated with TFCP, X-irradiated, then plated in the presence of cycloheximide. The results show (Table 2) that the plating efficiency
and the radiosensitivity were not modified by the presence of cycloheximide in the plating medium. However, the uncoupler does not
increase the survival of irradiated and cycloheximide treated cells.
Addition of leupeptine to the plating medium also prevents the increase of survival mediated by uncouplers in X-irradiated H_4 cells.

The influence of uncouplers on cell survival was related to
DNA metabolism. When the cells are irradiated in the presence of
CCCP or TFCP, the semi conversative DNA synthesis is less depleted
than in the case of cells irradiated in normal medium. Furthermore
the DNA repair synthesis is respectively 2 and 3 times higher in
irradiated cells treated with uncouplers than in control cells[6].

Incubation with uncouplers results in a transient decrease of
the cellular ATP (Table 3) determined by the luciferase assay[8].
Experiments were performed in order to determine if the survival
increase mediated by uncouplers could be related to this decrease
of cellular ATP concentration. In aerobic conditions where energy
is provided by oxidative phosphorylation[9], uncouplers decrease
the cellular energy and increase the survival. In hypoxia, energy

Table 2 : Effect of uncouplers and cycloheximide on the survival of X-irradiated H_4 cells.

CELL TREATMENT	SURVIVAL AFTER 1 000 RADS
Irradiation in MEM, no CX	0,55
Irradiation in MEM, CX	0,51
TFCP for 30 min, X-rays, no CX	16,10
TFCP for 30 min, X-rays, CX	0,61

The cells were incubated or not with TFCP ($2,5 \times 10^{-6}$M), X-irradiated, then cycloheximide (CX, 10μg/ml) was added or not for 6 hours in the plating medium.

is obtained by way of anaerobic glycolysis[9]. Metabolic inhibitors of glycolysis which decrease the ATP pool should modify the survival of cells irradiated in hypoxic conditions. Therefore hypoxic H_4 cells were incubated or not with 5-thio-d-glucose, then X-irradiated. The Do are plotted in Table 3. They show that decreasing the cellular ATP concentration corresponds to an increase of survival in X-irradiated cells.

Table 3 : Survival curve Parameters for H_4 cells X-irradiated in aerobic or hypoxic conditions.

CELL TREATMENT	PLATING EFFICIENCY(%)	Do (rad)	ATP concentration(%)
Aerobic Conditions			
MEM	75 - 85	152,8 ± 1,3	100
MEM + CCCP	73 - 84	269,0 ±12,8	10
MEM + TG	75 - 85	160,2 ± 2,4	90
Hypoxic Conditions			
MEM	72 - 85	466,2 ± 2,6	100
MEM + CCCP	76 - 87	460,2 ± 1,8	98
MEM + TG	75 - 85	686,5 ± 3,5	28

H_4 plateau cells were incubated for 30 min with CCCP (10^{-5}M) or for 60 min with 5-thio-d-glucose (TG, 1mg/ml) prior to X-irradiation in aerobic or hypoxic conditions.

The same influence of uncouplers on cell survival was observed in cells treated with methyl methane sulfonate, but they do not modify the survival of UV-irradiated cells. Although the results suggest that a relationship may exist between ATP concentration and cell survival, the problem is obviously complex. Uncouplers decrease the macromolecular synthesis at the time of irradiation and could act by one of these different pathways. They increase the mutation frequency and their influence on cell survival needs protein synthesis. These results suggest the existence of an error-prone repair process in X-irradiated mammalian cells.

ACKNOWLEDGEMENTS

The author thanks Professor M. TUBIANA for stimulating discussions. This work was supported by grants from CNRS and INSERM (CRL 78.5.151.3).

REFERENCES

1. D'Ambrosio, S.M., and Setlow, R.B. : Enhancement of postreplication repair in chinese hamster cells.
Proc. Natl. Acad. Sci.U.S.A., 73, 2 396, 1976.
2. Dasgupta U.B., and Summers, W.C. :UV-reactivation of herpes simplex virus is mutagenic and inducible in mammalian cells.
Proc. Natl. Acad. Sci., U.S.A., 75, 2 378, 1978.
3. Kennedy, A.R., and Little, J.B., : Protease inhibitors suppress radiation induced malignant transformation in vitro.
Nature, 276, 825, 1978.
4. Laval, F., and Little,J.B., : Enhancement of survival of X-irradiated mammalian cells by the uncoupler of oxidative phosphorylation m-chloro-carbonylcyanide phenylhydrazone.
Rad.Res., 71, 571, 1977.
5. Dehove, D., and Laval, F., : Influence of Daunorubicin on the survival of X-irradiated mammalian cells.
Int. J. Rad. Biol., 37,373, 1980.
6. Maher, V.M., and Wessel, J.E., :Mutations to azaguanine resistance induced in cultured diploid human fibroblasts by the carcinogen N-acetoxy-2-acetylaminofluorene.
Mut. Res., 28, 277, 1975.
7. Laval, F., : Effect of uncouplers on radiosensitivity and mutagenicity in X-irradiated mammalian cells.
Proc. Natl. Acad. Sci., U.S.A., 77, 2702, 1980.
8. Kimmich, G.A., Randles, J., and Brand, J.S., : Assay of picomole amounts of ATP, ADP, and AMP using the luciferase enzyme system.
Anal. Biochem., 69, 187, 1975.
9. Kim, J.H., Kim, S.H., Hahn, E.W., and Song, C.W. : 5-thio-d-glucose selectively potentiates hypethermic killing of hypoxic tumor cells.
Science, 200, 206, 1978.

CELL CYCLE PATTERNS OF MUTATION INDUCTION

AND THEIR RELATIONSHIP TO DNA REPAIR*

Regine Goth-Goldstein

Laboratory of Radiobiology
University of California
San Francisco, California 94143

ABSTRACT

From the finding that mutation induction by ethylnitrosourea is cell-cycle independent in synchronous Chinese hamster ovary cells and from the observation that these cells do not excise O^6-alkyl-guanine, it is concluded that enhanced mutagenesis at growing points, as observed in bacteria and yeast, is due to a repair system which removes mutagenic lesions at sites other than the DNA growing point, where immediate mutation fixation occurs.

INTRODUCTION

Alkylating agents, like N-methyl-N'-nitro-N-nitrosoguanidine (MNNG) and N-ethyl-N-nitrosourea (ENU), preferentially mutagenize in the Escherichia coli chromosome DNA sequences which are in the process of being replicated.[1,2] Different genetic markers are mutated at maximal rates at different times during a synchronous replication cycle, and the order in which markers are maximally mutated corresponds to the order in which they replicate. Similarly, MNNG induces mutations in synchronous yeast populations at much higher rates during a limited interval of the DNA synthesis period, individual genetic markers exhibiting maximum sensitivity to mutagenesis at different times of the S phase.[3,4]

As enhanced mutagenesis of the growing point is found not only in bacteria, but also in lower eukaryotes, it seemed likely that it would also occur in mammalian cells. This idea was tested in Chinese hamster ovary (CHO) cells.

─────────────
*This work was supported by the U.S. Department of Energy.

Cell Cycle Pattern of ENU-Induced Mutations in CHO Cells

A series of synchronous populations of CHO cells, spaced 1 h apart in the cell cycle and obtained by mitotic detachment, were treated with ENU for 30 min at $37°$. After an 8-day expression time cells were tested for mutation induction to 6-thioguanine resistance, ouabain resistance, and diphtheria toxin resistance (described in detail by Goth-Goldstein and Burki[5]). No increased mutation induction was observed for any of the three markers when cells were treated during the S phase; instead, the mutation induction was similar at all times of the cell cycle.

Removal of Ethylated Purines from DNA of CHO Cells

CHO cells were treated with 100 µg ml^{-1} $1-^{14}$C-ENU for 1 h at $37°$, a dose that is nontoxic to cells. The DNA was isolated either immediately after treatment or after cells had grown for another 22 h (for more details see Goth-Goldstein[6]). After DNA hydrolysis the purine bases were separated by Sephadex G10 chromatography. The amount of 3-ethyladenine (3 EtA), 7-ethylguanine (7 EtG), and 0^6-ethylguanine (0^6 EtG) was determined from the radioactivity eluted in the position of the appropriate marker and was expressed as a fraction of the parent base. Prelabeling of the cells with ^{3}H-thymidine and determination of the specific ^{3}H-activity of DNA permitted correction for dilution of ethylated bases resulting from DNA synthesis during the 22 h growth. The amounts of ethylated purine bases measured in two experiments are given in Table 1. 3 EtA and 7 EtG are easily depurinated from DNA due to the lability of their N-glycosidic bond. On the basis of their half-lives in DNA in vitro[7] one can calculate that during a 22 h interval 3 EtA would be reduced to 63% in CHO-DNA and 7 EtG to 93% due to spontaneous depurination alone. Actually, a much lower value was measured, so that one can conclude that the cell removes these derivatives by a repair process. In contrast, the amount of 0^6 EtG, which is a stable product in DNA, was virtually unchanged after 22 h. Therefore the CHO cell seems unable to excise this lesion.

0^6-alkylguanine has been implicated in mutagenesis: Its presence causes mutations in phages and miscoding by nucleic acid polymerases, and its formation and persistence in DNA of various organs after administration of N-nitroso carcinogens correlates with the incidence of tumors (for review see Pegg[8]). The deficiency in the removal of this lesion in CHO cells might therefore have important consequences for the mutation induction in these cells.

CONCLUSION

Enhanced mutagenicity at the growing point region of bacteria and yeast could originate either from enhanced reactivity of the mutagen with single-stranded regions of the replication fork or

Table 1. Loss of Ethylated Bases from CHO-DNA on Incubation of ^{14}C-ENU-Treated Cells

Experiment	Incubation Time (h)	3 EtA / A X 10^6 Value	% of 0 h Value	7 EtG / G X 10^6 Value	% of 0 h Value	O^6 EtG / G X 10^6 Value	% of 0 h Value
1	0	4.2		14.7		13.0	
	22	0.9*	22	9.7*	66	13.6*	105
2	0	3.9		20.5		15.8	
	21	0.4*	11	11.6*	57	14.5*	91

*Values corrected for dilution. The dilution factor (^{3}H-specific activity of DNA at 22 h versus that at 0 h) was 0.46 in the first experiment and 0.59 in the second experiment.

from an efficient repair system, which excises mutagenic lesions
before replication, so that only lesions at the growing point are
fixed as mutations. It is unlikely that an alkylating agent reacts
preferentially at growing points in yeast and bacteria, but not in
mammalian cells. It is known that Escherichia coli is very efficient
in the excision of 0^6-alkylguanine.[9,10] Therefore it is concluded
that growing point mutagenesis results from repair before DNA replication rather than enhanced reactivity. If the lesion cannot be
repaired by the cells, as for the 0^6 EtG in CHO cells, the mutation
frequency is independent of the time interval between formation of
the adduct in DNA and the replication of this part of the genome.

REFERENCES

1. E. Cerdá-Olmedo, P. C. Hanawalt, and N. Guerola, Mutagenesis of
    the replication point by nitrosoguanidine: Map and pattern
    of replication of the Escherichia coli chromosome, J. Mol.
    Biol. 33:705 (1968).
2. T. A. Hince and S. Neale, A comparison of the mutagenic action of
    the methyl and ethyl derivatives of nitrosamides and nitro-
    samidines on Escherichia coli, Mutation Res. 24:383 (1974).
3. T. W. Dawes and B. L. Carter, Nitrosoguanidine mutagenesis
    during nuclear and mitochondrial gene replication, Nature
    250:709 (1974).
4. W. Burke and W. Fangman, Temporal order in yeast chromosome
    replication, Cell 5:263 (1975).
5. R. Goth-Goldstein and H. J. Burki, Ethylnitrosourea-induced
    mutagenesis in asynchronous and synchronous Chinese hamster
    ovary cells, Mutation Res. 69:127 (1980).
6. R. Goth-Goldstein, Inability of Chinese hamster ovary cells to
    excise 0^6-alkylguanine, Cancer Res., in press.
7. R. Goth and M. F. Rajewsky, Molecular and cellular mechanisms
    associated with pulse-carcinogenesis in the rat nervous
    system by ethylnitrosourea: Ethylation of nucleic acids and
    elimination rates of ethylated bases from the DNA of different
    tissues, Z. Krebsforsch 82:37 (1974).
8. A. E. Pegg, Formation and metabolism of alkylated nucleosides:
    Possible role in carcinogenesis by nitroso compounds and
    alkylating agents, Adv. Cancer Res. 25:195 (1977).
9. P. D. Lawley and D. J. Orr, Specific excision of methylation
    products from DNA of Escherichia coli treated with N-methyl-
    N'-nitro-N-nitrosoguanidine, Chem.-Biol. Interact. 2:154
    (1970).
10. P. D. Lawley and W. Warren, Specific excision of ethylated purines
    from DNA of Escherichia coli treated with N-ethyl-N-nitro-
    sourea, Chem.-Biol. Interact. 11:55 (1975).

YEAST AS A MODEL FOR STUDYING DNA-ALTERATIONS AND BIOLOGICAL RESPONSE

INDUCED BY ALKYLATING ANTI-CANCER DRUGS: EFFECTS OF CYCLOPHOSPHAMIDE

Reinhard Fleer and Martin Brendel

Institut für Mikrobiologie
J.W.Goethe-Universität
Theodor-Stern-Kai 7, Haus 75A, D-6000 Frankfurt/M.

SUMMARY

Activated cyclophosphamide (CP) induces DNA interstrand cross-links and DNA strand-breakage in the chromosomes of yeast. In contrast to the post-incubation reaction of other polyfunctional alkylating agents, a further increase of cross-links after treatment with activated CP was hardly detectable. This is possibly due to the observed continued DNA fragmentation which decreases the sensitivity of the cross-link assay. Cell killing by non-activated CP depends on the function of DNA repair as well as on the mode of cellular energy metabolism. At concentrations 1000-fold higher than those used for activated CP it induces DNA single- as well as double-strand breaks. In contrast, no induction of interstrand cross-links could be detected.

INTRODUCTION

Recently we argued in favour of the genetically well known eucaryotic microbe Saccharomyces cerevisiae when testing for correlations between biological endpoints and DNA primary lesions[1,2]. Easily to handle, yeast cells are readily accessible to genetic manipulation, including defined blocking of one or more of the 60 genes affecting DNA dark repair[3]. Using strains of differing repair capacities, lethality as well as mutagenicity of a number of alkylating anti-cancer drugs were shown to depend substantially on the function of genes controlling repair processes[1,4]. This implicates that the lethal event induced by these agents involves some form of DNA damage. As one prominent type of DNA lesion, we demonstrated formation of DNA interstrand cross-links after treatment of cells with moderate doses of polyfunctional alkylating agents using a method that allows detection of less than one cross-link per chromosome[1,2,5]. In contrast to

bacteria and viruses, there is little agreement concerning the interrelation between DNA damage by alkylating anticancer drugs and cytotoxic effects induced in eucaryotic organisms. This is especially
true with regard to cyclophosphamide (CP)[6,7] which plays an outstan-
ding role in the clinical therapy of cancer. It is agreed that CP
exerts its cytotoxic action only after metabolic activation to
4-hydroxy-cyclophosphamide (4-OH-CP)[8] which further decomposes spontaneously in aqueous solutions yielding cleavage products with alkylating activity[9]. So far, however, no clear-cut correlation between
toxicity and alkylation of cellular constituents could be shown[10,11].
Therefore, it was concluded that DNA alkylation might not play an
important role for cytotoxicity[11,12]. In these investigations total
alkylation was used as a measure of biochemical induced lesions,
thereby covering cellular damages which differ considerably in their
lethal effects[13]. As it is known that toxicity and genetic activity
are due to qualitative rather than to quantitative aspects of DNA
alkylation we propose a more differential determination of DNA lesion
including their interrelation with repair processes[2]. So far, detailed information relating to specified DNA damage remains limited[6,?].
Some of our results obtained with CP and "activated" CP, which will be
reported in more detail elsewhere, will be summarized here.

MATERIAL AND METHODS

        Most of the procedures used have already been described in de-
tail[1,2,4,5]. Briefly, all treatments with alkylating agents have been
carried out with radioactively labeled, stationary haploid yeast cells
suspended in phosphate buffer (0,067M; pH 7) at 36°C. After stopping
the reaction by several washings, cells were either resuspended in
buffer for post-incubation experiments or immediately frozen in an
Eaton press at -70°C. Homogenization was performed in a crushing mixture containing 25% glycerol. The assay of DNA interstrand cross-links
was based on the technique of reversible denaturation: covalent
linkage of the DNA sister strands will allow renaturation of alkali-
denatured DNA upon neutralization, while molecules without cross-links
will remain single-stranded. Because of their different buoyant densities, double-stranded (renatured) and single-stranded (denatured) DNA
molecules are resolved into separate bands in ultracentrifugation.
Isopycnic centrifugation in CsCl as well as sedimentation analysis in
linear 15-30% sucrose gradients were parallely performed using un-
purified homogenates. After TCA precipitation, radioactivity of the
gradient fractions was assayed by standard LSC-techniques.

RESULTS AND DISCUSSION

Effects of "activated" CP

        CP has proved to be of great interest in both clinical therapy
and experimental laboratory work because of its relatively high can-

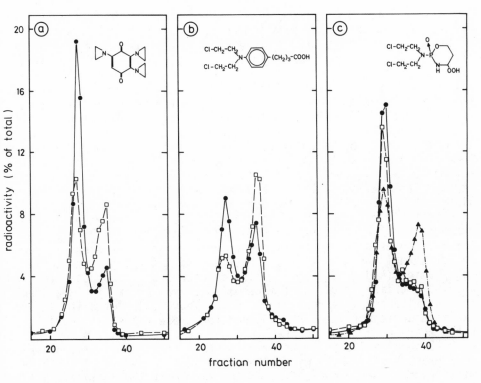

Fig. 1.   Effect of various bifunctional   alkylating agents on dena-
turability of DNA as measured in neutral CsCl equilibrium
density gradients: Influence of post-incubation treatment.
Right and left bands define positions of renatured and denatured DNA, respectively. (a) DNA from cells treated with
4.2×10^{-8}M triaziquone (12 min, 36^{0}C). Surviving fraction:
69%.●——●, without post-incubation;□----□, with 4 h
of post-incubation in phosphate buffer at 23^{0}C. (b) treatment with 5×10^{-3}M chlorambucil (2 h, 36^{0}C). Surviving fraction: 28%.●——●, without post-incubation;□----□, with
18 h of post-incubation at 23^{0}C. (c) treatment with 10^{-4}M
4-OOH-CP for various times at 36^{0}C.●——●, 4 h treatment,
surviving fraction 88%. □----□, additional 8 h of post-
incubation treatment at 36^{0}C. The banding of DNA after 4 h
of post-incubation was essentially the same (not shown). Surviving fraction 55%. ▲--·--▲ , 12 h treatment with 10^{-4}M
4-OOH-CP, without post-incubation. Surviving fraction: 9.5%.
The DNA distribution of the untreated control was similar to
control in Fig. 3.

cerostatic specificity. Essentially, two hypotheses have been proposed to explain this selective action of CP: (i) a reduced capacity

of tumor cells for enzymatic deactivation of the primary metabolite
4-OH-CP to non-toxic compounds[14], and (ii) a reversible fixation of
4-OH-CP proteins thus leading to a controlled toxication of this meta
bolite over a long period[9]. Both models imply release of the final
breakdown product(s) within the cell which in turn exert their toxic
effects by alkylation of target molecules. So far, however, virtually
nothing is known about the molecular mechanism by which the active
metabolite(s) of CP react with the cellular target site to cause letha
lity nor about the site of such action[7]. The nearly total loss of cytotoxic activity by monodechloroethylation and secondary oxydation of
4-OH-CP[14] supports the hypothesis that bifunctional alkylation of tar
get molecules is needed to block cell division. Recently, we demonstrated "activated" CP being able to induce DNA interstrand cross-
links[1] thus strengthening the hypothesis of DNA damage being involved
in cytostatic effects, as it is known that a single covalent linkage
between opposite DNA strands is lethal if not repaired[15]. More detailed studies of these processes revealed a dose-dependent increase
in renaturability of DNA after treatment with 4-hydroperoxy-cyclo-
phosphamide (4-OOH-CP)(Fig.1) which spontaneously decomposes to 4-OH-
CP[16], the primary product of metabolic activation of CP. Experiments
were carried out at a dose range conferring little lethality in the
excision deficient yeast strain used for these tests. It should be
noted that the applied 4-OOH-CP concentration is very similar to the
peak plasma level of phosphoramide mustard (PM) - the first breakdown product of 4-OH-CP with alkylating activity[9] - found after CP ad
ministration to humans[17]. At a concentration of 100 nmole/ml, a first
increase of renaturability of DNA (a measure for interstrand cross-
linking) is clearly detectable at a treatment time resulting in 88%
survival of cells. At about 12 h of incubation, renaturability reache
a maximum (Fig.1) and then drops to lower values upon further incubation (data not shown). Parallel measurements of sedimentation behaviour in neutral sucrose gradients indicated induction of DNA double-
strand breaks increasing with treatment time (data not shown). As the
sensitivity of each cross-link assay based on renaturability is closely related to the molecular weight of the DNA molecules[2], the decreas
of the renaturable fraction at prolonged 4-OOH-CP incubation may, at
least partially, be explained by the observed DNA fragmentation. So
far, it remains uncertain whether this strand breakage is due to chemical processes or to repair.

Considering reversibility of binding of 4-OH-CP to proteins which
leads to a slowed-down release of cytotoxic metabolites within the
cell, we anticipated a further increase in renaturability during a
post-incubation period (additional incubation of washed cells in
phosphate buffer). Such delayed formation of cross-links has already
been demonstrated for various anti-cancer drugs[2,18] and was either
attributed to different reaction kinetics of the two chloroethyl
residues[19] or to reversible non-covalent association to nuclear proteins[20]. However, in contrast to other alkylating agents (Fig.1), we
so far failed to measure a pronounced post-incubation effect on DNA
renaturability with 4-OOH-CP (Fig.1). We only observed a slight in-

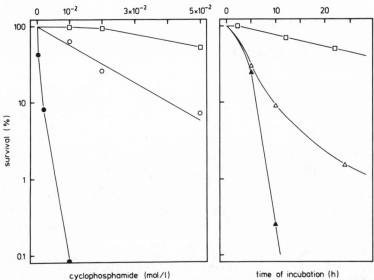

Fig. 2. Lethal effects of non-activated CP in yeast strains differing in capacity for DNA dark repair or mode of energy metabolism. Left panel: Surviving fractions as a function of CP concentration, treatment for 21 h at $36°C.\square$———$\square$, strain ZII, fully repair proficient wild-type; O———O, strain KC372, lacking function of gene RAD18; ●———●, strain 1111-1D, blocked in two non-epistatic genes affecting different repair pathways (RAD2, RAD18). Right panel: surviving fractions as a function of treatment time, CP concentration 10^{-1}M. $\triangle$———$\triangle$, respiration competent, excision-deficient strain KC370 lacking function of gene RAD2; ▲———▲, respiration-deficient isolate (rho^o) of the same strain; $\square$———$\square$, the repair-proficient, respiration-competent (rho$^+$) wild-type strain ZII.

crease which is reversed during a prolonged post-incubation time accompanied with further DNA fragmentation (data not shown). These findings may be explained assuming the release of 4-OH-CP from proteins being slower than the concurring degradation of already alkylated DNA. The latter process has been shown to lead to a decreased sensitivity of the cross-link assay[2].

Considering the close similarity between the types of DNA lesions produced by 4-OOH-CP and other mustards we propose the cancerospecific action of CP to be due to differences in the pharmacokinetic and pharmacodynamic properties of this drug (influencing bioavailability and uptake of biologically active metabolites and/or their precursors) rather than to differences in the molecular mode of action of the final metabolites. Additionally, we will have to take into account for differences in alkylation kinetics and possibly in repairability of DNA lesions: certainly it will be of importance whether DNA damage is induced rapidly - thus raising the probability of pre-replicational repair - or whether a continuous alteration of DNA takes place throughout the cell cycle.

Effects of non-activated CP

There is a considerable body of papers reporting CP to be inactive in its original form and to obtain its cytotoxic and genetic potential only after metabolic activation by mammalian enzymes[21]. Recently, however, it could be shown that CP is spontaneously degraded "in vitro" to give products which act genetically in bacterial[22] as well as in eucaryotic[23] organisms. The genetically active breakdown products are reported to accumulate over a period of 72 h when incubating CP in phosphate buffer (pH 7.4) at $37^{\circ}C$[22]. These findings are consistent with own experiments demonstrating increased cytotoxicity to yeast when pre-incubated CP solutions are used (unpublished data). As has been pointed out[22], no definite conclusions can yet be made as to the importance of the breakdown intermediates on an "in vivo" process such as damage to cellular DNA. Previously, we reported cytotoxic as well as mutagenic effects of non-activated CP "in vitro" being dependent on the function of some processes of DNA dark repair[4]. This implicates that DNA lesions may be regarded as a cause of cell death. Here, we present additional evidence of the importance of DNA repair processes for the cells' response to CP-treatment: induction of cell killing in a super-sensitive double mutant lacking function of RAD2

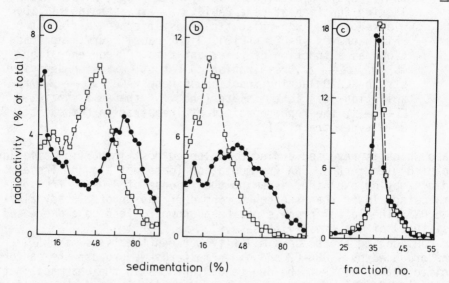

Fig. 3.   Influences of CP on molecular weight and renaturability of DNA. (a) Distribution of double-stranded and (b) of single-stranded DNA in neutral 15-30% sucrose gradients. (c) Banding of alkali denatured and re-neutralized DNA in isopycnic CsCl gradients. ●———●, control values; □----□, DNA of cells treated for 12 h with 10^{-1}M CP, surviving fraction 6%.

and RAD18 genes (belonging to two different repair pathways) requires only about 0.3% of the respective CP concentration when compared to the repair-proficient wild type strain (Fig.2). Apart from DNA repair, the mode of cellular energy metabolism can be demonstrated to exert a pronounced influence on cytotoxicity of CP: respiration-deficient isolates of a rad2 strain defective in excision repair were much more sensitive to CP treatment than was the respiration-competent parental strain (Fig.2). Similar results were reported by Mayer et al.[23]. Whether this effect may be related to a reduced capacity for DNA repair or to other metabolic processes remains unclear.

The hypothesis of DNA damage induced by CP - so far based on more indirect considerations - is now confirmed by a parallel assay of cytotoxicity, DNA renaturability, and sedimentation behaviour of single- as well as of double-stranded DNA (Fig.3). Whereas a CP treatment producing 95% cell killing (data not shown) induced both single- and double-strand breaks in a dose dependent manner, we failed to detect any effect on renaturability of DNA. Considering the reported CP breakdown under comparable conditions of pH and temperature[24] - finally leading to N-β-hydroxyethyl-β-chloroethylamine - the described effects of non-activated CP on DNA might be explained by monofunctional alkylation. This hypothesis would agree with our findings that in order to achieve the same level of cytotoxicity observed with nitrogen mustard we need a 200-fold higher concentration of the corresponding monofunctional half mustard[4]. Since treatment with 4-OOH-CP at comparable conditions of incubation time, cell killing, and DNA degradadation (data not shown) does produce renaturable DNA already at a concentration corresponding to 0.1% of that of CP, we conclude that the pathways of spontaneous degradation of CP differ from those of metabolic activation leading to different end products. In both cases, however, DNA damage causing cytotoxic as well as genetic effects is induced.

ACKNOWLEDGEMENTS

We wish to thank Dr.G.Peter, Gustav-Embden-Zentrum der Biologischen Chemie, for a gift of 4-OOH-CP and Dr.N.Brock, Asta-Werke, Brackwede, for supplying CP. This work was supported by a grant from the Deutsche Forschungsgemeinschaft.

REFERENCES

1. A. Ruhland, R. Fleer, and M. Brendel, Genetic activity of chemicals in yeast: DNA alterations and mutations induced by alkylating anti-cancer agents, Mutat.Res.,58:241 (1978)
2. R. Fleer and M.Brendel, Formation and fate of cross-links induced by polyfunctional anticancer drugs in yeast, Molec.gen.Genet., 176:41 (1979)
3. R. H. Haynes, B. J. Barclay, F. Eckhardt, O. Landman, B. Kunz, and J. G. Little, Genetic control of DNA repair in yeast, in: "Proc.XIV.Internat.Congr.Genetics", Moscow (USSR)(1978)in press

4. A. Ruhland and M. Brendel, Mutagenesis by cytostatic alkylating agents in yeast strains of differing repair capacities, Genetics, 92:83 (1979)

5. M. Kircher, R. Fleer, A. Ruhland, and M. Brendel, Biological and chemical effects of mustard gas in yeast, Mutat.Res., 63:273 (1979)

6. D. L. Hill,"A review of cyclophosphamide",Charles C. Thomas, Springfield (1975)

7. O. M. Friedman, A. Myles, and M. Colvin, Cyclophosphamide and related phosphoramide mustards, in: "Advances in Cancer Chemotherapy", A. Rosowsky, ed., Dekker, New York, Basel (1979)

8. N. Brock, and H. J. Hohorst, Über die Aktivierung von Cyclophosphamid in vivo und vitro, Arzneim.Forsch.(Drug Res.), 13:1021 (1963)

9. H. J. Hohorst, U. Draeger, G. Peter, and G. Voelcker: The problem of oncostatic specificity of cyclophosphamide (NSC-26271): Studies on reactions that control the alkylating and cytotoxic activity, Cancer Treat.Rep., 60:309 (1976)

10. G. P. Wheeler, and J. A. Alexander, Studies with mustards. V. In vivo fixation of C^{14} of labeled alkylating agents by bilaterally grown sensitive and resistant tumors, Cancer Res., 24:1331 (1964)

11. E. G. Trams, M. V. Nadkarni, and P. K. Smith, On the mechanisms of action of the alkylating agents. I. Interaction of alkylating agents with nucleic acids, Cancer Res., 21:560 (1961)

12. E. Liss, H. Schmidt, E. Schaumlöffel, and G. Palme, Alkylierung der DNA verschiedener Tumorzellarten nach in vivo Gabe von Cyclophosphamid, Z. Krebsforsch.Klin.Onkol., 80:239 (1973)

13. P. D. Lawley, J. H. Lethbridge, P. A. Edwards, and K. V. Shooter, Inactivation of bacteriophage T7 by mono- and difunctional sulphur mustards in relation to cross-linking and depurination of bacteriophage DNA, J.Mol.Biol., 39:181 (1969)

14. P. J. Cox, B. J. Phillips, and P. Thomas, Studies on the selective action of cyclophosphamide (NSC-26271): Inactivation of the hydroxylated metabolite by tissue-soluble enzymes, Cancer Treat.Rep., 60:321 (1976)

15. R. S. Cole, Inactivation of E.coli and bacteriophage lambda by psoralen plus 360 nm-light. Significance of DNA-crosslinks, J.Bacteriol., 107:846 (1971)

16. G. Voelcker, U. Dräger, G. Peter, and H. J. Hohorst: Studien zum Spontanzerfall von 4-Hydroxycyclophosphamid und 4-Hydroperoxycyclophosphamid mit Hilfe der Dünnschichtchromatographie, Arzneim.Forsch.(Drug Res.), 24:1172 (1974)

17. I. Jardine, C. Feselau, M. Appler, M.-N. Kan, R. B. Brundrett, and M. Colvin, Quantitation by gas chromatography - chemical ionization mass spectrometry of cyclophosphamide, phosphoramide mustard, and nor-nitrogen mustard in the plasma and urine of patients receiving cyclophosphamide therapy, Cancer Res., 38:408 (1978)

18. W. E. Ross, R. A. G. Ewig, and K. W. Kohn, Differences between melphalan and nitrogen mustard in the formation and removal of DNA cross-links, Cancer Res., 38:1502 (1978)

19. R. J. Rutman, E. H. L. Chun, and J. Jones, Observations on the mechanism of the alkylation reaction between nitrogen mustard and DNA, Biochim.Biophys.Acta, 174:663 (1969)

20. K. R. Harrap and E. W. Gascoigne, The interaction of bifunctional alkylating agents with the DNA of tumour cells, Eur.J. Cancer, 12:53 (1976)

21. G. R. Mohn and J. Ellenberger, Genetic effects of cyclophosphamide, ifosfamide and trofosfamide, Mutat.Res., 32:331 (1976)

22. D. Gatehouse, Detection of mutagenic derivatives of cyclophosphamide and a variety of other mutagens in a "microtitre" fluctuation test, without microsomal activation, Mutat.Res., 53:289 (1978)

23. V. W. Mayer, C. J. Hybner, and D. J. Brusick, Genetic effects induces in Saccharomyces cerevisiae by cyclophosphamide in vitro without liver enzyme preparations, Mutat.Res., 37:201 (1976)

24. H. Arnold and H. Klose, Über den hydrolytischen Abbau des hexacyclischen N-Lost-phosphamidesters B518 unter physiologischen Bedingungen, Arneim.Forsch.(Drug Res.), 11:159 (1961)

INDUCTION OF SISTER CHROMATID EXCHANGES

IS SISTER CHROMATID EXCHANGE FORMATION AN INDUCIBLE PROCESS ?

A. Gentil and G. Renault

Institut de Recherches Scientifiques sur le Cancer
B.P. n° 8, 94800 Villejuif, France

Sister chromatid exchange (SCE) formation is well correlated with the carcinogenic and mutagenic properties of DNA damaging agents[1]. However, the mechanism of induction of SCE is still unknown and this correlation is not well understood. At the present time many hypotheses have been advanced to explain this correlation but none is fully satisfactory. In particular many efforts have been made to link SCE formation and DNA repair, since treatment of cells with DNA damaging agents such as X-rays, UV light, alkylating or DNA cross-linking agents, increases considerably SCE frequency[2,3,4,5,6]. However, it does not seem that SCE induction depends on excision repair mechanisms. Indeed it has been shown that cells with different excision repair capacities exhibit the same level of SCE after treatment with DNA damaging agents. After an UV-irradiation at 254 nm, Xeroderma pigmentosum cells from the different complementation groups which excise pyrimidine dimers at various rates, exhibit the same SCE frequency[7]. There is not an inverse correlation between excision repair measured by unschedule DNA synthesis (UDS) and SCE frequency. Moreover it is not clear what kind of lesion is responsible for SCE induction after an UV-irradiation. Indeed some cells are able to monomerize pyrimidine dimers by a process called photoreactivation, and it has been shown that photoreactivation in protorous and in chick embryo cells does not affect SCE production[8]. It is therefore possible that lesions responsible for SCE formation are some minor photoproducts which are not excised, and not pyrimidine dimers.

It does not seem either that post-replication repair mechanisms are implicated in SCE formation since after a chemical or physical

DNA-damaging treatment a subsequent caffein treatment does not greatly modify SCE formation[9]. Furthermore in Xeroderma pigmentosum variant cells, which are deficient in post-replication repair, SCE
frequency is the same as in normal human cells[7].

Finally, in Bloom's syndrome which is an autosomal recessive
disorder as is Xeroderma pigmentosum, patients are also very sun
sensitive and present a high risk of malignancy. Cells from Bloom's
syndrome are characterized by a very high level of spontaneous and
UV-induced SCE frequency but no repair defect has been shown in these
cells, although an altered replication fork progression was remarked[10,11,12]. Moreover, it has been shown that the tumour promoter,
12-0-tetradecanoyl phorbol-13-acetate (TPA) is able to induce SCE
[13,14]. TPA is known to alter DNA synthesis[15] but is not known to
damage DNA. It is therefore difficult to correlate in a direct way
DNA repair processes and SCE induction.

SCE inducers, TPA included, inhibit or alter DNA synthesis and
although the mechanism of formation of SCE is still unknown some
evidence suggests that they occur during the S-phase of the cell cycle[16,17,18]. This and the previous considerations have allowed us
to propose a hypothesis in which SCE formation could be the consequence of either the alteration of the DNA synthesis by itself or
of an induced process initiated by DNA synthesis alteration. It is
well known that an inducible process called "SOS repair" exists in
bacteria and that this process is accompanied by pleitropic effects
such as mutagenicity, filamentation, prophage induction, etc.[19,20,
21]. This process is initiated by DNA synthesis inhibition, and it
has been recently shown using viral probes that such a mechanism
probably occurs in mammalian cells[22,23,24,25]. If SCE induction was
related to SOS functions the correlation between the formation of
SCE and the mutagenic and carcinogenic properties of DNA-damaging
agents could be understood. Indeed in that case the induction of SCE
by some mutagenic and carcinogenic compounds would be the consequence of the induction of such a mechanism which would be responsible
for their properties.

In order to give some support to this hypothesis we carried out
in vitro experiments to show that the inhibition of DNA synthesis
induces SCE. V79 chinese hamster cells were treated for 15 hours
with either various DNA synthesis inhibitors such as hydroxyurea
(5 mM), cycloheximide (5 µg/ml) or aphidicolin (1 µg/ml) or with a
shift of temperature to 21°C which also inhibits cell replication.
Cells were then allowed to enter into S-phase and 5 µg/ml of BrdU
was added for 6 hours. After two rounds of cell replication colchicine was added for 1.5 hour in order to accumulate metaphases. Cells
were then trypsinized and treated as described by Perry and Wolf[26]
which allows the recognition of SCE between differentially stained
chromatids. This method, in which only one DNA strand in one chromatid out of four is substituted with the thymidine analogue, allows

a "*a posteriori*" synchronisation since 100 % of the arrested metaphases are correctly labelled. 50 to 100 metaphases in each group
were scored for SCE frequency and variance analysis was performed
for statistical calculation. Results presented in table I show that
the four treatments inhibiting DNA synthesis induce significantly
SCE. These treatments probably induce side effects in the cells
and one cannot rule out the possibility that these may be responsible for the SCE formation observed. However, the side effects are
probably different from each other and it is unlikely that they all
result in SCE induction. It therefore seems more likely that SCE
formation is the consequence of the common denominator of these
treatments, that is to say the inhibition of the cellular DNA synthesis. Moreover two of the treatments, cycloheximide and hydroxyurea, have been shown to induce enhanced virus reactivation in mammalian cells[23] and our results show that they also increase SCE frequency. The mechanisms by which these treatments inhibit DNA synthesis are quite different and are not fully understood. Hydroxyurea
acts on the metabolic pathways of the precursor synthesis[27] leading
to a depletion of the pool of the precursors, and the inhibition by
cycloheximide is probably the consequence of the inhibition of protein synthesis by this compound. Aphidicolin is a tetracyclic diter-
penoïd which has been shown to inhibit DNA polymerase α but not DNA
polymerase β or γ[28], and the shift of temperature to 21°C also inhibits DNA polymerase α[29]. Dose fractionation treatment experiments
were also carried out in order to show whether a small UV-irradiation
may induce a mechanism able to increase SCE frequency when the cells
are further irradiated with a high UV-dose. Table II shows that a
pretreatment with a low dose of UV (2 Jm^{-2}) delivered 15 hours before
a higher dose of UV (5 Jm^{-2}) followed by a BrdU labelling for 26
hours and the usual processing, significantly increases SCE frequency. In some experiments UV_1 induces very few SCE but in that case
more SCE are formed in pretreated cells than the sum of the SCE induced by the separate treatments, although we could expect less SCE
than the sum since in a one hit experiment 7 Jm^{-2} induces less SCE
than the sum of SCE induced separately by 2 Jm^{-2} and 5 Jm^{-2} (not
shown). It is not impossible that the cell cycle is disturbed by
the low pretreatment and that the higher dose of UV is given at a
more sensitive phase of the cell cycle. However this probability
seems low since cells are arrested in metaphases for only 1.5 hour
and if the cell cycle of the pretreated cells was significantly
disturbed, metaphases of these cells would not be seen. Moreover in
a preliminary experiment cells were pulsed with tritiated thymidine.
The same percentage of labelled metaphases, detected by autoradiography, were observed in pretreated and in non pretreated cells.

Although it is clear that the results reported here are not
sufficient to allow any definite conclusions to be drawn, and that
more work is needed, the proposed hypothesis could explain the observations reported above.

TABLE I

SCE FREQUENCY AFTER TREATMENT WITH SOME DNA SYNTHESIS INHIBITORS

TREATMENT	SCE FREQUENCY ± SE/METAPHASE		
	EXP. I	EXP. II	EXP. III
CONTROLS	3.65 ± 0.22	5.28 ± 0.35	7.06 ± 0.41
CYCLOHEXIMIDE	6.40 ± 0.74	9.25 ± 0.97	-
APHIDICOLINE	-	10.7 ± 1.03	-
HYDROXYUREA	29.95 ± 3.30	34.7 ± 3.02	-
21°C	-	-	8.51 ± 0.32

Cells were treated for 15 hours with various DNA synthesis inhibitors then allowed to enter into S-phase. SCE were score after two rounds of cell replication.

TABLE II

SCE FREQUENCY AFTER A SPLIT DOSE OF UV

CONTROL	5.5 ± 0.5
UV_1 (2 Jm^{-2})	5.4 ± 0.5
UV_2 (5 Jm^{-2})	14.5 ± 1
UV_1 + UV_2	17.7 ± 0.9

Cells were or were not pre-irradiated with 2 Jm^{-2} of UV (UV_1). 15 hours later they were irradiated with 5 Jm^{-2} of UV (UV_2). SCE were scored after two rounds of cell replication.

REFERENCES

1. M. O. Bradley, I. C. Hsu and C. C. Harris, Relationship between sister chromatid exchanges and mutagenicity, toxicity and DNA damage, Nature 282 : 318 (1979).
2. P. Perry and H. J. Evans, Cytological detection of mutagen-carcinogen exposure by sister chromatid exchanges, Nature 258 : 121 (1975).
3. H. Kato, Spontaneous and induced sister chromatid exchanges as revealed by the BUdR-labeling method, Intern. Rev. of Cytol. 49 : 55 (1977).
4. H. Kato, Induction of sister chromatid exchanges by UV light and its inhibition by caffeine, Exp. Cell Res. 85 : 239 (1974).
5. H. Kato, Induction of sister chromatid exchanges by chemical mutagens and its possible relevance to DNA repair, Exp. Cell Res. 82 : 383 (1973).
6. S. Latt, Sister chromatid exchanges, indices of human chromosome damage and repair : detection by fluorescence and induction by mitomycine, Proc. Natl. Acad. Sci. USA 71 : 3162 (1974).
7. E. A. De Weerd-Kastelein, W. Keijzer, G. Rainaldi and D. Bootsma, Induction of sister chromatid exchanges in Xeroderma pigmentosum cells after exposure to ultraviolet light, Mutat. Res. 45 : 253 (1977).
8. S. Wolff, Relation between DNA repair, chromosome aberrations and sister chromatid exchanges, in "DNA repair mechanisms", P. C. Hanawalt, E. C. Friedberg and C. F. Fox, ed., Academic Press, New York (1978).
9. G. L. Chan, H. Nagasawa and J. B. Little, Induction and repair of lethal and oncogenic lesions and their relationship to cytogenic changes in UV-irradiated mouse 10 T 1/2 cells, in "Radiation Res.", S. Okada, M. Imamura, T. Terashima and H. Yamaguchi ed., Tokyo (1979).
10. R. S. K. Chaganti, S. Schonberg and J. German, A manyfold increase in sister chromatid exchanges in Bloom's syndrome lymphocytes, Proc. Natl. Acad. Sci. USA 71 : 4508 (1974).
11. R. Hand and J. German, A retarded rate of DNA chain growth in Bloom's syndrome, Proc. Natl. Acad. Sci. USA 72 : 758 (1975).
12. J. German, DNA repair defects and human disease, in "DNA repair mechanisms", P. C. Hanawalt, E. C. Friedberg and C. F. Fox, ed., Academic Press, New York (1978).
13. A. Kinsella and M. Radman, Tumor promoter induces sister chromatid exchanges : relevance to mechanism of carcinogenesis, Proc. Natl. Acad. Sci. USA 75 : 6149 (1978).
14. H. Nagasawa and J. B. Little, Effects of tumor promoter, protease inhibitors and repair processes on X-ray-induced sister chromatid exchanges in mouse cells, Proc. Natl. Acad. Sci. USA 76 : 1943 (1979).
15. W. M. Baird, J. A. Sedgwick and R. K. Boutwell, Effects of phorbol and four diesters of phorbol on the incorporation of tritiated precursors into DNA, RNA, and protein in mouse epidermis, Cancer Res. 31 : 1434 (1971).

16. B. A. Kihlmam, Sister chromatid exchanges in vicia faba. II. Effects of thiotepa caffeine and 8 ethoxycaffeine on the frequency of SCE'S, Chromosoma 51 : 11 (1975).

17. H. Kato, Mechanisms for sister chromatid exchanges and their relation to the production of chromosomal aberrations, Chromosoma 59 : 179 (1977).

18. H. Kato, Possible role of DNA synthesis in formation of sister chromatid exchanges, Nature 252 : 739 (1974).

19. M. Radman, Phenomenology of an inducible DNA repair which is accompanied by mutagenesis, in "Molecular mechanisms for repair of DNA", P. C. Hanawalt and R. B. Setlow, ed., Plenum Press, New York (1975).

20. R. Devoret, A. Goze, Y. Moule and A. Sarasin, Lysogenic induction and induced phage reactivation by aflatoxin B_1 metabolites, Colloques internationaux du C.N.R.S. 256 : 283 (1977).

21. E. M. Witkin, DNA repair and mutagenesis, Colloques internationaux du C.N.R.S. 256 : 203 (1977).

22. V. B. Das Gupta and C. Summers, Ultraviolet reactivation of herpes simplex virus is mutagenic and inducible in mammalian cells, Proc. Natl. Acad. Sci. USA 75 : 2378 (1978).

23. A. Sarasin and P. C. Hanawalt, Carcinogens enhance survival of UV-irradiated simian virus 40 in treated monkey kidney cells : induction of a recovery pathway ? Proc. Natl. Acad. Sci. USA 75 : 346 (1978).

24. A. Sarasin and A. Benoit, Induction of an error-prone mode of DNA repair in UV-irradiated monkey kidney cells, Mutat. Res. 70 : 71 (1980).

25. C. D. Lytle, J. Coppey and W. D. Taylor, Enhanced survival of ultraviolet-irradiated herpes simplex virus in carcinogen-pretreated cells, Nature 272 : 60 (1978).

26. P. Perry and S. H. Wolf, New giemsa method for the differential staining of sister chromatids, Nature 251 : 156 (1974).

27. C. H. Ockey and T. D. Allen, Distribution of DNA or DNA synthesis in mammalian cells following inhibition with hydroxyurea and 5-fluorodeoxyurdine, Exp. Cell Res. 93 : 275 (1975).

28. N. A. Berger, K. K. Kurohara, S. J. Petzold and G. W. Sikorski, Aphidicolin inhibits eukaryotic DNA replication and repair implications for involvement of DNA polymerase α in both processes, Biochem. Biophys. Res. Cons. 89 : 218 (1979).

29. T. Ooka, Thèse de Doctorat es Sciences, Université Claude Bernard, Lyon (1976).

INDUCTION OF SISTER CHROMATID EXCHANGES BY

NON-MUTAGENIC CARCINOGENS

K. Athanasiou and S.A. Kyrtopoulos

Biological Research Centre
The National Hellenic Research Foundation
48, Vassileos Canstantinou Avenue
Athens 501/1, Greece

The well known ability of many carcinogens to act as mutagens under suitable conditions constitutes strong evidence that interaction with, and chemical modification of DNA is an essential part of the process whereby such agents bring about neoplastic transformation (Ames, 1979; Hollstein et al., 1979). This view is further supported by the direct detection in many cases of covalent carcinogen-DNA adducts or other types of carcinogen-induced DNA modification (Grover, 1978).

A number of chemicals are known, however, which, although causing an increased incidence of tumours in animals to which they are administered, fail to act as mutagens or to cause DNA damage in a number of established test systems (Boyland, 1980). Such compounds are sometimes classified as "epigenetic carcinogens" to indicate that they may act by mechanisms other than DNA damage (Ashby et al., 1978).

In this communication we report the first results of a study in which a number of carcinogens which are not mutagenic in the Ames/Salmonella test system (McCann et et al.; 1975) are being examined for the ability to induce sister chromatid exchanges (SCE) in Chinese hamster ovary cells. Increased frequency of SCE is associated with the action of many carcinogens and DNA-damaging agents and it is increasingly used as a screening test for such compounds (Hollstein et al., 1979; Perry and Evans, 1975).

MATERIALS AND METHODS

Test chemicals

The chemicals examined were of the highest purity
commercially available and were used with no further
purification. DMSO, thioacetamide and 3-aminotriazole
were obtained from Serva, Heidelberg, sodium phenobarbital
and thiourea from Fluka, Switzerland and carbon tetrachloride and chloroform from BDH, England. Thiourea, thioacetamide and sodium phenobarbital were dissolved in HBSS
buffer (Flow Laboratories, Scotland) while the remaining
compounds were dissolved in DMSO.

Cell culture and chromosome preparation

The Chinese hamster ovary cell line used was kindly
donated by Dr. A. Guyalis of the Biological Research
Centre, The National Hellenic Research Foundation. The
cells were grown in Eagle's minimum essential medium
(modified) with Hank's salts (Flow Laboratories, Scotland)
and supplemented with 10% fetal calf serum (Flow Laboratories, Scotland). Twenty-four hours after the cells were
seeded in 75 ml culture flasks, BUdR (Gibco Bio-Cult,
Scotland) in a final concentration of 5 µg/ml and the
test chemical were added to the cultures in the dark. The
amount of solvent in each culture flask did not exceed
0.1 ml in the case of HBSS and 0.005 ml in the case of
DMSO. All cultures were kept in complete darkness at $37^{O}C$
for 26 hours and colchicine was added for the final 2
hours at a final concentration of 1×10^{-6}M. Cells were
collected by shaking and trypsinisation and chromosomes
prepared by standard methods. They were finally stained
for SCE with fluorochrome and Giemsa according to the
method of Perry and Wolff (1974) and scored for SCE. At
least 30 metaphases were scored for each treatment group.

RESULTS

The Table shows the results obtained for the six
chemicals examined. It can be seen that carbon tetrachloride, chloroform, sodium phenobarbital, thioacetamide
and thiourea all induce significant dose-dependent increases in the frequency of SCE (tested at the 5% significance
level by means of Student's t test). Of the compounds
examined only 3-aminotriazole failed to induce SCE even
at the highest dose used.

Table. Induction of SCE by non-mutagenic carcinogens

Agent	Dose (M)	SCE/cell (S.E.M.)	Range	t-test[a] (5%)
HBSS[b]		6.11 (0.20)	1-9	
DMSO[b]		6.52 (0.27)	1-9	
3-Aminotriazole	10^{-4}	5.90 (0.30)	2-10	−
	10^{-3}	5.78 (0.36)	2-10	−
Carbon tetrachloride	10^{-6}	7.10 (0.46)	4-10	−
	10^{-4}	12.35 (0.55)	7-16	+
	10^{-3}	12.80 (0.62)	8-20	+
Chloroform	10^{-6}	6.74 (0.41)	5-12	−
	10^{-5}	9.63 (0.51)	5-16	+
	10^{-4}	11.50 (0.62)	5-17	+
Na Phenobarbital	10^{-6}	7.40 (0.39)	4-13	+
	10^{-4}	9.97 (0.59)	5-18	+
	10^{-3}	9.93 (0.36)	6-16	+
Thioacetamide	10^{-5}	7.20 (0.34)	4-11	+
	10^{-4}	10.23 (0.57)	6-19	+
	10^{-3}	12.53 (0.72)	5-23	+
	10^{-2}	10.97 (0.43)	7-16	+
Thiourea	10^{-5}	5.50 (0.34)	3-10	−
	10^{-4}	7.34 (0.29)	5-11	+
	10^{-3}	9.74 (0.52)	7-16	+
	10^{-2}	10.47 (0.58)	6-18	+

[a]Compared with the mean value for the corresponding solvent control.
[b]Solvent controls.

In the cases of carbon tetrachloride and chloroform, significant numbers of chromosome aberrations were observed. These will be described in detail elsewhere.

DISCUSSION

The six chemicals tested are characterised by lack
of mutagenic activity in the well established Ames/Salmo-
nella system (McCann et al., 1975). None of them has, as
far as we are aware, been conclusively shown to form covalent adducts with DNA and, in addition, phenobarbital
does not damage DNA as judged by the alkaline elution
test (Swenberg et al., 1976). While it is possible that
all the above negative results may be due to methodolo-
gical weaknesses (Ashby and Styles, 1978), they may also
be interpreted as indicating that the carcinogenic activity of these compounds is not mediated by direct chemical
modification of DNA.

We have found that, of the six chemicals examined,
five give rise to increased numbers of SCE. This result
is in striking contrast to the negative results obtained
with other test systems mentioned above, and may be taken
as indicating a possible role for SCE in the mechanism
of cancer induction by these agents.

SCE are believed to be related in some way to DNA
damage and repair, although the details of the mechanism
of their induction are by no means understood (Wolff,
1977; Fornace et al., 1980). Consequently one cannot at
present decide whether their occurence indicates the
presence of carcinogen-induced damage which has as yet
gone undetected or whether they can be induced by "epigenetic" interactions of the carcinogens with molecules
other than DNA.

It has recently been shown by Kinsella and Radman
(1978) that the potent tumour promoter TPA (for which
no DNA-damaging activity is known) is also able to induce
SCE. Based on this discovery, Kinsella and Radman have
suggested that SCE may play an important role in tumour
promotion by allowing, by means of recombination, the
expression of a heterozygous, recessive trait which gives
rise to the cancer phenotype. It is interesting that
amongst the compounds which we have found to be active
in inducing SCE, phenobarbital and thioacetamide are
known tumour promoters, as well as being sole carcinogens
(Peraino et al., 1975; Cameron et al., 1978). Furthermore,
saccharin (a bladder cancer promoter) has also been
shown to induce SCE (Abe and Sasaki, 1977; Wolff and
Rodin, 1978). On the other hand, butylated hydroxytoluene,
also a tumour promoter, is inactive in this respect (Abe
and Sasaki, 1977). It appears, therefore, that of five

promoters so far examined, four can induce SCE. The possible significance of this finding for the mechanism of tumour induction and promotion may be better understood if a greated number of promoters is examined.

ACKNOWLEDGEMENTS

We acknowledge the financial support of the National Hellenic Research Foundation (to K.A.) and the Greek Agency for Scientific Research and Technology (to S.A.K.)

REFERENCES

Abe, S., and Sasaki, M., 1977, Chromosome aberrations and sister chromatid exchanges in Chinese hamster cells exposed to various chemicals, J. Natl. Canc. Inst., 58:1635.

Ames, B.N., 1979, Identifying environmental chemicals causing mutations and cancer, Science, 204:587.

Ashby, J. and Styles, J.A., 1978, Does carcinogenic potency correlate with mutagenic potency in the Ames assay? Nature, 271:452.

Ashby, J., Styles, J.A., Anderson, D., and Paton, D., 1978, Saccharin: an epigenetic carcinogen/mutagen? Food Cosmet. Toxicol., 16:95.

Boyland, E., 1980, History and future of chemical carcinogenesis, Brit. Med. Bull., 36:5.

Cameron, R., Lee, G., and Farber, E., 1978, Chemical mitogens as effective alternatives to partial hepatectomy in the new model for the sequential analysis of hepatocarcinogenesis, Proc. Am. Assoc. for Canc. Res., 19:56.

Fornace, A.J., Nagasawa, H., and Little, J.B., 1980, Relationship of DNA repair to chromosome aberrations, sister-chromatid exchanges and survival during liquid-holding recovery in X-irradiated mammalian cells, Mutat. Res., 70:323.

Grover, P.L., 1978, "Carcinogens and DNA", CRC Press, Boca Raton, Florida.

Hollstein, M., McCann, J., Angelosanto, F.A., and Nichols, W.W., 1979, Short-term screening tests for carcinogens and mutagens, Mutat. Res., 65:133.

Kinsella, A.R., and Radman, M., 1978, Tumour promo-
    ter induces sister chromatid exchanges -rele-
    vance to mechanisms of carcinogenesis, Proc.
    Natl. Acad. Sci. U.S.A., 75:6149.
McCann, J., Choi, E., Yamasaki, E., and Ames, B.N.,
    1975, Proc. Natl. Acad. Sci. U.S.A., 72:5135.
Peraino, C., Fry, M.R., Staffeld, E., and Christo-
    pher, J.P., 1975, Comparative enhancing effect
    of phenobarbital, amobarbital, diphenylhydan-
    toin, and dichlorodiphenyltrichloroethane
    on 2-acetylaminofluorene-induced hepatic tumo-
    rigenesis in the rat, Canc. Res., 35:2884.
Perry, P., and Evans, H.J., 1975, Cytological de-
    tection of mutagen-carcinogen exposure by
    sister chromatid exchange, Nature, 258:121.
Perry, P., and Wolff, S., 1974, New Giemsa method
    for the differential staining of sister chro-
    matids, Nature, 251:156.
Swenberg, J.A., Petzold, G.L., and Harbach, 1976,
    In vitro DNA damage/alkaline elution assay
    for predicting carcinogenic potential, Biochem.
    Biophys. Res. Commun., 72:732.
Wolff, S., 1977, Sister chromatid exchange, Ann.
    Rev. Genet., 11:183.
Wolff, S., and Rodin, B., 1978, Saccharin-induced
    sister chromatid exchanges in chinese hamster
    and human cells, Science, 200:543.

CHROMOSOME ABERRATIONS AND SISTER CHROMATID EXCHANGES IN PERSONS OCCUPATIONALLY EXPOSED TO MUTAGENS/CARCINOGENS

Helga Waksvik, Morten Boysen, Anton Brøgger and Olbjørn Klepp

Norsk Hydro's Institute for Cancer Research, The Norwegian Cancer Society and The Norwegian Radium Hospital, Montebello, Oslo 3, Norway.

INTRODUCTION

It is generally accepted that some 5% of all human cancer re-sults from occupational exposure to carcinogens. An increased in-cidence of respiratory tract cancer has been reported from nickel refineries in Wales, Russia, Canada and Norway[1]. Epidemiological studies from a Norwegian nickel refinery[2] show that workers exposed to various nickel compounds for more than 3 years have a ratio of observed versus expected respiratory tract cancer of 13.9. Chromo-some studies of humans exposed to nickel have not previously been reported.

Many cytostatic agents used in cancer chemotherapy are mutagens/carcinogens[3,4]. The possible induction of secondary neoplastic dis-ease in patients treated with cytostatic drugs is now being discussed[5]. These agents may also represent potential hazards to nurses and doc-tors handling them, but no study of this has yet been published.

It is generally believed that chromosome damage in somatic cells may give rise to precancerous cells by changing the normal cellular mechanisms that control and prevent cell multiplication.

Short-term tests such as the study of chromosome aberrations and sister chromatid exchange (SCE) can be used to detect _in vivo_ genetic damage in peripheral human lymphocytes. Since the lympho-cytes circulate throughout the body, they may be exposed to envir-onmental agents taken up by the subject.

The aim of the present investigation was to study whether humans occupationally exposed to various nickel compounds and cytostatic drugs, handling radiation equipment and anaesthetic gases show chromosome aberrations and/or increased level of SCE. We have therefore made chromosome analyses of nickel workers at Falconbridge Nikkelverk and nurses in a cancer hospital.

MATERIALS AND METHODS

Cell cultures were established of whole blood in Ham's F-10 medium supplemented with 20% fetal bovine serum, glutamine, streptomycin and phytohemagglutinin. Chromosome damage was scored as gaps and breaks in lymphocyte metaphases after 40-48 h culturing. Analysis of SCE was made in BrdU-labeled lymphocytes.

Nickel Workers

This study was concerned with three groups of individuals from a nickel refinery in Norway (Table 1).

1.   7 controls (44-59 years of age, mean 51). The controls worked in the same plant but had not been exposed to nickel compounds.

2.   9 workers (25-60 years of age, mean 44) from the roasting/ smelting department exposed to dry furnace dust containing principally nickel subsulphide and nickel oxide.

3.   10 workers (45-55 years of age, mean 52) from the electrolysis department exposed to aerosols of nickel sulphate and nickel chloride.

Since additional variables such as smoking, alcohol and drugs may influence results, we selected only non-smokers, non-drinkers and apparently healthy persons with no regular use of drugs for this study.

Nurses

Six groups were studied (Table 2).

1.   10 controls (24-60 years of age, mean 44). These were female office clerks working in the same hospital.

2.   10 nurses (30 - 58 years of age, mean 51) frequently handling cytostatic agents.

3.   11 nurses (25-61 years of age, mean 37) seldom handling cytostatic agents.

4.   8 nurses (27-61 years of age, mean 46) engaged in therapeutic or diagnostic radiology.

TABLE 1.   Chromosome aberrations and SCE in lymphocytes from persons
            occupationally exposed to nickel subsulphide, nickel oxide,
            nickel sulphate and/or nickel chloride, and controls.

Group/ Exposure	No. of subjects	Atm.conc. of Ni mg/m^3	Plasma conc.of Ni µg/l	% Chromosome[a] aberrations		SCE[b]	
				gap	break		Range
Controls	7		1	3.7	0.6	5.1	4.4-6.3
Ni subsulphide Ni oxide	9	0.5	4.2	11.9	0.9	4.8	4.4-5.2
Ni sulphate Ni chloride	10	0.2	5.2	18.3	1.3	4.9	4.4-5.3

a) Mean percentage. 100 cells from each person were counted.
b) Mean SCE of mean SCE per cell. 30 cells from each person counted.

5.   9 nurses (29-53 years of age, mean 35) exposed to anaesthetic
      gases such as halothane, enfluran and nitrous oxide in opera-
      ting theaters.

6.   7 nurses (24-56 years of age, mean 36) working in a post-
      operative ward.

All members of the groups were healthy at the time of blood sampling.
Previous diagnostic X-ray examinations, occasional use of drugs and
smoking habits were approximately equal in all groups. The cytostatic agents most frequently handled were cyclophosphamide, vin-
cristine, doxorubicine, methotrexate, vinblastine, mechlorethamine,
5-fluoro-uracil, actinomycin D, dacarbazine and bleomycin.

RESULTS AND DISCUSSION

     Table 1 shows the average results of the groups of nickel work-
ers. Only the percentage of gaps is greatly increased in the exposed groups compared to the controls (p < 0.001). The percentage
of breaks is low and the frequency of SCE is approximately in the
same range for all groups. The highest plasma concentration of
nickel is found in the group exposed to highly soluble aerosols of
nickel sulphate and nickel chloride. This group also showed the
highest percentage of gaps.

     Table 2 shows the average of the results for the nurses. The
percentage of gaps was significantly higher among the nurses most
frequently handling cytostatic agents (p < 0.001). The percentage
of breaks is low in all groups and within the limits usually found.
The frequency of SCE is significantly higher (p < 0.001) only among
the nurses handling cytostatic agents most frequently.

TABLE 2.   Chromosome aberrations and SCE in lymphocytes from nurses
           handling cytostatic agents, radiation equipment or ex-
           posed to anaesthetic gases, and nurses in ordinary practice

Group/Exposure	No. of subjects	% Chromosome[a] aberrations		SCE[b]	
		gap	break		range
Controls	10	9.3	0.3	6.5	4.5 - 7.9
Cytostatics frequently	10	18.5	0.8	8.4	5.5 -10.2
Cytostatics seldom	11	7.4	0.7	5.8	4.2 - 7.2
Radiation	8	13.9	1.0	6.2	4.3 - 8.8
Anaesthetic gases	9	7.8	1.5	6.1	4.6 - 7.6
Post-operative ward	7	10.3	0.6	5.1	4.6 - 6.1

a) and b) as in Table 1

In this study we found that chromosome aberrations such as
gaps (achromatic lesions) were increased after in vivo exposure to
various nickel compounds and cytostatic agents. Although the nature of the chromatid gap is unclear, dose-dependent responses to
clastogenic agents have previously been found[6], and therefore gaps
may be regarded as an indication of possible genetic damage.

The increase in SCE frequency in one of the groups handling
cytostatic agents points to these drugs as a possible occupational
health hazard and is a warning that proper measures - gloves, masks
and fume cabinets - should be used. The difference between the two
groups of nurses handling cytostatics may be due to the different
exposure time or to differences in the use of the protective measures.

REFERENCES

1. W. Sunderman, Ann. Clin. Lab. Sci. 7: 377 (1977)
2. E. Pedersen, A.C. Høgetveit and A. Anderson, Int. J. Cancer 2:
   32 (1973).
3. A. Schinzel and W. Schmid, Mutation Res. 40: 139 (1976).
4. B. Lambert, W. Ringborg, E. Harper and A. Lindblad, Cancer
   Treat. Rep. 62: 1413 (1978).
5. S.M. Sieber and R.H. Adamson, Cancer 40: 1950 (1977).
6. A. Brøgger, Mutation Res. 23: 353 (1974).

SISTER-CHROMATED EXCHANGES AS A MEASURE OF DAMAGE PRODUCED BY

VISIBLE LIGHT ON BrdUrd-SUBSTITUTED CHROMOSOMES[*],[**]

C. Gutiérrez and J.B. Schvartzman

Instituto de Biologia Celular

Velázquez, 144. Madrid-6, Spain

SUMMARY

We have studied the effect of visible light irradiation on BrdUrd-substituted chromosomes in <u>Allium cepa</u> L. meristem cells by measuring the yields of SCEs. The efficiency of this treatment in provoking SCEs is clearly dependent on the cell cycle time in which damage occurs, having obtained the highest frequencies when VL exposure coincides with the initiation of the S phase. Nitrogen atmosphere during irradiation reduces SCE frequencies, suggesting that some oxidizing radical species could be involved in the chemical reactions producing lesions that ultimately lead to SCE formation. These results are discussed with respect to the hypothesis that SCEs arise as a consequence of the passage of a damaged DNA through the replication process.

INTRODUCTION

In recent years the analysis of SCEs, which are provoked by a wide variety of physical and chemical agents, has become a useful tool in order to study not only the mechanism(s) by which they are

---

[*] Abbreviations used: SCEs = sister-chromatid exchanges. BrdUrd = 5-bromodeoxyuridine. VL = visible light. UV = ultraviolet. FPG = fluorescence plus Giemsa. XP = xeroderma pigmentosum. CHO = Chinese hamster ovary.

[**] This work has been partially supported by the "Comisión Asesora para la Investigación Cientifica y Técnica" and by the "Caja de Ahorros y Monte de Piedad de Madrid".

formed, but also in examining the potential mutagenic/carcinogenic activity of many substances[1]. Nevertheless, we only possess a poor understanding of the biological significance of SCEs.

It has been found that cells from human patients of several cancer-risk diseases, having defects in the repair of several types of lesions, show abnormal increases in SCE frequencies after DNA damaging treatments[2]. This has strengthened the suggestion that SCE phenomenon could be intimately associated with repair processes. However, only a few efforts have been made to characterize the relationship between the cell cycle time in which lesions are induced and the yields of SCEs.

In the present report we show that the efficiency of the same damaging treatment in increasing the yields of SCEs is strongly dependent on the period of time elapsed between induction of damage and DNA replication.

MATERIAL AND METHODS

Root meristems of Allium cepa L. were employed. BrdUrd-substitution was achieved in the dark and visible light irradiations were carried out by exposing roots to a Philips incandescent lamp at different moments throughout the second treatment cycle. BrdUrd-growth medium and irradiation conditions have both been previously described[3].

Once the treatment was finished, the roots were carefully washed with tap water and treated with 0.05% colchicine for the last 2.5 hours. Then root tips were washed, cut and transferred to an ethanol-acetic acid (3:1) fixative solution overnight. Cell cycle timing and the duration of the interphase periods were determined by labelling meristem population with 5 mM caffeine during 1 hour, according to González-Fernández et al[4]. Slides for cell cycle analysis were stained by the acetic-orcein method. Root tips in which SCEs were going to be studied were squashed and the FPG staining technique was performed as previously described[3].

RESULTS AND DISCUSSION

We have recently demonstrated[3] that VL illumination of BrdUrd-substituted DNA is able to increase the baseline frequencies of SCEs. Although the mechanism by which SCEs arise is still insufficiently understood, their formation seems to be closely associated with the DNA replication process[3].

Fig. 1 shows the relationship between the cell cycle period in which VL exposure takes place and the SCE frequencies detected in

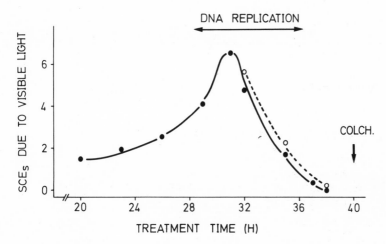

Fig. 1. SCE yields in second division chromosomes when VL irradiations take place at different times throughout the second
cycle when only the first one occurs in the presence of
BrdUrd (●——●), and when cells replicate the two consecutive cycles in the presence of the analogue (○---○).

second division chromosomes. Since even the chromosomes of the unirradiated cells exhibit some SCEs[5], we have expressed here only
the differences between the mean values in irradiated and unirradiated cells. It can be clearly seen that VL is able to increase the
number of SCEs regardless of the moment in which irradiation takes
place (except when it occurs in the last G_2 phase), being the increase maximum when VL irradiation conincides with the beginning
of the S phase. We examined the influence of post-replicated DNA
on the formation of VL-induced SCEs, by allowing cells to replicate during two consecutive cycles in the presence of BrdUrd,
occurring VL exposures after the second S period has started. Since
the increases were very similar in both set of experiments we can
conclude that only the pre-replicative BrdUrd-substituted DNA seems
to be the target for VL.

These results are in agreement with those previously reported
by Latt and Loveday[6] for the induction of SCEs by 8-methoxypsoralen
plus UV light. Nevertheless, the first part of the curve shows that
the efficiency of a damaging treatment in provoking SCEs is strongly dependent on the cell cycle time in which damage occurs. Although
we cannot rule out that some proteins associated with DNA during
non replicative periods might act as quenchers in photochemical
reactions[7], there exists to our knowledge no evidence that the
chromatin structure itselft might affect VL-lesions in BrdUrd-
substituted DNA.

Table 1. Effect of irradiation atmosphere on the yields of
          SCEs (25°)

Irradiation moment	G_1		S		G_2	
Atmosphere	Air	N_2	Air	N_2	Air	N_2
SCEs/chromosome	5.6	3.0	8.2	3.8	3.1	2.9

Dose   = 5×10^7 erg/mm^2
Control = 2.8 SCEs/chromosome

On the other hand, our results could be interpreted in relation
to the repair amount that a cell can perform before DNA replication.
Probably in our experiments we have modified the total repair capacity of the cell by varying the time between damaging treatment and
DNA replication: when this time diminishes, the number of DNA
lesions present during replication rises and, as a consequence, high
SCE frequencies are obtained. This last explanation is in agreement with previous observations of Cleaver[8], who postulated an inverse correlation between cell repair proficiency and SCE formation.
Further they support Wolff's report[9] in which he suggests that the
lifetime of the induced lesions could be an important factor determining the final yields of SCEs. Moreover, MacRae et al.[10] have
recently found a decrease in SCE frequencies in CHO cells deprived
of the amino acid arginine after UV-light exposure, and this be
another indication that UV-lesions could be repaired during cell
arrest.

However, the precise nature of the DNA lesions leading to
the formation of SCEs remains unknown. In fact, because SCEs can be
induced by a wide variety of physical and chemical agents[1], many
types of lesions probably lead to SCEs. Concerning the lesions
provoked by VL in BrdUrd-substituted DNA, it has been demonstrated
that a great amount of single-strand breaks are produced and the
formation seems to involve the production of a uracil radical[11].
Since radical formation is known to be strongly influenced by the
redox level, experiments in which cells were exposed to VL in a
nitrogen atmosphere, were carried out in order to test the possibility that these chemical species could be involved in the formation of SCEs. As can be clearly seen in Table 1, N_2 atmosphere
reduces the increase in SCE frequencies provoked by VL when irradiation occurs both in G_1 and in S period. As would be expected
in G_2-experiments, VL provoked no increase, neither in air nor in
N_2 atmosphere. Altogether these results strongly suggest that
radicals could induce the lesions that ultimately would lead to

the formation of an SCE, since a reducing environment prevents the breakage of the DNA backbone by stabilizing the uracil radicals.

In short, we have obtained new evidence supporting the idea that SCEs are formed by a mechanism operating in close association with DNA replication and SCE production could represent the consequence of the passage of a DNA bearing unrepaired lesions through the replication process.

REFERENCES

1. H. Kato, Spontaneous and induced sister chromatid exchanges as revealed by the BUdR-labeling method. Int. Rev. Cytol. 49:55 (1977).
2. C.F. Arlett and A.R. Lehmann, Human disorders showing increased sensitivity to the induction of genetic damage, Ann. Rev. Genet. 12:95 (1978).
3. J.B. Schvartzman, R. Postigo and C. Gutiérrez, Analysis of visible light-induced sister chromatid exchanges in 5-bromodeoxy-uridine substituted chromosomes. Chromosoma 74:317 (1979).
4. A. González-Fernández, G. Giménez-Martín and C. de la Torre, The duration of the interphase periods at different temperatures in root tip cells, Cytobiologie 3:367 (1971).
5. J.B. Schvartzman, F. Cortés, A. González-Fernández, C. Gutiérrez and J.F. López-Sáez, On the nature of sister-chromatid exchanges in 5-bromodeoxyuridine-substituted chromosomes, Genetics 92:1251 (1979).
6. S.A. Latt and K.S. Loveday, Chracterization of sister chromatid exchange induction of 8-methoxypsoralen plus near UV light, Cytogenet. Cell Genet. 21:184 (1978).
7. P.A. Cerutti, Repairable damage in DNA, in: DNA repair mechanisms, P.C. Hanawalt, E.C. Friedberg and C.F. Fox, eds., Academic Press, New York (1978).
8. J.E. Cleaver, DNA repair mechanisms and the generation of sister chromatid exchanges in human cell lines from xeroderma pigmentosum patients, in: Molecular Human Cytogenetics, R.S. Sparkes, D.E. Comings and C.F. Fox, eds., Academic Press, New York (1977).
9. S. Wolff, The induction of sister chromatid exchanges (SCE's) by lesions whose lifetimes are affected by the cell's excision repair capacity, Genetics Suppl. 91:5138 (1979).
10. W.D. MacRae, E.A. Mackinnon and H.F. Stich, The fate of UV-induced lesions affecting SCE's, chromosome aberrations and survival of CHO cells arrested by deprivation of arginine. Chromosoma 72:15 (1979).
11. L.A. Smet and J.J. Cornelis, Repairable and irrepairable damage in 5-bromouracil-substituted DNA exposed to ultra-violet radiation. Int. J. Radiat. Biol. 19:445 (1971).

DNA DAMAGE AND REPAIR IN RELATION TO CARCINOGENESIS

DNA DAMAGE AND CARCINOGENESIS

R. B. Setlow

Biology Department
Brookhaven National Laboratory
Upton, New York 11973 USA

INTRODUCTION:

A number of independent lines of evidence, other than cell biology ones, support the somatic mutation theory of cancer. They indicate that damage to DNA can lead to cancer and hence one should be concerned about environmental agents that react with DNA. Nevertheless, there are arguments against this point of view (1) and, even if damage to DNA is the important element in cancer initiation, one should always keep in mind the possibilities that the switch from normal to cancer cells may arise from faulty transcription and hence translation (as seems to be the case in the death of UV-irradiated arrested human fibroblasts). Moreover, promotion steps subsequent to initiation may be of overriding importance at the initiation doses received at low exposure-rate levels.

The following are direct reasons for associating initiating events in carcinogenesis with DNA damage:

1) There is an excellent correlation among those compounds that are mutagenic, when activated appropriately, and those that are carcinogenic. (Of course, a chemical that reacts with DNA will also readily react with RNA and protein.) (2)

2) In mammalian cell cultures there is an excellent quantitative correlation between the mutagenicity of metabolites of polycyclicaromatic hydrocarbons and transformation by these metabolites and also an excellent correlation between mutation and transformation by the activated metabolites of nitrosamines (3, 4).

3) If cells are treated in vitro with BrdU and long wavelength UV - a process known specifically to damage DNA - the resulting neoplastic transformation is correlated with the extent of DNA damage (5).

4) The UV-irradiation of thyroid cells of the fish Poecilia formosa gives rise to thyroid tumors when the cells are injected into isogenic recipients. If the cells are photoreactivated before injection - a treatment known to monomerize pyrimidine dimers in the cellular DNA - the number of tumors observed decreases by greater than ten fold (6).

5) In a number of human disorders, the affected individuals are cancer prone and their cells are more sensitive than normal to exogenous mutagens (7-9) (see previous chapters in this volume). Three of these disorders (Table 1) xeroderma pigmentosum (XP), ataxia telangiectasia (AT), and Fanconi's anaemia are associated with defects in DNA repair. However, the association is weak insofar as the causes of increased cell cytotoxicity compared to normal cells are not the same for all individuals with the same clinical disorder. Moreover the correlation between cytotoxicity of UV and excision repair deficiencies is not a good one (10), indicating either that there are other repair systems of importance or that cells die for reasons than the mere existence of damage to their DNA. Certainly, the time it takes to accomplish repair, before DNA replication and transcription take place, is important, as is the fidelity of replication and transcription on damaged templates. The etiologic agent is known for only XP. The distribution of cancer types in AT is different than that observed in the population exposed to atomic bombs, indicating that the enhanced cancer risk in the AT population does not arise from ionizing radiation (11). Morevover, AT cells are hypomutable to X-irradiation (8).

6) If human cells are held in a confluent state before replating, DNA damage decreases because of excision repair, and survival increases and mutations decrease. The changes in survival and mutation correlate well with the rate of excision repair in proficient and repair deficient human cells (12, 13). Transformation in human cells also decreases as a result of such a holding procedure (14) but various methods of enhancing transformation in Syrian hamster embryo cells do not affect known DNA repair processes (15).

Thus, it is well established that DNA damage and its repair play an important role in carcinogenesis but the quantitative aspects of its role are not clear. What is needed is a thorough understanding of the types of DNA damages that may result in the initiation of the carcinogenic process and the relative

Table I. Repair Deficient Diseases

	Xeroderma pigmentosum	Ataxia telangiectasi	Fanconi's anemia
frequency homozygotes heterozygotes	1/300,000 1/300	1/40,000 1/100	1,300,000 1/300
cancer probability* homozygotes less than age 20 yr.	skin cancer: > 0.5 (melanoma > 0.1)	0.1 (lymphoreticular:0.06, leukemia: 0.02)	> 10 fold normal
heterozygotes	5-fold normal in South	5-fold normal < 45 yr, 50% greater mortality than average.	like normal
etiologic agent	sunlight	?	?
cell sensitivity	UV and mimetics	x-rays, alkylating agents	cross linking agents
repair deficiencies	one or more of excision, photoreactivation, and postreplication (> 7 groups)	some cell strains defective defective in "X-ray" repair (> 3 groups)	some cell strains defective in crosslink repair

*Approximate average cancer probabilities
skin cancer prevalence: 0.005
melanoma incidence: 6 x 10⁻⁵/yr
lymphoreticular cancer: 13 x 10⁻⁵/yr
leukemia (t_{max} ~4 yr.): 42 x 10⁻⁶/yr.

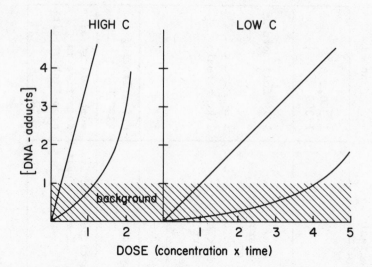

Fig. 1.   Possible dose-response curves for DNA adducts caused by
          physical or chemical agents at high or low dose rates or
          concentrations.

probability of initiation for one product versus another.
Sometimes DNA repair measurements themselves are used to
identify a crucial DNA adduct. For example, in new born rats
exposed in utero to ethylnitrosourea the repair of 0^6
ethylguanine in neuronal tissue is very low compared to that in
other tissues, but the major alkylation damage, N-7 ethylguanine
seems to be affected equally in all tissues(16). Since neuronal
tumors are the principal ones observed, these data are evidence
for the importance of the 0^6 ethylguanine in carcinogenesis and
the unimportance of the N-7 ethylguanine. A crude measurement of
DNA repair such as unscheduled DNA synthesis, or its equivalent,
would measure primarily the repair of the N-7 product.

QUANTITATIVE ASPECTS:

     An attractive feature of the strong implication that DNA
damages act as initiating events in carcinogenesis is the
possibility of using the implication to obtain good dosimetry at
the level of cellular DNA.   Thus, if one could identify
biological damages to DNA and develop ways of analyzing for them
in small numbers, as is now being done by the use of specific
antibodies or nucleases (17-19), one could measure the
accumulation of such products in experimental animals or people
exposed experimentally or environmentally.   It should be possible
to determine the relations between physical dose or chemical dose

(concentration x time) and the level of DNA adducts in terms of
adducts per unit length of DNA. It is important to determine the
dose response curve for such adducts – whether it is linear or
curvilinear – and how the relation depends on the concentration
or dose rate (see Fig. 1). The existence of activating and
deactivating enzyme systems for most chemical carcinogens, and
the existence of repair mechanisms for chemical damages, as well
as for physical damage, implies that the dose response relations
probably will depend on dose rate and will not be linear. The
determination of such dose response relations is within the realm
of technical capability. Of course, one of the first steps in
such determinations is the identification of the proper adduct to
measure.

Fig. 1 illustrates a second important consideration in
extrapolating experimental results on such simple systems to
humans. What is the background level of DNA adducts in the
absence of any man made contribution. An important question for
chemical carcinogenesis, as it is for radiation carcinogenesis,
is what is the level of man made adducts compared to the
background level? Even if one can do the dosimetry indicated in
Fig. 1, there is the near impossibility of using such data plus
carcinogenesis data on animal systems to extrapolate to low dose

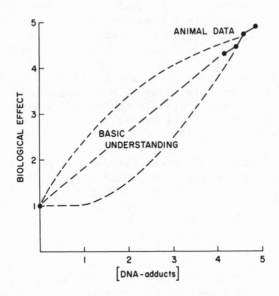

Fig. 2   Possible relations between DNA adducts and biological
         effects. The effects at zero adduct is deliberately
         shown as finite to emphasize that the effect could, in
         part, be independent of DNA or depend on other, not
         measured, adducts.

effects since most animal data are obtained at relatively high
levels of DNA adducts. The extrapolations to low levels, levels
that the general population might be exposed to, can only come
from a knowledge at the fundamental level relating adducts to
biological effects. Fig. 2 illustrates some of these possible
extrapolations. It is meant to emphasize the fact that human
epidemiological data at low levels of adducts do not exist and if
they did, might be confounded by genetic or physiological
variability in the population as well as the necessity for
considering steps in carcinogenesis other than initiation.
Nevertheless, a scientific goal should be the understanding of
the biological systems and their variabilities so as to make the
extrapolation based on sound biological theory. Part of this
theory has to do with the lifetime of adducts and the relations
between lifetime and other internal and external processes such
as replication, transcription, and promotion.

SKIN CANCER:

The experimental and epidemiological evidence indicates that
non-melanoma skin cancer arises from the cumulative exposure to
UV in sunlight (20). In the United States there are reasonable
data on skin cancer incidence over a wide range of latitudes
which, because of changes in light path through the stratospheric
ozone with latitude, means that there are data of skin cancer
incidence versus the average yearly UV exposure. At all ages
investigated, skin cancer incidence increases exponentially with
average yearly exposure (20, 21). I assume this rapid increase
means that most individuals are exposed to sufficient UV to bring
them well above any threshold level. (This would probably not be
the case for ambient chemicals.) In the United States, skin
cancer is the most common of all cancer. There are approximately
400,000 new cases per year (21).

The rate of increase of cancer incidence with UV exposure
depends critically on the wavelength band one considers to be the
effective one. Since the action spectrum for cytotoxicity for
normal and XP cells follows that for the production of pyrimidine
dimers in DNA (22) and such dimers are implicated in the lethal,
mutagenic, and tumorigenic effects of UV, and are poorly repaired
in most XP cells, it is reasonable to suppose that the effective
wavelengths in sunlight are those that affect DNA. Hence, one
should multiply the sun's spectrum by the DNA action spectrum in
mammalian cells, so as to obtain the average yearly UV exposure.
In arbitrary units of UV dose, the average exposure in the United
States is approximately 0.3 units per year, and the doubling dose
is 0.15 per year (20). Thus, a 50% increase in yearly dose would
double the cancer incidence rate*.

How do we extrapolate these data to assess the role of DNA repair in minimizing the effects of environmental agents in initiating carcinogenesis. Recall from Table 1 that XP individuals have a risk approximating unity of getting skin cancer before the age of 20. On the other hand, the average white person in the United Staes may be estimated, by a long extrapolation, to have a risk of somewhat less than 10^4 of getting skin cancer before that age (23). If one ascribes the differences between the average and XP individuals as solely the result of DNA repair, one must compare the doses that would give the same cancer incidence. Thus, the question is what UV dose would increase the average cancer prevelance before the age of 20 by a factor of 10^4? Since a two-fold increase arises from 0.15 units of dose, a 10^4 fold increase ($2^{13\ 3}$) would require 0.15 x 13.3 = 2 units of dose. Since the average exposure in the United States is approximately 0.3 units per year, these data indicate that there would have to be an approximately 7-fold increase in UV dose per year to give a skin cancer incidence in the average population equal to that observed in XP individuals at the present low level of flux. The 7-fold estimated increase obviously has large theoretical errors in it, as well as the practical one that most XP individuals tend to avoid direct sunlight, because of its more immediate deleterious effects, and hence are probably not exposed to more than 0.1 units of dose per year. If the latter number were correct, a twenty-fold increase in dose would be required to bring the average population up to the incidence observed in XP individuals. Thus, from these data on skin cancer one may conclude that in normal individuals repair processes act to remove between 6/7ths to 19/20ths of the UV damage and as a result decrease the skin cancer incidence rate by 10^4. Repair certainly seems to remove greater than 85% of the UV damage.

The population of XP individuals is very heterogeneous. Their repair deficiencies are not absolute for excision repair. They range from greater than 95% to 40% (24). If one considered the average deficiency, as approximately 85% one would conclude that this defect in excision repair leads to a 10^4 fold increase in skin cancer. Hence, one would also conclude that relatively small changes in DNA repair among the population could have large effects on initiation events for cancer. This conclusion is

---

*If a weighting factor similar to the erythemal spectrum were used, a two-fold increase in cancer incidence would arise from a 30% increase in UV dose (21). This number is lower than that obtained with a DNA spectrum because the longer wavelengths effective in erythema production are not effective in DNA damage.

reinforced by the observations that there is a rough correlation
between the severity of the clinical disease and the magnitude of
the excision repair defect observed on cells in culture.

CONCLUSIONS:

Although cancer may arise as a result of many different
types of molecular changes, there is little reason to doubt that
changes to DNA are one of the more important ones in cancer
initiation. Although DNA repair mechanisms seem able to
eliminate a very large fraction of deleterious changes to DNA, we
not only have little insight into the molecular mechanisms
involved in such repair, but have a negligible amount of
information to permit us to estimate the shape of dose response
relations at low doses. The case of skin cancer is a special
one, in that the average population is exposed to sufficient
solar UV so that the effects of small increments in UV dose may
be estimated. An approximate 85% reduction in DNA repair
increases skin cancer incidence 10^4 fold .

ACKNOWLEDGMENT:

This work was supported by the U. S. Department of Energy.

REFERENCES:

1. H. Rubin, Is somatic mutation the major mechanism of
    malignant transformation?, J. Natl. Canc. Inst. USA
    64:995 (1980).
2. B. N. Ames, Identifying environmental chemicals causing
    mutations and cancer, Science 204:587 (1979).
3. E. Huberman, Mutagenesis and cell transformation of
    mammalian cells in culture by chemical carcinogens, J.
    Environ. Pathol. Toxicol. 2:29 (1978).
4. C. A. Jones and E. Huberman, A sensitive
    hepatocyte-mediated assay for the metabolism of
    nitrosamines in mutagens for mammalian cells, Cancer
    Res.40:406 (1980).
5. J. C. Barrett, T. Tsutsui, and P. O. P. Ts'o,
    Neoplastic transformation induced by a direct
    perturbation of DNA, Nature 274:229 (1978).
6. R. W. Hart, R. B. Setlow, and A. D. Woodhead, Evidence
    that pyrimidine dimers in DNA can give rise to tumors,
    Proc. Natl. Acad. Sci. USA 74:5574 (1977).
7. R. B. Setlow, Repair deficient human disorders and
    cancer, Nature 271:713 (1978).

8.   C. F. Arlett and A. R. Lehmann, Human disorders showing
     increased sensitivity to the induction of genetic
     damage, Ann. Rev. Genet. 12:95 (1978).
9.   E. C. Friedberg, U. K. Ehmann, and J. I. Williams,
     Human diseases associated with defective DNA repair,
     Adv. Radiat. Biol. 8:85 (1979).
10. A. D. Andrews, S. F. Barret, and J. H. Robbins,
     Xeroderma pigmentosum abnormalities correlated with
     colony-forming ability after ultraviolet radiation,
     Proc. Natl. Acad. Sci. USA 75:1984 (1978).
11. D. G. Harnden, Mechanisms of Genetic Susceptibility,
     in: "13th Jerusalem Symposium on Carcinogenesis:
     Fundamental Mechanisms and Environmental Effects," D.
     Reidel, Dordrecht (1980).
12. B. Konze-Thomas, J. W. Levinson, V. M. Maher, and J.
     J. McCormick, Correlation among the rates of dimer
     excision, DNA repair replication, and recovery of human
     cells from potentially lethal damage induced by
     ultraviolet radiation, Biophys. J. 28:315 (1979).
13. V. M. Maher, D. J. Dorney, A. L. Mendrala, B.
     Konze-Thomas, and J. J. McCormick, DNA excision-repair
     processes in human cells can eliminate the cytotoxic
     and mutagenic consequences of ultraviolet irradiation,
     Mutat. Res. 62:311 (1979).
14. T. Kakunaga, Relationship between transformation and
     mutation in mammalian cells, in: "13th Jerusalem
     Symposium on Carcinogenesis: Fundamental Mechanisms
     and Environmental Effects," D. Reidel, Dordrecht
     (1980).
15. J. Doniger and J. A. DiPaolo, Excision and
     postreplication repair capacities, enhanced
     transformation, and survival of Syrian hamster cells
     irradiated by ultraviolet light, Cancer Res. 40:582
     (1980).
16. R. Goth and M. F. Rajewsky, Persistence of
     O^6-ethylguanine in rat brain DNA: Correlation with
     nervous system-specific carcinogenesis by
     ethylnitrosourea, Proc. Natl. Acad. Sci. USA 71:639
     (1974).
17. R. Mueller and M. F. Rajewsky, Sensitive
     radioimmunoassay for detection of
     O^6-ethyldeoxyguanosine in DNA exposed to the
     carcinogen ethylnitrosourea in vivo or in vitro, Zeit.
     f. Naturforsch. 33C:897 (1978).
18. M. C. Poirier, M. A. Dubin, and S. H. Yuspa, Formation
     and removal of specific acetylaminofluorene-DNA adducts
     in mouse and human cells measured by radioimmunoassay,
     Cancer Res. 39:1377 (1979).

19. M. C. Paterson, Use of purified Lesion-recognizing enzymes to monitor DNA repair in vivo, Adv. Radiat. Biol. 7:1 (1978).

20. E. L. Scott and M. L. Straf, Ultraviolet radiation as a cause of cancer, in: "Origins of Human Cancer," H. H. Hiatt, J. D. Watson, and J. A. Winston, eds., Cold Spring Harbor Laboratory, N. Y. (1977).

21. J. Scotto and T. Fears, quoted in: "Protection against depletion of stratosphere ozone by chlorofluorocarbons," Natl. Acad. Sci., Washington (1979).

22. G. J. Kantor, J. C. Sutherland, and R. B. Setlow, Action spectra for killing non-dividing normal human and xeroderma pigmentosum cells, Photochem. Photobiol. 31:459 (1980).

23. R. D. Rundel and D. S. Nachtwey, Skin cancer and ultraviolet radiation, Photochem. Photobiol. 28:345 (1978).

24. J. E. Cleaver and D. Bootsma, Xeroderma pigmentosum: Biochemical and genetic characteristics, Ann. Rev. Genet. 9:19 (1975).

SOME PARAMETERS AFFECTING DNA DAMAGE AND THEIR

RELEVANCE TO THE ACTION OF CHEMICAL CARCINOGENS

P.J. O'Connor

Paterson Laboratories
Christie Hospital and Holt Radium Institute
Manchester M20 9BX, U.K.

INTRODUCTION

The observation that chemical carcinogens will react with the
genetic material of cells has provided a rational basis on which to
build an understanding of their mutagenic action and possibly also of
their carcinogenic action. In this paper the interactions of carcinogenic methylating and ethylating agents with nucleic acids will be
considered in some detail. Over the past decade or so these compounds
have been investigated extensively. Firstly, because of the extent
to which they react with macromolecules, with the relative ease of
their detection by radioactive tracer techniques, and secondly, because the transfer of a simple alkyl group seemed to offer probably
the best long term possibilities for the interpretation of the actions
of these carcinogens at the molecular level[1]. Further, their potential as mutagens and the remarkable tissue-specificity of tumour formation by the N-nitroso compounds makes them very suitable as probes in
studies of the initiation of malignant change[2].

REACTION WITH NUCLEIC ACIDS

An important outcome of investigations with these agents which
are known to form a variety of products in RNA and DNA, has been the
demonstration[1] that their relative capacity to react with oxygen atoms
in nucleic acids, either in vitro or, in vivo correlates well with

Footnote: DEN, N,N-diethylnitrosamine; 1,2-DMH, 1,2-dimethylhydra-
zine; DMN, N,N-dimethylnitrosamine; DMPT, 3,3-dimethyl-1-phenyltria
zene; ENU, N-ethyl-N-nitrosourea; MMPT, 3-methyl-1-phenyltriazene;
MMS, methyl methanesulphonate; MNU, N-methyl-N-nitrosourea.

their overall mutagenic and carcinogenic action as originally proposed[3]. The pattern of reaction observed depends upon the agent itself, rather than the substrates involved[1]. When this data is summarised by reference to alkylation at the oxygen atom of guanine in DNA
it is evident that their increasing potency as carcinogens is para-
lleled by their relative capacity to form $\underline{O}^6$-alkylguanine[2]. On the
other hand, neither the capacity of these agents to react at the N-7
atom of guanine nor the total extent of reaction with DNA, can easily
be related to their carcinogenicity[1].

DISTRIBUTION OF REACTION IN THE BODY

The extent of reaction with the constituents of cells nevertheless, is an important feature in the interpretation of the biological
action of these agents. To a large extent this will depend upon the
need for metabolism in order to produce the chemically reactive alkylating species. For example when rats and hamsters are treated with
DMN, tissue differences can be of the order of 1000-fold and although
the overall pattern of distribution throughout the body tends to be
the same in the different species, liver, kidney and lung are consistently the organs most affected[2]. In general, tissues receiving very
low levels of alkylation by DMN or by DEN are not those in which tumours are commonly induced, rather they tend to occur in those tissues
which are alkylated more extensively[2,4]. In contrast, after treatment
with the nitrosoureas e.g. MNU or ENU which are hydrolysed spontaneously, the distribution of reaction is much more uniform throughout the
body and leads to a much wider spectrum of tumour sites[2,4]. In either
case, the site-specificity of tumour formation will depend upon a variety of factors such as the dose, the route of administration, the
treatment time schedule, the species treated and others (e.g. see
below).

PROCESSES WHICH MODIFY THE AMOUNT OF REACTION IN CELLS

Apparently then, it is not so much the gross extent of reaction
with nucleic acids that can be related to the appearance of tumours
but the amount of reaction of a particular type (i.e. with oxygen
atoms) which appears to be important. Given this premise, it follows
that any process leading to changes in the amounts of these products
might be expected to influence the initiation of tumours by this class
of carcinogen. Several biological processes are capable of altering
either, the amount or, the overall distribution of alkylation products
in cells and tissues. These include a) repair processes which eliminate modifications to normal structure (i.e. the primary or secondary
lesions) due to alkylation b) the replication of molecules, some of
which may already contain lesions c) recombinational events and d)

toxicity leading to partial cell lysis (nuclease action) or cell death. Only those processes concerned with the active removal of primary lesions from nucleic acids have been studied to any extent and the following summary is restricted to this aspect. In certain cases, however, (especially the 3- and 7-alkyl purines in DNA or phosphotriesters in RNA) an intrinsic chemical instability[1] also leads to changes, not only in the amounts of these products themselves, but concomitantly in the overall pattern of alkylation at the level of individual molecules within a cell.

REMOVAL OF SPECIFIC ALKYLATION INDUCED LESIONS IN NUCLEIC ACIDS

There are a few reports in which the repair of methylated bases in species of RNA have been considered as a possibility and others in which the evidence indicated that the level of alkylation declined simply as a function of natural RNA turnover. The repair of damage in long-lived species of RNA, however, could conceivably confer a biological advantage, but this has yet to be examined[2].

In the case of DNA there is now a good deal of information on the repair of lesions, much of it fragmentary but it is possible to attempt some degree of analysis. Early reports on the specific removal of alkylation products from DNA had considered that the major product 7-alkylguanine was most probably lost simply by depurination reactions due to the intrinsic instability of the N-glycosidic bond in the deoxynucleoside and that 3-alkyladenine, although much less stable in DNA, was subject to enzymic removal[1]. Studies with animals have indicated that all tissues examined apparently contain an enzyme capable of removing 3-alkyladenine[1,2,5] and a specific DNA-glycosylase[6] is most probably responsible for this. Some revision of earlier opinion is warranted in the case of 7-alkylguanine as the rate of removal of this base from hamster tissues[7,8] and from V79 cells[9] strongly implicates an enzymic mechanism. In the light of these later observations, perhaps rat liver is better regarded as deficient, rather than devoid of this enzyme[5]. By far the greatest volume of work has focussed upon aspects of the repair of O^6-alkylguanine which is known to be potentially promutagenic both in RNA and in DNA synthesising systems[10,11]. Since the early reports indicating the presence of an enzyme for its removal from DNA in bacteria[12] and in animals[13] and the suggestion that events leading to the development of tumours may be related to the efficiency of DNA repair systems for products such as O^6-alkylguanine[13], a number of studies have emerged in which it is possible to make such a correlation (see below). The enzyme has now been demonstrated in cell-free extracts of liver and of kidney[14] and its activity may be temporarily inhibited by high doses of carcinogens[14], or enhanced by pretreatment with low doses[15].

DNA SYNTHESIS AND CARCINOGENESIS

It is widely held that replication of DNA (induced either by repair reactions or, by cell division) on a template containing a miscoding lesion is important for the initiation of malignant change.
For convenience, studies seeking to relate the repair of $\underline{O}^6$-alkylgua-
nine with site-specificity of tumour formation can be considered in
relation to the turnover times of the tissues themselves. Very simply
these may be classified as a) fast proliferating b) slowly proliferating and c) very slowly proliferating or near-stationary tissues. Within these categories very considerable variations in both the amounts
and persistence of $\underline{O}^6$-alkylguanine in tissues can be anticipated without contravening the initial premise:
a) The combination of fast cell-turnover, especially when cells are
shed from the system, together with the efficient removal of promutagenic bases will quickly deplete a tissue DNA of potentially carcinogenic lesions and the number of cells remaining at risk will be very
small. Yet, if the repair system is deficient to some degree the statistical probability of replication over miscoding lesions is greatly
increased (Fig.1a). Data consistent with these two situations have
been described for the DNA of ileum and colon respectively, of rats
treated with 1,2-DMH[16]. Furthermore, such studies have measured only
the gross amounts of alkylation in the DNA that could be recovered
by extraction from tissues, so that the apparent removal rate is exag-
gerated by dilution of the DNA in situations of rapid cell turnover.
Parental DNA strands therefore, may contain higher levels of alkylation than indicated by the data and the actual rate of repair will be
very much slower. This situation is in contrast to the events in
liver (Fig 1a), in which 1,2-DMH produces many times (7-8 fold) more
alkylation than in colon[16,17] without inducing any tumours. However,
as a near-stationary tissue with an efficient repair system that may
be induced still further by a variety of agents[18] the chance of a promutagenic lesion surviving until a round of hepatic DNA replication
may be statistically very small.
b) In a more slowly dividing tissue, a slower (or more deficient) repair system might allow the level of promutagenic lesions in cells to
be maintained long enough for replication to occur (Fig 1b). An example would be the slower removal of $\underline{O}^6$-methylguanine from the DNA of
kidney which is correlated with the formation of kidney tumours in
protein-depleted rats treated with DMN. Again, this situation is in
contrast to the events observed in liver where there is initially a
much higher level of reaction[19].

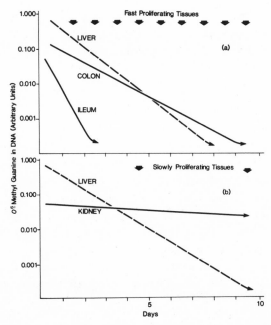

Fig.1.   Schematic representation of the relationship between the cap-
         acity for the repair of $\underline{O}^6$-methylguanine in the tissue DNA of
         carcinogen treated animals and the cellular renewal rates for
         individual tissues. Data consistent with the repair curves
         shown in (a) and (b) have been obtained experimentally (Ref
         Nos 16 and 19, respectively) but the arrows showing cell turn-
         over are indicative only of major differences in the rates of
         cellular renewal.

c)   In very slowly or virtually non-proliferating tissues promutagenic
lesions would be required to persist for much longer periods as typif-
ied by the early repair experiments with the nervous tissues of neo-
natal rats exposed to the neuro-oncogenic effects of ENU[20]. When repeated dose schedules are employed this deficiency of repair will lead
to a build up of $\underline{O}^6$-alkylguanine residues in the DNA of cells as e.g.
in brain DNA of young adult rats following repeated doses of MNU[21].

     There is clearly a need for a much more precise evaluation of
such repair studies in relation to the turnover of individual cell
types in order to take account of differences in the kinetics of tar-
get-cell turnover. For example, in the nervous tissue of rats where
single treatments with ENU[22], either transplacentally when the tissue is
especially sensitive[22] or to the neonate[20], or multiple treatments
with MNU to the young adult[21] all give rise to tumours of the nervous
system. In general terms, however, it is evident that the capacity
for the removal of DNA lesions can vary widely, not only between diff-

erent species but between different tissues within the same animal so
that the initial level of alkylation may not necessarily be as important as the capacity of a cell for repair[13]. More germane to the
process of initiation therefore, might be the amount and type of alkylation present in DNA at the time of replication[5] and depending upon
the turnover time of the cells concerned this may or may not be directly related to their capacity for the repair of lesions.

Some of the strongest support for the suggestion that the capacity for removal of O^6-alkylguanine from DNA might be a factor that is
related to tumour formation has come from the early reports that this
product was either persistent or accumulating in the brain DNA of rats
treated respectively either with single doses of ENU[20] or with multiple doses of MNU[21]. These observations now include a persistence in
brain DNA after single doses of MMPT[23] and single dose studies with
rats, Syrian and Chinese hamsters using DMN, 1,2-DMH and MNU have extended these observations to at least 4 other target tissues, namely,
kidney[19], colon[16,24], bladder[25] and liver[7,8]. In multiple dose experiments with rats treated with MNU, DMN and DMPT positive correlations were also observed for bladder[25], breast[26], kidney[27], and brain[28].

In contrast to these observations, chronic treatments, as in the
hepatocarcinogenic exposure of rats to low daily doses of DMN, lead
not to a delay but a marked enhancement of the capacity of liver for
the repair of O^6-methylguanine, as well as to an increased level of
hepatic DNA synthesis[15]. This situation may be more analogous therefore, to the formation of liver tumours after single doses of DMN
when these are given during the DNA synthesis that follows partial hepatectomy[29], treatment with carbon tetrachloride[30], or of renal tumours when given to new born rats[31]. The difficulty in inducing tumours of the liver in rats (but not in hamsters[8,32]) by single doses of
alkylating agents, except when administered during DNA synthesis emph-
asises the importance of cell proliferation and it is possible in the
case of rat liver that the inducible capacity for removal of O^6-alkyl-
guanine now observed after exposure to a variety of agents[18] might itself provide a selection pressure for initiation, particularly in situations where it is activated for long periods.

In experiments with mice[33,34] the situation is less clear and in
comparative experiments between strains of mice with differing tissue
sensitivities to the action of a carcinogen no obvious correlations
between removal of lesions and tumour formation could be made[2,5].
Here again, however, cell proliferation could be very important[2,5].
For example, GR and C3Hf mice (males) treated with DMN are susceptible

to liver tumour or to lung tumour induction, respectively and the proliferative response to the carcinogen was greatest in the target tissue of the susceptible strain[35], whilst in the livers of these animals
the initial extent of alkylation and subsequent loss of alkylated products was similar for both strains[36].

Relatively little work has been carried out in species other than
rats and mice. As noted above, positive correlations can be made for
the amount and retention of O^6-methylguanine and liver tumour induction in both Syrian[7] and Chinese[8] hamsters treated with DMN, but the
oncogenic effects of MNU in the Mongolian gerbil are quite different
from those in rats, in spite of a prolonged retention of O^6-alkylgua-
nine in brain DNA[37]. Although in this case melanomas are produced,
this tissue has a common origin with nervous tissue as it is also
derived from the neural crest during embryogenesis and for this reason
presents an interesting system for further study.

CONCLUSIONS

For practical reasons most of the work on the removal of specific
DNA lesions produced by the methylating and ethylating carcinogens has
been carried out with the alkylpurines, especially O^6-alkylguanine but
other products e.g. O^4-alkyl- and O^2-alkylpyrimidines are also known
to be potentially mutagenic[38,39,40]. More attention should now be
paid to the repair of these specific lesions[41] particularly since the
mutagenicity of a series of methylating agents could not be entirely
accounted for by the amounts of O^6-methylguanine formed in DNA, indicating that other sites may be contributing towards the observed mutation frequency[42]. As details of the mode of action of other classes
of chemical carcinogen begin to emerge it may prove more appropriate
to consider the repair of promutagenic lesions in a more general sense,
rather than of specific sites, in the context of the arguments made
above. The process of carcinogenesis from initiation through to the
appearance of tumours is an exceedingly complex sequence of events and
an attempt has been made to evaluate several factors that are apparently of importance in the very early stages of this process. Although a great deal of attention has been focussed recently on aspects
of the repair of DNA products much more information is needed particularly at the level of individual cells in order to assess its true
role and to place it in perspective as one of several factors that are
of importance in the process of initiation by this class of chemical
carcinogen. In conclusion, it should be remembered that the widely-
emerging relevance of DNA-carcinogen interactions does not necessarily
exclude other constituents of cells from an important role in malignant transformation.

ACKNOWLEDGEMENTS

    Financial support from the Cancer Research Campaign and the
Medical Research Council is gratefully acknowledged.

REFERENCES

1. G.P. Margison and P.J. O'Connor, Nucleic acid modification by
    N-nitroso compounds in: "Chemical Carcinogens and DNA",
    P.L. Grover ed. CRC Press, Boca Raton (1979) p111.
2. P.J. O'Connor, R. Saffhill and G.P. Margison, N-nitroso compounds:
    Biochemical Mechanisms of action in: "Environmental Carcinogen-
    esis", P. Emmelot and E. Kriek eds. Elsevier/North Holland Bio-
    medical Press, Amsterdam (1979) p73.
3. A. Loveless, Possible relevance of O-6 alkylation of deoxyguano-
    sine to the mutagenicity and carcinogenicity of nitrosamines and
    nitrosoamides, Nature (London) 223:206 (1969).
4. P.N. Magee, R. Montesano and R. Preussmann, N-nitroso compounds
    and related carcinogens in "Chemical Carcinogens" C.E. Searle
    ed. A.C.S. Monographs 173, Amer.Chem.Soc., Washington DC
    (1976) p491.
5. P.J. O'Connor, Interactions of chemical carcinogens with macro-
    molecules. J.Clin.Oncology and Cancer Res, in press 1980.
6. T. Lindahl, DNA glycosylases, endonucleases for AP-sites and
    base-excision repair. Prog.Nucleic Acid Res.Mol.Biol. 22:
    135 (1979).
7. G.P. Margison, J.M. Margison and R. Montesano, Methylated purines
    in the deoxyribonucleic acid of various Syrian golden hamster
    tissues after administration of hepatocarcinogenic doses of
    dimethylnitrosamine. Biochem.J. 157:627 (1976).
8. G.P. Margison, J.A. Swindell, C.H. Ockey and A.W. Craig, The eff-
    ect of a single dose of dimethylnitrosamine in the Chinese ham-
    ster and the persistence of DNA alkylation products in selected
    tissues. Carcinogenesis 1:91 (1980).
9. W. Warren, A.R. Crathorn and K.V. Shooter, The stability of meth-
    ylated purines and of methylphosphotriesters in the DNA of V79
    cells after treatment with N-methyl-N-nitrosourea. Biochim.
    Biophys.Acta, 563:82 (1979).
10. L.L. Gerschman and D.B. Ludlum, The properties of O^6-methylguanine
    in templates for RNA, polymerase. Biochim.Biophys.Acta, 308:310
    (1973).
11. P.J. Abbott and R. Saffhill, DNA synthesis with methylated poly-
    (dC-dG) templates. Evidence for a competitive nature of miscod-
    ing by O^6-methylguanine. Biochim.Biophys.Acta. 562:51 (1979).
12. P.D. Lawley and D.J. Orr, Specific excision of methylated products
    from the DNA of E.coli treated with MNNG. Chem.-Biol.Interactions
    2:154 (1970).

13. P.J. O'Connor, M.J. Capps and A.W. Craig, Comparative studies of the hepatocarcinogen N,N-dimethylnitrosamine in vivo: reaction sites in rat liver DNA and the significance of their relative stability. Br.J.Cancer. 27:153 (1973).

14. A.E. Pegg and G. Hui, Formation and subsequent removal of O^6-methylguanine from DNA in rat liver and kidney after small doses of dimethylnitrosamine. Biochem.J. 173:739 (1978).

15. R. Montesano, H. Brésil, G. Planche-Martel, G.P. Margison and A.E. Pegg, Effect of chronic treatment of rats with dimethylnitrosamine on the removal of O^6-methylguanine from DNA. Cancer Res. 40:452 (1980).

16. J.A. Swenberg, H.K. Cooper, J. Bücheler and P. Kleihues, 1,2-Dimethylhydrazine-induced methylation of DNA bases in various rat organs and the effect of pretreatment with disulfiram Cancer Res. 39:465 (1979).

17. K.J. Rogers and A.E. Pegg, Formation of O^6-alkylguanine by alkylation of rat liver, colon and kidney following administration of 1,2-dimethylhydrazine. Cancer Res. 37:4082 (1977).

18. P.J. O'Connor, G.P. Margison, R. Montesano, J.D. Buckley, Y-H, Chu, D.P. Cooper, R.A. Smith, N.J. Curtin, K. Snell, J. Ashby, C. Bradbrook and J.A. Styles, The enhanced repair of O^6-alkylguanine in mammalian systems (this volume).

19. J.W. Nicoll, P.F. Swann and A.E. Pegg, Effect of dimethylnitrosamine on the persistence of methylated guanines in rat liver and kidney DNA. Nature (London) 254:261 (1975).

20. R. Goth and M.F. Rajewsky, Persistence of O^6-ethylguanine in rat brain DNA: correlation with nervous system-specific carcinogenesis. Proc.Natl.Acad.Sci.U.S.A. 71:639 (1974).

21. G.P. Margison and P. Kleihues, Chemical carcinogenesis in the nervous system: preferential accumulation of O^6-methylguanine in rat brain DNA during repetitive administration of N-methyl-N-nitrosourea. Biochem.J. 148:521 (1975).

22. S. Ivankovic, Experimental prenatal carcinogenesis in "Transplacental Carcinogenesis" L. Tomatis and U. Mohr eds. IARC, Lyon 1973, p92.

23. G.P. Margison, A.J. Likhachev and G.F. Kolar, In vivo alkylation of foetal, maternal and normal rat tissue nucleic acids by 3-methyl-1-phenyltriazene. Chem.-Biol.Interactions 25:345 (1979).

24. A.J. Likhachev, G.P. Margison and R. Montesano, Alkylated purines in the DNA of various rat tissues after administration of 1,2-dimethylhydrazine. Chem.-Biol.Interactions 18:235 (1977).

25. R. Cox and C.C. Irving, Selective accumulation of O^6-methylguanine in DNA of rat bladder epithelium after intravesical administration of N-methyl-N-nitrosourea. Cancer Lett. 3:265 (1977).

26. R. Cox and C.C. Irving, O^6-methylguanine accumulates in the DNA of mammary glands after administration of N-methyl-N-nitrosourea to rats. Cancer Lett. 6:273 (1979).

27. J.W. Nicoll, P.F. Swann and A.E. Pegg, The accumulation of O^6-methylguanine in the liver and kidney of rats treated with dimethylnitrosamine for a short and a long period. Chem.-Biol. Interactions 16:301 (1977).

28. H.K. Cooper, E. Hauenstein, G.F. Kolar and P. Kleihues, DNA alkylation and oncogenesis by 3,3-dimethyl-1-phenyltriazene. Acta Neuropathol(Berl) 43:105 (1978).

29. V.M. Craddock, Induction of liver tumours in rats by a single treatment with nitroso-compounds given after partial hepatectomy. Nature (London) 245:386 (1973).

30. A.W. Pound and T.A. Lawson, Partial hepatectomy and toxicity of dimethylnitrosamine and carbontetrachloride in relation to the carcinogenic action of dimethylnitrosamine. Br.J.Cancer 32:596 (1975).

31. B. Terracini and P.N. Magee, Renal tumours in rats following injection of dimethylnitrosamine at birth. Nature (London) 202:502 (1964).

32. L. Tomatis and F. Cefis, The effects of multiple and single administration of dimethylnitrosamine to hamsters. Tumori 53:447 (1967).

33. P.D. Lawley, Methylation of DNA by carcinogens: some applications of chemical analytical methods in "Screening Tests in Chemical Carcinogenesis" R. Montesano, H. Bartsch and L. Tomatis eds. IARC, Lyon 1976, p181.

34. J.V. Frei, D.H. Swenson, W. Warren and P.D. Lawley, Alkylation of DNA in vivo in various organs of C57Bl mice by the carcinogens N-methyl-N-nitrosourea, N-ethyl-N-nitrosourea and ethylmethanesulphonate in relation to the induction of thymic lymphoma. Biochem.J. 174:1031 (1978).

35. H.K. DeMunter, L. Den Engelse, and P. Emmelot, Studies on lung tumours IV. Correlation between $[H^3]$-thymidine labelling of lung and liver cells and tumour formation in GRS/A and C3Hf/A male mice following administration of dimethylnitrosamine. Chem.-Biol.Interactions 24:299 (1979).

36. L. Den Engelse, The formation of methylated bases in DNA by dimethylnitrosamine and its relation to differences in the formation of tumours in the livers of GR and C3Hf mice. Chem.-Biol.Interactions 8:329 (1974).

37. P. Kleihues, S. Bamborschke and G. Doerjev, Persistence of alkylated bases in the Mongolian gerbil (Meriones unguiculatus) following a single dose of methylnitrosourea. Carcinogenesis 1:111 (1980).

38. P.J. Abbott and R. Saffhill, DNA synthesis with methylated poly-
    (dA-dT) templates: possible role of $\underline{O}^4$-methylthymidine as a
    promutagenic base. Nucleic Acids Res 4:761 (1977).
39. R. Saffhill and P.J. Abbott, Formation of $\underline{O}^2$-methylthymine in
    poly(dA-dT) on methylation with N-methyl-N-nitrosourea and
    dimethyl sulphate. Evidence that $\underline{O}^2$-methylthymine does not mis-
    code during DNA synthesis. Nucleic Acids Res. 5:1971 (1978).
40. B. Singer, H. Fraenkel-Conrat and J.T. Kuśmierek, Preparation and
    template activities of polynucleotides containing $\underline{O}^2$- and $\underline{O}^4$-
    alkyluridine. Proc.Natl.Acad.Sci.U.S.A. 75:1722 (1978).
41. W.J. Bodell, B. Singer, G.H. Thomas and J.E. Cleaver, Evidence
    for removal at different rates of $\underline{O}$-ethylpyrimidines and ethyl-
    phosphotriesters in two human fibroblasts. Nucleic Acids Res
    6:2819 (1979).
42. R.F. Newbold, W. Warren, A.S.C. Medcalf and J. Amos, Mutagenicity
    of carcinogenic methylating agents is associated with a specif-
    ic DNA modification. Nature (London) 283:596 (1980).

EFFECT OF A TUMOR PROMOTER ON DNA REPAIR AFTER TREATMENT OF C3H/10T½ CELLS WITH A CARCINOGEN

Johan R. Lillehaug and Ruth K. Kleppe

Department of Biochemistry
University of Bergen
Bergen, Norway

SUMMARY

The mechanism by which tumor promoters act is still unknown. Since the initiating carcinogens are believed to interact with the cellular DNA finally resulting in mutations, tumor promoters could act through interference with DNA repair. In this work we investigated the effect of 12-O-tetradecanoyl-phorbol-13-acetate (TPA) on repair synthesis after 20-methylcholanthrene (MCA) treatment. Both in early and in late logarithmic growth phase of C3H/10T1/2 cells an inhibition of repair synthesis was observed when the cells were exposed to 0.17 µM TPA. The damage introduced by 2 µM MCA exposure for 24 hrs was present 20 hrs later at nearly 0-hrs level in TPA treated cells, while repair synthesis was reduced to 60% in cells not exposed to TPA. At least one of the effects of TPA therefore seems to be an inhibition of DNA repair.

INTRODUCTION

The mechanism by which tumor promoters enhance the yield of transformation resulting from treatment of target cells with small doses of carcinogens is still poorly understood. Several theories have been put forward but conclusive evidence in support of one of these has so far not been produced. (1-6). Gaudin et al. (7-8) suggested that inhibition of DNA repair could be an essential feature of tumor promoters. Their conclusions was later challenged by Poirier et al. (9) and by Cleaver and Painter (10). They pointed out that tumor promoters inhibited DNA repair and DNA replication in a similar manner. The inhibition of DNA repair therefore, seemed to be nonspecific. However, one of the characteristics of tumor promotion is the fact that tumor promoters must be made available

to the target cells after initiation has taken place, and furthermore, the tumor promoters have to be administrated to the cells
for a prolonged period of time.

Since the above mentioned reports (7-10) all were short term
experiments, the longest lasting less than 6 hrs, and in one report the tumor promoters to be tested were added prior to the
treatment of initiator, it may well be that the effects studied do
not relate to tumor promotion. The observation by Teebor et al.
(11) that a 24 hrs TPA treatment inhibited thymidine dimer excision in HeLa cells supported the original findings of Gaudin et al.
(7). The availability of a mouse embryo fibroblast cell line,
C3H/10T1/2, in which the twostage carcinogenesis process can be
studied in vitro (12) offers a good opportunity to reinvestigate
the effect of TPA on DNA repair. The effect of tumor promoters on
the DNA synthesis in the C3H/10T1/2 cells has been studied by Pe-
terson et al. (13). They concluded that TPA only had a transient
effect on DNA synthesis. An initial inhibition was observed 6-12
hrs after TPA administration. At the end of this period the cells
recovered and resumed normal or near to normal growth rates as
well as DNA synthesis. Thus, if DNA repair is monitored 12 hrs or
more after addition of TPA, the effect on DNA replication should
be negligible.

MATERIALS AND METHODS

The C3H/10T1/2 C18 cells were obtained from C. Heidelberger,
USC, Los Angeles. Basal medium Eagle (BME) and foetal calf serum
(FCS) were purchased from Gibco. [methyl-^{3}H]thymidine (^{3}H-TdR) at
52 mCi/mmol was obtained from Amersham. Hydroxyurea(HU) and thymidine were from Sigma. 20-methylcholanthrene (MCA) was purchased from
Koch-Light Laboratories Ltd. and 12-0-tetradecanoyl-phorbol-13-
acetate (TPA) was from P.L.Biochemicals Inc.

The growth medium was either BME supplemented with 10% FCS or
BME with 10% heatinactivated, dialyzed FCS. The cells were grown
in 55 mm plast petridishes (Nunc). The number of cells in each assay was determined in duplicate, parallell dishes using a Coulter
counter.

Inhibition of replication was obtained by incubating the cells for
3 hrs in the presence of 10 mM hydroxyurea. It is essential that
the hydroxyurea treatment is carried out in the presence of extensively dialyzed FCS.

Repair synthesis was measured by incorporation of ^{3}H-TdR into cold
5% TCA precipitable material in the presence of 10 mM HU. The incubation mixture was BME enriched with 10% heat-inactivated, dia-
luzed FCS, 10 mM HU, 1 µM thymidine (TDR) and 5 µC/ml 3H-TdR. At
the times indicated the radioactive medium was removed and the

petridishes were left at -20° over night. The cells were scraped off the dishes in 1.5 ml PBS buffer, 10% cold TCA was added and the DNA was coprecipitated with 200 µg calf thymus DNA. The precipitates were collected onto glassfiber filters, washed three times with 5 ml 5% cold TCA and twice with 5 ml ethanol. The filters were dried in an vacuum oven and the radioactivity was measured by liquid scintillation.

20-methylcholanthrene treatment. The MCA was dissolved in acetone and added to the fresh medium in the concentrations desired. Control cultures received medium containing acetone. The acetone concentrations was always less than 0.5%. The cells were exposed to MCA containing medium for 24 hrs.

RESULTS

Inhibition of DNA replication. The background incorporation of ^{3}H-TdR into DNA was reduced by the addition of hydroxyurea to the growth medium. To inhibit the DNA replication 95% or more 10 mM HU had to be employed as shown in Table I. This concentration effectively inhibited the replication over a 7 hrs period while at 1 mM HU the inhibition was overcome by the cells 4 to 5 hrs after addition of the hydroxyurea. Analysis on alkaline sucrose showed that HU at 50 mM caused single-stranded breaks in the DNA. Using freshly prepared solutions of HU at 10 mM, single-strand breaks were however not observed (data not shown).

Induction of DNA repair synthesis with MCA. MCA is a powerful carcinogen which is used as the standard polycyclic hydrocarbon initiator in the C3H/10T1/2 two-stage carcinogenesis assay (13). When added to the cultures of C3H/10T1/2 cells, DNA repair, as measured by incorporation of ^{3}H-TdR, increases as shown in Figure 1. The amount of repair synthesis increased with dose of MCA relative to the control cultures receiving acetone only. At a concentration of 0.4 µM of MCA and with approximately 90% cell survival, the rate of DNA repair was one half of that caused by 4 µM MCA. At 4 µM MCA the cell plating efficiency was between 20 and 30% of acetone control. A 24 hrs treatment of the C3H/10T1/2 cells with 2 µM MCA also led to a decrease in the average molecular weight of cellular DNA as revealed by alkaline sucrose gradient centrifugation analysis of uniformed labelled DNA (data not shown).

Early effects of TPA. A 6 hrs treatment of the C3H/101/2 cells with 0.17 µM TPA reduced the rate of ^{3}H-TdR incorporation both in MCA treated and in control cultures, as shown in Figure 2. The inhibition by TPA was approximately 50% in both cases. The DNA replication is also reduced by 15 - 30% during the first 6 hrs of TPA treatment. From Figure 2, left panel, it is apparent that in the control cell cultures where the total DNA synthesis is reduced to 3% by 10 mM HU, 0.17 µM TPA caused an additional 50% inhibition

Table 1.   Effect of Hydroxyurea on Incorporation
           of ^{3}H–TdR

Concentration of hydroxyurea:

	0 mM	1.0 mM	10 mM
Time:	cpm/10^{5}cells (%)	cpm/10^{5}cells (%)	cpm/10^{5}cells (%)
0.5 h	70966 (100)	2224 (3.1)	721 (1.0)
1 hrs	107466 (100)	8058 (7.5)	1703 (1.6)
2 hrs	165533 (100)	13947 (8.4)	4318 (2.6)
4 hrs	255452 (100)	56932 (22.3)	8539 (3.3)

C3H/10T1/2 cells were treated with different concentrations
of hydroxyurea for 3 hrs. At the end of this period ^{3}H–TdR
was added to the medium, still containing the same concentration of HU, and assayed for ^{3}H–TdR incorporation into
DNA at times indicated. The cell density was 155.000 cells/
plate.

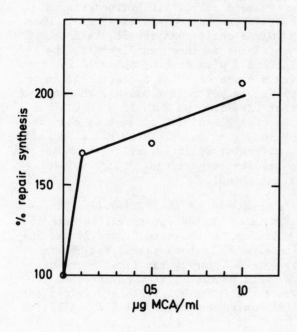

Fig. 1.
Repair synthesis after
24 hrs treatment with
MCA. Logarithmically
growing C3H/10T1/2 cells
were treated with MCA at
different concentrations
for 24 hrs. Repair syn-
thesis was then measured
by 90 minutes incorpora-
tion of ^{3}H–TdR as des-
cribed in Materials and
Methods. The cell density
was 339.700 cells/plate.
Each point represents
two measurements.

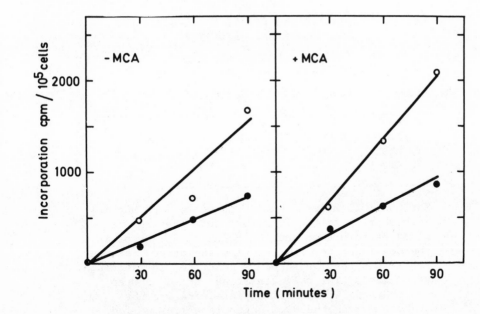

Fig. 2.   Effect of short time exposure to TPA.
C3H/10T1/2 cells in early logarithmic growth phase
were treated with 2 μM MCA for 24 hrs, then the MCA
was removed and the cells were exposed to 0.17 μM
TPA for 3 hrs. DNA replication was inhibited by
further 3 hrs incubation with 10 mM HU in the presence
of 0.17 μM TPA. Thymidine incorporation was measured
as described in Materials and Methods. The cell density was 17.000 cells/plate for the untreated cells
and 14.900 cells/plate in the case of the MCA treated
cells. Open circles; cells not exposed to TPA.
Closed circles; cells treated with 0.17 μM TPA.

of this rest DNA synthesis. This suggests that the HU-insensitive
DNA synthesis is more susceptible to TPA than the overall DNA replication.

TPA retards the rate of DNA repair in MCA treated cells. The inhibition of DNA repair found early during TPA treatment suggested
that the half lives of MCA adducts in DNA as well as other damages
in the cellular DNA due to the MCA treatment increased in cells
exposed to TPA compared to unexposed cells. To test this hypothesis logarithmic phase cultures of C3H/10T1/2 cells were treated
with 2 μM MCA for 24 hrs, after the MCA was removed 0.17 μM TPA
was added while the control cultures received appropriate amounts
of acetone. DNA repair was then determined at different times after the MCA removal. The results are shown in Table 2. The DNA

Table 2. Relative DNA Repair Synthesis After
          Long Time Exposure to TPA

Time	Without TPA	With 0.17 µM TPA
0 hrs	100	
20 hrs	60.0	99.4
40 hrs	47.3	48.3

The C3H/10T1/2 cells were exposed to 2 µM MCA for
24 hrs prior to the TPA treatment. The TPA was
added at zero time and the cultures were kept in
the CO_2 incubator for the time indicated. The DNA
repair was then measured as described in Materials
and Methods. The cell densities varied between
443.000 and 541.000 cells/plate. TPA was not
present during the 90 minutes labelling period.

repair synthesis in the TPA treated cells remained essentially unchanged the first 20 hrs while a decrease of approximately 40% was
observed in the control cultures. After 40 hrs of treatment with
TPA, the remaining DNA repair synthesis was nearly the same in the
two groups. It is also interesting to note close to 40% of the
initial repair synthesis still remained 40 hrs after removal of
the MCA.

DISCUSSION

     In the present report we have shown that the carcinogen MCA
introduces defects into cellular DNA of the normal mouse embryo
fibroblast cell line C3H/10T1/2 C18 that result in DNA repair.
The influence of the tumor promoter TPA has also been studied under conditions similar to those used in two-stage carcinogenesis
experiments.

     The inhibition of DNA repair found is in good agreement with
previous findings (7,8,9,10). This inhibition is not due to toxic
effects of TPA since a concentration of 0.17 µM TPA did not effect
the plating efficiency of the cells. This concentration of TPA is
commonly used in the C3H/10T1/2 C18 two-stage carcinogenesis assay.
The early effect of TPA may be related to inhibition of the overall DNA synthesis as pointed out by Poirier et al. (9), since we
also observed a reduction in DNA replication in control cultures
not exposed to MCA. This can not, however, be the case for remaining DNA repair synthesis after MCA treatment which was determined 20 hrs after initiating the TPA treatment. At this time,

DNA replication was equal to that found in control cultures. In some cases a slight increase in total DNA synthesis could be observed in cells exposed to 0.17 μM TPA for 20 hrs.

Since TPA causes both an inhibition (early) and an enhancement (late)(13) of DNA synthesis in C3H/10T1/2 C18 cells, the effect on DNA synthesis is not easily related to DNA replication. To obtain a better understanding of the mechanism by which TPA influences DNA repair we are now studying the removal of labelled MCA adducts from DNA in vivo. The data presented in this report support the findings of Teebor et al. (11) that TPA slows down the DNA repair processes. From the extensive data published on the properties of TPA we may conclude that it is highly unlikely that the tumor promoter activity of TPA is related directly to one specific property. The promoting capability may well be the sum of several properties and the inhibition of DNA repair could well be one of those essential in the promoting process.

ACKNOWLEDGEMENT

This work was supported by the Norwegian Research Council for Science and Technology (NTNF).

We wish to thank Ms. Karin Wiig Johannessen for excellent technical assistance.

REFERENCES

1. I. Bernblum, Sequential aspects of chemical carcinogenesis, in: "Skin Cancer". Vol. 1 p. 323, F. F. Becker, ed., Plenum Publishing Corp. New York (1975)
2. R. K. Boutwell, The function and mechanism of promoters of carcinogenesis. CRC Critical Rev. Toxicol., 2:419 (1974).
3. S. Mondal, Two stage carcinogenesis: Possible role of promoters, in: "Genetic differences in chemical carcinogenesis", R. Kouri, ed., CRC Critical Rev. Toxicol. (in press)(1979).
4. I. B. Weinstein, M. Wigler, P. B. Fisher, E. Siskin, and C. Pietropolo, Cell culture studies on the biologic effects of tumor promoters, in: "Carcinogenesis" Vol. 2, p. 313, ed., T.J. Slaga, A. Sivak, and R. K. Boutwell, Raven Press, New York (1978).
5. A. R. Kinsella, and M. Radman, Tumor promoter induces sister chromatid exchanges: Relevance to mechanisms of carcinogenesis, Proc. Natl. Acad. Sci. U.S.A., 75:6149 (1978).
6. E. Hecker, Isolation and characterization of the co-carcinogenic principles from Croton oil, in: "Methods in Cancer Research", Vol. 6, p. 439, ed., H. Buch, Academic Press, New York (1971).
7. D. Gaudin, R. S. Gregg, and K. L. Yielding, DNA repair inhibition: A possible mechanism of action of co-carcinogens,

Biochem. Biophys. Res. Com., 45:630 (1971).

8. D. Gaudin, R. S. Gregg, and K. L. Yielding, Inhibition of DNA repair by co-carcinogens, Biochem. Biophys. Res. Com., 48: 945 (1972).

9. M. C. Poirier, B. T. De Cicco, and M. W. Lieberman, Nonspecific inhibition of DNA repair synthesis by tumor promoters in human diploid fibroblasts damaged with N-Acetoxy-2-acetylaminofluorene, Cancer Research, 35:1392 (1975).

10. J. E. Cleaver, and R. B. Painter, Absence of specificity in inhibition of DNA repair replication by DNA-binding agents, co-carcinogens and steroids in human cells, Cancer Research, 35:1773 (1975).

11. G. W. Teebor, N. J. Duker, S. A. Ruacan, and K. J. Zachary, Inhibition of thymine dimer excision by the phorbol ester, phorbol myristate acetate, Biochem. Biophys. Res. Com., 50:66 (1973).

12. S. Mondal, D. W. Brankow, and C. Heidelberger, Two stage chemical oncogenesis in cultures of C3H/10T1/2 cells, Cancer Research, 36:2254 (1976).

13. A. R. Petersen, S. Mondal, D. W. Brankow, W. Thon, and C. Heidelberger, Effects of promoters on DNA synthesis in C3H/10T1/2 mouse fibroblasts, Cancer Research, 37:3223 (1977).

ENHANCED SURVIVAL OF ULTRAVIOLET-DAMAGED PARVOVIRUS LuIII AND

HERPES VIRUS IN CARCINOGEN PRETREATED TRANSFORMED HUMAN CELLS

Magali Gunther, René Wicker, Sylvie Tiravy and
Jacques Coppey

Institut Curie, Section de Biologie
75231 PARIS Cedex 05, France

An increased survival of UV (254 nm)-damaged DNA viruses
(Herpes, Adenovirus and SV40) occurs in mammalian cells
treated prior to infection by low doses of physical (UV and
X-rays) or chemical carcinogens[1-6]. This increase is more
pronounced when the treatment is given a few days before
infection. The rate of forward mutations (Herpes virus)[7] and
of backward mutations (SV40)[8] is significantly higher in the
reactivated than in the control viruses, indicating that the
underlying process could be, as in bacteria a manifestation
of an error-prone mode of DNA repair[9-10]. In order to study
such an inducible repair process in mammalian cells without
any interference with excision repair mechanisms,we analysed
the reactivation of an UV-damaged single-stranded DNA virus
(parvovirus LuIII) in human cells treated by different carcinogens. For comparison the reactivation of UV-damaged Herpes
virus (HSV) was examined in parallel cultures.

RESULTS

We measured the virus yields of one-cycle growth of intact
or UV-damaged Lu III virus in NBK cells (Table 1). A 50 J.m^{-2}
dose of UV gave a virus survival of 10^{-3} when tested in control
cells. This survival was three times enhanced when tested in
cells which were UV-irradiated (5 J.m^{-2}) immediately or 12 h
before infection. In such irradiated cells the production of
intact virus was also slightly increased. In fig. 1a, results
are expressed as the virus production in UV-irradiated cells
relative to that of intact cells. The values obtained with UV-
irradiated Lu III are higher than those obtained with intact
virus, so that reactivation factor (RF), as defined in the

Table 1 : Lu III production in SV40-transformed newborn
human kidney cells (NBK)

		Intact Lu III ($\times 10^{-7}$ PFU/ml)	UV-damaged Lu III ($\times 10^{-4}$ PFU/ml)
Control cells		2.4	3.3
UV-damaged cells	48 h	1.25	3.4
	36 h	1.65	4.0
	24 h	1.6	3.15
	12 h	2.9	10.3
	0 h	3.8	11.3

Control cells or cells irradiated at different times
before infection with an unique dose of UV-light
(5 J.m^{-2}) were infected (0 time) either with intact
Lu III (0.5 PFU per cell) or with Lu III irradiated at
an UV-fluence of 50 J.m^{-2}. Twenty four hours later,
i.e. after a viral one-step growth, infected cells
were collected and the produced virus titrated in NBK
cells. Results are expressed in plaque forming units
per ml of harvested cells (PFU/ml).

caption of fig. 1, is always superior to one. It is maximum
when cells are irradiated 12 h before infection. Such a pattern
of reactivation of UV-irradiated virus was found in 6 separate
experiments. The same results have been obtained using Kilham
rat virus (KRV), parvovirus growing in rat cells (data not
shown).

The reactivation curves of Lu III and HSV in UV-irradiated
NBK were compared (fig. 1). As for Lu III, UV (450 J.m^{-2})-
irradiated HSV I was reactivated in these cells, but kinetics
of the phenomenon were different. A time interval of 36 h
between cellular irradiation and viral infection was necessary
to observe the highest RF. Cells irradiated just before infection produced 3 times less intact virus and 10 times less
UV-irradiated virus than control cells.

In other experiments (fig. 2) we treated NBK cells for
1 h at different times before infection, with N-acetoxy-
acetyl-aminofluorene (N-AAAF) or N-methyl-N'-nitro-N-nitro-
soguanidine (MNNG). RF curves of Lu III were similar whatever
the carcinogen used, and analogous to that observed in UV-
irradiated NBK cells. Especially biphasic profiles were constantly found. HSV, again behaved differently, with a maximal
RF value when cells were treated 36 h before infection, and
a fall down at the 0 time.

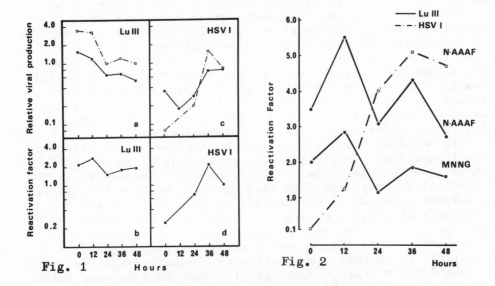

Fig. 1.

Fig. 2

Fig. 1.   Enhanced survival of UV-damaged Lu III and HSV I
            in UV-irradiated NBK cells. Experiments were per-
            formed as described in the caption of table 1,
            except that HSV I was irradiated at an UV-fluence
            of 450 J.m^{-2} and virus production titrated
            in CV1 cells. a and c : results are expressed as
            the viral production in UV-irradiated cells relative
            to that of control cells. NBK cells infected with :
            a,■———■intact Lu III or □— · —□UV-irradiated Lu III;
            c,■———■intact HSV I or □—·—□UV-irradiated HSV I.
            b and d : RF of both viruses are expressed as the
            ratio of relative viral productions of UV-irradiated
            virus and of intact virus.

Fig. 2.   RF obtained with UV-damaged Lu III and HSV I when
            NBK cells are treated with either 0.2 ug/ml or
            N-AAAF or 1 ug/ml   MNNG for 1 h at different times
            before infection. ●————●, ■————■ Lu III and
            of □— · —□ HSV I.

DISCUSSION

        The survival of UV-damaged Lu III parvovirus is increased
when host cells are pretreated with physical (UV) or chemical
(N-AAAF, MNNG) carcinogens. Survival of UV-damaged KRV, is
similarly increased (data to be published). This confirms and
extends a preliminary result obtained with UV-irradiated KRV
in UV-exposed rat cells[3]. Moreover the production of intact

parvovirus is also constantly increased in carcinogen pretreated host cells but to a less extent than that of UV-
irradiated virus. Curves of reactivation factors as a function
of time-intervals between cell treatment and virus infection
show two-waves profiles for both systems studied (rat fibroblasts and NBK cells) whatever the carcinogen used. A maximal
reactivation is reached when cells are treated 12 h before
infection. Similar result has been obtained by J. Rommelaere
using MVM parvovirus in mouse cells synchronised by an
isoleucine deficiency (personal communication).

In parallel NBK cultures, there is a low but significant
reactivation of UV-damaged HSV (survival 10^{-5}) as already
shown in other human tumor cells[2]. But the time-course of HSV
reactivation contrasts with that obtained with Lu III, since
a maximum enhancement of survival is reached for time-
intervals of 36-48 h between cell treatment and infection, as
already described in monkey cells[4] and in 30 human skin
fibroblasts lines (Coppey and Menezes, in preparation).
Similarly an optimal reactivation of UV-damaged SV40 is
obtained for a time-interval greater than 24 h between treatment and infection[5,6].

As the kinetics of reactivation are quite different for
both viruses, it must be emphasized that the two viruses differ
by their dependence upon cellular enzymes and their genome
length and structure. The UV doses to Lu III and to HSV
leaving comparable survivals in control NBK cells are about
50 and 300 $J.m^{-2}$ respectively. Since the molecular weight (MW)
of Herpes DNA is 60 times greater than that of Lu III, we can
roughly estimate that the amount of UV-photoproducts in the
DNA of HSV is at least 3 logs greater than in the DNA of
Lu III. Thus the major fraction of UV-photoproducts in Herpes
DNA could be removed by the constitutive excision repair
systems of host cells. On the contrary single-stranded DNA of
parvoviruses does not provide any fitting substrate for
excision repair enzymes. Let us remind that an UV dose of
2000 $J\ m^{-2}$ is necessary to elicite, in CV-1 cells, a 10^{-3}
survival of SV40[6], which has a double-stranded DNA of length
close to that of Lu III. In conclusion the use of single-
stranded DNA viruses appears to be more suitable for studying
the time-course of the expression of a repair system different
from excision processes. The two waves of Lu III reactivation
at 0-12 and 36 h after treatment would indicate that this
repair system is acting in parallel with normal DNA synthesis.
Experiments using synchronized cells are in progress in order
to clarify this point.

ACKNOWLEDGEMENTS

We thank Dr Günter SIEGL for having kindly provided us
Lu III virus and NBK cells. This work was supported by grant

ATP 77.79.109.6 and contract CRL 78.4.042.3 from the Institut National de la Santé et de la Recherche Médicale.

REFERENCES

1. L.E. Bockstahler and C.D. Lytle, Radiation enhanced reactivation of nuclear replicating mammalian viruses, Photochem. Photobiol., 25:477 (1977).
2. C.D. Lytle, S.G. Benane and L.E. Bockstahler, Ultraviolet-enhanced reactivation of Herpes virus in human tumor cells, Photochem. Photobiol., 20 : 91 (1974).
3. C.D. Lytle, Radiation-enhanced virus reactivation in mammalian cells, Nati. Cancer Inst. Monogr., 50 : 145 (1978).
4. C.D. Lytle, J. Coppey and W.D. Taylor, Enhanced survival of ultraviolet-irradiated herpes simplex virus in carcinogen-pretreated cells, Nature (London), 272:60 (1978).
5. A. Sarasin and P.C. Hanawalt, Carcinogens enhance survival of UV-irradiated simian virus 40 in treated monkey kidney cells : Induction of a recovery pathway?, Proc. Nati. Acad. Sci. (U.S.A.), 75:346 (1978).
6. A. Sarasin and P.C. Hanawalt, Simian virus 40 as a probe for studying DNA repair pathways in mammalian cells, in : "DNA repair mechanisms", P.C. Hanawalt, E.C. Friedberg and C.F. Fox, Ed., Academic Press, New York, p 547 (1978).
7. U.B. Das Gupta and W.C. Summers, Ultraviolet reactivation of herpes simplex virus is mutagenic and inducible in mammalian cells, Proc. Nati. Acad. Sci. (U.S.A.), 75:2378 (1978).
8. A. Sarasin and A. Benoit, Induction of an error-prone mode of DNA repair in UV-irradiated monkey kidney cells, Mutation Res. 70:71 (1980).
9. M. Radman, Phenomenology of an inducible mutagenic DNA repair pathway in Escherichia Coli : SOS repair hypothesis, in: "Molecular and environmental aspects of mutagenesis", L. Praskash, F. Sherman, M.W. Miller, C.W. Lawrence and H.W. Taber, Eds., C.C. Thomas Publ., Springfield III, p 128 (1974).
10. E.M. Witkin, Ultraviolet mutagenesis and inducible DNA repair in Escherichia Coli, Bacteriol. Rev. 40:869 (1976)

AUTHOR INDEX